Edible Food Packaging
Materials and Processing Technologies

Contemporary Food Engineering

Series Editor

Professor Da-Wen Sun, Director
Food Refrigeration & Computerized Food Technology
National University of Ireland, Dublin
(University College Dublin)
Dublin, Ireland
http://www.ucd.ie/sun/

Edible Food Packaging: Materials and Processing Technologies, *edited by Miguel Ângelo Parente Ribeiro Cerqueira, Ricardo Nuno Correia Pereira, Óscar Leandro da Silva Ramos, José António Couto Teixeira, and António Augusto Vicente* (2016)

Light Scattering Technology for Food Property, Quality and Safety Assessment, *edited by Renfu Lu* (2016)

Advances in Heat Transfer Unit Operations: Baking and Freezing in Bread Making, *edited by Georgina Calderon-Dominguez, Gustavo F. Gutierrez-Lopez, and Keshavan Niranjan* (2016)

Innovative Processing Technologies for Foods with Bioactive Compounds, *edited by Jorge J. Moreno* (2016)

Edible Food Packaging: Materials and Processing Technologies, *edited by Miquel Angelo Parente Ribeiro Cerqueira, Ricardo Nuno Correia Pereira, Oscar Leandro da Silva Ramos, Jose Antonio Couto Teixeira, and Antonio Augusto Vicente* (2016)

Handbook of Food Processing: Food Preservation, *edited by Theodoros Varzakas and Constantina Tzia* (2015)

Handbook of Food Processing: Food Safety, Quality, and Manufacturing Processes, *edited by Theodoros Varzakas and Constantina Tzia* (2015)

Edible Food Packaging: Materials and Processing Technologies,
edited by Miquel Angelo Parente Ribeiro Cerqueira, Ricardo Nuno Correia Pereira, Oscar Leandro da Silva Ramos, Jose Antonio Couto Teixeira, and Antonio Augusto Vicente (2015)

Advances in Postharvest Fruit and Vegetable Technology,
edited by Ron B.H. Wills and John Golding (2015)

Engineering Aspects of Food Emulsification and Homogenization,
edited by Marilyn Rayner and Petr Dejmek (2015)

Handbook of Food Processing and Engineering, Volume II: Food Process Engineering, *edited by Theodoros Varzakas and Constantina Tzia* (2014)

Handbook of Food Processing and Engineering, Volume I: Food Engineering Fundamentals, *edited by Theodoros Varzakas and Constantina Tzia* (2014)

Juice Processing: Quality, Safety and Value-Added Opportunities, *edited by Víctor Falguera and Albert Ibarz* (2014)

Engineering Aspects of Food Biotechnology, *edited by José A. Teixeira and António A. Vicente* (2013)

Engineering Aspects of Cereal and Cereal-Based Products, *edited by Raquel de Pinho Ferreira Guiné and Paula Maria dos Reis Correia* (2013)

Fermentation Processes Engineering in the Food Industry, *edited by Carlos Ricardo Soccol, Ashok Pandey, and Christian Larroche* (2013)

Modified Atmosphere and Active Packaging Technologies, *edited by Ioannis Arvanitoyannis* (2012)

Advances in Fruit Processing Technologies, *edited by Sueli Rodrigues and Fabiano Andre Narciso Fernandes* (2012)

Biopolymer Engineering in Food Processing, *edited by Vânia Regina Nicoletti Telis* (2012)

Operations in Food Refrigeration, *edited by Rodolfo H. Mascheroni* (2012)

Thermal Food Processing: New Technologies and Quality Issues, Second Edition, *edited by Da-Wen Sun* (2012)

Physical Properties of Foods: Novel Measurement Techniques and Applications, *edited by Ignacio Arana* (2012)

Handbook of Frozen Food Processing and Packaging, Second Edition, *edited by Da-Wen Sun* (2011)

Advances in Food Extrusion Technology, *edited by Medeni Maskan and Aylin Altan* (2011)

Enhancing Extraction Processes in the Food Industry, *edited by Nikolai Lebovka, Eugene Vorobiev, and Farid Chemat* (2011)

Emerging Technologies for Food Quality and Food Safety Evaluation, *edited by Yong-Jin Cho and Sukwon Kang* (2011)

Food Process Engineering Operations, *edited by George D. Saravacos and Zacharias B. Maroulis* (2011)

Biosensors in Food Processing, Safety, and Quality Control, *edited by Mehmet Mutlu* (2011)

Physicochemical Aspects of Food Engineering and Processing, *edited by Sakamon Devahastin* (2010)

Infrared Heating for Food and Agricultural Processing, *edited by Zhongli Pan and Griffiths Gregory Atungulu* (2010)

Mathematical Modeling of Food Processing, *edited by Mohammed M. Farid* (2009)

Engineering Aspects of Milk and Dairy Products, *edited by Jane Sélia dos Reis Coimbra and José A. Teixeira* (2009)

Innovation in Food Engineering: New Techniques and Products, *edited by Maria Laura Passos and Claudio P. Ribeiro* (2009)

Processing Effects on Safety and Quality of Foods, *edited by Enrique Ortega-Rivas* (2009)

Engineering Aspects of Thermal Food Processing, *edited by Ricardo Simpson* (2009)

Ultraviolet Light in Food Technology: Principles and Applications, *Tatiana N. Koutchma, Larry J. Forney, and Carmen I. Moraru* (2009)

Advances in Deep-Fat Frying of Foods, *edited by Serpil Sahin and Servet Gülüm Sumnu* (2009)

Extracting Bioactive Compounds for Food Products: Theory and Applications, *edited by M. Angela A. Meireles* (2009)

Advances in Food Dehydration, *edited by Cristina Ratti* (2009)

Optimization in Food Engineering, *edited by Ferruh Erdoğdu* (2009)

Optical Monitoring of Fresh and Processed Agricultural Crops, *edited by Manuela Zude* (2009)

Food Engineering Aspects of Baking Sweet Goods, *edited by Servet Gülüm Sumnu and Serpil Sahin* (2008)

Computational Fluid Dynamics in Food Processing, *edited by Da-Wen Sun* (2007)

Edible Food Packaging

Materials and Processing Technologies

Edited by
**Miguel Ângelo Parente Ribeiro Cerqueira
Ricardo Nuno Correia Pereira
Óscar Leandro da Silva Ramos
José António Couto Teixeira
António Augusto Vicente**

CRC Press is an imprint of the
Taylor & Francis Group, an **informa** business

CRC Press
Taylor & Francis Group
6000 Broken Sound Parkway NW, Suite 300
Boca Raton, FL 33487-2742

© 2016 by Taylor & Francis Group, LLC
CRC Press is an imprint of Taylor & Francis Group, an Informa business

No claim to original U.S. Government works

Printed on acid-free paper
Version Date: 20160208

International Standard Book Number-13: 978-1-4822-3416-9 (Hardback)

This book contains information obtained from authentic and highly regarded sources. Reasonable efforts have been made to publish reliable data and information, but the author and publisher cannot assume responsibility for the validity of all materials or the consequences of their use. The authors and publishers have attempted to trace the copyright holders of all material reproduced in this publication and apologize to copyright holders if permission to publish in this form has not been obtained. If any copyright material has not been acknowledged please write and let us know so we may rectify in any future reprint.

Except as permitted under U.S. Copyright Law, no part of this book may be reprinted, reproduced, transmitted, or utilized in any form by any electronic, mechanical, or other means, now known or hereafter invented, including photocopying, microfilming, and recording, or in any information storage or retrieval system, without written permission from the publishers.

For permission to photocopy or use material electronically from this work, please access www.copyright.com (http://www.copyright.com/) or contact the Copyright Clearance Center, Inc. (CCC), 222 Rosewood Drive, Danvers, MA 01923, 978-750-8400. CCC is a not-for-profit organization that provides licenses and registration for a variety of users. For organizations that have been granted a photocopy license by the CCC, a separate system of payment has been arranged.

Trademark Notice: Product or corporate names may be trademarks or registered trademarks, and are used only for identification and explanation without intent to infringe.

Visit the Taylor & Francis Web site at
http://www.taylorandfrancis.com

and the CRC Press Web site at
http://www.crcpress.com

Contents

Series Preface ix
Preface xi
Acknowledgments xiii
Series Editor xv
Editors xvii
Contributors xxi

Chapter 1 Edible Packaging Today 1

Miguel Ângelo Parente Ribeiro Cerqueira, José António Couto Teixeira, and António Augusto Vicente

Chapter 2 Edible Film and Packaging Using Gum Polysaccharides 9

Marceliano B. Nieto

Chapter 3 Protein-Based Films and Coatings 81

Pedro Guerrero and Koro de la Caba

Chapter 4 Edible Coatings and Films from Lipids, Waxes, and Resins 121

Jorge A. Aguirre-Joya, Berenice Álvarez, Janeth M. Ventura, Jesús O. García-Galindo, Miguel A. De León-Zapata, Romeo Rojas, Saúl Saucedo, and Cristóbal N. Aguilar

Chapter 5 Production and Processing of Edible Packaging: Stability and Applications 153

Nuria Blanco-Pascual and Joaquín Gómez-Estaca

Chapter 6 Mass Transfer Measurement and Modeling for Designing Protective Edible Films 181

Valérie Guillard, Carole Guillaume, Mia Kurek, and Nathalie Gontard

Chapter 7 Edible Packaging: A Vehicle for Functional Compounds 215

Joana O. Pereira and M. Manuela Pintado

Chapter 8	Antimicrobial Edible Packaging	243

*Lorenzo M. Pastrana, Maria L. Rúa, Paula Fajardo,
Pablo Fuciños, Isabel R. Amado, and Clara Fuciños*

Chapter 9	Nanotechnology in Edible Packaging	287

Marthyna Pessoa de Souza and Maria G. Carneiro-da-Cunha

Chapter 10	Nanostructured Multilayers for Food Packaging by Electrohydrodynamic Processing	319

María José Fabra, Amparo López-Rubio, and José M. Lagaron

Chapter 11	How to Evaluate the Barrier Properties for Edible Packaging of Respiring Products	333

Maria José Sousa-Gallagher

Chapter 12	Edible Packaging for Fruits and Vegetables	353

*Marta Montero-Calderón, Robert Soliva-Fortuny,
and Olga Martín-Belloso*

Chapter 13	Edible Packaging for Dairy Products	383

*Óscar Leandro da Silva Ramos, Ricardo Nuno Correia Pereira,
Joana T. Martins, and F. Xavier Malcata*

Chapter 14	Edible Coatings and Films for Meat, Poultry, and Fish	413

*Eveline M. Nunes, Ana I. Silva, Claúdia B. Vieira,
Men de Sá M. Souza Filho, Elisabeth M. Silva, and
Bartolomeu W. Souza*

Index 431

Series Preface

CONTEMPORARY FOOD ENGINEERING

Food engineering is a multidisciplinary field of applied physical sciences combined with the knowledge of product properties. Food engineers provide the technological knowledge transfer essential to the cost-effective production and commercialization of food products and services. In particular, food engineers develop and design processes and equipment to convert raw agricultural materials and ingredients into safe, convenient, and nutritious consumer food products. Food engineering topics are continuously undergoing changes to meet diverse consumer demands, and the subject is being rapidly developed to reflect market needs.

In the development of food engineering, one of the many challenges is to employ modern tools and knowledge, such as computational materials science and nanotechnology, to develop new products and processes. Simultaneously, improving food quality and safety continues to be a critical issue in food engineering studies. New packaging materials and techniques are being developed to provide more protection to foods, and novel preservation technologies are emerging to enhance food quality and safety. Additionally, process control and automation are among the top priorities identified in food engineering. Advanced monitoring and control systems are developed to facilitate automation and flexible food manufacturing. Furthermore, energy saving and minimization of environmental problems continue to be important food engineering issues, and significant progress is being made in waste management, efficient utilization of energy, and reduction of effluents and emissions in food production.

The Contemporary Food Engineering Series addresses some of the recent developments in food engineering. The series covers advances in classical unit operations in engineering applied to food manufacturing, as well as such topics as progress in the transport and storage of liquid and solid foods; heating, chilling, and freezing of foods; mass transfer in foods; chemical and biochemical aspects of food engineering and the use of kinetic analysis; dehydration, thermal processing, nonthermal processing, extrusion, liquid food concentration, membrane processes, and applications of membranes in food processing; shelf life and electronic indicators in inventory management; sustainable technologies in food processing; and packaging, cleaning, and sanitation. The books in this series are aimed at professional food scientists, academics researching food engineering problems, and graduate-level students.

The editors of these books are leading engineers and scientists from different parts of the world. All the editors were asked to present their books to address the market's needs and pinpoint cutting-edge technologies in food engineering.

All chapters have been contributed by internationally renowned experts who have both academic and professional credentials. All authors have attempted to provide critical, comprehensive, and readily accessible information on the art and science of a relevant topic in each chapter, with reference lists for further information. Therefore, each book can serve as an essential reference source to students and researchers in universities and research institutions.

Da-Wen Sun
Series Editor

Preface

The edible packaging industry has had remarkable growth in recent years and is expected to have an important impact on the food market in the following years. This growth is a consequence of increasing knowledge developed on edible films and coatings technology, achieved through research and product development, as well as of advances in material science and processing technology. Simultaneously, the impact on sustainability and the interest in using renewable resources further contribute to reinforce growth perspectives. The edible packaging field is related to materials and food sciences, where materials' properties and processing and food characteristics are considered simultaneously to guarantee the safety and quality of food products. In the last 20 years, a great number of scientific and opinion articles have been published as a result of research projects undertaken by the academic and research communities as well as the food industry. These advances in edible packaging technology justify the publication of a book revisiting existing edible packaging technologies, while adding the most recent developments in that field and providing an insight on new applications.

The book is divided into 14 chapters, providing a comprehensive review on aspects such as materials used, their structure–function relationship, and new processing technologies for the application and production of edible coatings and films; the application of functional compounds using edible packaging and the development of active and smart edible packaging; nanotechnology applied to edible packaging; the most important properties of packaging to be applied on foods; and examples of edible packaging application to food products, commercialization, and regulatory aspects.

This book will be useful as a reference for all aspects of edible packaging development and characterization. Undergraduate and postgraduate students, researchers, and industries will find the topics covered in this book an invaluable contribution to their learning and work. This book is an important source of information for all those who aim to do research and development in edible packaging technology.

Acknowledgments

Miguel Ângelo Parente Ribeiro Cerqueira, Ricardo Nuno Correia Pereira, and Óscar Leandro da Silva Ramos gratefully acknowledge their postdoctoral grants (SFRH/BPD/72753/2010, SFRH/BPD/81887/2011, and SFRH/BPD/80766/2011, respectively) from Fundação para a Ciência e Tecnologia (FCT, Portugal). The authors thank the FCT Strategic Project of UID/BIO/04469/2013 unit, the project RECI/BBB-EBI/0179/2012 (FCOMP-01-0124-FEDER-027462), and the project "BioInd—Biotechnology and Bioengineering for improved Industrial and Agro-Food processes," REF. NORTE-07-0124-FEDER-000028 cofunded by the Programa Operacional Regional do Norte (ON.2 – O Novo Norte), QREN, FEDER.

Series Editor

Born in southern China, Dr. Da-Wen Sun is a world authority in food engineering research and education; he is a member of the Royal Irish Academy (RIA), which is the highest academic honor in Ireland; he is also a member of Academia Europaea (The Academy of Europe) and a fellow of the International Academy of Food Science and Technology. His main research activities include cooling, drying, and refrigeration processes and systems, quality and safety of food products, bioprocess simulation and optimization, and computer vision/image processing and hyperspectral imaging technologies. His many scholarly works have become standard reference materials for researchers in the areas of computer vision, computational fluid dynamics modeling, vacuum cooling, among others. Results of his work have been published in over 800 papers, including more than 400 peer-reviewed journal papers (Web of Science h-index = 71), among them, 31 papers have been selected by Thomson Reuters' *Essential Science Indicators^{SM}* as highly-cited papers, ranking him No. 1 in the world in Agricultural Sciences (December 2015). He has also edited 14 authoritative books. According to Thomson Reuters *Essential Science Indicators^{SM}* based on data derived over a period of 10 years from the ISI Web of Science, there are about 4500 scientists who are among the top 1% of the most cited scientists in the category of agriculture sciences. For many years, Dr. Sun has consistently been ranked among the top 50 scientists in the world (he was at the 20th position in December 2015), and has recently been named Highly Cited Researcher 2015 by Thomson Reuters.

He earned a first class BSc Honours and MSc in mechanical engineering, and a PhD in chemical engineering in China before working in various universities in Europe. He became the first Chinese national to be permanently employed in an Irish university when he was appointed college lecturer at the National University of Ireland, Dublin (University College Dublin [UCD]), in 1995, and was then continuously promoted in the shortest possible time to senior lecturer, associate professor, and full professor. Dr. Sun is now the professor of Food and Biosystems Engineering and the director of UCD Food Refrigeration and Computerised Food Technology.

As a leading educator in food engineering, Dr. Sun has significantly contributed to the field of food engineering. He has trained many PhD students who have made their own contributions to the industry and academia. He has also delivered lectures on advances in food engineering on a regular basis in academic institutions internationally and delivered keynote speeches at international conferences. As a recognized authority in food engineering, he has been conferred adjunct/visiting/consulting professorships from 10 top universities in China, including Zhejiang University, Shanghai Jiaotong University, Harbin Institute of Technology, China Agricultural University, South China University of Technology, and Jiangnan University. In recognition of his significant contribution to food engineering worldwide and for his outstanding leadership in the field, the International Commission of Agricultural and Biosystems Engineering (CIGR) awarded him the CIGR Merit Award twice in 2000 and in 2006, the Institution of Mechanical Engineers based in the United Kingdom named him Food Engineer of the Year 2004. In 2008, he was awarded the CIGR Recognition Award in honor of his distinguished achieve- ments as the top 1% of agricultural engineering scientists in the world. In 2007, he was presented with the only AFST(I) Fellow Award by the Association of Food Scientists and Technologists (India), and in 2010, he was presented with the CIGR Fellow Award; the title of Fellow is the highest honor in CIGR and is conferred to individuals who have made sustained, outstanding contributions, worldwide. In March 2013, he was presented with the You Bring Charm to the World award by Hong Kong–based Phoenix Satellite Television with other award recipients including Mr. Mo Yan—the 2012 Nobel Laureate in Literature and the Chinese Astronaut Team for Shenzhou IX Spaceship. In July 2013, he received the Frozen Food Foundation Freezing Research Award from the International Association for Food Protection for his significant contributions to enhancing the field of food freezing technologies. This is the first time that this prestigious award was presented to a scientist outside the United States, and in June 2015 he was presented with the IAEF Lifetime Achievement Award. This IAEF (International Association of Engineering and Food) award, highlights the lifetime contribution of a prominent engineer in the field of food.

He is a fellow of the Institution of Agricultural Engineers and a fellow of Engineers Ireland (the Institution of Engineers of Ireland). He is also the editor-in-chief of *Food and Bioprocess Technology—An International Journal* (2012 impact factor = 4.115), former editor of *Journal of Food Engineering* (Elsevier), and editorial board member for a number of international journals, including the *Journal of Food Process Engineering, Journal of Food Measurement and Characterization*, and *Polish Journal of Food and Nutritional Sciences*. He is also a chartered engineer.

At the 51st CIGR General Assembly held during the CIGR World Congress in Quebec City, Canada, on June 13–17, 2010, he was elected incoming president of CIGR, became CIGR President in 2013–2014, and is now a CIGR Past President. CIGR is the world's largest organization in the field of agricultural and biosystems engineering.

Editors

Miguel Ângelo Parente Ribeiro Cerqueira graduated with a degree in chemical and biological engineering from the University of Minho (UM), where he earned two scholar merit awards (2002 and 2003) and a scholarship merit award (2005) during graduation. He earned his PhD in December 2010, and received an award for the best PhD thesis in 2011 from the School of Engineering of UM. Since April 2011, he has been a postdoctoral researcher at UM. To date, he has published more than 45 papers in peer-reviewed journals, 1 patent, 5 papers in nonpeer-reviewed journals, 12 book chapters, 12 extended abstracts in international conferences, 52 abstracts in international conferences, and 18 abstracts in national conferences. He was selected as one of the winners of the Young Scientist Award at the 17th IUFoST World Congress of Food Science and Technology, 2014 (Canada). In his career, he has performed several short-term missions (Chemistry Department, University of Aveiro [Pt]; Department of Process and Chemical Engineering, University College Cork [Ie]; Department of Analytical Chemistry and Food Science, University of Vigo [Es]; Department of Food Engineering, University of Campinas [Br]; and Novel Materials and Nanotechnology Group of the Institute of Agrochemistry and Food Technology [IATA] of the Spanish Council for Scientific Research, Valencia [Es]). He has maintained contact with the food industry as a consultant (implementation of ISO 22000 and HACCP) and in 2013, he cofounded Improveat, Lda, a spin-off company of UM.

Ricardo Nuno Correia Pereira graduated in 2003 with a degree in food engineering from the Portuguese Catholic University in Porto, Portugal, and earned his PhD in chemical and biological engineering in 2011 at the University of Minho in Braga, Portugal. During his career as a young researcher, he has participated in several research projects, both nationally and internationally, in the food industry. He has worked as a consultant (professionally and educationally) for a number of private and public sector clients related to food technology. Before earning his PhD, he earned an MSc in biotechnology–bioprocess engineering in 2007 from the University of Minho in Braga, Portugal.

His main research interests at the moment are as follows:

- Thermal food processing by ohmic heating (namely, the study of the effects of electric currents on biomolecules, cells, and microorganisms)
- Extraction of value-added compounds from food
- Nanotechnology applied to food technology (development of nanoparticles and nanogels, all from food-grade materials)
- Novel food processing technologies (i.e., pulsed electric fields and high-pressure processing)
- Food packaging technology, food safety, and quality control

He has published approximately 20 research articles in international peer-reviewed journals and four book chapters in international books. He is currently a postdoctoral researcher at the Centre for Biological Engineering (CEB) in the Department of Biological Engineering of the University of Minho in Braga, Portugal.

Óscar Leandro da Silva Ramos earned a BSc in microbiology from the Portuguese Catholic University, Porto, Portugal, and a PhD in technological and engineering sciences with a specialization in biochemical engineering from the New University of Lisbon, Lisboa, Portugal. He is now a postdoctoral researcher at the Centre for Biological Engineering (CEB) of Minho University, Braga, Portugal, and at the Department of Chemical Engineering, University of Porto, Porto, Portugal. He has authored or coauthored more than 20 papers in peer-reviewed journals in the field of food science and technology and 6 chapters in edited books. He also has participated in more than 30 international conferences and workshops worldwide and done more than 40 oral and poster presentations. He has cosupervised 10 graduate, 2 MSc, and 1 PhD students and has collaborated in the R&D of several independent projects. He has a national patent (No. 105852-11/02/2013) developed during his PhD program, titled "Revestimento comestível para alimentos." His research domains include the following:

- The development of edible film and coating formulations for food products
- Production of value-added foods using by-products of the food industry
- Application of natural antimicrobial and antioxidant agents in food products
- Development of functional foods
- Design of innovative food-grade nanostructures for the controlled delivery and release of nutraceuticals on food products

Editors

José António Couto Teixeira is a full professor and currently the head of the Centre for Biological Engineering (CEB) of Minho University. He is responsible for scientific research and advanced formation in the areas of biological and chemical engineering and carries out research in industrial and food biotechnology and bioengineering.

His research activities have been focused on two main topics: fermentation technology (multiphase bioreactors, in particular) and food technology. He is also interested in food nanotechnology as well as in the production of bioactive compounds for food and medical applications. He has been the scientific coordinator of 26 research projects, including two Alfa networks. He (co)authored 400 peer-reviewed papers and is the coeditor of books *Reactores Biológicos-Fundamentos e Aplicações* (in Portuguese), *Engineering Aspects of Milk and Dairy Products*, and *Engineering Aspects of Food Biotechnology*.

António Augusto Vicente graduated with a degree in food engineering in 1994 from the Portuguese Catholic University in Porto, Portugal, and earned his PhD in chemical and biological engineering in 1998 at the University of Minho. He received his habilitation in chemical and biological engineering from the University of Minho in 2010. From early in his career, he has kept close contact with the food industry and he is involved in several research projects, both national and international, together with industrial partners either as a participant or as a project leader. Currently, he has collaborations established in countries such as Brazil, France, Germany, Ireland, Italy, Mexico, New Zealand, Slovakia, Spain, Sweden, United Kingdom, and the United States. His main research interests are

- Food processing by ohmic heating/moderate electric fields (namely, the study of the effects of electric currents on biomolecules and cells)
- Edible films and coatings for food products (chemical, physical, and functional characterization)
- Nanotechnology applied to food technology (nanomultilayered films and coatings, nanoparticles, and nanogels, all from food-grade materials)
- Fermentation technology (including design and operation of bioreactors)

He has supervised 23 PhD theses and also several MSc theses and 15 postdoctoral fellows. He has published approximately 150 research articles in international peer-reviewed journals and approximately 20 book chapters in international books.

Contributors

Cristóbal N. Aguilar
Food Research Department
Universidad Autónoma de Coahuila
Saltillo, Coahuila, Mexico

Jorge A. Aguirre-Joya
Food Research Department
Universidad Autónoma de Coahuila
Saltillo, Coahuila, México

Berenice Álvarez
Food Research Department
Universidad Autónoma de Coahuila
Saltillo, Coahuila, México

Isabel R. Amado
Departamento de Química Analítica
  e Alimentaria
Universidade de Vigo
Ourense, Spain

Nuria Blanco-Pascual
Products Department
Institute of Food Science, Technology
  and Nutrition (ICTAN-CSIC)
José Antonio Novais, Madrid, Spain

Maria G. Carneiro-da-Cunha
Department of Biochemistry
Universidade Federal de Pernambuco
Recife, Pernambuco, Brazil

Miguel Ângelo Parente Ribeiro Cerqueira
Centre of Biological Engineering
University of Minho
Braga, Portugal

Koro de la Caba
BIOMAT Research Group
University of the Basque Country
  (UPV/EHU)
Donostia-San Sebastián, Spain

Miguel A. De León-Zapata
Food Research Department
Universidad Autónoma de
  Coahuila
Saltillo, Coahuila, México

Marthyna Pessoa de Souza
Department of Biochemistry
Universidade Federal de Pernambuco
Recife, Pernambuco, Brazil

María José Fabra
Novel Materials and Nanotechnology
  Group
IATA-CSIC
Paterna, Valencia, Spain

Paula Fajardo
ANFACO-CECOPESCA
Vigo, Spain

Men de Sá M. Souza Filho
Embrapa Tropical Agroindustry
R Dra Sara Mesquita
Fortaleza, Ceará, Brazil

Clara Fuciños
Centre of Biological Engineering
University of Minho
Braga, Portugal

Pablo Fuciños
Department of Health, Food and Environment
International Iberian Nanotechnology Laboratory
Braga, Portugal

Jesús O. García-Galindo
Food Research Department
Universidad Autónoma de Coahuila
Saltillo, Coahuila, México

Joaquín Gómez-Estaca
Products Department
Institute of Food Science, Technology and Nutrition (ICTAN-CSIC)
Madrid, Spain

Nathalie Gontard
Joint Research Unit: Agropolymers Engineering and Emerging Technologies
Montpellier, France

Pedro Guerrero
BIOMAT Research Group
University of the Basque Country (UPV/EHU)
Donostia-San Sebastián, Spain

Valérie Guillard
Joint Research Unit: Agropolymers Engineering and Emerging Technologies
Montpellier, France

Carole Guillaume
Joint Research Unit: Agropolymers Engineering and Emerging Technologies
Montpellier, France

Mia Kurek
Joint Research Unit: Agropolymers Engineering and Emerging Technologies
Montpellier, France

José M. Lagaron
Novel Materials and Nanotechnology Group
IATA-CSIC
Valencia, Spain

Amparo López-Rubio
Novel Materials and Nanotechnology Group
IATA-CSIC
Valencia, Spain

F. Xavier Malcata
LEPABE—Laboratory of Process Engineering, Environment, Biotechnology and Energy
Department of Chemical Engineering
University of Porto
Porto, Portugal

Olga Martín-Belloso
Department of Food Technology
University of Lleida—Agrotecnio Center
Lleida, Spain

Joana T. Martins
Centre of Biological Engineering
University of Minho
Braga, Portugal

Marta Montero-Calderón
Postharvest Technology Laboratory
Agronomic Research Center
University of Costa Rica
San Pedro, Costa Rica

Marceliano B. Nieto
TIC Gums, Inc.
Belcamp, Maryland

Eveline M. Nunes
Department of Food Technology
Federal University of Ceará
Ceará, Brazil

Contributors

Lorenzo M. Pastrana
Department of Health, Food and Environment
International Iberian Nanotechnology Laboratory
Braga, Portugal

Joana O. Pereira
Centro de Biotecnologia e Química Fina
Escola Superior de Biotecnologia
Universidade Católica Portuguesa
Porto, Portugal

Ricardo Nuno Correia Pereira
Centre of Biological Engineering
University of Minho
Braga, Portugal

M. Manuela Pintado
Centro de Biotecnologia e Química Fina
Escola Superior de Biotecnologia
Universidade Católica Portuguesa
Porto, Portugal

Óscar Leandro da Silva Ramos
Centre of Biological Engineering
University of Minho
Braga, Portugal

and

Laboratory for Engineering of Processes, Environment, Biotechnology and Energy
Porto, Portugal

Romeo Rojas
Research Center and Development for Food Industries (CIDIA)
Universidad Autónoma de Nuevo León
General Escobedo, Nuevo León, Mexico

Maria L. Rúa
Departamento de Química Analítica e Alimentaria
Universidade de Vigo
Pontevedra, Spain

Saúl Saucedo
División de Metrología Industrial
Universidad Politécnica de Ramos Arizpe
Ramos Arizpe, Coahuila, Mexico

Ana I. Silva
Department of Fisheries Engineering
Federal University of Ceará
Ceará, Brazil

Elisabeth M. Silva
Department of Food Technology
Federal University of Ceará
Ceará, Brazil

Robert Soliva-Fortuny
Department of Food Technology
University of Lleida—Agrotecnio Center
Lleida, Spain

Maria José Sousa-Gallagher
Process and Chemical Engineering
University College Cork
Cork, Ireland

Bartolomeu W. Souza
Department of Fisheries Engineering
Federal University of Ceará
Ceará, Brazil

José António Couto Teixeira
Centre of Biological Engineering
University of Minho
Braga, Portugal

Janeth M. Ventura
Food Research Department
Universidad Autónoma de Coahuila
Saltillo, Coahuila, Mexico

António Augusto Vicente
Centre of Biological Engineering
University of Minho
Braga, Portugal

Claúdia B. Vieira
Department of Fisheries Engineering
Federal University of Ceará
Ceará, Brazil

1 Edible Packaging Today

*Miguel Ângelo Parente Ribeiro Cerqueira,
José António Couto Teixeira, and
António Augusto Vicente*

CONTENTS

Abstract 1
1.1 Definition of Edible Packaging 1
1.2 Social, Commercial, and Scientific Interest of Edible Packaging 2
1.3 Legislation and Regulatory Assessment 4
1.4 Future Perspectives 5
Acknowledgments 6
References 6

ABSTRACT

Edible packaging has taken a rebirth in the past 20 years with the use of coatings and films in food applications. In this chapter, the growing interest in this technology is enlightened and aspects such as consumers and environmental needs, commercial interest, and new scientific findings are presented as the promoters of the recent development of edible packaging. In addition, the definition as well as legislation and regulatory aspects of edible packaging is discussed. Future perspectives on using edible packaging are presented, together with the potential use of this technology by academic and industrial players.

1.1 DEFINITION OF EDIBLE PACKAGING

Packaging has been defined as a socioscientific discipline that operates in society to ensure the delivery of goods to the ultimate consumer in the best condition intended for their use (Lockhart, 1997). In the particular case of foods, a package can be used for a great number of applications where the main goals are to serve as a container for the protection and preservation of a perishable product and as a way of communication with the consumer. According to Zepf (2009, p. 1296), a package is defined as "a metal can, glass bottle, plastic bag, or pouch which serves the functions of containing and protecting the product, as well as providing convenience and communicating to the consumer."

Edible packaging can be included in the presented packaging definition; however, in this case, all the materials used should be edible according to the legislation, both in the initial (packaging ingredients) and in the final (packaging) forms. Two types of edible packaging are usually described: coatings and films. It is considered a coating when the film-forming solution is applied directly on the food product (by immersion, spray, brushing, or other) and is left to dry on the food surface to form a thin film, which will perform the desired function. A film is the dried film-forming solution that is used and applied as a self-standing material on the food product. In order to be considered a film, the value of thickness should be below 254 μm (Robertson, 2012).

1.2 SOCIAL, COMMERCIAL, AND SCIENTIFIC INTEREST OF EDIBLE PACKAGING

Interest in edible packaging for food applications has been growing in recent years, due to consumers' demand for higher-quality and safe foods with extended shelf lives and also their desire for natural and biodegradable materials instead of synthetic and nonbiodegradable materials. The use of edible, biodegradable, and renewable materials to replace (partially or totally) petroleum-based packaging materials has increased the interest of the global market for edible packaging solutions.

It is known that edible packaging is only one of the parts of this global bio-based packaging material option; it is expected that the global bioplastic packaging demand may reach to 884,000 tons by 2020, the compounds' annual growth rate being 24.9% from 2010 to 2015 and 18.3% from 2015 to 2020. In 2010, the group of water-soluble polymers and starch- and cellulose-based bioplastics represented 44.3% of this market. According to Smithers Pira (2013), the bioplastics change from biodegradable and compostable polymers to bio-based packaging from renewable and sustainable materials. One of the evidences regarding the use of edible packaging is the number of registered inventions related to this topic, where edible coatings and films are distinguished based on their possible innovative applications.

One of the first information of commercial interest regarding edible coatings appeared in 1933, in the United States, through a patent submitted by the Food Machinery Corporation, where a new way to apply a wax coating in fresh fruit (mainly citrus fruits) was presented (De Ore, 1933). In the same year, the Griffith Laboratories presented an invention aiming at using gelatin-based coatings for the preservation of foods such as meats, fruits, vegetables, and packaged foods (Griffith and Hall, 1933). Since then, a great number of inventions concerning edible coatings have been presented, and a great number of applications have been registered. Some of the examples are dried fruit (Brekke and Watters, 1959; Durkee et al., 1962), meat and frozen meat (Allingham, 1949; Cadwell, 1943; Cornwell, 1951), shrimp and seafood (Hussey, 1986; Toulmin, 1956), coffee beans (Verena and Specht, 1959), and cheese (Van et al., 2001; Wood, 1939).

Today, most of the inventions submitted regarding edible coatings are related to their application in fresh-cut fruits. Fruitsymbiose Inc. submitted an invention for the preservation of at least one organoleptic property of fruits and vegetables, in which the coating composition comprises a polysaccharide solution and a cross-linking agent solution (Girard, 2013).

Inventions regarding edible films for food applications appeared much later when compared with edible coatings, the first one being from the 1990s (Fennema et al., 1987, 1990; Lazard et al., 1997). This invention used polysaccharides and proteins as main materials, with the possibility of using plasticizers and lipids during the production. In the last 5 years, several inventions appeared dealing with new film compositions, such as a film based on purple potato starch (Shaoxiao et al., 2014) and mixtures of carboxymethyl cellulose with improved properties (Brown and Verrall, 2010), and new films used as sausage casings and wrappers for hams and other cured meat products (Macquarrie, 2014).

From these inventions and other research studies, several companies actually present edible coatings in their products. Table 1.1 presents examples of companies marketing commercial coatings for food industry.

It is clear that the eco-friendly label of this kind of packaging and potential applications for several food products leads to a growing interest by the industry for the development of new solutions. According to Attila Pavlath (Bernstein and Woods, 2013), the market in edible coatings for food products, in particular for fresh fruits and vegetables, presented in 2013 more than 1000 companies that exceed $100 million in annual sales.

From a scientific point of view, the use of bio-based materials for the production of edible coatings and films integrates the knowledge and contributions of a significant number of scientific areas (e.g., polymer, chemistry, food science, and biology). Besides the potential application in food and pharmaceutical industries (proved by a great number of publications and patents), the great interest of these systems regarding more in-depth scientific works is due to the following aspects:

1. Different materials used such as polysaccharides, proteins, lipids, and waxes, and the interaction between these and other materials used in their production (e.g., plasticizers and surfactants).
2. Processing techniques and different techniques of application.
3. Transport, thermal and mechanical properties, influenced by the materials used and processing techniques, and also by the external factors (e.g., relative humidity, temperature, gases, moisture, and light).
4. Possible use as a vehicle for bioactive and functional compounds.
5. Use of nanotechnology for the formation of coatings and multilayers at the nanoscale and the incorporation of nanostructures in films, potentially bringing new and interesting findings.

TABLE 1.1
Companies with Commercial Edible Packaging Materials for Food Applications

Company Name	Product	Main Use
AgriCoat Nature Seal, Ltd.	Semperfresh	Fresh-cut fruits and vegetables
Fruitsymbiose Inc.	Pürbloom	Fresh-cut fruits and vegetables
BASF	FreshSeal	Tomatoes and melon
De Leye Agro	Bio-Fresh	Pears, apples, and nectarine
Improveat	BioCheeseCoat	Cheese
Caragum International	FibreCoat	Fried food
Loders Croklaan	Durkex 500	Dried fruits and nuts
Mantrose-Haeuser	Crystalac	Chocolates and candies
Wikifoods	Wikipearl	Wrapper for foods and beverages

1.3 LEGISLATION AND REGULATORY ASSESSMENT

Global regulation is perhaps one of the most difficult challenges for the food industry. Today, with a global economy where the amount of commercialized products between countries increases daily, it is a very hard task for the food industry to comply with the legislations of different countries. In European countries, the United States, Canada, and Australia, food regulations existed for decades; therefore, it is difficult to create a unique document for all countries. Aiming at a global legislation, the Food and Agriculture Organization of the United Nations and World Health Organization established their own guidelines in an attempt to render uniform food regulations in all countries and create international food standards, the Codex Alimentarius (2014).

Edible packaging regulation should meet the legislation of the country where the packed food product is sold to the consumer. It is known that such legislation should guarantee the safety of the consumers, and it is also important to enforce a correct labeling and clear information regarding the ingredients used, processing technologies used, and final functionality of the food product. This information should be provided to the consumers to guarantee the total transparency of the product being sold.

Edible packaging materials should be regarded as food additives once they are intended for consumption with the food product. In Europe, edible packaging materials are included in the Regulations EC 1331/2008 and EU 234/2011 for food additives, enzymes, and flavorings. The materials used for the production of edible packaging must be part of this list and be used according to the specific food where they are allowed (e.g., fruits, cheese, or bread actually have different additives that are allowed in each case). In the United States, the Food and Drug Administration (FDA) lists the additives that can be used in food products in a document that also presents the allowed specific applications of such additives as coatings (only the mentioned ingredients can be used) (FDA, 2013). The particular section "Indirect Food Additives: Adhesives and Components of Coatings" presents specific ingredients that can be used in the production of edible packaging (FDA, 2011).

One of the interesting points in the United States is that fresh fruits and vegetables having a coating should be labeled to inform the consumers that the product that they are buying is coated (FDA, 2012). This is undoubtedly a way to supply all the information to consumers and win their confidence.

One of the potential applications of edible packaging is their use as a wrap, thus in some cases, they are not supposed to be eaten by the consumer. When this is the final application, the legislation that should be considered is the one applied for materials intended for food contact.

Current legislation is clear regarding the ingredients and materials that can be added to foods and which of those can be eaten, and the materials that are approved as food contact materials. As a rule of thumb, packaging materials used as films/wraps should follow food contact material guidelines, and in the specific case of packaging that is expected to be eaten by the consumer, the rules for food additives should be used instead.

1.4 FUTURE PERSPECTIVES

Despite the great number of research works and products already available in the market, the use of edible packaging to guarantee the quality and safety of food products is still under development, and investment is still needed. Edible packaging is without any doubt one of the emergent technologies that meets consumers' demands for sustainability, use of biodegradable materials, and replacement of synthetic additives by natural compounds. The use of new materials, their modification to enhance their properties, the evaluation of their combination with other processing and conservation techniques, as well as their use as biodegradable packaging are some of the trends of edible packaging that are expected to increase their possible applications and, consequently, commercial interest in the near future.

Several new materials have appeared in recent years as good candidates for the development of edible packaging solutions. However, a full characterization is always needed, along with attention to the regulatory issues regarding the approval of new ingredients (to guarantee consumer safety). The use of agriculture feedstock, animal, and marine food processing industry wastes is very attractive once such materials can, in principle, be used directly in packaging of foods while adding value to materials that otherwise are treated as a subproduct or as a waste.

New approaches to enhance transport, mechanical, and thermal properties, as well as to address the control of moisture in edible packaging are needed, despite the great number of studies regarding these topics, success cases are limited. Chemical structure modification (e.g., creation of hydrophobic groups) and the interaction between molecules of different materials (e.g., protein–polysaccharide blends, nanocomposite incorporation, and cross-linking) are some of the methodologies used to obtain edible packaging materials with enhanced properties (Abdollahi et al., 2013; Martins et al., 2012; Rhim et al., 2013; Su et al., 2010); nevertheless, in these cases, the edibility of the corresponding packaging materials has not always been guaranteed.

Today a great number of methods are used in the processing and preservation of food products. From ultraviolet or gamma radiation sterilization to refrigeration and high-pressure processing, there are several techniques that need to be used in combination with the packaging materials. This implies evaluating their behavior during the application of those processing and preservation techniques, including the possibility of materials modification and their interaction with food products. This calls for more studies to understand how edible packaging could be used in combination with these techniques and thus guarantee the quality and safety of the products.

When edible packaging is presented as biodegradable packaging and focused as one of the solutions for the replacement of synthetic packaging, the market and food applications increase. However, the still underperforming properties (when compared with petroleum-based synthetic materials), the adaptation of processing devices, and the cost of the materials used are some of the problems faced in the development of biodegradable packaging. In the past years, several authors and companies have presented this possibility, replacing synthetic- and petroleum-based materials with bio-based, biodegradable materials (e.g., Ecoflex® and Mater-Bi®).

One of the further potential applications of edible packaging is their use as vehicles for bioactive and functional compounds (active and smart packaging), where the release of such compounds can be controlled by the materials used in the packaging production. Besides the diffusion control of bioactive compounds into the matrix (depending on the materials used and on the interaction between materials), the encapsulation of these bioactive compounds followed by their entrapment in the packaging matrix can also be of great interest, once it is possible to control their release rates under different external environments. The use of nanostructures such as nanohydrogels, nanoemulsions, and nanoparticles for the incorporation of antifungals, vitamins, and antioxidants, respectively, should increase the potential applications of active and smart edible packaging (Cerqueira et al., 2014; Woranuch and Yoksan, 2013).

The future of edible packaging materials is promising, and their use for very specific applications is already viewed as an innovative solution close to market. Still open is their use to replace synthetic nonbiodegradable packaging, where great efforts are needed regarding the investment in research (both public and private), which can turn the global use of edible packaging into a reality during the next decade.

ACKNOWLEDGMENTS

Miguel Ângelo Parente Ribeiro Cerqueira (SFRH/BPD/72753/2010) is a recipient of a fellowship from the Fundação para a Ciência e Tecnologia (FCT, Portugal). The authors acknowledge the FCT Strategic Project PEst-OE/EQB/LA0023/2013 and the project "BioInd–Biotechnology and Bioengineering for Improved Industrial and Agro-Food processes," REF.NORTE-07-0124-FEDER-000028, cofunded by the Programa Operacional Regional do Norte (ON.2—O Novo Norte), QREN, FEDER.

REFERENCES

Abdollahi, M., Alboofetileh, M., Behrooz, R., Rezaei, M., and Miraki, R. 2013. Reducing water sensitivity of alginate bio-nanocomposite film using cellulose nanoparticles. *International Journal of Biological Macromolecules*, 54, 166–173.

Allingham, W. J. 1949. Preservative coatings for foods. United States Patents Office US2470281 A.

Bernstein, M., and Woods, M. 2013. *Edible Coatings for Ready-to-Eat Fresh Fruits and Vegetables*. Washington, DC: American Chemical Society. Accessed on July 10, 2014.

Brekke, J. E., and Watters, G. G. 1959. Coating of raisins and other foods. United States Patents Office US2909435 A.

Brown, S. E., and Verrall, A. P. 2010. Carboxymethyl cellulose-based films, edible food casings made therefrom, and method of using same. World Intellectual Property Organization, Patents WO2009018503 A1. Switzerland: Monosol, LLC.

Cadwell, L. L. 1943. Meat treatment. United States Patents Office US2337645 A. United States: Industrial Patents Corporation.

Cerqueira, M. A., Costa, M. J., Fuciños, C., Pastrana, L. M., and Vicente, A. A. 2014. Development of active and nanotechnology-based smart edible packaging systems: Physical–chemical characterization. *Food and Bioprocess Technology*, 7(5), 1472–1482.

Cornwell, R. T. K. 1951. Protective coating composition for hams. United States Patents Office US2558042 A. United States: American Viscose Corp.

Codex Alimentarius. 2014. Codex Alimentarius. Available at http://www.codexalimentarius. org. Accessed on July 10, 2014.

De Ore, M. R. 1933. Process of treating fruit for the market. United States Patents Office US1900295 A. United States: FMC Corporation.

Durkee, E. L., Edison, L., and Hamilton, W. E. 1962. Process for coating food products. United States Patents Office US3046143 A.

FDA. 2011. Indirect food addittives: Adhesives and components of coatings. *21CFR175*. U.S. Food and Drug Administration.

FDA. 2012. Raw produce: Selecting and serving it safely. U.S. Food and Drug Administration.

FDA. 2013. Food additives permitted for direct addition to food for human consumption. *21CFR172*. U.S. Food and Drug Administration.

Fennema, O. R., Kamper, S. L., and Kester, J. J. 1987. Edible film barrier resistant to water vapor transfer. World Intellectual Property Organization WO 1987003453. Switzerland: Wisconsin Alumni Research Foundation.

Fennema, O. R., Kamper, S. L., and Kester, J. J. 1990. Method for making an edible film and for retarding water transfer among multi-component food products. United States Patents Office US 4915971 A. United States: Wisconsin Alumni Research Foundation.

Girard, G. 2013. Composition and methods for improving organoleptic properties of food products. World Intellectual Property Organization WO 2013049928 A1. Switzerland: Fruitsymbiose Inc.

Griffith, E. L., and Hall, L. A. 1933. Protective coating. United States Patents Office US1914351 A. United States: Griffith Laboratories.

Hussey, E. S. 1986. Process for preserving seafood. United States Patents Office US 4585659 A.

Lazard, L., Doreau, A., and Nadison, J. 1997. Edible films. E. P. Office, Patents EP0547551 B1. Belgium: National Starch and Chemical Investment Holding Corporation.

Lockhart, H. E. 1997. A paradigm for packaging. *Packaging Technology and Science*, *10*(5), 237–252.

Macquarrie, R. 2014. Edible film compositions for processing of meat products. United States Patents Office US8728561 B2. United States: Living Cell Research Inc.

Martins, J. T., Cerqueira, M. A., Bourbon, A. I., Pinheiro, A. C., Souza, B. W. S., and Vicente, A. A. 2012. Synergistic effects between κ-carrageenan and locust bean gum on physico-chemical properties of edible films made thereof. *Food Hydrocolloids*, *29*(2), 280–289.

Pira, S. 2013. The future of bioplastics for packaging to 2020: Global market forecasts. Available at http://www.smitherspira.com/products/market-reports/packaging/innovations-and-technologies/the-future-of-bioplastics-for-packaging-to-2020. Accessed on December 10, 2015.

Rhim, J.-W., Park, H.-M., and Ha, C.-S. 2013. Bio-nanocomposites for food packaging applications. *Progress in Polymer Science*, *38*(10–11), 1629–1652.

Robertson, G. L. (Ed.) 2012. Edible, biobased and biodegradable food packaging materials. In *Food Packaging: Principles and Practice* (3rd ed., pp. 49–90). Boca Raton, Florida: CRC Press.

Shaoxiao, Z., Yunyun, L., Longtao, Z., Yi, Z., Baodong, Z., and Xuhui, H. 2014. Purple potato starch edible film and preparation method thereof. TSIPOOTPSRO China, Patents CN103739883 A. China: Fujian Agriculture and Forestry University.

Su, J.-F., Huang, Z., Yuan, X.-Y., Wang, X.-Y., and Li, M. 2010. Structure and properties of carboxymethyl cellulose/soy protein isolate blend edible films crosslinked by Maillard reactions. *Carbohydrate Polymers*, *79*(1), 145–153.

Toulmin, J. H. A. 1956. Method of preserving shrimp. United States Patents Office US 2758929 A. United States: The Commonwealth Engineering Company of Ohio.

Van, M. P. J. J., Kosters, H. A., Van, D. H. G. A., and Smegen, J. 2001. Bio-degradable coating composition for food and food ingredients, in particular cheese. European Patent Office EP1299004 A2. The Netherlands: CSK Food Enrichment BV.

Verena, C. S., and Specht, K. 1959. Wax treatment of coffee beans. United States Patents Office US 2917387 A. United States: Tobeler, P.O.

Wood, R. 1939. Cheese coating. United States Patents Office US 2172781 A.

Woranuch, S., and Yoksan, R. 2013. Eugenol-loaded chitosan nanoparticles: II. Application in bio-based plastics for active packaging. *Carbohydrate Polymers*, *96*(2), 586–592.

Zepf, P. 2009. Glossary of packaging terminology and definitions. In K. L. Yam (Ed.), *Encyclopedia of Packaging Technology* (pp. 1287–1304). New Jersey: John Wiley & Sons, Inc.

2 Edible Film and Packaging Using Gum Polysaccharides

Marceliano B. Nieto

CONTENTS

Abstract 10
2.1 Introduction 10
    2.1.1 WikiCells and WikiPearls 11
    2.1.2 Edible, Dissolvable Packaging 12
    2.1.3 Edible Films, Wrappers, and Food Coatings 13
2.2 Film-Forming Properties of Gum Polysaccharides 14
2.3 Thickening, Gelling, and Film-Forming Properties 22
2.4 Method of Manufacture and Properties of Edible Gum Film 23
2.5 Source, Chemical Structure, and Film-Forming Properties of Gum Polysaccharides 25
    2.5.1 Nonionic Gum Polymers with Extended Twofold Helix Structures 25
        2.5.1.1 Methylcellulose (E461; CAS#99638-59-2; 21CFR§182.1480; FEMA 2696) 26
        2.5.1.2 Hydroxypropyl Methylcellulose (E464; CAS#9004-65-3; 21CFR172.874) and Hydroxypropyl Cellulose (E463; CAS#9004-64-2) 28
        2.5.1.3 Ethylcellulose (CAS#9004-57-3; E462) 30
        2.5.1.4 Fenugreek (CAS#977155-29-5; FEMA 2484; GRAS as Botanical Substance to Supplement Diet) 31
        2.5.1.5 Guar Gum (E412, CAS#9000-30-0; 21CFR184.1339; FEMA#2537) 32
        2.5.1.6 Tara Gum (E417; CAS#39300-88-4; GRAS) 34
        2.5.1.7 Locust Bean Gum (E410; CAS#9000-40-2; 21CFR184.1343; FEMA#2648) 35
        2.5.1.8 Konjac Gum (E425; CAS#37220-17-0) 38
    2.5.2 Ionic Gum Polymers with Extended, Twofold Helix Structures 40
        2.5.2.1 Carboxymethylcellulose or Cellulose Gum (E466; CAS#:9004-32-4; 21CFR 182.1745) 41
        2.5.2.2 Sodium Alginate (E401; CAS#9005-38-3; 21CFR184.1724; FEMA 2015) 42

         2.5.2.3 Propylene Glycol Alginate (E405; CAS#9005-38-3;
                 21CFR172.858; FEMA#2941)... 45
         2.5.2.4 Chitosan (CAS#9012-76-4)... 47
   2.5.3 Gums Polymers with Extended, Threefold Helix Structures............. 49
         2.5.3.1 Agar (E406; CAS#64-19-7; 21CFR 182.101;
                 FEMA#2006) 49
         2.5.3.2 Carrageenan (E407; CAS#9000-07-1; 21CFR 172.620;
                 FEMA#2596)... 51
         2.5.3.3 Gellan (E418; CAS#7101052-1; 21CFR172.665) 54
   2.5.4 Gum polymer with Extended, Fivefold Helix Structure.................... 56
         2.5.4.1 Xanthan Gum (E415; CAS#11138-66-2; 21CFR172.695) 56
   2.5.5 Gum Polymers with Extended, Sixfold Helix Structures................... 59
         2.5.5.1 Curdlan (CAS#54724-00-4; E424; 21CFR172.809)............. 59
   2.5.6 Gum Polymers with Complex Conformations................................... 61
         2.5.6.1 Pectin (E440; CAS#7664-38-2; 21CFR184-69-5)............... 61
         2.5.6.2 Pullulan (E1204; CAS#9257-02-7; GRAS) 64
2.6 Composite Films.. 67
   2.6.1 Rapid Melt Film Composite... 70
   2.6.2 Slow Melt Film Composite 70
   2.6.3 Insoluble Composite Films 70
2.7 Conclusions.. 71
References... 72

ABSTRACT

Developing cost-effective biodegradable films and rigid or semirigid packages that match the properties of synthetic plastics have received growing interest in the last decade. Progress of bio-based polymers using renewable resources has been made as follows: (i) using natural bio-based polymers with partial modification to meet the requirements (e.g., starch); (ii) producing bio-based monomers by fermentation/conventional chemistry followed by polymerization (e.g., polylactic acid, polybutylene succinate, and polyethylene); and (iii) producing bio-based polymers directly by bacteria (e.g., polyhydroxyalkanoates). In addition to the given examples of materials, there are many gum polysaccharides that are commercially produced today that meet the above requirements of being bio-based, natural, partially modified, or produced from bacterial fermentation, and these gum polymers are discussed individually in this chapter.

2.1 INTRODUCTION

The European Bioplastics reported that today there are three groups of bioplastics in the market: (1) bio-based or partially bio-based polyethylene (PE), polyethylene terephthalate, polypropylene (PP), polytrimethylene terephthalate, thermoplastic ether–ester elastomer (TPC-ET); (2) new bio-based and biodegradable plastics such as polylactic acid (PLA) and polyhydroxyalkanoates (PHA); and (3) biodegradable plastic based on fossils such as polybutylene adipate-*co*-therepthalate and polycaprolactone.

When it comes to food packaging, group 2 bioplastics such as the PLA and PHA are currently being produced commercially. Food bioplastic such as cellophane, a biodegradable cellulose acetate film, was invented a long time ago in 1908, and cellulose-based food casings for sausages, hotdogs, salami, bologna, and deli meats were invented in 1926 but these have not really competed with plastics. Then, NatureWorks, LLC, owned by Cargill Inc., was the first to manufacture polylactide or PLA from fermentation of corn starch; and now Walmart uses biopolymers by employing polylactide in packing fresh-cut produce (Bastioli 2005). However, bioplastics are currently more expensive than most petroleum-based polymers; hence, substitution has increased packaging cost that has become the major deterrent for its use as multipurpose packaging material. PLA for instance commands 30%–40% price premium compared to conventional films but a lot of efforts have been spent to reduce this price to a single digit parity as reported by BI-AX International (http://www.plasticsnews.com/article/20141105/NEWS/141109968/firm-cuts-cost-of-pla-packaging-film). Another bioplastic being commercially produced via bacterial fermentation of sugars or lipids is the PHAs, the polyhydroxybutyrate being the most popular PHA used in food packaging (Liu 2006).

What do gum polymers offer as alternative packaging materials? Gum polysaccharides are structural materials that come from renewable resources of plant, animal, or microbial origins. They are food grade and are used in almost any food product that are sold and consumed all over the world. In addition to providing thickening, gelling, foaming, emulsifying, and other functionalities in food systems, they also form films that can potentially be used as edible and biodegradable packaging materials. At their current production capacities and supply volume, they are not as cost-effective as petroleum-based material; however, they are renewable and can be farmed extensively or produced in a fermentation plant to whatever demand volume is required to make them competitive.

While it is still not economically feasible to replace plastic as packaging for food products, there are many specific food applications where edible films and packaging can replace the synthetic plastic polymers, which are described in the following sections.

2.1.1 WikiCells and WikiPearls

WikiCells is a food packaging idea created back in 2009 by Dr. David Edwards and colleague Francois Azambourg based on the principle of enclosing food and drink inside a soft skin or hard shell using edible film (Figure 2.1). Edible film is key, and gum polysaccharides play a potential role in creating this skin. Gums from seaweed sources like the alginate from brown algae can easily cross-link with Ca^{2+} to create this skin. Alternatively, gelling gums such as agar and carrageenan can be used that will form a gel around the cold food on contact. WikiPearls, on the contrary, are iterations of WikiCells that are bite-size foods, that is, the size of meatballs (made of ice cream, yogurt, juice, desserts, etc.) that are surrounded

FIGURE 2.1 WikiPearl bite-size food portions with outer membrane made with edible, food-grade gum polysaccharides. (With permission, copyright 2015. WikiFoods, Inc.)

by gel-like edible skin made of the same seaweed gum polymers. WikiPearl bite-size yogurts are now being marketed by Stonyfield in many whole food stores in New England and expanding in New York, New Jersey, and Connecticut.

2.1.2 Edible, Dissolvable Packaging

Dissolvable packaging comes to mind for many different types of food that are now packed in plastic or paper packages, such as liquid sauces for dinner entrees, liquid flavors for various soups, high intensity sweeteners, pasta sauce, powder flavor mixes, etc. (Figure 2.2). Similar to the concept of soluble detergent packs, ManoSol, LLC plans to extend to dissolvable food packaging and wrapping that

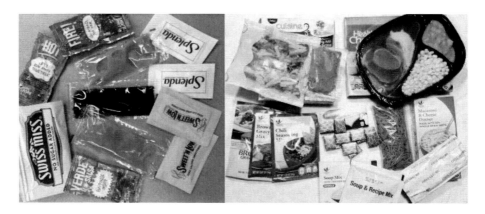

FIGURE 2.2 Examples of food packaging that can be potentially replaced by dissolvable, transparent packaging materials, for example, individually packaged chocolate mix, various sauces, high intensity sweeteners, gravy/soup flavors, bouillon cubes, soup mixes, cheese flavor, and macaroni individual packs, plastic film cover of dinner entrees, individually packages veggies, and sauce or gravy.

is sure to have a bigger impact in the food industry. The new packaging technology, composed of Food and Drug Administration (FDA)-approved edible polymers that have an undetectable taste and smell, is ideal for items we dissolve in water anyway, and there are plenty of these materials in the form of gum polysaccharides that we will be discussing in this chapter. Monosol is expected to hit the market soon, so it may not be long before we start to see these dissolvable wrappers and packages in our grocery store for products like hot chocolate, drink mix, instant, oatmeal, pasta, and more (http://www.fastcompany.com/1812661/monosol-creates-innovative-dissolvable-packaging-combat-waste).

Biodegradable, edible, and dissolvable wrappers that will replace Saran wrap are also an attractive option to reduce packaging wastes. These wrapping material will be used like Saran wrap, and end users have the option of either reheating or recooking the food together with the wrapper that can be made to dissolve when heated, or use it exactly like the plastic wrap that are thrown away after use, but have the benefit of being biodegradable.

2.1.3 Edible Films, Wrappers, and Food Coatings

Gum polymers can hold insoluble particulates, soluble solutes, drugs, flavors, and other active ingredients in the film matrix as shown in Figure 2.3. Other polymers, such as proteins and lipids, can be used in conjunction with the gums to alter the solubility, water vapor, and oxygen permeabilities of the edible film. Plasticizers—such as glycerol, propylene glycol, and sugar alcohols—can also be used to increase the plasticity and elasticity of the film. There is a wide choice of gum polymers, including their various combinations, that can be used in film applications such as breath film, sore throat, and cold–flu film strips, fresh produce coatings, confection coatings and glazes, etc. When one gum polymer or a combination of gums is used,

FIGURE 2.3 Example of edible films made with various gum polysaccharides including flavored film, pepper film, various fast dissolving edible films as carrier for drugs and actives, sushi wrap, edible glitters.

the property of the final film is dictated by both the gum polymers used and the composite effects of all ingredients used to make the film.

With the versatility and established safety of most gum polymers, novel applications for gum films seem almost limitless. The concept of drug, vitamin, and mineral delivery films will likely expand and while there are more and more of these novelty film products appearing in stores, other creative concepts are likely to evolve, including cheese, ketchup, or sour cream films, or various spice films that food scientists, product formulators, or chefs could create. The convenience and microbial stability of dried forms of food films are the biggest selling point for developing these products. The market for edible films has already experienced noteworthy growth, from ~$1 million in 1999 to more than $100 million in 2006 (http://www.amfe.ift.org/cms/?pid=1000355). In 2013, annual sales for edible film is still estimated at around $100 million per year but that the biggest potential application is the ready-to-eat, cut fruits and vegetables that now account for about 10% of all produce sales, with sales exceeding $10 billion annually (http://www.foodproductdesign.com/News/2013/09/Health-Conscious-Consumers-Fuel-Edible-Film-Growt.aspx).

Active ingredients, flavors, and colors can be incorporated directly into these recipes before the film is cast. These active ingredients, which can comprise up to 30% of the film by weight, become locked into the film matrix, and must remain stable until product consumption. Examples of actives used in film strips include ingredients for oral hygiene such as mint, alertness such as caffeine, as well as nutrients and botanicals.

2.2 FILM-FORMING PROPERTIES OF GUM POLYSACCHARIDES

Gums are complex polysaccharides of different molecular weights (MWs) and conformation that are linkages of one type of sugar monomer, or two or more different sugars. They have an abundance of free OH groups that are polar and water loving, and technically they belong to a family of polymers called hydrocolloids. In addition, they can have anionic and cationic groups that increase their polarity, and steric groups such as a sugar substitution on its linear backbone, or longer side chains or branching, or ester groups that are naturally present in the polymer, or ether groups with different polarity that are chemically added to the parent polymer structure. All these features contribute to their varying functional properties when used in foods, pharmaceutical, and industrial applications. Their main functions include thickening, gelling, film forming, and foaming/emulsification. Thickening, gelling, and film formation are gum properties that are related to each other, specifically to the structures the gum polymers form when they dissolve in water. When gums are hydrated in water, their free OH groups interact with water by forming hydrogen bonds. At the same time, the polymer strands unwind and assume a conformation that is stabilized by intramolecular hydrogen bonds and intermolecular hydrogen bonds between polymer chains.

Figure 2.4 shows how gums interact with water during the dissolution process. The free OH groups from the sugar molecules making up the gum polymers form hydrogen bonds with water that open up the dried polymer chains and allow them to unwind. If polymer chains are long and extended, intermolecular hydrogen

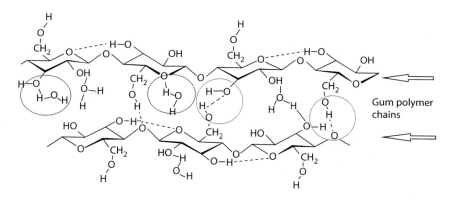

FIGURE 2.4 Interaction of gum polymers with water showing hydrogen bonding between OH groups of the sugar component of the gum and water molecule, as well as intermolecular hydrogen bonding between polymer chains of the gum.

bonding occurs as shown by the dark dotted lines. When this happens, water is trapped in a 3-D network or matrix, resulting in the immobilization of the water molecules as manifested by thickening of the water, or in other instances gelling. Sudo (2011) showed using hydroxypropyl cellulose (HPC) that the motion of free water was restricted by HPC molecules, when the dielectric relaxation parameters were monitored and this was attributed to the formation of a cholesteric-phase structure of the HPC molecules; while Patra et al. (2012) determined that relaxation dynamics of water slows down with a progressive increase in HPC content, indicating restriction of the relaxation pathway of water molecules that translates to thickening. When it comes to film formation, the intermolecular hydrogen bonds that are formed between aligned polymer chains are the structures that create the film. These structures are preserved when gum solutions are cast on a surface and dried.

Different gum polysaccharides come in varying MWs naturally or induced by processing. Linear gum polymers with MW of 50,000 or higher have shown to form good film structures. Many gum polymers have MW over 1 million such as konjac, guar, xanthan, curdlan, and pullulan; however, their film-forming property vary in much the same way as they vary in thickening properties. Factors that affect their film-forming properties are as follows:

a. *Structural conformation*: Gum polysaccharide form ordered conformations based on measurements after they are fully hydrated and dried into a film. From the physical properties observed and measured when hydrating gums in water at room temperature (RT) or from heating and cooling their solutions, it is widely shown that gum polysaccharides assume ordered structures corresponding to their lowest energy state that give them varying mechanical and rheological properties. These ordered conformations depend on the sugar residues making up the molecule and the glycosidic linkages joining the monomers together.

They can be broadly grouped as either linear or branched, the linear gum polymers being of particular interest in this chapter, because these are the types that produce good film. Linear gums assume conformations ranging from a twofold helix to a sixfold helix (Figure 2.5) that determine how extended the gum polymer is and how much their polymer chains can associate with each other in forming: (1) intramolecular hydrogen bonds that help stabilize their conformations, and (2) intermolecular hydrogen bonds that create networks that trap and immobilize water resulting in thickening, gelling, or film formation. The more stretched the helix is, as in the case of gums forming twofold helices such as konjac, galactomannans, and cellulose derivatives, the longer or the more extended the hydrated polymer chains are. These form better network from interpolymer association and therefore, better thickening and film structure. However, linear gums have other structural features such as varying MWs, presence of ionic groups, and steric groups (such as sugar substitution on the linear backbone) that reduce or restrict interpolymer association ultimately affecting their inherent thickening and film-forming properties. There is a good reason why a 1% konjac gum gives the highest viscosity (up to 25,000 cP at 25°C and 20 RPM) of all the nongelling gums at the same concentration. Konjac is a high MW, neutral polymer that forms twofold helices that are extended or elongated and hence has a very high hydrodynamic volume; the absence of surface charge also allows better intermolecular associations to form water trapping networks.

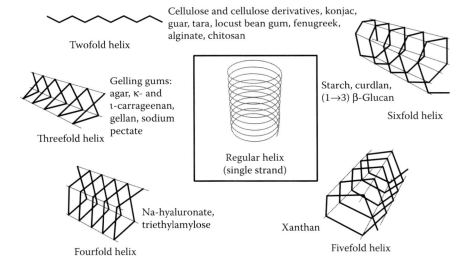

FIGURE 2.5 Representation of structural conformation of extended, helical gum polymer backbones showing two- to sixfold helices. Sugar monomers, dimers, or tetramers make up each side of various helices and are called repeating units. Side chains and other steric groups (not shown) will be protruding outside of the helix axis.

b. *Molecular weight*: For the same gum with different MWs, the lower MW grade will always hydrate and achieve full viscosity faster. Unwinding of a large molecule is expected to take longer during the hydration process. Therefore, a low MW carboxymethylcellulose (CMC) with degree of polymerization (DP) of ~100 will achieve full viscosity faster than the high MW CMC with a DP of 1000; however, at the same concentration, the low MW grade will give a lower final viscosity because its polymer strands are shorter and less efficient in forming water-immobilizing network. The high viscosity achieved with a large MW gums indicates better structures are formed and, therefore, stronger films.

c. *Ionic charge on the molecule*: For different linear gums with the same MW, anionic gums will always hydrate faster than the linear, neutral gums because the ionic groups make their molecules more polar and more readily able to create hydrogen bond with water. Therefore, linear anionic gums such as CMC, alginate, λ-carrageenan, and xanthan will hydrate faster than neutral gums such as guar and konjac because of the higher polarity of these anionic polymers. This means they will form weaker intermolecular hydrogen bonds and weaker film structure if they are not protonated with mineral ions like Na^+, K^+, or Ca^{2+} to neutralize charge repulsion. Both alginate and CMC are produced commercially as sodium salt; hence, their name sodium alginate and NaCMC, and carrageenans and xanthan also come naturally with mineral ions. When dissolved in water and cast into film, they form film with decent tensile strength that dissolves rapidly in water compared to films from nonionic gum polymers.

d. *Steric groups*: Sugar monomers making up the gum polymer theoretically have three free OH groups. In some gums, some of these OH sites are esterified with methyl, acetyl, glyceryl, or sulfate esters naturally as in the case of pectin, konjac, xanthan, gellan, and carrageenans, or synthetically as in case of cellulose derivatives that can be produced with various ether substitution. These groups alter the intermolecular hydrogen bonding between polymer chains of the gums due to steric hindrance and hence, film-forming property. Some of these steric groups are nonpolar such as the methyl and ethyl groups that change the hydrophilic–lipophilic balance value of the gum that affect the melting/sealing property of the film and also give them foaming property. Other gums like the galactomannans have galactose substitution on the linear mannose chain that protrudes out of the helix backbone to help open the molecule. They reduce intermolecular hydrogen bonding due to steric hindrance, and hence, depending on how much galactose substitution is present, the film-forming property of galactomannans varies.

Table 2.1 summarizes the structural features of various gum polysaccharides and their key chemical and physical properties, showing the diversity of gum polysaccharides and the variability of their sugar compositions, conformation, and physical characteristics.

TABLE 2.1
Comparison of Relative Viscosities at 1% Gum in Water, Sugar Composition, Conformation, and Film-Forming Properties of Various Gum Polysaccharides

Gum Polysaccharide	Relative Viscosity, cP (1%, 20 RPM, 25°C)	Sugar Composition, Others	Other Structural Features	Ionic Charge, etc.	Film-Forming and Other Properties
Methylcellulose, DP ~1000 (A4M grade)	50–150	Glucose with only methyl ether substitutions, DS 1.64–1.92	Extended, twofold helix	Neutral, modified	Cold water soluble (CWS), foams heavily, 3% w/w to make film, strong film that dissolves slowly, available in different viscosity grades
HPMC (K4M grade)	200–400	Glucose with both methyl (DS 1.12–2.03) and hydroxypropyl (DS 0.07–34) substitutions	Extended, twofold helix	Neutral, modified	CWS, foams heavily, 3% gum in water to make film, strong film that dissolves slowly, available in different viscosity grades
HPC (EF grade)	6–10	Glucose with only hydroxypropyl substitutions (DS not more than 4.6)	Extended, twofold helix	Neutral, modified	CWS, foams heavily, 3%–5% gum in water to make film; strong film that dissolves slowly, available in different viscosity grades
Ethylcellulose (N100 grade)	1800–2400	Glucose with ethyl ether substitutions, typical DS 2.6	Extended, twofold helix	Neutral, modified, hydrophobic	Dissolves in organic solvent, insoluble in water, 5% in IPA to make film, very strong film, heat sealable, shrink wrap properties, available in different viscosity grades
Fenugreek	1000–3400	Mannose, galactose (~1:1 ratio)	Extended, twofold helix	Neutral, natural	CWS dissolves up to 3% for ease of casting, brittle film without plasticizer, needs thicker cast
Std. grade, high MW guar	5000–6500	Mannose, galactose (~2:1 ratio)	Extended, twofold helix	Neutral, natural	CWS, slow hydrating, dissolves at 3% ease of casting, brittle film without plasticizer, available in different viscosity grades

(Continued)

TABLE 2.1 (Continued)
Comparison of Relative Viscosities at 1% Gum in Water, Sugar Composition, Conformation, and Film-Forming Properties of Various Gum Polysaccharides

Gum Polysaccharide	Relative Viscosity, cP (1%, 20 RPM, 25°C)	Sugar Composition, Others	Other Structural Features	Ionic Charge, etc.	Film-Forming and Other Properties
Tara gum	3500–5200	Mannose, galactose (~3:1 ratio)	Extended, twofold helix	Neutral, natural	Swells and partially thicken in cold water, 3% gum in water, full hydration at ~65°C, brittle film without plasticizer
Locust bean	2500–3500	Mannose, galactose (~4:1 ratio)	Extended, twofold helix	Neutral, natural	Sparingly soluble in cold water, requires 82°C to full hydrate, brittle film without plasticizer, stronger film versus guar gum
Konjac	13,000–25,000	Glucose and mannose with ~13% acetyl substituted groups	Extended, twofold helix	Neutral, natural	Dissolves slowly in cold water, heating hastens hydration, 1.0%–1.5% gum concentration to make film, needs thicker cast due to low gum solids, strong film
Pullulan	5–10	Glucose	Extended, step ladder conformation	Neutral, natural	CWS; 20%–25% in water to make film; strong film that dissolves fast, film heat sealable and stretchable
Chitosan	4–9	Glucose with amine and acetyl amine substitution	Extended, twofold helix	Cationic, natural	Dissolves in acidic solution like acetic acid and becomes cationic; strong film
High MW CMC	4000–7000	Glucose with carboxymethyl substitutions, DS 0.7–1.2	Linear, twofold helix	Anionic, modified	CWS, 1.5%–3% in water to cast into film, flexible moderately strong film compared to other cellulose derivatives, not heat sealable, available in different viscosity grades and DS

(*Continued*)

TABLE 2.1 (Continued)
Comparison of Relative Viscosities at 1% Gum in Water, Sugar Composition, Conformation, and Film-Forming Properties of Various Gum Polysaccharides

Gum Polysaccharide	Relative Viscosity, cP (1%, 20 RPM, 25°C)	Sugar Composition, Others	Other Structural Features	Ionic Charge, etc.	Film-Forming and Other Properties
Sodium alginate (std grade)	400–755	Guluronic acid, mannuronic acid	Linear, twofold helix	Anionic, natural	Very soluble in water and tends to form lumps with insufficient shear, dissolves up to 8% depending on grade, produced mainly as a sodium salt, moderately strong film
Propylene glycol alginate	100–200	Guluronate/mannuronate ester of PG, guluronic acid, mannuronic acid	Linear, twofold helix	Slightly anionic, modified	CWS, foams, 4% gum in water to make film, moderately strong film
Agar	Insoluble	D-galactose, 3,6-anhydro-L-galactose	Extended, threefold helix	Neutral, natural	Regular grade requires boiling to dissolve; quick soluble grade requires 180°F, 4%–6% gum in water to make film, strong clear film, does not heat seal on its own
κ-Carrageenan	800–1600	D-galactose, 3,6-anhydro-D-galactose with sulfate ester substitutions	Extended, threefold helix	Anionic, natural	Swells in cold water but requires 82°C to fully hydrate, 4%–6% to make film, cast hot solution, strong clear film, does not heat seal on its own
ι-Carrageenan	1000–2000	D-galactose, 3,6-anhydro-D-galactose with sulfate ester substitutions	Extended, threefold helix	Anionic, natural	Thickens cold but requires ~65°C or higher to fully hydrate, 5%–8% gum in water to make film, moderately strong clear film
λ-Carrageenan	2000–3000	D-galactose with sulfate ester substitutions	Extended, twofold helix	Anionic, natural	CWS, dissolves up to 8%, smooth flow, easy to cast, moderately strong clear film, does not heat seal on its own

(Continued)

TABLE 2.1 (Continued)
Comparison of Relative Viscosities at 1% Gum in Water, Sugar Composition, Conformation, and Film-Forming Properties of Various Gum Polysaccharides

Gum Polysaccharide	Relative Viscosity, cP (1%, 20 RPM, 25°C)	Sugar Composition, Others	Other Structural Features	Ionic Charge, etc.	Film-Forming and Other Properties
Low acyl gellan	4000–5500	Rhamnose, glucose, glucuronic acid with minimal glyceryl and acetyl ester groups on the glucose	Extended, threefold helix	Anionic, natural	Swells in cold water; full hydration at 82°C, 3% gum in water to make film, moderately strong film
Xanthan	1200–1600	Glucose, glucuronic acid, mannose with and without acetyl ester and pyruvate groups	Extended, fivefold helix	Anionic, natural	CWS, 3% gum in water to make film, weak film, needs thicker cast, moderately fast dissolving film
Curdlan	500–750	Glucose	Extended, sixfold single to triple helix	Neutral, natural	CWS, gels on heating, 3% gum in water to make film, slow dissolving film
Unstandardized HM pectin	25–100	Methylgalacturonate, galacturonic acid, xylose, rhamnose, arabinose, galactose, apiose	Homogalacturonan threefold helix, branched hairy regions	Anionic, natural	CWS but requires heat to fully dissolve, 6%–8% gum in water to make film, slow dissolving film

2.3 THICKENING, GELLING, AND FILM-FORMING PROPERTIES

Thickening of water, gelation, and film formation are direct results of structures created when gum polymers are dissolved in water with or without heating. Cross-linking agents are not required to create these structures and these structures are easily manipulable or improved using other food-grade cross-linkers, salts, or plasticizers. How are they related? If a gum solution is more viscous at the same gum concentration, this means that there are better 3-D structures created as related to intermolecular hydrogen bonding or interpolymer chain association. In some extreme examples, as in the case of gelling gum polymers, aggregation of their polymer chains is so efficient that they can solidify water into a gel. Thickening and gelling are all indication of 3-D structures that are forming when gum is hydrated in water, and these same structures are responsible for film formation.

As earlier pointed, gelation is an exceptional case of gum structure creation when gum is dissolved in water. Only a few gums will naturally form a gel in aqueous system. Examples are agar, κ-carrageenan, ι-carrageenan, and gellan, and all these gum polymers form threefold helices that aggregate on cooling of their heated solutions resulting in gelation. They vary in gel strengths which is a direct result of how much aggregation of their helices happens. Agar forms the strongest gel structure because its molecule is neutral and hence, association of the helices is not restricted by charge repulsion unlike in the anionic molecules of κ- and ι-carrageenans which carry a negative charge from the sulfate groups. Therefore, agar forms stronger films compared to carrageenans. Gellan, on the contrary, has steric effect from the acetyl and glyceryl ester groups. Other gelling gums include: (1) alginate and low methoxy pectin that can form gel by cross-linking with a bivalent ions like Ca^{2+}; (2) those that are structurally synergistic with each other like the blend of xanthan–konjac or xanthan–locust bean gum (LBG); and (3) the high ester pectin that gels when the required dissolved sugar solids and pH are met as a result of both hydrophilic and hydrophobic associations, as the pectin polymer chains are brought close to each other. In all of these examples, film strength is greatly improved when they form a gel or when their polymer chains are cross-linked as a result of more structures being created.

The solubility of the film follows that same rule as the solubility of the parent gum. The rule of thumb is gums that are soluble in cold water will form films that are also soluble in cold water. Conversely, gums that dissolve in boiling water or hot water will produce film that will dissolve in boiling or hot water and will be insoluble in cold water until heated. Corollary to this, gum powders of CMC, alginate, λ-carrageenan, and xanthan, being anionic and more polar, will dissolve faster than the neutral gums such as guar and konjac in RT water; therefore, their films will also dissolve faster in the mouth when used in edible oral thin films with the thickening effect of the parent material. The low MW grades of these gums form weaker film compared to the high MW grades but with less thickening when their films dissolve in the mouth. Gums, like agar or κ-carrageenan that requires boiling or heating to 180°F (82°C) to fully dissolve, will also require more or less the same heating conditions for their films to dissolve.

As one can imagine, gum films or dissolvable packaging with different melting and textural attributes when eaten or when they dissolve in water can easily be tailored to the end application. To name a few examples, applications, like breath film, drug delivery film, chewable nicotine film, edible candy wrapper, sushi wrap, etc., will require different gum films to achieve acceptable textural properties and solubility. Breath films and drug delivery films need to be readily soluble in the mouth with minimal thickening of the saliva. This means that low viscosity gums with good film-forming properties are suitable like pullulan and the anionic, highly water soluble gums such as low viscosity CMC, low MW alginate, or a gum blend. Edible wrapper for candies also requires low viscosity and fast disintegration. Sushi wrap, on the other hand, requires an insoluble gum film similar to dried seaweed sushi wrap. There are several gum films that will work in this application like the insoluble agar film, cross-linked calcium–alginate film, or to some extent κ-carrageenan film. Thick gums like the high viscosity CMC, xanthan, alginate, carrageenan, guar gum, konjac, and methylcellulose, although good film formers are worse in terms of mouthfeel. When their films dissolve in the mouth, they tend to thicken the saliva which is unacceptable in most edible film applications. Films made with nonionic gums tend to be chewy and dissolve slowly. However, these gum films will find suitable application when used as dissolvable packaging material, where the films are meant to dissolve and provide thickening to give a sauce or soup consistency during heating.

2.4 METHOD OF MANUFACTURE AND PROPERTIES OF EDIBLE GUM FILM

Gum polysaccharides are very unique hydrocolloids. After dissolving them in water, thickening happens as a result of structure formation when the molecule unwinds and imbibes water. When cast on a surface and dried, they leave films that vary in tensile strength, clarity, and solubility. Nieto (2009) compared these gum films to each other, and his results can be summarized as follows:

   a. Linear high-MW, nonionic gum polymers generally formed the stronger films that peeled off in one piece even at casting thickness of 10 mils (254 microns) at low gum concentration of 1.5%–3% w/w as shown by konjac and methylcellulose.
   b. Branched gums such as gum arabic, gum ghatti, and larch gum, on the contrary, formed weak, noncohesive films that flaked rather than peeled and are excluded from the list of gum films discussed in this chapter. Linear, anionic gums (carrageenans, alginate, and CMC) produced good films but with tensile strengths lower than konjac and methylcellulose.
   c. Linear, neutral gums such as fenugreek, guar, and LBG with varying galactose substitution on the linear mannose backbone produced better films compared to branched gums but weaker compared to the linear, anionic gums indicating that the size of the side chain poses a greater steric hindrance to intermolecular association and film structure.

d. Inulin, a linear, neutral oligosaccharide, did not form film because its molecule is too small to create the film network.
e. Film clarity and solubility also depended on the characteristics of the parent gum. All cellulose derivatives and refined grades of agar, carrageenans, and alginates produced clear films and seed gums such as fenugreek, guar, LBG, xanthan gum produced hazy films. In terms of solubility, agar, LBG, and κ-carrageenan produced insoluble to slow dissolving film in RT water and in the saliva; konjac, guar, and methylcellulose films dissolved slowly in the saliva; while the anionic gum films dissolved fast with varying thickening effects.
f. All films showed very good oil barrier properties, as expected. Likewise, altering solubility and disintegration of the gum films were achieved by a proper combination of these gum polysaccharides.

Figure 2.6 shows a simple lab process for making films from gum polysaccharides, consisting of dissolving the gum in water, casting to the right thickness, and drying. Concentration depends on the gum type and grade to achieve an optimal casting viscosity of 6000–10,000 cP. Casting can be done on a smooth surface such as a glass plate, a piece of stainless steel, or a plastic (*mylar*) sheet. Casting thickness can vary between 10 mils (0.254 mm) or thicker depending on the gum and gum concentration and target tensile strength. The film is then dried to the proper moisture or humidity in an oven or ambient air, and then peeled. Overdrying makes gum films stick strongly to the plate and will make peeling difficult. Applying a very thin layer of release agent such as potassium stearate (5% solution) or oil-based

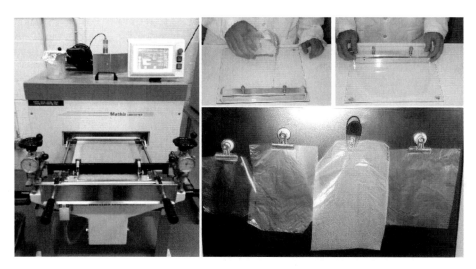

FIGURE 2.6 A simple lab process for making gum films, using a lab coater (left) or manual (right) both showing a casting knife spreading a thin layer of gum solution on stainless plate or glass plate. The process on left involves drying the plate with gum coat inside the machine, and the process on right involves air drying in either an oven or ambient air, then film is peeled manually.

material such as nonstick spray (e.g., PAM) will mitigate this problem. Alternatively, a lab coater/dryer, such as the one shown in Figure 2.6, can be used to cast the gum solution at desired thickness and suitable casting speed depending on the viscosity of gum solution so that a uniform layer of gum coat is spread on the surface. The plate is then sent into the machine for drying where air flow, air temperature, dwell time, and surface temperature can be programmed and controlled. The gum film is then peeled manually.

Nieto (2009) also provided a comprehensive guideline for the use of gum polysaccharides in film applications to achieve the desired attributes, such as choosing the gum or gum combination, the concentration of the gum that will provide optimal viscosity for casting, pointers on how to dissolve the gum in water to achieve desired attributes, the use of release agent so that gum films peel off easily, etc. In a commercial setting, the production of composite films and flakes requires an elaborate setup of machines for automated processing on an industrial scale, and details of this commercial process for the manufacture of edible films are described in detail by Rossman (2009).

2.5 SOURCE, CHEMICAL STRUCTURE, AND FILM-FORMING PROPERTIES OF GUM POLYSACCHARIDES

All gum polymers discussed in this chapter come from renewable resources and can be produced to be cost competitive with oil-based materials for food packaging, if deemed to be a potential film and packaging material. Raw material sources such as cellulose, various plants, shellfish wastes, and fermentation products offer many options for polymeric materials that are GRAS and approved for use in foods. This section discusses the source of the gum polymers, their chemical structure, their film-forming properties, and film studies done on each one of them.

2.5.1 Nonionic Gum Polymers with Extended Twofold Helix Structures

There are many gum polymers that fall into this group of structures of being nonionic, extended twofold helices which are very good film formers. They vary in solubility due to the presence of steric groups in the sugar monomers making up the helix backbone which are either naturally present or chemically attached during the modification of the parent material. For instance, cellulose derivatives are chemically modified celluloses, where different groups are substituted for the –OH groups of the sugar, such as methyl, hydroxypropyl, or ethyl groups. These groups add nonpolar character to the polymer, and depending on the degree of substitution (DS), the gum derivative can have both polar and nonpolar behavior for foaming or could be transformed into a nonpolar polymer that only dissolves in organic solvent but with improved sealability as packaging material. Other gum polymers, such as galactomannans (fenugreek, guar, tara, and LBG), naturally have varying amount of sugar substitutions (galactose) on its mannose backbone that similarly gives these polymers different solubilities in cold water. Another example of a nonionic, twofold helical gum polymer is konjac, a glucomannan that has naturally occurring acetyl substitution on its sugar monomers. These gum polymers vary in MWs relative to each other; furthermore, for any of these gums, commercial grades are available differing mainly in MWs that directly correlate to viscosity, hence, film-forming property.

2.5.1.1 Methylcellulose (E461; CAS#99638-59-2; 21CFR§182.1480; FEMA 2696)

The main raw materials for the manufacture of methylcellulose are pine trees/wood pulp and cotton linters, as all the other cellulose derivatives. Its manufacture consists of treating cellulose (α-cellulose, cotton linters, or wood pulp) with 50% aqueous sodium hydroxide solution at 60°C and pressure of 2.26×10^4 MPa (http://www.lotus-biotech.com/en/shownews.asp?id=188), then reacting alkali cellulose with methylchloride, under controlled conditions, purifying, drying, and milling to a fine powder. The resulting polymer, which has the same backbone as cellulose, is soluble in cold water unlike the parent material. Depending on the raw material or the MW of the starting cellulose, different viscosity grades of methylcellulose are created. Average MW of methylcellulose grades range from 13,000 to 140,000 or DP of 50–750 (http://www.brenntagspecialties.com/en/downloads/Products/Multi_Market_Principals/Aqualon/Benecel_Brochure.pdf).

Chemically, methylcellulose is an ether of cellulose with a methyl DS of 1.6–1.9 on O6, O2, or O3 positions of glucose units (Figure 2.7). Like cellulose, the glucose

FIGURE 2.7 Molecular structure of methylcellulose showing four glucose units in a tight twofold helix conformation, where the glucose units are in $^4C_1$ conformation (upright and inverted to show each turn of the helix) and with methyl ether substitution or DS of 1, 2, 3, and 0 (no substitution) on each of the four glucose units shown or average DS 1.5. Also shown are pictures of its film, cast from a 3% solution and thickness of 10 mils (0.01 inch or 254 μm) and 20 mils (0.02 inch or 508 μm). Tensile strength of dried film measured across a 2.5-cm wide film strip using the TA.XT.PlusTexture analyzer, yielding tensile strengths of 760 and 1400 g and puncture strengths of 350 and 830 g, respectively. (With kind permission from Springer Science+Business Media: *Edible Films and Coatings for Food Applications*, Structure and function of polysaccharide gum-based edible films and coatings, 2009, pp. 57–112, Nieto, M.)

units are connected by β-(1 → 4) glycosidic linkages. It forms an extended twofold helix conformation, but unlike cellulose, it is capable of dissolving in cold water and forming viscous solutions. The addition of the methyl ether groups to the cellulose backbone opens up its structure as a result of the steric effects of the $-CH_3$ groups, thus preventing the polymer strands from reassociating into microfibrils when redried into powder. Although substitution reduces the number of free hydroxyl groups on the glucose molecules, the level of substitution still allows hydrophilic interactions of the polymer with water with the remaining free OH groups, as well as interpolymer hydrogen bonding. In addition, there are intermolecular hydrophobic interactions between the nonpolar methyl groups, and both interactions contribute to the film structure of methylcellulose and to the ability of methylcellulose to foam when mixed in water. Methylcellulose contains 27.5%–31.5% methoxyl, or a DS of 1.64–1.92, a range that yields maximum water solubility. A lower DS gives products with lower cold water solubility, leading to products that are only soluble in caustic solutions; while higher degrees of substitution with nonpolar methyl groups produce methylcellulose products that are soluble only in organic solvents. The level of substitution of 27.5%–31.5% imparts optimum functionality to commercial grades of methylcellulose as a hydrocolloid.

Methylcellulose is unique in the sense that, although it completely dissolves in cold water and forms a clear solution, its solution gels upon heating to a temperature ranging between 48°C and 64°C depending on the grade and concentration. The gelling phenomenon is a stage in the flocculation process, where the methylcellulose molecules are losing their water of hydration and hydrophilic interactions, and switching to hydrophobic interactions of its nonpolar methyl groups. Depending on the concentration, however, the heated methylcellulose solution can either hold the gel or can completely fall out of solution and settle. For a 2% solution of high viscosity methylcellulose in water (~6000–8000 cP), the heated solution will hold an intact gel but will show some amount of syneresis or weeping. At lower concentrations, or in the presence of competing solutes or interfering insoluble solids, the methylcellulose can completely fall out of solution during heating. The presence of polar and nonpolar groups in methylcellulose makes it foam when mixed in water. The presence of nonpolar methyl groups also makes it difficult to disperse in RT water due to its tendency to float. It is best to disperse methylcellulose in hot water, 65–82°C or 150–180°F, depending on the grade, and then cool the solution with very slow stirring until it thickens into a clear solution without whipping and creating foam. Furthermore, methylcellulose solution is stable over the pH range of 3–11 (Murray 2000).

Methylcellulose film dissolves in cold water providing some thickening effect. When used as dissolvable food packaging for soup flavor, sauce, instant coffee/tea, or similar products, it can provide a case for the product and a thickening/texturizing effect after it dissolves. When used as an edible packaging on its own, the film, however, is not heat sealable and requires another layer of a melting component to form a seal.

There is hardly any study on methylcellulose as an edible packaging. One study by Debeaufort and Voilley (1997) characterized several edible films based on methylcellulose and polyethylene glycol 400 (PEG400) that showed that tensile strength of the films strongly depended on relative humidity or water content, more than on PEG400 content; on the contrary, their elongation is dependent both on water and PEG400.

2.5.1.2 Hydroxypropyl Methylcellulose (E464; CAS#9004-65-3; 21CFR172.874) and Hydroxypropyl Cellulose (E463; CAS#9004-64-2)

Hydroxypropyl methylcellulose (HPMC) and HPC are also derived from cellulose. The production of HPC starts with same alkali cellulose. The addition of the alkali, in combination with water, activates the cellulose matrix by disrupting its ordered crystalline structure, thereby increasing the access of OH groups by the alkylating agent to promote the etherification reaction. This activated matrix is called alkali cellulose (Kirk-Othmer 1993). To make HPC, the alkylating agent used to induce the etherification process is propylene oxide, under controlled conditions. HPMC, on the contrary, has two types of ether substitution such as methyl and hydroxypropyl. The current process involves a two-step etherification process where methyl groups are grafted first, then a second derivatization using propylene oxide is done to graft hydroxypropyl groups onto the cellulose. In another commercial scenario, a one-step etherification process can also be done where the methylchloride and propyne oxide reactions happen at the same time (http://www.lotus-biotech.com/en/shownews.asp?id=188). MW of commercial HPMC ranges from 13,000 to 200,000, while MW of HPC is from 30,000 to 1,000,000 based on JECFA specifications. These correspond to various viscosity grades of each cellulose ether that is commercially sold.

HPMC and HPC have the same sugar backbone as cellulose and composed of substituted glucose units that are glycosidically linked by β-(1 $\rightarrow$ 4) linkages. They assume a tight twofold helix conformation as shown in Figure 2.8, similar to methylcellulose and differ from methylcellulose only in the type of ether substitution. For HPMC, JECFA allows a methoxyl content of not less than 19% and not more than 30% on dry basis or DS of 1.12–2.03, and hydroxypropyl content of not less than 3% and not more than 12% or DS of 0.07–0.34. In contrast to methylcellulose, a 2% HPMC solution flocculates or gels at higher temperature (~82°C vs. 48–64°C for methylcellulose). For HPC, JECFA allows not more than 80.5% of hydroxypropyl content, equivalent to DS of not more than 4.6, where maximum DS for methylcellulose is 3.0. In the case of HPC, a DS of greater than 3 is possible, since there is a free OH in the hydroxypropyl group that can chain with another hydroxypropyl forming a side chain with more than one mole of combined propylene oxide. HPC has a critical solution temperature of 40–45°C at which the gum polymer will precipitate or fall out of solution, and below this critical temperature, HPC redissolves and forms viscous solution like HPMC and methylcellulose. Its solution viscosity remains unchanged as pH is varied over the range of 2–11 (http://www.ashland.com/Ashland/Static/Documents/ASI/PC_11229_Klucel_HPC.pdf).

Using a grade of HPC that gives 840 cP at 5% concentration in water at 20 RPM, the resulting HPC film cast 20 mils (0.002 inch thick) gave longitudinal tensile strength of 972 g and transversal tensile strength of 933 g, with some stretchability like the other cellulose derivatives. The film as shown in Figure 2.8 is very clear, and

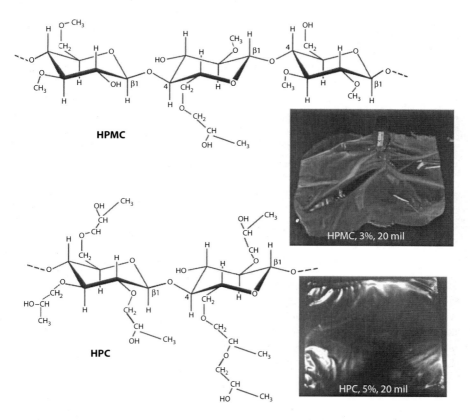

FIGURE 2.8 Idealized twofold helix structure of HPMC (top) with methyl substitution, DS of 1.67 and hydroxypropyl substitution, DS of 0.33 as drawn; and the twofold helix of HPC (bottom) with a DS of 3.0. Also shown are pictures of HPMC film, cast from a 3% solution and thickness of 20 mils (0.02 inch or 508 µm) and HPC film cast from a 5% solution and at the same thickness. Tensile strengths of dried HPMC film as measured across a 2.5-cm wide film strip using the TA.XT.PlusTexture analyzer was ~1100 g and puncture strength was ~590 g. (With kind permission from Springer Science+Business Media: *Edible Films and Coatings for Food Applications*, Structure and function of polysaccharide gum-based edible films and coatings, 2009, pp. 57–112, Nieto, M.)

like the methylcellulose and HPMC films, it dissolved rather slowly in water and in the mouth.

Bilbao-Sainz et al. (2010) studied how to improve the mechanical strength and moisture barrier properties of HPMC edible film by reinforcing it with microcrystalline cellulose (MCC) and lipid coating (LC). They found that the HPMC/MCC and HPMC/LC-MCC films increased in tensile strength by up to 53% and 48%, respectively, in comparison with unfilled HPMC films. Furthermore, addition of unmodified MCC nanoparticles reduced the moisture permeability by up to 40%, and the use of LC-MCC reduced this value by up to 50%; and finally, they concluded that water vapor permeability (WVP) was mainly influenced by the differences in water solubility of different composite films.

Reddy et al. (2012) studied the miscibility of HPC and PEG to produce film. They reported that the materials are miscible when HPC is more than 40% of the polymer blends. They theorized that the compatibility of the polymer composition is due to the formation of hydrogen bonding between hydroxyl groups of the HPC and etheric atom of the PEG molecules. Reddy et al. (2013) also studied the miscibility of HPC and polyvinyl alcohol (PVA) and concluded that HPC/PVA blends are miscible at all compositions due to the strong intermolecular hydrogen bonding interactions between the two polymers. These results open new insight on how the film properties of biofilms can be improved for use as food packaging. Alyanak (2004) compared HPC film with NaCMC film and found that it produced film with lower water vapor sorption capacity or high water vapor barrier ability and higher tensile strength. This result is expected since it is nonionic and less polar compared to CMC. This also means that its film will dissolve slower when added to water. Whether it is used as film coating or dissolvable packaging, this slower solubility may be good for some specific situations.

2.5.1.3 Ethylcellulose (CAS#9004-57-3; E462)

The raw materials for the production of ethylcellulose are the same as the other cellulose derivatives such as wood pulp and cotton pulp. It starts with alkali cellulose which is reacted with ethyl chloride, under controlled conditions. Ethyl chloride and pulp are the only reactants in the process to make ethylcellulose, but processing aids are used to activate the pulp which is then removed to meet acceptable residual limits. The reacted material undergoes filtration, precipitation, washing and neutralization, centrifugation, drying and milling to mesh specification (http://www.fda.gov/ucm/groups/fdagov-public///www.fao.org/ag/agn/jecfa-additives/specs/monograph13/additive-178-m13.pdf). The MW of ethylcellulose depends on the MW of the starting material, but the MW of commercially produced ethylcellulose by DOW Wolff Cellulosics ranges from 44,900 to 232,200 Da (http://www.dow.com/dow-wolff en/pdf/192-00818.pdf).

The chemical structure of ethylcellulose is a β-(1 → 4)-glucose linked, twofold helix where most of the free OH groups are substituted with ethoxyl groups (Figure 2.9). DS for ethylcellulose is 2.6 on average, leaving a minimal OH groups free to interact with water. The molecule is, therefore, hydrophobic in contrast to the other cellulose derivatives where polar and nonpolar groups are in good balance. This explains the low dielectric constant ethylcellulose which ranges from 2.5 to 4.0 at 25°C and 60 Hz and is only soluble in organic solvents but forms film that softens at 133–138°C, and melts at 165–173°C (Majewics et al. 2002).

Ethylcellulose is an excellent film former. At 10 mils, longitudinal tensile strength of ethylcellulose is 1229 g (average) and transversal tensile strength is 1405 g (average) with extensibility is 1.59 and 2.89 mm. At 50 mils (1270 μm), average longitudinal tensile strength is 5052 g and transversal tensile strength is 5809 g, with extensibility of 1.86 and 3.59 mm. The film is heat sealable and flexible especially at 10 mil. Viscosity of the ethylcellulose used in this experiment was 5700 mPa s at 3% in isopropyl alcohol.

According to DOW (http://www.dow.com/dowwolff/en/pdf/192-00818.pdf), ethylcellulose yields a greater volume of film-forming solids than any other cellulose

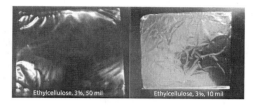

FIGURE 2.9 Molecular structure of ethylcellulose showing four monomer units in a tight twofold helix conformation, where the glucose units are in $^4C_1$ conformation (upright and inverted to show each turn of the helix) and with ethyl ether substitution or DS of 2.5. Also shown are pictures of its film cast from a 3% solution in isopropyl alcohol and thickness of 10 mils (0.01 inch or 254 μm) and 50 mils (0.05 inch or 1270 μm). Tensile strength of dried film measured across a 2.5-cm wide filmstrip using the TA.XT.PlusTexture analyzer.

derivatives; its films are highly flexible and retain their flexibility at temperatures well below freezing, compatible with most common plasticizers and polymers, and excellent water barrier and absorb little moisture either on exposure to humidity or after long immersion in water.

2.5.1.4 Fenugreek (CAS#977155-29-5; FEMA 2484; GRAS as Botanical Substance to Supplement Diet)

Fenugreek (*Trigonella foenum-graecum*) is a renewable agricultural resource. It is indigenous to India, Middle East, and Southern Europe and currently is cultivated in Middle East, North Africa, Southern Europe, and North America (Mathur and Mathur 2005). Zimmer (1984) reported that fenugreek was first grown commercially in Western Canada. It requires ~120 days to produce mature seeds (Basu 2006). Today, it is now grown in various quantities all over the world. It is estimated that 80% of the world's fenugreek is produced in India (http://panhandle.unl.edu/at_the_center_72). Fenugreek galactomannan is produced by dehusking the seeds, swelling with 25% water, flaking, and then milling to powder. To remove the spice aroma of the seeds, a further purification is done using solvent like hexane to extract oil-based flavor, then the gum is dissolved in water and precipitated with alcohol (http://www.chemtotal.com/Fenugreek%20gum%20material.pdf; U.S. Patent 587109).

Fenugreek gum, the polysaccharide reserve in the seed endosperm, was commercially produced in 1993 (Mathur and Mathur 2005). Reported MW for fenugreek gum is 500,000–950,000 (http://www.chemtotal.com/Fenugreek%20gum%20material.pdf).

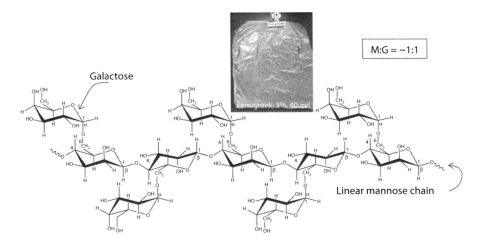

FIGURE 2.10 Idealized structure of fenugreek showing a twofold helix conformation with six mannose units shown, each mannose has a galactose substitution at C6 position. Also shown is picture of its film cast from a 3% solution and thickness of 40 mils (0.04 inches or 1016 μm). Tensile strength measured across a 2.5-cm wide film strip using the TA.XT. PlusTexture analyzer, yielding tensile strength of 2330 g and puncture strength of 230 g. (With kind permission from Springer Science+Business Media: *Edible Films and Coatings for Food Applications*, Structure and function of polysaccharide gum-based edible films and coatings, 2009, pp. 57–112, Nieto, M.)

Fenugreek gum is a galactomannan consisting of a linear chain of mannose linked by β-(1 → 4) glycosidic bonds, with galactose substitution at the C6 position of almost every mannose (Figure 2.10); hence, the ratio of mannose to galactose is ~1:1. Brummer et al. (2003) reported a mannose to galactose ratio of 1.02:1 to 1.14:1. A linear mannan polymer is completely insoluble in water, just like the native cellulose; however, the galactose substitution in fenugreek prevents polymer chains from associating intimately during drying and acting as spacer between polymer chains, making it completely soluble in cold water. In fact, fenugreek is the most soluble seed galactomannans.

2.5.1.5 Guar Gum (E412, CAS#9000-30-0; 21CFR184.1339; FEMA#2537)

Guar gum is the ground endosperm of the guar seed of the leguminous shrub *Cyamopsis tetragonoloba* that requires a growing season of 14–16 weeks. The guar bean is principally grown in India, Pakistan, United States, China, Australia, and Africa. India produces 250,000 to 350,000 metric tons of guar gum annually, making it the largest producer with about 80% of world production (http://en.wikipedia.org/wiki/Guar_gum). According to Trostle, in 2013, ~200 million pounds of guar is used in North America (or over 90,000 metric tons), and much of the guar gum consumed in the United States is imported from India and Pakistan. The industry expects the 2014 guar gum exports to be around 550,000 to 600,000 tons due to the revived demand from the food sector and the growing usage of this gum in the oil drilling sector (http://articles.economictimes.indiatimes.com/2014-02-06/

news/47089814_1_guar-seed-guar-gum-rajesh-kedia), unless the fracking industry demand for guar diminishes due to lower oil prices.

Processing of guar gum starts with the pod. First, the pods are dried in sunlight, and then seeds are manually separated for further processing. Seeds are dehusked by a mechanical process of roasting, differential attrition, sieving, and polishing. This break the seeds, germ is separated from the endosperm, and two halves of the endosperm are obtained from each seed that are called guar splits. Processing of the guar splits into guar gum requires soaking in water to swell the splits prior to flaking into thin ribbons. This flaking process disrupts the ordered packing of the guar molecule, enabling the finished guar powder to hydrate and open up completely for hydrogen bonding with the water and interpolymer chain association to give optimal viscosity and thickening. Various viscosity grades of guar gums are also produced commercially that correlate to various MWs of the gum polymer.

Guar gum is a galactomannan similar to fenugreek, tara, and LBG. It consists of a linear mannose chain linked by β-(1 → 4) glycosidic bonds, with galactose (G) substitution at the C6 position of approximately every other mannose (M) unit, or a M:G ratio of ~2:1 (Figure 2.11). It assumes a twofold helix conformation that is extended or elongated that favors thickening through intermolecular hydrogen bonding. Guar gum is a nonionic polysaccharide and consists of longer chain molecules (average DP 10,000 for standard grade of guar) than those found in LBG. Higher galactose substitution of guar gum compared to LBG (Petkowicz et al. 1998) increases the solubility of this galactomannan; hence, like fenugreek, guar achieves full hydration

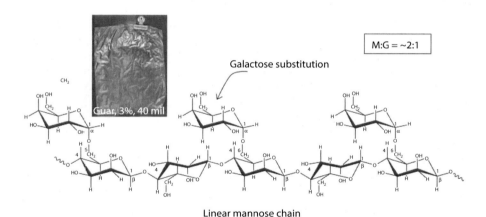

FIGURE 2.11 Idealized structure of guar gum showing five mannose units in a twofold helix conformation with a galactose substitution at every other mannose; two mannoses on the backbone make up one helix turn plus a galactose side chain. Also shown is guar gum film produced using the standard grade of guar gum, cast from a 3% solution and thickness of 40 mils (0.04 inches or 1016 μm). Tensile strength measured across a 2.5-cm wide film strip using the TA.XT.PlusTexture analyzer, yielding tensile strength of 4580 g and a puncture strength of 655 g. (With kind permission from Springer Science+Business Media: *Edible Films and Coatings for Food Applications*, Structure and function of polysaccharide gum-based edible films and coatings, 2009, pp. 57–112, Nieto, M.)

in cold water; however, guar dissolves slower. The standard grade of guar gum has a 1% viscosity in water of ~4500–5000 cP at 20 RPM.

Mikkonen et al. (2007) prepared films from guar gum and enzymatically modified guar gum to obtain structurally different galactomannans. They found that cohesive and flexible films were formed from guar galactomannan plasticized with glycerol, showing higher tensile strength of 13 mPa at a ratio of guar (1%):glycerol (0.2%) versus ~6 mPa for guar (1%):glycerol (0.6%), but elongation at break was the reverse. They also reported that modified guar gum with lower galactose side chains produced films with higher elongation at break and tensile strength, and modified guar galactomannans with approximately 6 galactose units per 10 mannose backbone units resulted in films with two peaks in loss modulus spectra, whereas films from galactomannans with approximately 2 galactose groups per 10 mannose units behaved as a single phase in dynamic mechanical analysis.

2.5.1.6 Tara Gum (E417; CAS#39300-88-4; GRAS)

Tara gum is the ground endosperm of the seed of *Caesalpinia spinosa* shrub or tree that can reach a height of 8 m with spreading, gray-barked leafy branches. The fruit is a flat oblong reddish pod which contains 4–7 large round black seeds composed of endosperm (22% by weight), germ (40%), and hull (38%) (http://en.silvateam.com/Products-Services/Food-Ingredients/Tara-gum). The tara tree is native to the Cordillera region of Peru and Bolivia in South America, where the fruit grows from April to December. Tara is also cultivated in Morocco and East Africa. Mature pods are usually harvested by hand and typically sun dried before processing. Tara seeds are separated from the pod by threshing, sieving to remove impurities, roasting to break the hull, and removing the germ, then tara gum in the form of splits, like guar, is soaked in water and milled to produce tara gum in powder form (http://en.silvateam.com/Products-Services/Food-Ingredients/Tara-gum/Production-process).

In the European Union (EU), only two grades of tara gum are traded—a high viscosity grade that is 6000 mPa s at 1% and medium viscosity grade with 4500 mPa s at 1% (http://www.cbi.eu/system/files/marketintel_documents/2014_natural_thickeners_-_product_factsheet_tara_gum.pdf).

Tara gum is another galactomannan comprised of approximately 25% galactose and 75% mannose. Like the other galactomannan, tara assumes a twofold extended helix conformation with a linear mannose backbone and a galactose substitution occurring on approximately every third mannose unit, or M:G ratio of 3:1 (Figure 2.12). Compared to guar gum, it has less galactose substitution which translates to more intermolecular association and reduced solubility. Tara hydrates in RT water slower than guar and requires heating to fully dissolve and achieve full viscosity. Like LBG, tara gum forms a composite gel with xanthan and also exhibits gelling synergy with both agar and carrageenan.

Antoniou et al. (2015) successfully produced tara gum films with the inclusion of bulk chitosan or chitosan nanoparticles at various concentrations and reported that the incorporation of chitosan nanoparticles improved mechanical, physicochemical, and barrier properties of tara gum films. Tensile strength was increased by 35.73 MPa, while the elongation was decreased by 7.21%. Water solubility and WVP were reduced by 74.3% and 22.7%, respectively. Additionally, the microstructure of

Edible Film and Packaging Using Gum Polysaccharides

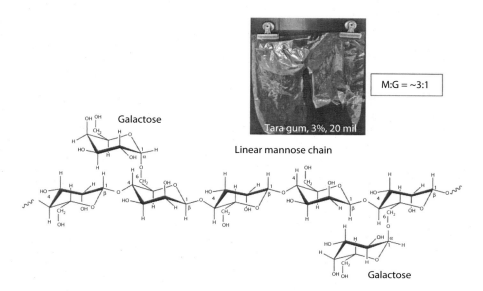

FIGURE 2.12 Idealized structure of tara gum showing five mannose units in a twofold helix conformation with a galactose substitution at C6 position of every third mannose. Also shown is a picture of its film cast from a 3% solution and thickness of 20 mils (0.02 inch and 508 μm). Tensile strength measured across a 2.5-cm wide film strip using the TA.XT. PlusTexture analyzer, yielding tensile strength of 1365 g and puncture strength of 390 g. (With kind permission from Springer Science+Business Media: *Edible Films and Coatings for Food Applications*, Structure and function of polysaccharide gum-based edible films and coatings, 2009, pp. 57–112, Nieto, M.)

the films showed that the nanoparticles were distributed homogenously within the structure but it increased the roughness of the surface. Furthermore, they reported that tara gum films with bulk chitosan exhibited better antimicrobial activity.

2.5.1.7 Locust Bean Gum (E410; CAS#9000-40-2; 21CFR184.1343; FEMA#2648)

Locust bean gum, or carob bean gum, is derived from the endosperm of carob seeds from the tree *Ceratonia siliqua* that grows in semiarid environments of the Mediterranean region. The tree starts bearing fruits/pods after 5–7 years and from flowering it takes 6–8 months for the pods to mature before they are harvested. The seeds usually comprise 8%–10% of the pod by weight, and the LBG is basically the endosperm portion of the seed that account for about half of the whole seeds (http://www.fao.org/docrep/v9236e/v9236e06.htm). According to FAO Corporate Document Repository for seed gums (http://www.fao.org/docrep/v9236e/v9236e06.htm), the processing of LBG involves the following steps:

1. Pods are taken to the kibbling factories, where they are left to dry for about a month. They are then crushed and broken in the kibbling machines, which are usually of the hammer mill type, and put through a series of sieves

which sorts the broken pieces according to size, and the seeds are further separated from pieces of pod of the same size by blowing air through the mixture.
2. The next step involves removal of the seed hull. This is achieved either by mechanical abrasion or by chemical treatment. In one method, the seeds are roasted, which loosens the hull and enables it to be removed from the rest of the seed; the remaining part is cracked and the crushed germ, which is more friable than the endosperm, is sifted off from the unbroken endosperm halves. An alternative method is to treat the whole seed with acid at an elevated temperature; this carbonizes the hull, which is removed by a washing and brushing operation, and the dried germ/endosperm is then processed as before. Efficient removal of the hull prior to separation of the germ and endosperm is important, since residual specks of it will detract from the quality and value of the final product. The pieces of endosperm are then ground to the required particle size to furnish LBG.
3. To make the clarified LBG grade, additional steps such as extraction with hot water, centrifugation to remove insolubles, and precipitation with alcohol are done, then milling to fine powder.

El Batal and Hazib (2013) optimized the extraction step with water; and based on the RSM analysis, optimum conditions for LBG processing included a temperature of 97°C at 36 min and the water to endosperm ratio of 197:1. Robbins (1988) estimated total world exports of LBG at about 12,000 tons year^{-1} (or ~26 million pounds) over a 6-year period of 1979–1985 with the EU being the biggest market, while the United States imports averaged 2300 tons year^{-1}. The main gum-producing countries are Spain, Italy, and Portugal contributing about 5000, 3000, and 1500 tons, respectively, out of 12,000 tons. The remaining 2500 tons were accounted for mainly by Morocco, Greece, Cyprus, and Algeria. Turkey, Israel, India, and Pakistan produce locust bean but were not, then, believed to be significant traders of this gum. The demand for LBG seems to be more or less the same today. In 2011, United States imported ~2.9 tons with Spain as the largest source of imports (roughly 1800 tons), Italy (670 tons), and Morocco (400 tons) (http://en.wikipedia.org/wiki/Locust_bean_gum).

Molecular weight of LBG ranges from 50,000 to 3,000,000 based on JECFA specifications. Different grades of LBG are sold that vary in purity from powder with visible brown specks from the husk to fine mesh LBG without visible specks. Clarified LBG is also commercially being offered.

Locust bean gum is another galactomannan that consists of a linear backbone of mannose linked by β-(1 → 4)-glycosidic with the galactose substitution occurring at the C6 position of every fourth mannose unit or M:G ratio of 4:1 on average (Figure 2.13). LBG structure is a twofold extended helix. There are regions of the molecule that are densely substituted and regions where galactose substitution occurs only on every 10–11 mannose units; hence, the galactose content of LBG is reported to be between 17% and 26% (Wielinga 2000). The linear mannose chain is quite similar to cellulose, and if not for the galactose substitution, this gum would be insoluble in water. Heating to 82°C is required to fully activate LBG; thus, when

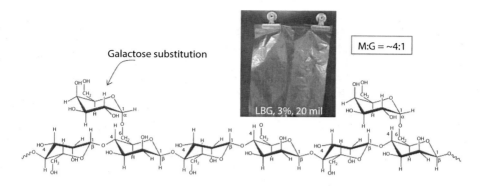

FIGURE 2.13 Idealized structure of locust bean gum showing six mannose units in a two-fold helix conformation with a galactose substitution at C6 position of every fourth mannose. Also shown is a picture of its film, cast from a 3% solution and thickness of 20 mils (0.02 inch and 508 μm). Tensile strength measured across a 2.5-cm wide film strip using the TA.XT. PlusTexture analyzer, yielding tensile strength of its film was 2000 g and puncture strength was 460 g. (With kind permission from Springer Science+Business Media: *Edible Films and Coatings for Food Applications*, Structure and function of polysaccharide gum-based edible films and coatings, 2009, pp. 57–112, Nieto, M.)

its solution is redried, LBG reverses to being almost insoluble in cold water like the original material. Therefore, if used in a film application, LBG is one gum polymer that can be used to make film that will sparingly dissolve in cold water but dissolves and thicken on heating.

A 1% fully hydrated LBG has a viscosity of 2500–3300 mPa s at 1% w/w. It exhibits strong synergy with xanthan gum, and a 60/40 to 50/50 blend of LBG/xanthan yields the most synergistic ratio (Sworn 2000). At concentrations of up to 0.4% of the total gum blend, its heated solution shows a very pseudoplastic, gel-like behavior. Between 0.4% and 1.0% total gum concentrations, it is a soft gel; at 1% total gum concentration or higher, it forms an elastic gel that is not prone to syneresis. LBG is also synergistic with κ-carrageenan with the optimum ratio being around 20/80 LBG to carrageenan. LBG makes the carrageenan gel more elastic. This synergy also gives stability to the carrageenan gel, allowing its use in acidic applications with pH values between 3.0 and 5.0. Similar synergy is exhibited by LBG and agar.

Mikkonen et al. (2007) prepared films from LBG:glycerin at 1:0.2% and 1:0.6% ratios. Tensile strength was higher with less glycerin but elongation at break was weaker. Aydinli et al. (2004) showed that WVP of LBG films with PEG 200 and PEG 600 increased. However, when PEG 1000 was used, WVP of the films decreased which they explained was due to the solid state of PEG 1000 at RT. Bozdemir and Tutas (2003) found that the WVP of LBG films with PEG 200 and sorbitol gave the lowest WVP values, whereas films containing glycerol gace the highest WVP. Chen and Nussinovitch (2001) studied WVP, oxygen permeability, and CO_2 permeability of wax films blended with LBG and guar gum compared to commercial coatings with polyethylene and shellac. The incorporation of LBG and guar gum in wax coatings decreased the permeability to oxygen and CO_2. They also studied LBG and guar gum in traditional wax formulations applied to citrus fruits and showed that LBG-wax coating produced juice with the best taste (Chen and Nussinovitch 2000a,b).

Bozdemir and Tutas (2003) found that when LBG was blended with stearoptene and beeswax, the WVP of films gave lower WVP when compared to control LBG films. They also showed that if PEG 200 and sorbitol were used as plasticizer, LBG film with stearoptene gave lower WVP than LBG film with beeswax. More recently, Martins et al. (2011) studied different ratios of LBG and κ-carrageenan on their film properties. The films composed of 40/60 carrageenan/LBG showed a synergistic effect presenting enhanced water vapor barrier and mechanical properties.

2.5.1.8 Konjac Gum (E425; CAS#37220-17-0)

According to The Konjac Association of Chinese Society for Horticultural Science (http://www.konjac.org/english/readnews.asp?rid=449), konjac, known also as elephant's foot of devil tongue, represents a group of perennial herbal plants in *Amorphophallus* genus of Araceae family. There are altogether about 170 species of konjac in the world, distributed mainly in Asia and Africa. China is rich in the resources of konjac germplasm and boasts of 20 odd species, of which 13 can be found only in this country.

Briefly, the process for making konjac glucomannan (KGM) involves the following steps: (1) the harvested fresh konjac tuber is first washed, cleaned, and peeled, followed by slicing; (2) drying is then achieved in a special type of oven dryer in which the konjac slices are subjected to high temperature to stabilized color, followed by low temperature drying; and (3) then the dried slices are broken and made into fine powder and the starch, fiber, and other impurities removed.

Konjac is grown in China, Korea, Taiwan, Japan, and Southeast Asia for its large starchy corms that are used to create a flour and jelly of the same name. It is also used as a vegan substitute for gelatin. KGM comprises 40% by dry weight of the roots, or corm, of the konjac plant. China and Japan are the main producing countries of konjac flour with annual output of 5000 ton (or 11 million pounds) and export of 2500 ton for China, and 3000–4000 ton for Japan (http://www.hnxcfood.com/cs2_e.asp). Based on JECFA specifications, konjac has a MW of 200,000–2,000,000 Da; however, actual MW of KGM depends on the konjac variety, processing method, and storage time of the raw material.

Konjac is a heteropolysaccharide consisting of glucose and mannose monosaccharide units in the ratio of 5:8 (Shimahara et al. 1975b), connected by β-(1 → 4) glycosidic bonds (Figure 2.14). Physical images, based on x-ray data revealed that the backbone conformation of the chain is a twofold helix stabilized by intramolecular O3...O5 hydrogen bonds (Yui et al. 1992). This report is in agreement with KGM molecular chain parameters determined using the Mark–Houwink equation by Li et al. (2006). This conclusion, however, is contrary to the molecular structure proposed for konjac mannan in earlier studies by Shimahara et al. (1975a), Kato (1973), and Kato and Matsuda (1980), which proposed slight branching on every 50–60

Edible Film and Packaging Using Gum Polysaccharides

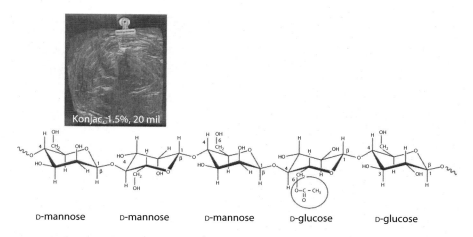

FIGURE 2.14 Idealized structure of konjac mannan linear chain showing the repeating units. (Adapted from Maeda, M., Shimahara, H., and Sugiyama, N. 1980. *Agricultural and Biological Chemistry*, 44:245–252.) Other proposed repeating structural units are G-G-M-M-M-G-M or G-G-M-G-M-M-M-M (Kato and Matsuda 1969) and G-G-M-M-G-M-M-M-M-M-G-G-M (Maeda et al. 1980; Takahashi et al. 1984). Also shown is a konjac film, cast from a 1.5% solution and thickness of 20 mils (0.02 inches or 508 μm) measured across a 2.5-cm wide film strip using the TA.XT.PlusTexture analyzer, yielding tensile strength of ~805 g and puncture strength of ~630 g. (With kind permission from Springer Science+Business Media: *Edible Films and Coatings for Food Applications*, Structure and function of polysaccharide gum-based edible films and coatings, 2009, pp. 57–112, Nieto, M.)

backbone sugar units. In addition, konjac mannan also contains approximately one acetyl ester group per 19 sugar residues (Maekaji 1978). The MW of konjac mannan depends on the species, or even variety, of *Amorphophallus* from which it is extracted and on the extraction method employed. Sugiyama et al. (1972) reported average MW of 0.67–1.9 million, depending on the *Amorphophallus* variety used, while Li et al. (2006) reported a average MW of 1.04 million Daltons. The idealized molecular structure of konjac is shown in Figure 2.14.

Konjac solution gels if it is heated after treatment or exposure to alkali. This gelation occurs as a result of the hydrolysis of the acetyl groups, which no longer hinder intermolecular hydrogen bonding or association of the polymer chains (Maekaji 1978). A further interesting characteristic of konjac gum lies in its synergy with other hydrocolloids. Takigami (2000) reported synergy between konjac mannan and xanthan such as the formation of an elastic gel and between konjac mannan–carrageenan and konjac mannan–agar as a significant increase in gel strength. Furthermore, konjac mannan is unique in that it produces a viscosity as high as 25,000 cP for a 1% solution, the highest viscosity ever reported for any nongelling gum. This is attributed to its extended or elongated twofold helix conformation, the absence of ionic charge on the molecule and its high MW. Konjac is a slow hydrating gum and much less prone to lumping. In solution, konjac gum is also quite heat stable similar to xanthan.

Konjac, like other gums with a linear structure and high MW, produces strong film. The acetyl ester groups pose a steric hindrance that, in this case, serves to somehow open the molecule by acting as spacer between polymer chains, enabling the konjac powder to dissolve in RT water, although much slower compared to guar gum. This means that konjac can form film structure even without heating its solution but heating will be advantageous. As could be expected, the removal of acetyl ester groups will strengthen the intermolecular association of konjac molecules in solution and will result in the formation of a gel and stronger films, but this will also make konjac powder insoluble in RT water and would require heating to dissolve, similar to agar powder. A potential disadvantage of konjac mannan in film application is that it is too viscous. Optimally, the concentration of a gum polymer and other solids in a film application should be at least 25% for processing efficiency. Konjac, however, can only be dissolved in water at concentrations of 1.5% maximum before its solution becomes too viscous to cast.

Wu et al. (2012) made edible blend films of KGM and curdlan by a solvent-casting technique using different ratios of the two polymers. The results showed formation of strong intermolecular hydrogen bonds between KGM and curdlan, and the interaction was much higher when the KGM content in the blend films was around 70 wt% with maximum tensile strength of 42.93 ± 1.92 MPa. Furthermore, the blend films displayed excellent moisture barrier properties.

2.5.2 IONIC GUM POLYMERS WITH EXTENDED, TWOFOLD HELIX STRUCTURES

Anionic gums exist as a polymers of sugar acids (e.g., alginate); sugar units biosynthesized with anionic substituents such as sulfate groups (e.g., carrageenans); or carboxyl groups that are substituted through a chemical reaction (e.g., CMC). The presence of these anionic groups increases the polarity and water solubility of gum polymers, though the inherent charge weakens intermolecular associations between polymer chains due to repulsion. However, these anionic gums are manufactured in the form of a salt, most commonly as sodium salt that protonates and neutralizes the anionic charge. When cast and dried into film, they form films with superior clarity and decent tensile strength. The negative charge on molecules prevents gums from forming an excessively tight fiber structure in the dried state, which characteristic allows the polymers to imbibe water readily during the hydration process without foaming. Depending on the extent and distribution of charge on polymer molecules, the gum is either completely soluble in cold water, as in the case of alginate, λ-carrageenan and the highly substituted CMC; or only partially soluble in water such as κ-carrageenan, ι-carrageenan, and pectin which are anionic gums but with threefold helix conformation that tend to aggregate; and therefore, require heat activation to fully hydrate the polymer molecules.

On the contrary, there are only few cationic gums that are currently produced and sold in commerce. Chitosan is one of them, a polymer of glucose amine where the amino group confers positive charge when the gum is dissolved in acidic solution. Several other grades of gums such as low methoxy, amidated pectin, and cationically modified guar that contain $-NH_2$ groups are also available in the market.

2.5.2.1 Carboxymethylcellulose or Cellulose Gum (E466; CAS#:9004-32-4; 21CFR 182.1745)

Carboxymethylcellulose or CMC is first commercialized in 1946 by Hercules Inc. It is a cellulose ether produced by reacting the alkali cellulose with sodium monochoroacetate, under controlled conditions. There are many degrees of substitution producing different properties of CMC; hence, many grades of CMC are in the market to perform different functions, that is, as an all purpose thickener, suspending system, stabilizer, binder, and film former in a wide variety of uses. Average MW of commercial CMC grades ranges between 90,000 for the low viscosity and 700,000 for the high viscosity grade, corresponding to DP of 400–3200 (http://www.brenntagspecialties.com/en/downloads/Products/Multi_Market_Princpals/Aqualon/Aqualon_CMC_Booklet.pdf). Viscosity grades ranging from 20–50 cP for a 2% concentration up to as high as 13,000 cP at 1% concentration in water are available, as measured at 30 RPM and 25°C. For general thickening applications, high viscosity grades are chosen for economic reasons, because it allows lower usage to achieve the same target viscosity. However, in other applications like flavored syrups, the rheology of high MW CMC and the presence of graininess or grit in solution due to unsubstituted and insoluble regions in the molecule are sufficient reasons to use lower viscosity grades which are less prone to this textural problem. In film applications, low viscosity CMC is also preferred over the viscous grades because higher gum concentrations can be achieved in casting solutions, which means less water to dry.

The CMC structure involves carboxymethyl substitution of the native cellulose polymer at the C2, C3, or C6 positions of anhydroglucose units (Figure 2.15). The DS

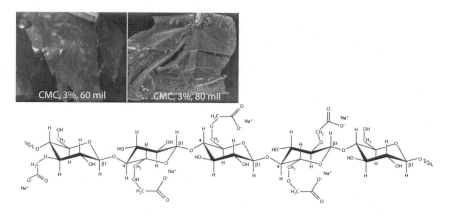

FIGURE 2.15 Idealized structure of carboxymethylcellulose showing five glucose units in a twofold helix conformation with an average DS of 1.0 (the fourth glucose from the left having two substitutions and the fifth having 0). Also shown is picture of its film cast from a 3% solution with thicknesses of 60 mils (0.06 inch or 1524 µm) and 80 mils (0.08 inch or 2032 µm). Tensile strength measured across a 2.5-cm wide film strip using the TA.XT. PlusTexture analyzer, yielding tensile strength of 1920 g and puncture strength of 1170 g. (With kind permission from Springer Science+Business Media: *Edible Films and Coatings for Food Applications*, Structure and function of polysaccharide gum-based edible films and coatings, 2009, pp. 57–112, Nieto, M.)

is generally in the range of 0.65–0.95, but it can be as high as 1.2. Higher DS means higher solubility for CMC. However, the uniformity of the substitution along the cellulose backbone also influences the solubility or smoothness of CMC solutions. The glycosidic linkage of CMC is the same as the parent cellulose and is composed of β-(1 → 4) glucose linkages that assume a twofold helix, extended conformation with lengths varying depending on the MW. The carboxylate ends of the carboxymethyl groups are free and therefore, CMC is anionic. In addition to the free –OH groups remaining in the sugar residues, it is very polar and has good solubility in water. CMC is produced as sodium salt or NaCMC, and when it is dissolved in water, the –COONa groups ionize to Na^+ and $–COO^-$, creating a negative charge on the gum molecule. At acidic pH levels of 3.0 or less, CMC becomes insoluble due to the protonation of the –COOH and loss of solubility. Figure 2.15 shows the structure of CMC and its film. It forms a clear film, water soluble with thickening property, but it does not heat seal.

Alyanak (2004) studied the film-forming properties of NaCMC and HPC and found that NaCMC film produced good mechanical properties but has a higher water vapor sorption capacities or low water vapor barrier properties. This is expected since CMC is anionic and more polar, whereas HPC is nonionic. If used in edible dissolvable packaging, this means that CMC film will dissolve faster when added to water or in the saliva when put in the mouth. Tongdeesontorn et al. (2011) studied the effect of CMC on properties of cassava starch films and found that CMC increased its tensile strength, reduced elongation at break, and decreased water solubility of the blended films. They further determined that FT-IR spectra indicated intermolecular interactions between cassava starch and CMC in blended films by shifting of carboxyl (C = O) and OH groups. Shekarabi et al. (2014) made composite films based on plum gum and CMC plasticized with glycerol and found that WVP of the films increased when glycerol is increased from 5% to 20%, and decreased tensile strength but this improved flexibility of the films and elongation at break.

2.5.2.2 Sodium Alginate (E401; CAS#9005-38-3; 21CFR184.1724; FEMA 2015)

Alginate is present in the cell walls of brown algae as calcium, magnesium, and sodium salts of alginic acid. Commercially, alginate is mostly sold as sodium alginate although potassium alginate, ammonium alginate, or ammonium–calcium alginate grades are also produced to some extent. Alginic acid has limited functionality and requires conversion to its salt form to improve its solubility and stability. Briefly, alginate processing starts with wet chopped seaweeds that are subjected to alkaline extraction, where the residue is removed and the alginate in solution is processed further to produce the final sodium alginate in two ways (http://www.fao.org/docrep/006/y4765e/y4765e08.htm):

1. Alginic acid process: Acid is added to the alginate extract to form alginic acid which forms a gel. To remove excess water from the gel, alcohol is added to the alginic acid extract, followed by sodium carbonate which converts the alginic acid into sodium alginate. Sodium alginate does not dissolve in the mixture of alcohol and water and is therefore precipitated out, pressed to remove excess water, and then dried and milled to an appropriate particle size.

Edible Film and Packaging Using Gum Polysaccharides

2. Calcium alginate process: In this method, calcium chloride is added to the alginate extract to form fibrous calcium alginate gels. The calcium alginate gel is then treated with acid to convert it to alginic acid. These fibrous alginic acid fibers are then reacted with sodium carbonate gradually, until all the alginic acid is converted to sodium alginate. The paste of sodium alginate is sometimes extruded into pellets that are then dried and milled.

Annual production of alginate has not really increased much from 33 million tons in 2001 (http://www.fao.org/docrep/006/Y4765E/y4765e08.htm) to 38 million tons in 2008 (Helgerud et al. 2010). Based on JECFA specification, commercial sodium alginate has average MWs ranging from 10,000 to 600,000 and these correspond to different viscosity grades.

Alginates are linear, unbranched polymers, containing β-$(1 \rightarrow 4)$-linked D-mannuronic acid (M) and α-$(1 \rightarrow 4)$-linked L-guluronic acid (G) units, and are therefore highly anionic and very polar polymers. Alginates are not strictly random copolymers but are instead block copolymers. Based on the study by Haug et al. (1966, 1969), alginates consist of blocks of both similar (or homopolymeric) and alternating (or heteropolymeric) sugar units (i.e., MMMMMM, GGGGGG, MG, and GM blocks) as shown in Figure 2.16, with each of these blocks having different conformation behaviors (Sabra and Deckwer 2005). The G block forms a twofold helix with a pitch of 8.7 Å, which is 1.7 Å shorter than in cellulose or mannan (Atkins et al. 1973b). The guluronic acid residues are in the $^1C_4$ conformation and are, therefore, diaxially linked along the polymer chain. This gives the structure of the G-block polymer a buckled, as opposed to flat, conformation. This helix is stabilized by O2...O61 hydrogen bonds, and adjacent helices are bridged by water molecule which are present one

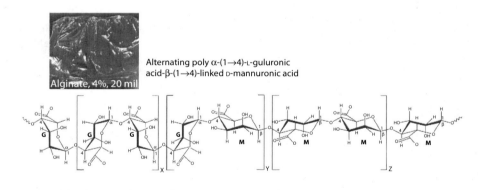

FIGURE 2.16 Idealized structure of alginate showing polyguluronic (G), polymannuronic (M), and alternating guluronic–mannuronic (GM) fractions. Also shown is a picture of its film, cast from a 4% solution and thickness of 20 mils (0.02 inch or 508 µm) measured across a 2.5-cm wide film strip using the TA.XT.PlusTexture analyzer, yielding tensile strength of ~1500 g and puncture strength of ~830 g. (With kind permission from Springer Science+Business Media: *Edible Films and Coatings for Food Applications*, Structure and function of polysaccharide gum-based edible films and coatings, 2009, pp. 57–112, Nieto, M.)

per guluronate residue (Rao et al. 1998). The structure of the polymannuronic acid segments is very similar to that of cellulose and mannan, especially the O3...O5 hydrogen bonds (Atkins et al. 1973a,b). The mannuronic acid residues are in the $^4C_1$ conformation and consequently it is diequatorially linked but also assumes a twofold helix or extended conformation. The linkages give the polymer segments containing polymannuronic acid a flattened, ribbon-like structure. It has been proposed that this structure is further stabilized by the formation of hydrogen bond to atom O3 in one ring with the ring oxygen of an adjacent residue. Another hydrogen bond, between the carboxyl group's hydroxyl and the oxygen atom attached to C3 of a parallel chain causes the polymannuronic acid chains to bond into sheets of anti-parallel residues. Regularly alternating poly(MG)n, on the contrary, has a sinusoidal conformation.

Although alginates from brown seaweed sources may exist predominantly as high G or high M, all three blocks are present within a single alginate molecule. For example, the M/G ratio of alginate from *Macrocystis pyrifera* is about 1.6:1, whereas that from *Laminaria hyperborea* is about 0.45:1. Alginates may also be prepared with a wide range of average chain lengths (50–100,000 DP) to suit the application. Commercial grades that are high in guluronic acid are usually labeled HG.

There are two structural blocks in alginate molecule that respond differently to calcium or bivalent ions, such as the G-block and M-block. The G-blocks respond to calcium cross-linking faster than the M-blocks, because of its three-dimensional "egg box" molecular conformation, which structure accommodates Ca^{2+} ions to form salt bridges. The M-blocks also associate through Ca^{2+} salt bridges, forming more elastic gels with good heat stability (Donati et al. 2005). Different viscosity grades corresponding to varying MWs of either high G or high M alginates are commercially produced.

Sodium alginate forms a clear film as shown in Figure 2.16. It does not stretch and does heat seal like all polar gum films. Cross-linking the alginate with calcium increases the tensile strength of the resulting film, indicating that film structure forms better without the charge repulsion. Cross-linked calcium–alginate always yields a strong, clear film that is insoluble in water compared to sodium alginate film; this combination has been used to microencapsulate probiotics, such as *Lactobacillus acidophilus*, to increase its survival under gastric conditions. Based on a study by Chandramouli et al. (2004), the viability of the bacterial cells within calcium–alginate microcapsules increased as both capsule size and alginate concentration increased.

One important film application for sodium alginate is sausage casing. As shown in Figure 2.17, the meat and alginate can be coextruded, where the meat is extruded in the inner tube and the alginate solution in the outer tube. As the sausage comes out of the tube, it showered immediately with calcium chloride solution to form the calcium–alginate casing. The use of alginate as sausage casing has attracted the attention of many sausage manufacturers and has already started commercial productions in the United States, Europe, and Canada. Other film applications for sodium alginate have been published. Russo et al. (2007) made three films of sodium alginate, with different amounts of guluronic fraction and found that increasing the fraction of guluronic units promoted chain-to-chain interaction through Ca^{2+} cross-linking and appreciable change in the physical properties, and increased free volume during the cross-linking process took place. Tapia et al. (2007) studied film made

FIGURE 2.17 Use of sodium alginate as sausage casing. Shown in the picture is the coextrusion process where meat (inner tube) and sodium alginate solution (outer tube) are coextruded together; alginate casing is formed by cross-linking with Ca^{2+} as sausage comes out the opening and showered with calcium chloride solution.

with alginate (2% w/v) containing glycerol (0.6% to 2.0%), N-acetylcysteine (1%), and/or ascorbic acid (1%) and citric acid (1%) and then used these to coat fresh-cut apple and papaya cylinders. WVP was significantly higher in alginate films than in the gellan films, and fresh-cut apple and papaya cylinders were successfully coated with 2% (w/v) alginate or gellan film-forming solutions containing viable bifidobacteria. This work demonstrated the feasibility of alginate- and gellan-based edible coatings to carry and support viable probiotics on fresh-cut fruits.

On the contrary, da Silva et al. (2009) manufactured composite biofilms of alginate and LM-pectin cross-linked with calcium ions using a two-step contact with Ca^{2+}—initially a low-structured prefilm is formatted which is further cross-linked in a second contact with a more concentrated Ca^{2+} solution containing plasticizer such as glycerol from 1% to 15% (w/v). The results indicated that increasing the glycerol concentration of the cross-linking solution increased film solubility in water, moisture content, volumetric swelling, and flexibility but decreased tensile strength. Transparent alginate and pectin composite films with acceptable mechanical properties, low solubility, and limited degree of swelling were obtained with 10% glycerol in the second contact solution.

2.5.2.3 Propylene Glycol Alginate (E405; CAS#9005-38-3; 21CFR172.858; FEMA#2941)

Propylene glycol alginate (PGA) is chemically derived by exposure of the parent alginate to propylene oxide gas. This treatment introduces a propylene glycol ester group, as bonded to the carboxyl groups of guluronic and mannuronic units. The result is the reduction in negative charge, and this changes the behavior of the polymer in many ways compared to standard alginate. Commercial PGAs are esterified up to ~90%. The esterification makes PGA much less calcium sensitive; it thickens to some extent in the presence of calcium ions but does not gel. PGA also tolerates low pH conditions and has good emulsifying and foaming properties relative to native alginate. Based on JECFA specifications, commercial PGA has MW ranging from 10,000 to 600,000 similar to sodium alginate, but annual production is much less, at 3100 tons in 2001 (http://www.fao.org/docrep/006/Y4765E/y4765e08.htm) to about 10,000 tons currently.

Structurally, the backbone of PGA has the same glycosidic linkages between the G units, M units, and GM or MG units as the parent alginate. The main difference is the presence of propylene glycol (also called hydroxypropyl) ester substitutions on C6, where the carboxyl groups of the guluronic and mannuronic acids are esterified. These groups add nonpolar properties to PGA and the presence of both polar and nonpolar groups make PGA foam and also give it both thickening and emulsifying properties. The reduction of the anionic charge on the PGA molecule, however, does not confer additional tensile strength to the PGA film compared to sodium alginate. This is likely because the propylene glycol ester group is bulkier (relative to the free carboxyl group of native alginate) and impedes to a greater extent the interpolymer chain association that is responsible for the thickening and formation of film structure. PGA film is shown in Figure 2.18.

Edible film studies based on PGA are scarce. One study by Rhim et al. (1998) used different levels (5%, 10%, 15%, 17.5%, or 20% w/w of solid) of propyleneglycol alginate (PGA) incorporated into soy protein isolate (SPI) films to form biodegradable composite films with modified physical properties. Color of the SPI films was significantly affected by the incorporation of PGA; tensile strength increased with addition of PGA up to 17.5%, while the percentage elongation at break decreased with incorporation of PGA of higher levels; WVP and water solubility also decreased

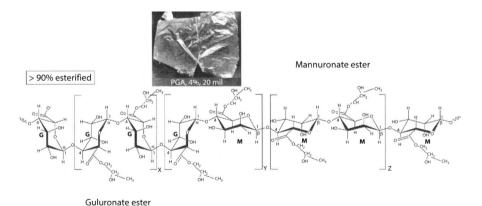

FIGURE 2.18 Idealized structure of propylene glycol alginate showing propylene glycol ester substitution at position C6, where the free carboxyl groups of the guluronic and mannuronic acids are located. Also shown is a picture of its film, cast from a 4% solution and thickness of 40 mils (0.04 inches or 1016 μm) using the TA.XT.PlusTexture analyzer, yielding tensile strength of 990 g and puncture strength was 445 g. (With kind permission from Springer Science+Business Media: *Edible Films and Coatings for Food Applications*, Structure and function of polysaccharide gum-based edible films and coatings, 2009, pp. 57–112, Nieto, M.)

by adding PGA up to 10%, but further addition of PGA increased values for these properties; and results suggest that the site of reaction with PGA on the protein chain may become saturated with PGA at the 10% level.

2.5.2.4 Chitosan (CAS#9012-76-4)

Chitosan is produced commercially by deacetylation of chitin, which is the structural element in the exoskeleton of crustaceans (such as crabs, shrimps, oysters, and other mollusk shells) and cell walls of fungi. Nearly all chitin and chitosan produced commercially are chemically extracted from crab, shrimp, and prawn exoskeleton wastes (Roberts 1997), and it seems that this is still the case currently. Chitin can also be produced from shell waste by fermentation with microorganisms or with the aid of enzymes (Rao et al. 2000). Enzymatic deacetylation of chitin to chitosan has also been accomplished in lab scale by Win et al. (2000). The production of chitosan involves many steps such as: shell wastes are crushed, washed, and decalcified using dilute acid (HCl), then washed again and deproteinized using dilute alkali (NaOH) to get chitin. The final chemical step is deacetylation with concentrated NaOH, then the product is dried and milled (http://www.emergingkerala2012.org/pdf/project-under-msme/manufacture-of-chitin-chitosan.pdf). The quality of chitosan produced from shrimp shells depends on the condition of the chemical extraction process (Lertsutthiwong et al. 2000, 2002), a 4% NaOH is suitable for deproteination at RT, and 4% HCl is used for decalcification to get chitin with low ash content, followed by deacetylation using concentrated alkaline.

The degree of deacetylation (% DD) in commercial chitosans ranges between 60% and 100% and on average, the MW of commercially produced chitosan is between 3800 and 20,000 Da (http://en.wikipedia.org/wiki/Chitosan). Hwang et al. (2002) reported that chitosan is widely depolymerized during processing in a range from 1100 to 100 kDa and deacetylated from 67.3% to 95.7% by NaOH alkaline treatment.

Transparency Market Research has published a new report titled "Chitosan Market—Global Industry Analysis, Size, Share, Growth, Trends and Forecast, 2014–2020." According to the report, the global chitosan market was valued at U.S.$1.35 billion in 2013 and is likely to reach U.S.$4.22 billion in 2020, expanding at a CAGR of 17.7% between 2014 and 2020 (http://www.transparencymarketresearch.com/pressrelease/chitosan-market.htm). According to Global Industry Analyst, Inc., worldwide demand for chitosan will exceed $21 billion by 2015 (http://kytosanusa.com/News/KUSA_Press_Release_12072013.pdf). The global market for chitosan was estimated at 13,700 metric tons for 2010, with a positive projection of 21,400 metric tons expected by 2015. The Asia-Pacific region (including Japan) was the leading chitosan market with an estimated 7800 metric tons in 2010 and a projected 12,000 metric tons by 2015. The United States represented the second biggest market for chitosan, with an estimated market size of 3600 metric tons in 2010 (http://www.nutraceuticalsworld.com/contents/view_online-exclusives/2010-12-02/the-global-chitosan-market-/#sthash.xEamFdmC.dpuf).

Chitosan is poly-β-(1 → 4)-2-amino-deoxy-D-glucopyranose. Its idealized structure is shown in Figure 2.19. In solution, chitosan forms micelle-like aggregates from the fully acetylated segments of the polysaccharide chains, interconnected by

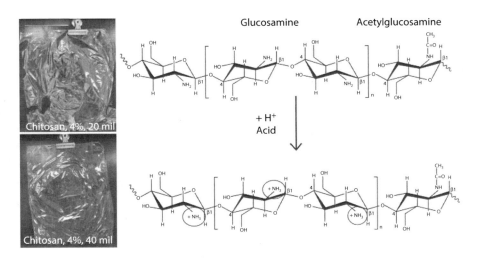

FIGURE 2.19 Idealized structure of chitosan showing the cationic glucosamine units and acetylglucosamine. Also shown are pictures of its film, cast from a 4% solution in acetic acid and thicknesses of 20 mils (0.02 inch or 508 μm) and 40 mils (0.04 inch or 1016 μm), yielding tensile strengths of 1780 and 2970 g, and puncture strengths of 1500 and 2035 g, respectively. (With kind permission from Springer Science+Business Media: *Edible Films and Coatings for Food Applications*, Structure and function of polysaccharide gum-based edible films and coatings, 2009, pp. 57–112, Nieto, M.)

blocks of almost fully deacetylated polysaccharide, stretched by electrostatic repulsion (Pedroni et al. 2003). Chitosan in the free amine form is insoluble in water at neutral pH. However, it is soluble in glacial acetic acid and dilute HCl, but insoluble in dilute sulfuric acid at RT. Chitosan carries a large number of amino groups along its chain and is, thus, capable of forming multiple complexes. At acid pH, the protonation of $-NH_2$ groups converts them to $-NH^{3+}$, which can associate with polyanions to form complexes and bind to anionic sites on bacterial and fungal cell wall surfaces. At higher pH levels, greater than 4, chitosan can form complexes with colorants and heavy metals. These appealing features make chitosan widely applicable in wound healing, production of artificial skin, food preservation, cosmetics, and wastewater treatment (Juang and Shao 2002; Risbud et al. 2000).

When a plastic wrap made from chitosan was used for storage of fresh mangoes, the shelf life of the produce was extended up to 18 days, without any microbial growth and off flavor (Srinivasa et al. 2002). Cardenas and Miranda (2004) also made chitosan films with glycerol, tween, and beeswax and concluded that these composite films are good alternatives to existing food storage materials due to important similarities with commercial PPs, not to mention their environmental advantages as biodegradable. Park et al. (2002) prepared unplasticized films using 3 MW chitosans, that is, 37,000, 79,000, and 92,000 Da. Tensile strength increased with MW of chitosan. Using acetic acid as solvent produced the toughest film; WVP was not influenced significantly and ranged from 0.3 to 0.7 ng m m^{-2} s^{-1} Pa^{-1}. Oxygen permeability of chitosan film using malic acid was the lowest, followed by acetic, lactic, and citric acid. Kujawa et al. (2007) made films via layer-by-layer assembly of hyaluronan and modified chitosan and found that HA/PC–CH films were stable over a wide pH range (3.0–12.0), exhibiting a stronger resistance against alkaline conditions as compared to HA/unmodified CH films. They concluded that given the versatility of the PEM methodology, HA/PC–CH films are attractive tools for developing biocompatible surface coatings of controlled mechanical properties.

2.5.3 Gums Polymers with Extended, Threefold Helix Structures

There are gum polymers that naturally form threefold helical structures when hydrated in water, in contrast to other gelling gums that form gels by cross-linking their polymer chains with salt or bivalent ions. What is amazing with these structures is that their helices can organize into aggregates, completely immobilize water, and convert it to a gel that you can hold in your hand. There are only a few examples of these naturally occurring gum polysaccharides that gel water, that is, agar, κ- and ι-carrageenans from red algae, and gellan gum from bacterial fermentation. Pectin is also an example of a threefold helix but that its structure is composed of mixed branched polysaccharides (hairy region) and the gelling threefold helical region of homogalacturonan, and is much more complex that a straight threefold helical gums.

2.5.3.1 Agar (E406; CAS#64-19-7; 21CFR 182.101; FEMA#2006)

Agar is extracted from two major commercial sources of red seaweed such as *Gelidium* and *Gracilaria* that customers normally would specify when ordering agar grade. In the seaweed, it consists of a mixture of agarose, a gelling fraction that is void of sulfate, and agaropectin, the nongelling fraction which bears sulfate and other charged groups. In the commercial manufacture of food-grade agar, most of the agaropectin is removed during processing, hence, commercial agars are mainly composed of the agarose fraction.

Tseng (1944) defines agar as the dried amorphous, gelatin-like, nonnitrogenous extract from Gelidium and other red algae, a linear galactan sulfate insoluble in cold but soluble in hot water, and a 1%–2% solution of which upon cooking solidifies to a firm gel at 35–50°C and melting at 90–100°C. The chemical nature of agar varies according to the seaweed source, the environment where the seaweeds grow and on the method of preparation of the agar. Meer (1980) recognized two types of agar, the Gelidium and Gracilaria agars.

Processing of agar starts with the seaweed. Cleaning of the seaweed is better done in the field when still fresh using seawater because it will be easier to remove attached shells, sand and algae contaminants other than *Gracilaria* or *Gelidium*. Seaweeds are then dried under the sun which also bleaches the seaweed. Alternatively, bleaching is done by sprinkling lime on seaweed during drying under the sun. Prior to extraction, Gracilaria agar is first treated with alkali to produce agar with higher gel strength; whereas Gelidium agar does not undergo akali treatment.

The next step is agar extraction using hot water. The difference between Gelidium and Gracilaria extraction is that Gelidium is extracted under pressure or temperature between 105°C and 110°C for 2–4 h, whereas the Gracilaria agar is extracted with 95–100°C water for the same length of time (http://www.fao.org/docrep/006/Y4765E/y4765e06.htm#TopOfPage). Extract is filtered to remove residues and cooled to gel and broken in pieces for further refining and drying which are done in 2 ways:

1. *Freezing and thawing*: The agar gels are frozen slowly so that large ice crystals are formed, then thawed. Structure of the agar gel is retained during freezing making it insoluble again while the water thaws out and drains away. The agar is then dried and milled to mesh specification.
2. *Gel press*: The agar gels are placed in a cloth and then pressed under pressure squeezing the water out of the gel that drains through the pores of the cloth. The pressed cake is then dried and milled to mesh specification.

Annual productions of dried agar seaweeds (mostly Gracilaria and Gelidium) was 55,650 tons in 2001 (http://www.fao.org/docrep/006/Y4765E/y4765e05.htm#TopOfPage). In 1976, Yamada estimated an annual output of agar from over 20 countries of 6000 tons and this number remains the same more than a decade after where total world production of finished agar was estimated at 7000–10,000 tons in 1987 (http://www.fao.org/docrep/field/003/AB728E/AB728E12.htm). Today, annual volume is still within this estimate. In 2012, a total of 1429 tons of agar was imported by the United States based on NOAA Current Fishery Statistics No. 2012-2 (http://www.st.nmfs.noaa.gov/st1/trade/documents/TRADE2012.pdf). MW of commercial agars was between 10,400 and 243,500 Da based on the study by Mitsuiki et al. (1999).

Agarose is a linear polymer with a MW of about 120,000 Da and is composed of a repeating dimer of D-galactosyl and 3,6-anhydro-L-galactosyl units connected via alternating α-(1 → 3) and β-(1 → 4) glycosidic linkages (Labropoulos et al. 2002). Its structural formula is depicted in Figure 2.20. The fundamental difference in covalent structure between agarose and the gelling carrageenans is the inversion of anhydrogalactose residues from D to L (Rao et al. 1998). With both monosaccharides in the dimer being in the preferred $^4C_1$ chair conformations, Arnott et al. (1974b) have proposed a conformation for agar as a threefold, left-handed, half-staggered, parallel double helix with pitch of 19 Å. Only van der Waals forces stabilize the double helix but in the central cavity of the helix, water molecules can mediate interchain hydrogen bonding with the oxygen atoms, O2 of galactose unit and O5 of the anhydrogalactose unit. Similar to the gelling carrageenans, the presence of anhydrogalactose unit in the B-residue of the dimer allows the molecule to coil into a threefold helix with three dimers making each turn of the helix. Gelation occurs when these helices aggregate upon cooling of the agar solution.

Commercial agars are not completely void of agaropectin and other charged groups such as pyruvic acid in addition to the sulfate ester, even after the refining process. Hirase (1957) found the presence of pyruvic acid in the agar of *Gelidium amansii* as a ketal attached to C4 and C6, while Araki (1965) found that the D-galactose residues are 6-O-methylated to certain degrees that can impart foaming property to some agar grades.

Most applications for agar are based on its gelling ability. It is differentiated from the gelling carrageenans by its better gel stability at low pH and high temperature conditions. Standard agar requires boiling to fully dissolve and gelation results from

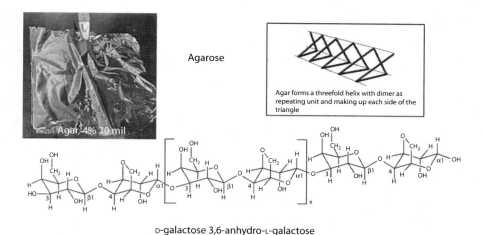

D-galactose 3,6-anhydro-L-galactose
dimer

FIGURE 2.20 Idealized molecular structure of agarose showing the dimer composed of β-(1 → 4)-linked D-galactose-(3,6)-anhydro-L-galactose and α-(1 → 3) linkages between dimers. The dimer turns to form a threefold helix as shown, where three dimers or six monomers make up each turn of the helix. Also shown is a picture of its film, cast from a 4% solution and thickness of 20 mils (0.02 inch or 508 μm) measured across a 2.5-cm wide film strip using the TA.XT.PlusTexture analyzer, yielding tensile strength of 2540 g and puncture strength of 2150 g of force. (With kind permission from Springer Science+Business Media: *Edible Films and Coatings for Food Applications*, Structure and function of polysaccharide gum-based edible films and coatings, 2009, pp. 57–112, Nieto, M.)

the formation of a network of threefold agarose double helices upon cooling. These double helices are stabilized by the presence of water molecules bound inside the double helical cavity (Labropoulos et al. 2002). Exterior hydroxyl groups allow for aggregation of these double helices via intermolecular hydrogen bonds. In this aggregation process, suprafibers consisting of up to 10,000 double helices are formed, and contribute to the strength of agar films. Agar solutions gel at temperatures of 32–40°C (90–103°F). In making agar film, the solution and casting surface needs to be maintained above these temperatures to avoid premature gelation. Sousa et al. (2010) reported that agar/glycerol films and coating have a protective function, reduced moisture, oxygen, and flavor transfers between food and surroundings. Compared to κ- and ι-carrageenans, agar biofilms were significantly less hygroscopic which constitutes a great advantage; they also gave higher tensile strength and high strain at break, their oxygen permeability was identical to synthetic polymers used in food packaging and they were effective in extending shelf life and gloss of cherry tomatoes.

2.5.3.2 Carrageenan (E407; CAS#9000-07-1; 21CFR 172.620; FEMA#2596)

Carrageenan is a collective term for sulfated polysaccharides extracted from certain species of red seaweed of the family, Rhodophyceae. Major commercial sources are *Eucheuma spinosum*, *Eucheuma cottonii*, *Gigartina* spp., *Chondrus crispus*, and *Hypnea* spp. In terms of grades, four commercial types of carrageenan extracts are available: kappa I, kappa II, iota, and lambda types. Different seaweeds produce different carrageenans fractions, with one type being more predominant in any species. A purer κ- and ι-carrageenan can be produced following potassium chloride precipitation, where κa- and ι-fractions are made insoluble, and lambda is removed in the soluble phase. Physical separation of different species of seaweed, or picking of seaweed contaminants, is also a good way to obtain high quality fractions.

There are two different methods of producing carrageenan, based on different principles (http://www.fao.org/docrep/006/y4765e/y4765e0a.htm), such as: (1) the original method—the only one used until the late 1970s–early 1980s—where the carrageenan is extracted from the seaweed into an aqueous solution, the seaweed residue is removed by filtration and then the carrageenan is recovered from the solution by precipitation, eventually as a dry solid containing little else than the refined carrageenan. This recovery process is difficult and expensive; and (2) the second method, where the carrageenan is not actually extracted from the seaweed, but rather just washed with alkali and water or alcohol to remove soluble color and smell, leaving the carrageenan and other insoluble matter behind. This insoluble residue, consisting largely of carrageenan and cellulose, is then dried and sold as semirefined carrageean. Because the carrageenan does not need to be recovered from solution, the process is much shorter and cheaper.

Commercial (food-grade) carrageenans have a weight average MW ranging from 400,000–600,000 Da with a minimum of 100,000 Da (van de Velde and de Ruiter 2002). In 1976, the U.S. FDA defined food-grade carrageenan as having a water viscosity of no less than 5 mPa s (5 cP) at 1.5% concentration and 75°C, which corresponds to the above mentioned MW of 100 kDa. The minimum viscosity is set in response to reports of caecal and colonic ulceration induced by highly degraded carrageenan.

According to CyberColloids Carrageenan Industry Report 2012 (http://www.cybecoloids.net/sites/default/files/private/downloads/Carrageenan%20market%20report%202012%20contents.pdf), the carrageenan market worldwide is estimated to be around $640M. Imports into EU/United States from China have increased to nearly 6000 tons, whereas imports from the Philippines were 5800 tons in 2011. While Bixler and Porse (2011) estimated that world carrageenan production exceeded 50,000 tons in 2009 with a value of over U.S.$527 million.

Carrageenans are made up of alternating galactopyranosyl dimer units linked by alternating β-(1 → 4) and α-(1 → 3) glycosidic bonds. The sugar units are sulfated either at C2, C3, or C6 of the galactose or C2 of the anhydrogalactose unit (Figure 2.21). They are similar in structure to agar, except for the presence of 3,6-anhydro-D-galactose (B unit of the dimer) rather than 3,6-anhydro-L-galactose units found in agar, and also the presence of sulfate ester groups in all carrageenans. Kappa I, II, and iota are gelling carrageenans. Similar to agarose, gelation is a result of the formation of threefold helices that can only occur in repeat structures where the B unit is in a $^1C_4$ conformation in the form of anhydrogalactose. Lambda carrageenan does not have the anhydrogalactose bridges and has both its sugar residues in a $^4C_1$ conformation and hence, it does not coil into a threefold helix and does not gel. All the gelling types of carrageenan contain a 3,6-anhydrogalactose bridge on the B unit which forces the sugar to flip from a $^4C_1$ conformation to a $^1C_4$ conformation that can then coil into a threefold helix and aggregate to form cross-linked networks and

Edible Film and Packaging Using Gum Polysaccharides

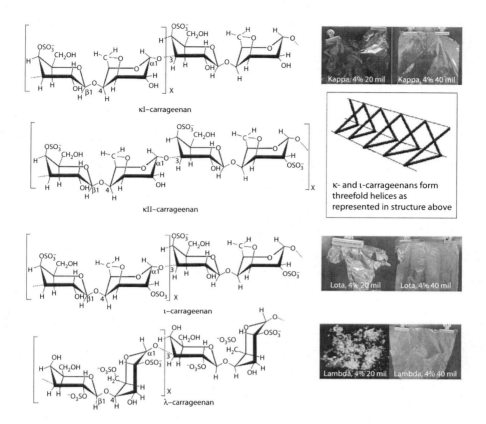

FIGURE 2.21 Idealized structures of various carrageenan fractions showing the anhydrogalactose (B unit) for κI, κII, and ι in $^1C_4$ conformation, which enable the dimer to turn and form a triangular helix structure, while λ-carrageenan is void of anhydrogalactose. Also shown are pictures of κI-, ι-, and λ-carrageenan films, cast from 4% solutions and thicknesses of 20 mils (0.02 inch or 508 μm) and 40 mils (0.04 inches or 1016 μm) measured across a 2.5-cm wide film strip using the TA.XT.PlusTexture analyzer, yielding tensile strengths of 1940 g for κ-carrageenan, 944 g for the ι, and 659 g for the λ films, and almost similar puncture strengths of 955–1040 g. (With kind permission from Springer Science+Business Media: *Edible Films and Coatings for Food Applications*, Structure and function of polysaccharide gum-based edible films and coatings, 2009, pp. 57–112, Nieto, M.)

gel. According to Arnott et al. (1974a) both ι-carrageenan and κ-carrageenan form threefold, right-handed, parallel, half-staggered double helices. The ι-carrageenan double helix is stabilized by six interchain hydrogen bonds, while the κ-carrageenan double helix is stabilized by only three interchain hydrogen bonds. Falshaw et al. (2001) studied the differences between the four types of carrageenan in the degree of sulfation—kappa I being the least sulfated with 25% ester sulfate content, and the lambda being the most with 35%. These differences in sulfation levels give carrageenans varying degrees of anionic charge and solubility in water. Lambda carrageenan is completely and highly cold water soluble, while κ-carrageenan is only partially cold water soluble and requires heating to 82°C for full activation. The ι-carrageenan has solubility in between that of λ and κ types.

Refined carrageenans produce clear solutions and therefore, clear films. Kappa 1, being the least anionic and least soluble in cold water, will swell in RT water and thicken the water with a grainy texture that disappears on heating. On cooling and when shear is removed, its hydrated molecules or threefold helices are better able to aggregate with each other because of less repulsion. The presence of mineral ions such as K^+ or Na^+ also reduces the negative charge repulsion. The result is better aggregation of helices, better gelling and stronger film formation upon drying compared to other carrageenans, but weaker compared to agar. The dried κ-carrageenan film is also only partially soluble in cold water, similar to the parent carrageenan, which requires heating to ~50–80°C to dissolve. λ-carrageenan, on the contrary, being the most anionic and most soluble in cold water, produces a slightly weaker film than both kappa and iota. As might be expected, the lambda film dissolves in cold water or thickens the saliva when film is eaten.

Abdou and Sorour (2014) studied the mechanical properties of starch/carrageenan films with glycerol as plasticizer. The results show that the mechanical property of the films increased with increasing carrageenan content, but so as the WVP. Another study on edible films made with κ-carrageenan and rice starch was reported by Larotonda et al. (2014) that also showed that carrageenan increased the hygroscopic property of the film, but improved extensional property and Young's modulus. Wu (1999) compared edible composite films and coatings based on carrageenan, chitosan, wheat gluten, and soy protein for their effectiveness in controlling moisture loss and lipid oxidation of precooked ground beef patties and found that wheat gluten, soy protein, carrageenan and chitosan films and coatings had different effects on maintaining the quality of precooked patties after 3-day storage at 4°C. All coatings were as effective as polyvinyl chloride film in reducing patty moisture loss. Coating with wheat gluten, carrageenan, or soy protein and wrapping with carrageenan films were effective in lowering patty thiobarbituric acid reactive substance (TBARS) values and hexanal values, with wheat gluten coating being the most effective which resulted in about 28% and 48% reduction in TBARS and hexanal, respectively. Incorporating fatty acids, especially 30%–40% of palmitic acid and/or stearic acid, had a significant impact on water barrier properties and mechanical properties of the films.

2.5.3.3 Gellan (E418; CAS#7101052-1; 21CFR172.665)

Kaneko and Kang discovered gellan in 1978, the exocellular polysaccharide produced by the bacterium *Sphingomonas elodea*, formerly called *Pseudomonas elodea*. Gellan was approved for use in Japan in 1988 and by the U.S. FDA in 1992. The manufacturing process for gellan is very similar to xanthan gum (http://www.biopolymer-international.com/manufacturing-process/). Gellan gum is produced by aerobic submerged fermentation using the bacterium *P. elodea*. There is a multistage bacterial inoculum build up from agar plate to shake flasks to small seed fermentation vessels to large final fermentation vessel. At all stages of production, fermentation equipment is thoroughly cleaned and sterilized before use and strict aseptic techniques are followed to ensure that culture is pure and without contamination. The fermentation medium is comprised of glucose syrup derived from maize or wheat, inorganic nitrogen (ammonium or nitrate salts), an organic nitrogen source (protein), and trace elements. Once the final fermentation is complete, the contents of the vessel are pasteurized to kill all the bacterial cells used in the initial culture and optimize the conformation of the polymers. The gums are recovered from the fermentation broth by addition of alcohol (usually isopropyl alcohol) causing precipitation of gellan fibers. The resulting fibers are then treated to remove the excess alcohol, dried, milled into a powder, and packaged in a controlled environment.

According to CP Kelco's Gellan brochure (http://www.appliedbioscience.com/docs/Gellan_Book_5th_Edition.pdf), high acyl gellan has MW of 1–2 million, while low acyl gellan has between 200,000 and 500,000 Da. JECFA specifies MW for gellan to be approximately 500,000.

Gellan is a linear, anionic polymer of about 50,000 DP, consisting of a tetrasaccharide repeat of $(1 \rightarrow 4)$-α-L-rhamnopyranosyl-$(1 \rightarrow 3)$-β-D-glucopyranosyl-$(1 \rightarrow 4)$-D-glucuronopyranosyl-$(1 \rightarrow 4)$-β-D-glucopyranosyl-$(1\rightarrow)$. The $(1 \rightarrow 3)$-linked glucose unit possesses L-glyceryl ester groups attached through C2 and acetyl at C6 (Chandrasekaran and Radha 1995) as shown in Figure 2.22. Gellan gum may or may

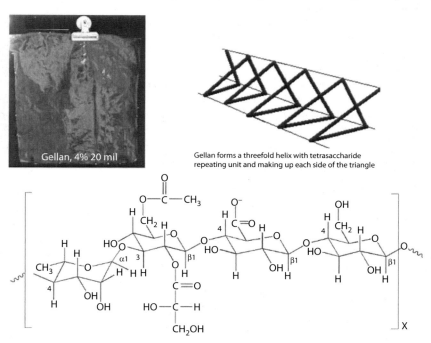

Gellan forms a threefold helix with tetrasaccharide repeating unit and making up each side of the triangle

FIGURE 2.22 Idealized structure of gellan gum showing the basic tetramer repeating unit that make up each side of a threefold helix, consisting of rhamnose, glucose with acetyl and glyceryl substitutions, glucuronic acid, and another glucose. Also shown is a picture of its film, cast from a 4% solution and thickness of 50 mils (0.05 inches or 1270 μm) measured across a 2.5-cm wide film strip using the TA.XT.PlusTexture analyzer, yielding tensile strength of ~700 g and puncture strength of ~250 g. (With kind permission from Springer Science+Business Media: *Edible Films and Coatings for Food Applications*, Structure and function of polysaccharide gum-based edible films and coatings, 2009, pp. 57–112, Nieto, M.)

not be deesterified by alkali treatment to produce the low acyl and high acyl grades, respectively, and they can be highly refined to produce clear grades of gellan, or not so refined with opaque solutions in water. Rao et al. (1998), as discussed in their book, said that the best model for gellan conformation is that of a threefold, left-handed, half-staggered, parallel double helix with a pitch of each chain being 56.3 Å. The helix repeating unit is a tetrasaccharide, instead of a dimer for agar and the gelling carrageenans, therefore, forming a bigger threefold helix than can accommodate an acyl group in the central cavity for stabilization of its structure. Geremia and Rinaudo (2005) determined that glyceryl groups interact inside the double helix and its hydrolysis leads to the destabilization of the helical structure, whereas taking off the acetyl groups does not modify the stability of the double helix and favors double helix interactions with improved gelation.

Gellan contains glucuronic acid in its polymer chain giving it a partial negative charge. In addition, some of the glucose units contain either a glyceryl ester or acetyl ester groups, and these substitutions pose steric hindrance and affect the relative solubility, gelling property and film-forming property of different gellan grades. Low acyl and high acyl gellans, like κ-carrageenan, require heat to fully hydrate. The low acyl gellan is less soluble and disperses well in RT water, while the high acyl gellan is more soluble and tend to form lumps when added to RT water with insufficient shear. Either grades, however, tend to drastically lose their viscosities with increase in temperature, a characteristic shared by all gelling gums where their structural conformation transitions to a disordered state or random coil.

Tapia et al. (2007), in addition to studying alginate film, also made gellan edible film (0.5%) containing glycerol (0.6%–2.0%), N-acetylcysteine (1%), and/or ascorbic acid (1%) and citric acid (1%), and then used this film to coat fresh-cut apple and papaya cylinders. WVP of gellan films was significantly lower compared to alginate films, and that addition of 0.025% (w/v) sunflower oil decreased its WVP. The gellan coatings and films exhibited better water vapor properties in comparison with the alginate coatings. Both gellan and alginate films maintained >10^6 CFU g^{-1} of Bifidobacterium *lactis* Bb-12 for 10 days during refrigerated storage of fresh-cut fruits, demonstrating the feasibility of alginate- and gellan-based edible coatings to carry and support viable probiotics on fresh-cut fruit. Leon and Rojas (2011) also studied edible gellan films as carrier for L-(+)-ascorbic acid (AA) for nutritional purposes and antioxidant effect on foods. Initial AA retention was around 100%, and half-lives were 36, 26, and 11 days, respectively. Xiao et al. (2011) made edible gellan membranes with good tensile strength, moisture resistance, and excellent gas-barrier properties after drying at temperatures between 60°C and 70°C.

2.5.4 Gum polymer with Extended, Fivefold Helix Structure

2.5.4.1 Xanthan Gum (E415; CAS#11138-66-2; 21CFR172.695)

Xanthan gum is a microbial gum polysaccharide that is produced by the bacterium *Xanthomonas campestris* from a fermentation process involving a carbohydrate substrate and other growth-supporting nutrients. CP Kelco pioneered the production of xanthan gum in 1964 which has become the most studied gum thickener and stabilizer in innumerable applications. Xanthan gum manufacturing process is described in the Cargill link https://www.cargillfoods.com/na/en/products/hydrocolloids/xanthan-gum/manufacturing-process/index.jsp and the CP Kelco link http://www.bisi.cz/cmsres.axd/get/cms$7CVwRhc3USVqgzxkKF96gI$2BChNrXcTq$2BOUdiEtz5TfYA$2B1dJUbBlKfluXdoDfiqojVRVU$2FkQ343xA$3D. The process consists of the following:

1. The culture of *Xanthomonas campestris* that is preserved in a freeze-dried state is activated by inoculation into a nutrient medium containing a carbohydrate, a nitrogen source, and mineral salts. The growth culture is then used to inoculate successive fermenters. Throughout the fermentation process, pH, aeration, temperature, and agitation are monitored and controlled. Once the carbohydrate is exhausted, the broth is pasteurized, then the fermenter is emptied, cleaned, and sterilized before the next fermentation takes place.
2. Xanthan gum is recovered by precipitation in alcohol (isopropyl or ethanol). The coagulum obtained is separated, rinsed, pressed, dried, and ground before quality control testing.

Total global production in 2012 is expected to be in excess of 110,000 tons. China became the world's largest producer of xanthan gum in 2005 and now produces and exports about two thirds of the world's supply. Other major producers and exporters are the United States, France, Austria, and Japan (http://spendmatters.com/2012/09/24/xanthan-gum-another-possibly-sticky-situation/). Xanthan gum was approved for food use in the United States in 1969 and in 1982 in Europe. MW of xanthan is approximately 2 million Daltons based on weight average but the distribution can be as wide as 1 million to 10 million (Viebke 2005). Based on nine published data, as cited by Viebke (2005), average molar mass values quoted for xanthan gum single strand are between 2 and 9 million.

Xanthan gum is anionic polymer with a molecular structure that consists of a linear glucose chain linked by β-(1 $\rightarrow$ 4) glycosidic bonds like cellulose; it also possesses a trisaccharide side chain attached through O3 of alternate glucose units of the main chain (Jansson et al. 1975; Melton et al. 1976). The trisaccharide side chain is composed of an inner mannose unit, partially acetylated at the C6 position, between 60% and 70% (Born et al. 2002); a glucuronic acid unit; and a terminal mannose unit, partially substituted with pyruvic acid, between 30% and 40% (Born et al. 2002), in the form of a 4,6-cyclic acetal (Figure 2.23). The glucuronic acid and the pyruvate groups confer a negative charge on the xanthan molecule. The repeating unit of xanthan, therefore, is a pentamer. Each molecule of xanthan consists of about 7000 pentamers. Xanthan gum is actually an excretion to protect the bacterial cells when the pH of the fermentation medium becomes too low and unfavorable for growth. It is likely because of this native protective function that xanthan gum fold into a conformation that impart better stability over other gums. It is stable over a wide range of pH, from 1 to 13, very resistant to enzyme hydrolysis, tolerant of high salt, high sugar, and high alcohol systems, stable at boiling temperatures, and tolerates retort processing better than most other gums.

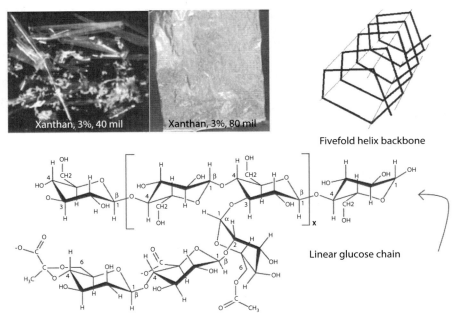

FIGURE 2.23 Idealized structure of xanthan gum (Adapted from Jansson, P. E., Keene, L., and Lindberg, B. 1975. *Carbohydrate Research*, 45:275–282 and Melton, L. D., Mindt, L., Rees, D. A., and Sanderson, G. R. 1976. *Carbohydrate Research*, 46:245–257.) showing the pentamer repeating unit composed of two glucose units in the main backbone and the trisaccharide side chains consisting of a mannose with an acetyl ester group, glucuronic acid, and terminal mannose with a pyruvate substitution. The presence of the trisaccharide side chains pulls the otherwise twofold helix of β-(1 → 4) glucan to a fivefold helix with two glucose units making up each side of the pentagon, while the side chains protrude outward of the five corners. Also shown are pictures of its film, cast from a 3% solution, and thicknesses of 40 mils (0.04 inch or 1016 μm) and 80 mils (0.08 inch or 2032 μm) measured across a 2.5-cm wide film strip using the TA.XT.PlusTexture analyzer.

The three dimensional structure of xanthan has been proposed based on fiber pattern by Okuyama et al. (1980) as antiparallel right-handed fivefold helix; the only one gum polymer that has this structural conformation, which is stabilized by four intramolecular and one intermolecular hydrogen bond. The β-(1 → 4)-glucose linkages that make up the backbone was similar to cellulose but instead of a twofold helix, the trisaccharide substitution at every second glucose pulls this otherwise twofold helix into a fivefold helix structure. When hydrated in water, xanthan gum forms an ordered structure of an extended but wide fivefold helices, a conformation that is different from other linear gums that are capable of hydrogen bonding with water and intermolecular hydrogen bonding between polymer chains. The unique conformation of xanthan gives its solution a very unique rheology, that is, very pseudoplastic, a property that is very desirable for suspension of particulates. A vast number of studies have been published regarding the order-disorder conformation of xanthan solutions as reviewed by Viebke (2005), and evidence suggests that it is not easy to convert xanthan to a disordered conformation. When this happens, xanthan lose its viscosity and shear thinning behavior, and to obtain this form, the xanthan gum solution needs to be heated to 80°C for an extended time (Kawakami et al. 1991) or by dissolving xanthan gum in 4 M urea and heating to 95°C for 3 h (Southwick et al. 1982).

A 1% concentration of standard xanthan grade gives a viscosity between 1200 and 1600 cP at 20 RPM. Its solution is highly pseudoplastic and gives a much higher viscosity at low shear or at rest. It hydrates rapidly in cold water compared to many other gum polymers, even in the presence of high levels of sugar, salt, or alcohol. The consistent water-holding ability of xanthan may be used for control of syneresis and to retard ice recrystallization in freeze-thaw situations as shown by Giannouli and Morris (2000).

Xanthan gum is capable of synergistic interactions with glucomannan like konjac gum and with galactomannans such as guar, tara, and LBGs boosting its viscosity, hence, intermolecular hydrogen bonding and structure formation between polymer chains. With guar, a boost in water binding capacity and viscosity is achieved at all ratios, with the optimum ratio being between 75/25 and 80/20 guar to xanthan. With LBG, optimal synergy and gelation is achieved at 50/50 LBG to xanthan (Goycoolea et al. 2001). Tara gum exhibits synergy with xanthan intermediate of that achieved with guar and LBGs. An extreme synergy between xanthan and konjac mannan takes place at an optimal ratio of 80:20 konjac to xanthan, respectively, with a viscosity build-up as high as 161,000 cP at a 1% total gum concentration (Takigami 2000). This noted synergy allows for very low usage levels for this gum combination in food applications.

When de Melo et al. (2011) made films from nine starch/xanthan/nanoclay combinations, containing glycerol as plasticizer, scanning electron microscopy (SEM) of the starch-xanthan extruded films showed reticulated surface and smooth interior, indicating that the gum was mostly concentrated on the surface of the films, while starch/xanthan/nanoclays films showed a more homogeneous surface, suggesting that the introduction of nanoclays provided a better biopolymeric interaction. In general, nanoclays addition (2.5–5.0 wt%) generated more transparent and resistant films, with lower WVPs and lower water sorption capacities, and xanthan gum addition improved the elongation of starch films. Veiga-Santos et al. (2005) studied the effect of deacetylated xanthan gum, additives (sucrose, soybean oil, sodium phosphate, and propylene glycol) and pH modifications on mechanical properties, hydrophilicity, and water activity of cassava starch–xanthan gum films. Sucrose addition had the highest effect on cassava starch films elongation at break; while deacetylated xanthan gum increased elongation at break of the cassava starch film better than the acetylated gum, but both sucrose and deacetylated xanthan gum addition resulted in a slight hydrophilicity increase.

2.5.5 Gum Polymers with Extended, Sixfold Helix Structures

2.5.5.1 Curdlan (CAS#54724-00-4; E424; 21CFR172.809)

Curdlan is a linear β-glucan, a high-MW polymer of glucose that was discovered by Harada et al. in 1966. In 1989, curdlan was approved for food use in Japan, then in the United States in 1996. This polymer is produced by nonpathogenic bacteria such as

Agrobacterium biovar., formerly called *Alcaligenes faecalis* var. *myxogenes*, during a fermentation process where the bacterial culture is grown in a medium containing a carbon source (maltose and sucrose), nitrogen and phosphate. Many studies have focused on optimizing processing conditions for curdlan (Kim et al. 2000; Lee et al. 1999) where fermentation temperature, pH, agitation, aeration, and nutrients needed were studied to increase yield. More recent study by Kalyanasundaram et al. (2012) focused on a mutant strain of *Agrobacterium* that produced higher levels of curdlan 66 g/L vs. 42 g/L *Agrobacterium* sp. *ATCC 31750*; while that of Zhang et al. (2012) optimized dissolved oxygen and pH. The production of curdlan by *Alcaligenes faecalis* has been developed for use in gel production as well where they showed a 43.8 g/L yield or a 30% increase.

Based on published studies, curdlan has an average DP of ~450 and is unbranched (as cited by Lee 2004), equivalent to MW of ~73,000. Nakata et al. (1998), on the other hand, reported an average MW for curdlan to be in the range of 53,000 to 2 million Daltons.

Curdlan consists of β-(1 → 3)-linked glucose residues that form elastic gels upon heating in aqueous suspension. There are two types of curdlan gels (Miwa et al. 1994)—low set curdlan that gels at 40–60°C and high set curdlan that gels at 80°C or higher. Both gels are reversible like agar and gelatin gels. According to Stone and Clark (1992) curdlan gelation begins at 54°C forming what is termed as a low set gel, an additional change occurs at 95°C to form what is termed a high set gel. The high set gel has the properties of being much stronger and more resilient than the low set gel. This change is explained by the hypothesis that microfibrils dissociate at 60°C as the hydrogen bonds are broken, but then reassociate at higher temperatures as hydrophobic interactions between the curdlan molecules occur (Harada et al. 1979).

Studies on structural conformation or 3D structure of curdlan differentiate native curdlan (called Curdlan I), hydrated form after annealing at 140°C (Curdlan II) and after drying the annealed curdlan (Curdlan III). It is shown that native curdlan, based on the observation of meridional reflection on the sixth layer line, is a sixfold, right-handed single helix (Okuyama et al. 1991) with a pitch of the single helix of 22.8 Å, and that the chain conformation is stabilized by a weak hydrogen bond between O4H...O5. The helices are oriented in a parallel fashion. Figure 2.24 shows the idealized structure of curdlan.

On the contrary, Chuah et al. (1983) showed that Curdlan II, or the hydrated form of Curdlan, as analyzed by fiber x-ray diffraction is organized as a triple helical structure or three strands of sixfold helices intertwined in parallel right-handed fashion. The core of the triple helix exhibits triad of interchain hydrogen bonds (O2H...O2) and six water molecules have been located in the unit cell that bridge adjacent helices through hydrogen bonds. Similarly, Deslandes et al. (1980) reported that Curdlan III, or the dehydrated form of curdlan, is structurally well organized and that the layer line spacing is consistent with a sixfold parallel, triple helix whose pitch is 17.61 Å. The core of the triple helix exhibits triad of interchain hydrogen bonds (O2H...O2) occurring at successive levels separated by distance of 2.94 Å.

Curdlan film studies include that of Han et al. (2007) where they prepared curdlan composite films and determined their properties in order to select the most appropriate setting methods, moisture barrier materials, and viscoelasticity enhancing

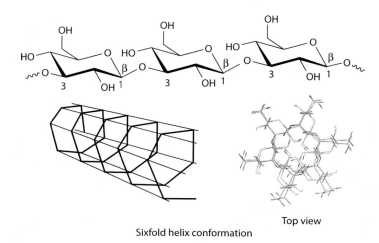

FIGURE 2.24 Idealized structure of curdlan I showing three glucose monomers in a β-(1 → 3) linkages and the top view of the sixfold helix conformation. (Adapted from Okuyama, K., Otsubo, A., Fukuzawa, Y., Ozawa, M., Harada, T., and Kasai, N. 1991. *Journal of Carbohydrate Chemistry*, 10(4):645–656.)

materials. They reported that high set curdlan films with PEG showed higher tensile strength and moisture barrier properties than low set films; films with oleic acid as a moisture barrier material had greater tensile strength, elongation and moisture barrier properties than films with acetylated monoglyceride; while the films using polyisobutylene as a viscoelasticity enhancing material showed higher elongation than films with polybutene. Ren (2007) also made composite films with SPI and curdlan and used this film to keep carambola fruits fresh, and found that when the film-forming solution was adjusted to pH 8 and heated at 75°C for 20 min, the film had better physical and moisture barrier properties—tensile strength was 6.32 MPa, elongation was 62.08%, and WVP property was greatly reduced.

2.5.6 Gum Polymers with Complex Conformations

2.5.6.1 Pectin (E440; CAS#7664-38-2; 21CFR184-69-5)

Pectin is mainly produced commercially from citrus peels and apple pomace. Sugar beet pectin has gained some commercial attention in the last several years but still has not gotten any good traction as a commercial pectin grade. About 85.5% of pectin sold today are from citrus, 14% from apple, 0.5% from sugar beet, and annual production is estimated at 42,000 metric tons in 2009 (https://conference.ifas.ufl.edu/citrus09/htms09/pdfs/Thursday/0905%20Staunstrup.pdf). The process for the manufacture of pectin is described in the following websites: http://www.ippa.info/commercial_production_of_pectin.htm by the International Pectin Producers Association), http://en.silvateam.com/Products-Services/Food-Ingredients/Pectin/Production-process by the Silvateam, and https://www.cargillfoods.com/lat/en/products/hydrocolloids/pectins/manufacturing-process/index.jsp by Cargill Foods.

Process details vary between different companies, but the general process is as follows:

1. The pectin factory receives apple pomace or citrus peel from a number of juice producers. In most cases, these starting materials have been washed and dried, so it can be transported and stored without spoilage. If wet citrus peel is used, it has to be used immediately as received as it deteriorates very rapidly.
2. The raw material is added to hot water containing a mineral acid, as water alone will only extract a limited amount of pectin. Enzymes could also be used in place of the acid.
3. After extraction of the pectin, the remaining solids are separated, and the solution clarified and concentrated by centrifugation or other means. The solution is then filtered again to clarify it if necessary.
4. Either directly, or after some further holding time to modify the pectin, the concentrated liquid is mixed with an alcohol to precipitate the pectin. The pectin can be partly deesterified at this stage, or earlier or later in the process.
5. The precipitate is separated, washed with more alcohol to remove impurities, and dried. The alcohol wash may contain salts or alkalis to convert the pectin to a partial salt form (sodium, potassium, calcium, ammonium). The alcohol (usually isopropanol) is recovered and reused.
6. Before or after drying, the pectin may be treated with ammonia to produce an amidated pectin if required. Amidated pectins are preferred for some applications for its calcium reactivity and better gelation.
7. The dry solid is ground to a powder, tested, and blended with sugar or dextrose to a standard gelling power or other functional property such as viscosity. Pectins are also sold blended with other approved food additives for use in specific products.

The properties of pectins vary according to the plant origin, raw material, season, as well as extractions conditions. Viscosity measurements are often employed to determine the MW of pectins because this method gives more accurate results. Based on viscometry, the average MW falls within range 50,000–200,000 (http://polysac3db.cermav.cnrs.fr/discover_pectins.html), whereas other results report MW of ~1 million or higher due to aggregation of pectin polymer strands.

Pectin is probably the most complex polysaccharide structure of all gum polysaccharides. The chemical structure of pectins has been the subject of scientific investigations for many decades, and because of its heterogeneity and variability from one source to another, it is not surprising that only recently that a consensus description of the primary structure of pectins was proposed. Figure 2.25 shows the idealized structure of pectin molecule. It is generally agreed that pectin is not a regular chain, but has a backbone made up of $\alpha(1 \rightarrow 4)$-linked D-galacturonic acid residues, and a greater proportion of these galacturonic acid residues is methyl esterified naturally (Pérez et al. 2000, 2003). This homogalacturonans block is also called smooth region and can have a size of about 72–100 units as obtained from apple,

Edible Film and Packaging Using Gum Polysaccharides 63

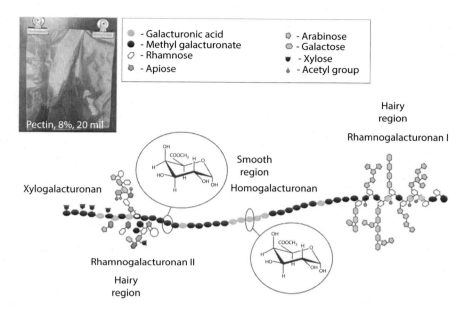

FIGURE 2.25 Idealized structure of pectin showing the five fractions—The homogalacturonan (smooth region), RG-I and RG-II (hairy regions), and xylogalacturonan. Also shown is picture of its film, cast from an 8% solution and thickness of 20 mils (0.02 inch or 508 μm) measured across a 2.5-cm wide film strip using the TA.XT.PlusTexture analyzer, yielding tensile strength of 965 g and puncture strength of 28 g. (With kind permission from Springer Science+Business Media: *Edible Films and Coatings for Food Applications*, Structure and function of polysaccharide gum-based edible films and coatings, 2009, pp. 57–112, Nieto, M.)

beet, and citrus pectins after controlled acid degradation. According to Walkinshaw and Arnott (1981), the x-ray diffraction pattern of sodium pectate shows that this polymer forms a threefold helix with a pitch of 13.36 Å; the diaxial link and the $^4C_1$ conformation of the galacturonate unit generates right-handed single helices that form antiparallel arrangement with intrachain hydrogen bonds stabilizing each helix. Association of helices is through direct hydrogen bonds involving the carboxyl groups.

The homogalacturonan moieties are interrupted with $(1 \rightarrow 2)$-linked α-L-rhamnopyanose residues in adjacent or, more likely, in alternate positions. The amount of rhamose is typically 1%–4%, and many of them carry side chains made up of neutral sugars. These regions are also called hairy or branched regions and are made up of two blocks such as rhamnogalacturonan (RG)-I and RG-II. Their structures have been elucidated by Engelsen et al. (1996) and Pérez et al. (2003). The frequency of rhamnose occurrence remains to be firmly characterized, although it has been suggested that the α-L-rhamnosyl units are concentrated in rhamnose-rich areas interposing relatively long galacturonan segments having typical DP around 50. In the branched block of the molecule, both arabinan and galactan chains are attached to rhamnoses, with further arabinan segments on the galactan chain. Arabinans are branched polysaccharides with a backbone of $(1 \rightarrow 5)$ linked α-L-arabinofuranosyl residues; other α-L-arabinofuranosyl units being attached to varying numbers of residues at the O2 and/or O3 position. Some structures of these arabinans can generate comb-like types of arrangements; some can imply a more ramified type of structures. There may be different types of branched blocks in pectins from one cell wall or even within a single pectin macromolecule. There is also a region in the galacturonan backbone, where xylose side chains occur, hence, a fourth component called xylogalacturonan as shown in Figure 2.25. In addition, pectins also carry noncarbohydrate substituents, acetyl groups, borate, and amide groups that are incorporated during processing of LM amidated pectin.

There are many published studies on pectin edible films alone and in combination with other ingredients. Galus et al. (2012) studied the sorption and wetting properties of pectin edible films and found that water content increases slowly for these films from 0 to 0.75 water activities, but increases sharply from 0.75 to 0.90. Mariniello et al. (2003) studied whole soy flour and apple pectin as raw materials for producing hydrocolloid edible films. The best ratio between the two components (2:1 pectin–soy flour) was determined to obtain films which could be easily handled during use. The addition of transglutaminase, an enzyme able to produce isopeptide bonds along the soy polypeptide chains, gave a smoother homogeneous surface, increased strength but reduced flexibility. Liu et al. (2008) studied antimicrobial edible films prepared from natural fiber of pectin and other food hydrocolloids for food packaging or wrapping by extrusion followed by compression, or blown film method. Microscopic analysis revealed a well-mixed integrated structure of extruded pellets and an even distribution of the synthetic hydrocolloid in the biopolymers. The resultant composite films possess mechanical properties that are comparable to films cast from most natural hydrocolloids that are consumed as foods or components in processed foods. The inclusion of poly(ethylene oxide) alters the textures of the resultant composite films and therefore, demonstrates a new technique for the modification of film properties. The composite films were produced in mild processing conditions, thus the films are able to protect the bioactivity of the incorporated nisin, as shown by the inhibition of *Listeria monocytogenes* bacterial growth by a liquid incubation method. Perez Espitia et al. (2014) prepared edible films based on pectin and açai with infused apple skin polyphenols and thyme essential oils. The polyphenol increased mechanical resistance of the film, while thyme had a plasticizing effect at optimum levels of 6.07% and 3.1%, respectively. Giosafatto et al. (2014) studied pectin edible film and reported that mechanical properties and barrier properties to CO_2, O_2, and water vapor revealed that these films possess technological features comparable to those possessed by commercial plastics and that these characteristics are maintained even following storage of the films at 4°C or −20°C, suggesting that these bioplastics can be tailored to protect food at low temperature.

2.5.6.2 Pullulan (E1204; CAS#9257-02-7; GRAS)

Pullulan is a water-soluble, extracellular polysaccharide produced by certain strains of the polymorphic fungus *Aureobasidium pullulans*, formerly known as *Pullularia pullulans*, an organism that was first described by De Bary in 1866. It is produced by fermentation of substrate containing sucrose or other carbohydrate as carbon source and other growth nutrients. Mishra and Vuppu (2012) described the following process for pullulan production:

1. To obtain the pure biopolymer from the fermentation broth, cell harvesting, removal of the melanin pigments that are formed during fermentation and precipitation of the polymer are essential. After fermentation, the broth is centrifuged (800 RPM for 30 min), and the greenish black sludge containing the melanin pigments is discarded. The supernatant is then subjected to heating (80°C for 1 h) to inactivate the pullulanase enzymes formed during fermentation, followed by further demelanization by adsorption on activated charcoal, or by use of solvent or solvent/salt combinations. The pure pullulan is then precipitated out with alcohol (IPA), dried, and milled.
2. The average MW of pullulan ranges from 150,000 to 10 million based on published studies (Kim et al. 2000) using optimized conditions. However, based on the report by Ivan Stankovic to JECFA, commercially available pullulan has a purity of 90% or more, and its average MW is about 200,000 (http://www.fao.org/fileadmin/templates/agns/pdf/jecfa/cta/65/pullulan.pdf).

Pullulan has a very different structural conformation not found in any other gum polysaccharides. It is not nearly as complex as pectin, but very unique in its own way. It is a homopolysaccharide composed of only glucose monomer. The basic structure of pullulan was elucidated from the works of several researchers (Bender and Wallenfels 1961; Bernier 1958; Bouveng et al. 1962, 1963; Sowa et al. 1963; Ueda et al. 1963; Wallenfels et al. 1961, 1965). From the results of these studies, pullulan was characterized as linear polymer of maltotriose subunits, glucose joined by α-$(1 \rightarrow 4)$ linkages connected via α-$(1 \rightarrow 6)$ glycosidic linkages (Figure 2.26). Catley and coworkers subsequently established the occurrence of a minor percentage of randomly distributed maltotetraose subunits in pullulan (Catley 1970; Catley and Whelan 1971; Catley et al. 1966; Carolan et al. 1983) and reported that these subunits were also joined by α-$(1 \rightarrow 6)$ linkages. The regular occurrence of α-$(1 \rightarrow 6)$ linkages kinks the structure of pullulan into a step ladder conformation and interrupts what would otherwise be a linear, sixfold helix starch amylose chain. The unique pattern of α-$(1 \rightarrow 6)$ linkages between maltotriose subunits gives the pullulan polymer distinctive physical properties, such as structural flexibility and high water solubility, resulting in distinct film- and fiber-forming characteristics. Leathers (2003) reported that pullulan films and fibers resemble certain synthetic polymers, such as plastics derived from petroleum, and possess the following unique functional characteristics: (1) oxygen impermeability in contrast to other polysaccharide films; (2) edible and biodegradable, although more expensive than plastics; and (3) highly soluble in water. Pullulan solutions are of relatively low viscosity compared to most gums of the same MW. Depending on grade, a 25% solution of pullulan could be 1800–22,000 cP at 20 RPM and 25°C, which is much more viscous than a solution of gum Arabic at the same concentration of ~50–100 cP but is significantly less viscous compared to other gums like guar, xanthan, konjac, and many others. Comparing

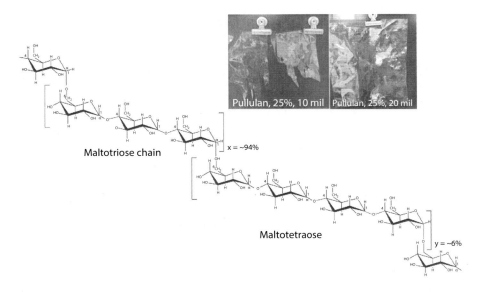

FIGURE 2.26 Idealized structure of pullulan showing both α-(1 → 4) and α-(1 → 6) glycosidic linkages in a step ladder conformation for the maltotriose and maltotetraose repeating blocks. Also shown are pictures of its film, cast from a 25% solution and thickness of 10 mils (0.01 inches or 254 μm) and 20 mils (0.02 inches or 508 μm) measured across a 2.5-cm wide film strip using the TA.XT.PlusTexture analyzer, yielding tensile strength of 14,000 g and a puncture strength of 2200 g. (With kind permission from Springer Science+Business Media: *Edible Films and Coatings for Food Applications*, Structure and function of polysaccharide gum-based edible films and coatings, 2009, pp. 57–112, Nieto, M.)

the structures of pullulan and gum Arabic, there is really no branching in the structure of pullulan; hence, polymer chains are still able to associate and form a much stronger film than gum Arabic. The kink caused by the 1 → 6 linkages that happens quite regularly at every maltotriose, or every maltotetraose is a key to pullulans low viscosity and high water solubility but uniquely so that it can still form thick solutions that are water clear at 25% concentration. Its MW is sufficiently large enough (200,000 to several millions) and quite extended to induce intermolecular association and film structure formation.

Pullulan films are highly oxygen-impermeable and have excellent mechanical properties according to Yuen (1974). Diab et al. (2001) used pullulan as edible film and coating to preserve fruit and found that the application of pullulan/sorbitol/SE coating on strawberries resulted in large changes in internal fruit atmosphere composition which were beneficial for extending the shelf life of this fruit; the coated fruit showed much higher levels of CO_2, a large reduction in internal O_2, better firmness and color retention, and reduced rate of weight loss. At Penn State College of Agricultural Sciences, pullulan films containing essential oils derived from rosemary, oregano, and nanoparticles were used to wrap meat and showed that bacterial pathogens were inhibited significantly by the use of the antimicrobial films (http://www.science20.com/news_articles/edible_pullulan_films improve_the_microbiological_safety_of_meat-135412).

More recently, Liu et al. (2013) showed that the addition of pullulan dramatically improved the elongation at break of edible film, when the amount was 50% by weight and the tensile strength and the luminousness reached the best. Glycerol both increased the elongation at break and optical properties of edible film, and when the amount was 20%, the tensile strength was highest. Choudhary et al. (2011) made rapid dissolving films using pullulan in combination with xanthan and showed excellent film-forming capacity along with tensile strength of 5.56 N mm^{-2}, disintegration time of 22 s, and dissolution time of 42 s.

2.6 COMPOSITE FILMS

It is very common to see that gum polymers are blended together to make composite films with improved mechanical properties and with altered solubility to meet a specific packaging requirement. Dissolvable food packaging is one specific film application, where it is critical to tailor the solubility of the film to the food system inside the package. Table 2.2 summarizes the results of a study done by Nieto and Grazaitis (unpublished), comparing dissolution time of various film composites (without the flavor, color, or active ingredients). The films were prepared using sodium alginate and CMC in combination with other gums and a plasticizer. Twelve composite film recipes with different tensile strengths and solubilities are shown. The solubility of the alginate–CMC film is greatly improved by the addition of gums that do not form cohesive films (weak gums), such as gum Arabic, inulin, resistant maltodextrin, and polydextrose, as well as the insoluble MCC. This phenomenon is mainly due to the weakening of the film matrix structure, leading to faster disintegration of the film. On the contrary, glycerol, a plasticizer, strengthens and increases the dissolution time of the resulting film.

Many published studies on edible films are already combining different gum polymers, proteins, plasticizers, insoluble particulates, and actives. Many of these studies are already presented under specific gum heading. Additionally, composite film studies summarized below were done basically with same goal, that is, to improve film properties of the composite films to make them suitable as edible film, coating, or as a food packaging.

Chambi and Grosso (2011) made composite films with methylcellulose, glucomannan, and pectin and found that a ratio of 1:4:1 produced film with best tensile strength of 72.63 Pa and elongation of 9.85%. The addition of gelatin to this film at pH 5 improved WVP. Rao et al. (2010) prepared composite films by casting method using chitosan and guar gum in different ratios. The concentration of guar gum ranged from 0% to 50% (v/v). Addition of guar gum in varied proportions to chitosan solution led to changes in transparency and opacity of films. The water vapor transmission rate did not change significantly upon addition of guar gum. Films containing 15% (v/v) guar gum showed very low oxygen permeability, good tensile, and puncture strength. The antimicrobial activity of films containing 15% (v/v) guar gum was comparable to chitosan films against *Escherichia coli* and *Staphylococcus aureus*. Wu et al. (2012) studied a series of novel edible blend films made with KGM and curdlan by a solvent-casting technique with different blending ratios of the two polymers. The results showed that strong intermolecular hydrogen bonds or interactions

TABLE 2.2
Tensile, Puncture Strength, and Dissolution Time of Edible Film Composites

	Composite Film Recipe	%	Tensile Strength (g)	Puncture Strength (g)	Dissolution Time (s)
A	Water	91.1	1168.7	468.9	40.25
	Sodium alginate	4.0			
	LV CMC	4.0			
	Glycerin	0.9			
B	Water	89.1	1360.4	192.9	24.5
	Sodium alginate	4.0			
	LV CMC	4.0			
	MCC	2.0			
	Glycerin	0.9			
C	Water	89.1	1701.7	356.9	34.75
	Sodium alginate	4.0			
	LV CMC	4.0			
	Maltodextrin	1.0			
	Glycerin	0.9			
D	Water	95.0	735.9	189.5	83
	Sodium alginate	2.5			
	LV CMC	1.3			
	κ-Carrageenan	1.3			
E	Water	87.6	1171.3	1175.5	48.5
	LV CMC	8.0			
	κ-Carrageenan	3.5			
	Glycerin	0.9			
F	Water	91.1	834.1	341.2	31
	Pectin	4.0			
	Sodium alginate	4.0			
	Glycerin	0.9			
G	Water	95.0	834.1	341.2	31
	Sodium alginate	2.5			
	LV CMC	2.5			
H	Water	95.0	713.7	290.2	20
	Sodium alginate	2.5			
	LV CMC	1.3			
	Gum Arabic	1.3			
I	Water	95.0	456.1	102.0	16
	Sodium alginate	2.5			
	LV CMC	1.3			
	Inulin	1.3			
J	Water	95.0	804.3	107.5	12.6
	Sodium alginate	2.5			
	LV CMC	1.3			
	Polydextrose	1.3			

(Continued)

TABLE 2.2 (Continued)
Tensile, Puncture Strength, and Dissolution Time of Edible Film Composites

	Composite Film Recipe	%	Tensile Strength (g)	Puncture Strength (g)	Dissolution Time (s)
K	Water	95.0	761.9	122.1	14.2
	Sodium alginate	2.5			
	LV CMC	1.3			
	Maltodextrin	1.3			
L	Water	95.0	355.7	86.9	15.4
	Sodium alginate	2.5			
	LV CMC	1.3			
	MCC	1.3			

Source: Nieto, M. and Grazaitis, D. 2006. Edible films with unique properties using gums. Presented at IFT in Orlando, Florida, June 2006 (unpublished).

Note: Tensile strength measured across a 2.5-cm wide film strip using the TA.XT.PlusTexture analyzer; dissolution time measured by a trained panel on 1 cm^2 film, 0.05 mm thick.

took place between konjac and curdlan, and the interaction of the blend film was much greater than that of the other films, when konjac was around 70 wt%, resulting in excellent miscibility. In addition, the blend films displayed excellent moisture barrier properties, which had a potential application in food packaging. Wu (1999) studied edible composite films and coatings based on polysaccharides, proteins, and/or lipids and tested them for their effectiveness in controlling moisture loss and lipid oxidation of precooked ground beef patties. Wheat gluten, soy protein, carrageenan, and chitosan films and coatings had different effects on maintaining the quality of precooked patties after 3-day storage at 4°C. All coatings were as effective as polyvinyl chloride film in reducing patty moisture loss. Composite films of soy protein and PGA showed improved moisture barrier properties and mechanical properties by adding up to 10% and 17.5%. During 6-day storage at 4°C, starch–alginate–stearic acid–based films were effective in limiting moisture loss from precooked patties. Tocopherol-treated starch–alginate films and starch–alginate–stearic acid films were effective in inhibiting the lipid oxidation in patties by lowering TBARS and other oxidation products. Martins et al. (2011) prepared composite films with 80:20 ratio of κ-carrageenan and LBG, versus 100% LBG. The WVP decreased significantly in the composite film, increased tensile strength but gave lower value elongation at break. These authors recommended based on principal component analysis that composite films composed of 40/60% (w/w) of κ-carrageenan/LBG could be the best choice to be applied on food systems due to their good water vapor barrier and mechanical properties. Martins et al. (2012, 2013) studied physical and antimicrobial properties of biodegradable films composed of mixtures of κ-carrageenan and LBG with modified clay dispersed in the biopolymer matrix. The increase in clay concentration caused a decrease in WVP and an increase in CO_2 permeability; while O_2 permeability did not change. Films with 16% clay exhibited the highest tensile strength and elongation at break. The films exhibited an inhibitory effect only against *L. monocytogenes* and not against *E. coli* and *Salmonella*.

Luceniusa et al. (2014) prepared all-polysaccharide composite films from native, unmodified cellulose nanofibrils mixed with various natural water-soluble polysaccharides like carboxymethyl cellulose, galactoglucomannan, xyloglucan, and guar gum. Composite films were manufactured by pressurized filtration and hot pressing. It was found that all the tested polysaccharides increased the strength and toughness of the dry composite films at 2 wt% addition. After soaking the samples for 24 h in water, wet strength of film with 2 wt% CMC films diminished, while the uncharged polysaccharide like guar gum improved the wet strength. Young's modulus of film with 2% guar gum increased by a factor of 1.3, the tensile strength by a factor of 2.8, and the toughness by a factor of 3.4.

2.6.1 Rapid Melt Film Composite

In general, a weak film dissolves and undergoes rapid disintegration in water. A strong gum film can be weakened using fillers like maltodextrin, sugar, dextrose, sugar alcohols, cellulose, gum Arabic, etc., to make it melt rapidly. In applications such as breath film, cheese film, and spice film, fast dissolution is a desirable attribute. However, the gum polymer or blend needs to provide adequate tensile strength for machinability during processing. Any of the recipes from H to L in Table 2.2 can be used or further manipulated to produce films that dissolve faster in the mouth. There are many options for making rapid melt films using various polymers as matrix materials. Aliginate, pectin, λ-carrageenan, and methylcellulose can be used together to provide the primary film structure, while other ingredients can be added to weaken the film for rapid melting.

2.6.2 Slow Melt Film Composite

There are many gums or gum combinations that will meet the criterion of slow melting, and this is also easily manipulated by adjusting casting thickness. As shown in the same study by Nieto and Grazaitis (unpublished), the doubling of the casting thickness for recipes H to L (Table 2.2) increased dissolution time by 1.4–3.3 times faster. The use of methylcellulose, HPMC, and konjac in a gum film blend will increase the tensile strength of the resulting films, allowing them to dissolve more slowly. If meant as a dissolvable packaging, slow melting films will dissolve faster when heated, and in this case the dissolution of the film is improved. In addition, they provide the thickening effect.

2.6.3 Insoluble Composite Films

Insoluble films with water and oil barrier properties are required in applications such as a film barrier between the dough and pizza filling and similar applications like dinner entrees where flavors are packed in insoluble film then dissolves when heated. The best choice of gum is agar, a good film former whose film does not dissolve until heating to 100°C. κ-Carrageenan and LBG to some extent can be used especially if formulated with oil and suitable plasticizers. From the same study by Nieto and Grazaitis (unpublished), κ-carrageenan and glycerol produced a film that is only sparingly soluble in cold water. Cross-linking alginate with calcium also yields a strong, clear film that is insoluble in water, and this combination has been used to microencapsulate probiotics, such as *L. acidophilus*, to increase survival of the bacterial cells under gastric conditions. Based on a study by Chandramouli et al. (2004), the viability of the bacterial cells within calcium–alginate microcapsules increased as both capsule size and alginate concentration increased. Agar and gellan may also be used in this same application.

For dissolvable packaging that possesses high mechanical strength, this group of gum polymers has the most potential in packaging flavors for soups and sauces that would normally be heated.

2.7 CONCLUSIONS

The use of gum films as antimicrobial films, encapsulating material, carrier for actives, flavors or color, or as casing for sausages have gained research successes and commercial applications. As food packaging, gum polymer films have shortcomings that need to be addressed to make them suitable as packaging materials to replace plastic, such as

1. Heat sealability and melting property: All gum films presented in this chapter do not have good melting property and heat sealability except for ethylcellulose. Most of them are polar and therefore, their films do not melt. Ethylcellulose is predominantly nonpolar and consequently possesses very good melting property and heat sealability. Research focusing on colayering of these gum films with a heat sealable component would be very helpful in improving the functionality of the gum bioplastics.
2. Increasing tensile strength and elongation at break: More research required to optimize tensile strength and stretchability of gum films to make them suitable as plastic packaging replacement. They are not nearly as stretchable as Saran wrap, and therefore, more research focusing on improving elongation at break of the gum films to make them comparable to plastic films are needed.
3. According to Lin and Zhao (2007), despite significant benefits from using edible coatings for extending product shelf life and enhancing the quality and microbial safety of fresh and minimally processed fruits and vegetables, commercial applications on a broad range of edible coating of fruits and vegetables are still very limited. Of the many improvements that are needed for polysaccharides to be used more extensively as edible film and coating materials, the most important is improving the water resistance of gum films. The creation of a nanocomposite, incorporation of lipid to the gum polymer matrix, or formation of nanolaminates can provide solutions to this challenge. Weiss et al. (2006), in their scientific status summary

on materials in food nanotechnology, describe how a polysaccharide–clay nanocomposite can be formed. A carbohydrate is pumped together with clay layers, through a high shear cell, to produce a film that contains exfoliated clay layers. The nanocomposite carbohydrate film has substantially reduced WVP. Similarly, chitosan containing exfoliated hydroxyapatite layers maintains functionality in humid environments, with good mechanical properties.

4. The incorporation of hydrophobic ingredients such as lipids for improving the moisture barrier, while maintaining the desirable function of resistance to vapor, gas, or solute is another research area that can be pursued. Decher and Schlenoff (2003) created a nanolaminate based on the layer-by-layer deposition technique in which charged surfaces are coated with interfacial films that consist of multiple nanolayers of different materials. Nanolaminates can give food scientists some advantages for the preparation of edible coatings and films over conventional technologies, because they can be more efficient moisture, lipid, and gas barriers.

REFERENCES

Abdou, E. S., and Sorour, M. A. 2014. Preparation and characterization of starch/carrageenan edible films. *International Food Research Journal*, *21*(1):189–193.

Alyanak, D. 2004. Water Vapor Permeable Edible Membranes. Master's thesis. IZMIR Institute of Technology.

Antoniou, J., Liu, F., Majeed, H., and Zhong, F. 2015. Characterization of tara gum edible films incorporated with bulk chitosan and chitosan nanoparticles: A comparative study. *Food Hydrocolloids*, *44*:309–319.

Araki, C. 1965. Some recent studies on the polysaccharides of agarophytes. *Proceedings of International Seaweed Symposium*, *5*:3–17.

Arnott, S., Fulmer, A., Scott, W. E., Dea, I. C. M., Moorhouse, R., and Rees, D. A. 1974b. The agarose double helix and its function in agarose gel structure. *Journal of Molecular Biology*, *90*:269–284.

Arnott, S., Scott, W. E., Rees, D. A., and McNab, C. G. A. 1974a. ι-Carrageenan: Molecular structure and packing of polysaccharide double helices in oriented fibers of divalent cation salts. *Journal of Molecular Biology*, *90*:253–267.

Atkins, E. D. T., Nieduzynski, I. A., Mackie, W., Parker, K.D., and Smolko, E. E. 1973a. Structural components of alginic acid I. The crystalline structure of poly-β-D-mannuronic acid. *Biopolymers*, *12*:1865–1878.

Atkins, E. D. T., Nieduzynski, I. A., Mackie, W., Parker, K. D., and Smolko, E. E. 1973b. Structural components of alginic acid II. The crystalline structure of poly-α-L-guluronic acid. *Biopolymers*, *12*:1879–1887.

Aydinli, M., Tutas, M., and Bozdemir, O. A. 2004. Mechanical and light transmittance properties of locust bean gum based edible films. *Turkish Journal of Chemistry*, *28*:163–171.

Bastioli, C. 2005. *Handbook of Biodegradable Polymers*. Toronto-Scarborough, Ontario, Canada: ChemTec Publishing, 553 p.

Basu, S. K. 2006. Seed Production Technology for Fenugreek (*Trigonella foenum-graecum* L.) in the Canadian Prairies. Master's thesis, University of Lethbridge, Alberta, Canada.

Bender, H., and Wallenfels, K. 1961. Investigations on pullulan II. Specific degradation by means of a bacterial enzyme. *Biochem Z*, *334*:79–95.

Bernier, B. 1958. The production of polysaccharides by fungi active in the decomposition of wood and forest litter. *Canadian Journal of Microbiology, 4:*195–204.

Bilbao-Sainz, C., Avena-Bustillos, R. J., Wood, D. F., Williams, T. G., and McHugh, T. H. 2010. Composite edible films based on hydroxypropyl methylcellulose reinforced with microcrystalline cellulose nanoparticles. *Journal of Agricultural and Food Chemistry*, 58(6):3753–3760.

Bixler, H. J., and Porse, H. 2011. A decade of change in the seaweed hydrocolloids industry. *Journal of Applied Phycology 23*: 321–335.

Born, K., Langendorff, V., and Boulenguer, P. 2002. Xanthan. In: Steinbüchel, A., Vandamme, E. J., and De Baets, S. (Eds.), *Biopolymers* (vol. 5, pp. 261–291). Weinheim, Germany: Wiley-VCH.

Bouveng, H. O., Kiessling, H., Lindberg, B., and McKay, J. 1962. Polysaccharides elaborated by *Pullularia pullulans*. I. The neutral glucan synthesized from sucrose solutions. *Acta Chemica Scandinavica, 16*:615–622.

Bouveng, H. O., Kiessling, H., Lindberg, B., and McKay, J. 1963. Polysaccharides elaborated by *Pullularia pullulans*. II. The partial acid hydrolysis of the neutral glucan synthesised from sucrose solutions. *Acta Chemica Scandinavica, 17*:797–800.

Bozdemir, O. A., and Tutas M. 2003. Plasticizer effect on water vapour permeability properties of locust bean gum edible films. *Turkish Journal of Chemistry*, 27:773–782.

Brummer, K., Cui, W., and Qang, Q. 2003. Extraction, purification and physicochemical characterization of fenugreek gum. *Food Hydrocolloids*, 17(3):229–236.

Cardenas, G., and Miranda, S. P. 2004. FTIR and TGA studies of chitosan composite films. *Journal of Chilean Chemical Society*, 49(4):291–295.

Carolan, G., Catley, B. J., and McDougal, F. J. 1983. The location of tetrasaccharide units in pullulan. *Carbohydrate Research, 114*:237–243.

Catley, B. J. 1970. Pullulan, a relationship between molecular weight and fine structure. *FEBS Letters 10*:190–193.

Catley, B. J., Robyt, J. F., and Whelan, W. J. 1966. A minor structural feature of pullulan. *Biochemical Journal 100*: 5P–8P.

Catley, B. J., and Whelan, W. J. 1971. Observations on the structure of pullulan. *Archives of Biochemistry and Biophysics, 143*:138–142.

Chambi, H. N. M., and Grosso, C. R. F. 2011. Mechanical and vapor permeability properties of biodegradable films based on methylcellulose, glucomannan, pectin and gelatin. *Ciencia e Tecnologia de Alimentos Campinas*, 31(3):739–746.

Chandramouli, V., Kailasapathy, K., Peiris, P., and Jones, M. 2004. An improved method of microencapsulation and its evaluation to protect *Lactobacillus* spp. in simulated gastric conditions. *Journal of Microbiological Methods*, 57:27–35.

Chandrasekaran, R., and Radha, A. 1995. Molecular architectures and functional properties of gellan gum and related polysaccharides. *Trends in Food Science*, 6:143–148.

Chen, S., and Nussinovitch, A. 2000a. The role of xanthan gum in traditional coatings of easy peelers. *Food Hydrocolloids*, 14(4):319–326.

Chen, S., and Nussinovitch, A. 2000b. Galactomannans in disturbances of structured wax-hydrocolloid-based coatings of citrus fruit (easy peelers). *Food Hydrocolloids*, 14:561–568.

Chen, S., and Nussinovitch, A. 2001. Permeability and roughness of wax-hydrocolloid coatings and their limitations in determining citrus fruit overall quality. *Food Hydrocolloids*, 15:127–137.

Choudhary, D. R., Patel, V., Patel, H., and Kundawala, A. J. 2011. Exploration of film forming properties of film formers used in the formulation of rapid dissolving films. *International Journal of ChemTech Research*, 3(2):531–533.

Chuah, C. T., Sarko, A., Deslandes, Y., and Marchessault, R. H. 1983. Packing analysis of carbohydrates and polysaccharides. Part 14. Triple-helical crystalline structure of curdlan and paramylon hydrates. *Macromolecules*, 16(8):1375–1382.

da Silva, M. A., Bierhalz, A. C. K., and Kieckbusch, T. G. 2009. Alginate and pectin composite films crosslinked with Ca^{2+} ions: Effect of the plasticizer concentration. *Carbohydrate Polymers*, 77(4):736–742.

Debeaufort, F., and Voilley, A. 1997. Methylcellulose-based edible films and coatings: 2. Mechanical and thermal properties as a function of plasticizer content. *Journal of Agricultural and Food Chemistry*, 45(3):685–689.

Decher, G., and Schlenoff, J. B. 2003. *Multilayer Thin Films: Sequential Assembly of Nanocomposite Materials* (p. 543), Weinheim, Germany: Wiley VCH.

de Melo, C., Garcia, P. S., Grossmann, M. V. E., Yamashita, F., Dall'Antônia, L. H., and Mali, S. 2011. Properties of extruded xanthan–starch–clay nanocomposite films. *Brazilian Archives of Biology and Technology*, 56(4). Published online at http://www.scielo.br/scielo.php?script=sci_arttext&pid=S1516-89132013000400014.

Deslandes, Y., Marchessault, R. H., and Sarko, A. 1980. Triple-helical structure of (1 → 3)-β-D-glucan. *Macromolecules*, 13(6):1466–1471.

Diab, T., Biliaderis, C. G., Gerasopoulos, D., and Sfakiotakis, E. 2001. Physicochemical properties and application of pullulan edible films and coatings in fruit preservation. *Journal of Science and Food Agriculture*, 81:988–1000.

Donati, I., Holtan, S., Mørch, Y. A., Borgogna, M., Dentini, M., and Skjåk-Bræk, G. 2005. New hypothesis on the role of alternating sequences in calcium-alginate gels. *Biomacromolecules*, 6:1031–1040.

El Batal, H., and Hazib, A. 2013. Optimization of extraction process of carob bean gum purified from carob seeds by response surface methodology. *Chemical and Process Engineering Research*, 12:1–8.

Engelsen, S. B., Cros, S., Mackie, W., and Perez, S. 1996. A molecular builder for carbohydrates: Application to polysaccharides and complex carbohydrates. *Biopolymers*, 39(3):417–433.

Falshaw, R. Bixler, H. J., and Johndro, K. 2001. Structure and performance of commercial kappa-2 carrageenan extracts I. Structure analysis. *Food Hydrocolloids* 15:441–452.

Galus, S., Turska, A., and Lenart, A. 2012. Sorption and wetting properties of pectin edible films. *Czech Journal of Food Science*, 30(5):446–455.

Geremia, R., and Rinaudo, M. 2005. Biosynthesis, structure and physical properties of some bacterial polysaccharrides. In: *Polysaccharides—Structural Diversity and Functional Versatility*, Dumitriu, S. (ed.), Marcel Dekker, New York, pp. 411–430.

Giannouli, P., and Morris, E. R. 2000. Cryogelation of xanthan. *Food Hydrocolloids*, 17:495–501.

Giosafatto, C. V., Di Pietro, P., Gunning, P., Mackie, A., Porta, R., and Mariniello, L. 2014. Characterization of citrus pectin edible films containing transglutaminase-modified phaseolin. *Carbohydrate Polymers*, 106:200–208.

Goycoolea, F. M., Milas, M., and Rinaudo, M. 2001. Associative phenomena in galactomannan-deacetylated xanthan systems. *International Journal of Biological Macromolecules*, 29:181–192.

Han, Y. J., Roh, H. J., and Kim, S. S. 2007. Preparation and physical properties of curdlan composite edible films. *Korean Journal of Food Science and Technology*, 39(2):158–163.

Harada, T., Koreeda, A., Sato, S., and Kasai, N. 1979. Electron microscopic study on the ultrastructure of curdlan gel: Assembly and dissociation of fibrils by heating. *Journal of Electron Microscopy*, 28(3):147–153.

Haug, A., Larsen, B., and Smidsrod, O. 1966. A study of the constitution of alginic acid by partial acid hydrolysis. *Acta Chemica Scandinavica*, 20:183–190.

Haug, A., Larsen, B., Smidsrod, O., and Painter, T. 1969. Development of compositional heterogeneity in alginate degraded in homogenous solution. *Acta Chemica Scandinavica*, 23:2955–2962.

Helgerud, T., Gaserod, O., Fjaereide, T., Andersen, P. O., and Larsen, C. K. 2010. Alginates. In Imeson, A. (Ed.), *Food Stabilizers, Thickeners and Gelling Agents* (Chapter 4) (pp. 50–72). Iowa, USA: Blackwell Publishing Limited, Oxford United Kingdon and Ames.

Hirase, S. 1957. *Pyruvic Acid as a Constituent of Agar-Agar.* Mem Fac Ind Arts Kyoto Tech Univ Sci Technol, Kyoto, Japan, pp. 17–29.

Hwang, K. T., Jung, S. T., Lee, G. D., Chinnan, M. S., Park, Y. S., and Park, H. J. 2002. Controlling molecular weight and degree of deacetylation of chitosan by response surface methodology. *Journal of Agricultural Food Chemistry*, 50(7):1876–1882.

Jansson, P. E., Keene, L., and Lindberg, B. 1975. Structure of the exocellular polysaccharide from *Xanthomonas campestris*. *Carbohydrate Research*, 45:275–282.

Juang, R. S., and Shao, H. J. 2002. A simplified equilibrium model for sorption of heavy metal ions from aqueous solutions on chitosan. *Water Research*, 36(12):2999–3008.

Kalyanasundaram, G. T., Doble, M., and Gummadi, N. 2012. Production and downstream processing of (1 → 3)-β-D-glucan from mutant strain of *Agrobacterium* sp ATCC 31750. AMB Express. 2:31. http://www.amb-express.com/content/pdf/2191-0855-2-31.pdf.

Kato, K. 1973. Isolation of oligosaccharides corresponding to the branching-point of konjac mannan. *Agricultural and Biological and Chemistry*, 37(9):2045–2051.

Kato, K., and Matsuda, K. 1980. Studies on chemical structure of konjac mannan. I. Isolation and characterization of oligosaccharides from partial hydrolysis of mannan. *Agricultural and Biological and Chemistry*, 33:1446–1453.

Kawakami, K., Okabe, Y., and Norisuye, T. 1991. Dissociation of dimerized xanthan in aqueous solution. *Carbohydrate Polymers*, 14:181.

Kim, M. K., Lee, I. Y., Kim, K. T., Rhee, Y. H., and Park, Y. H. 2000. Residual phosphate concentration under nitrogen-limiting conditions regulates curdlan production in *Agrobacterium* sp. *Journal of Industrial Microbiology and Biotechnology*, 25:180–183.

Kirk-Othmer. 1993. Cellulose ethers. In: Kroschwitz, J. (Ed.), *Encyclopedia of Chemical Technology*, 4th edition (pp. 541–561). New York: John Wiley and Sons.

Kujawa, P., Schmauch, G., Viitala, T., Badia, A., and Winnik, F. M. 2007. Construction of viscoelastic biocompatible films via the layer-by-layer assembly of hyaluronan and phosphorylcholine-modified hitosan. *Biomacromolecules*, 8:3169–3176.

Labropoulos, K. C., Niesz, D. E., Danforth, S. C., and Kevrekidis, P. G. 2002. Dynamic rheology of agar gels: Theory and experiment. Part I. Development of a rheological model. *Carbohydrate Polymers*, 50:393–406.

Larotonda, F. D. S., Hilliou, L., Sereno, A. M. C., and Goncalves, M. P. 2014. Green edible films obtained from starch-domestic carrageenan mixtures. *2nd Mercosur Congress on Chemical Engineering and 4th Mercosur Congress on Process Systems Engineering*, Costa Verde, Rio de Janeiro, Brazil.

Leathers, T. D. 2003. Biotechnological production and applications of pullulan. *Applied Microbiology and Biotechnology*, 62:468–473.

Lee, I. Y. 2004. Curdlan. In: Steinbuchel, A. (Ed.), B*iopolymers: Polysaccharides I-Polysaccharides from Prokaryotes* (vol. 5, pp. 135–155). Weinheim, Germany: Wiley-VCH Publications.

Lee, J. H., Lee, I. Y., Kim, M. K., and Park, Y. H. 1999. Optimal pH control of batch processes for production of curdlan by *Agrobacterium* species. *Journal of Industrial Microbiology and Biotechnology*, 23:143–148.

Leon, P. G., and Rojas, A. M. 2011. Gellan gum films as carriers of L-(+)-ascorbic acid. *Procedia Environmental Sciences*, 8:756–763.

Lertsutthiwong, P., How, N. C., Chandrkrachang, S., and Stevens, W. F. 2002. Effect of chemical treatment on the characteristics of shrimp chitosan. *Journal of Metals, Materials and Minerals*, 12(1):11–18.

Lertsutthiwong Rao, M. S., Munoz, J., and Stevens, W. F. 2000. Critical factors in chitin production by fermentation of shrimp biowaste. *Applied Microbiology and Biotechnology*, 54:808–813.

Li, B., Xie, B., and Kennedy, J. F. 2006. Studies on the molecular chain morphology of konjac glucomannan. *Carbohydrate Polymers*, 64(4):510–515.

Lin, D., and Zhao, Y. 2007. Innovations in the development and application of edible coatings for fresh and minimally processed fruits and vegetables. *Comprehensive Reviews in Food Science and Food Safety*, 6(3):60–75.

Liu, L. 2006. Bioplastic in Food Packaging: Innovative Technologies for Biodegradable Packaging. Packaging Engineering, San Jose State University. http://www.iopp.org/files/public/sanjoseliucompetitionfeb06.Pdf.

Liu, L. S., Liu, C. K., Jin, T., Hicks, K., Mohanty, A. K., Bhardwaj, R., and Misra, M. 2008. A preliminary study on antimicrobial edible films from pectin and other food hydrocolloids by extrusion method. *Journal of Natural Fibers*, 5:366–382.

Liu, X. Y., Zhang, L. Q., Zhao, S. F., and Li, X. F. 2013. Effect of pullulan, glycerol blend on the properties of gelatin based edible film. *Advanced Materials Research*, 779–780:136.

Luceniusa, J., Parikkab, K., and Osterberg, M. 2014. Nanocomposite films based on cellulose nanofibrils and water-soluble polysaccharides. *Reactive and Functional Polymers*, 85:167–174.

Maekaji, K. 1978. A method for measurement and kinetic analysis of the gelation process of konjac mannan (kinetic study on the gelation of konjac mannan). *Nippon Nogeikagakukaishi*, 52:251–257.

Majewics, T. G., Erazo-Majewicz, P. E., and Podlas, T. J. 2002. Cellulose ethers. In: *Encyclopedia of Polymer Science and Technology*. New York: John Wiley and Sons Inc.

Mariniello, L., Di Pierro, P., Esposito, C., Sorrentino, A., Masi, P., and Porta, R. 2003. Preparation and mechanical properties of edible pectin soy flour films obtained in the absence or presence of transglutaminase. *Journal of Biotechnology*, 102:191.

Martins, J. T., Bourbon, A. I., Pinheiro, A. C., Souza, B. W. S., Cerqueira, M. A., and Vicente, A. A. 2013. Biocomposite films based on k-carrageenan/locust bean gum blends and clays: Physical and antimicrobial properties. *Food and Bioprocess Technology*, 6:2081–2092.

Martins, J. T., Cerqueira, M. A., Bourbon, A. I, Pinheiro, A. C., and Vicente, A. A. 2011. Edible films based on k-carrageenan/locust bean gum-effects of different polysaccharide ratios on film properties. In: *11th International Congress on Engineering*, Athens, Greece, May 22–26, 2011. (www.icef11.org/content/papers/aft/AFT998.pdf).

Martins, J. T., Cerqueira, M. A., Bourbon, A. I., Souza, B. W. S., and Vicente, A. A. 2012. Synergistic effects between κ-carrageenan and locust bean gum on physicochemical properties of edible films made thereof. *Food Hydrocolloids*, 29(2):280–289.

Mathur, V., and Mathur, N. K. 2005. Fenugreek and other lesser known legume galactomannan-polysaccharides: Scope for development. *Journal of Scientific and Industrial Research*, 64:475–481.

Meer, W. 1980. Agar. In Davidson, R. (Ed.), *Handbook of Water Soluble Gums and Resins* (pp. 7-1 to 7-19). New York: McGraw-Hill.

Melton, L. D., Mindt, L., Rees, D. A., and Sanderson, G. R. 1976. Covalent structure of the polysaccharide from Xanthomonas campestris: Evidence from partial hydrolysis studies. *Carbohydrate Research*, 46:245–257.

Mikkonen, K. S., Rita, H., Helén, H., Talja, R. A., Hyvönen, L., and Tenkanen, M. 2007. Effect of polysaccharide structure on mechanical and thermal properties of galactomannan-based films. *Biomacromolecules*, 8:3198–3205.

Mishra, B., and Vuppu, S. 2012. A study on downstream processing for the production of pullulan by *Aureobasidium pullulans*-SB-01 from the fermentation broth. *Research Journal of Recent Science*, 2(ISC-2012):16–19.

Mitsuiki, M., Mizuno, A., and Motoki, M. 1999. Determination of molecular weight of agars and effect of the molecular weight on the glass transition. *Journal of Agricultural and Food Chemistry*, 47(2):473–478.

Miwa, M., Nakao, Y., and Nara, K. 1994. Food applications of curdlan. In: Nishinari, K., and Doi, E. (Eds.), *Food Hydrocolloids: Structures, Properties, and Functions* (pp. 119–124). New York: Plenum Press.

Murray, J. C. F. 2000. Cellulosics. In: Phillips, G. O., and Williams, P. A. (Eds.), *Handbook of Hydrocolloids* (pp. 219–229). Boca Raton, FL: Woodhead Publishing.

Nakata, M., Kawaguchi, T., Kodaky, Y., and Kono, A. 1998. Characterization of curdlan in aqueous sodium hydroxide. *Polymer Science*, 39:1475–1481.

Nieto, M. 2009. Structure and function of polysaccharide gum-based edible films and coatings. In Embuscado, M. E., and Huber, K. C. (Eds.), *Edible Films and Coatings for Food Applications* (pp. 57–112). New York: Springer.

Okuyama, K., Arnott, S., Moorhouse, R., Walkinshaw, M. D., Atkins, E. D. T., and Wolf-Ulish, C. H. 1980. Fiber diffraction studies of bacterial polysaccharides. In: *Fiber Diffraction Methods*. French, A. D., and Gardner, K. C. H. (eds), ACS Symposium Series 141. Washington, DC, 411 pages.

Okuyama, K., Otsubo, A., Fukuzawa, Y., Ozawa, M., Harada, T., and Kasai, N. 1991. Single-helical structure of native curdlan and its aggregation state. *Journal of Carbohydrate Chemistry*, 10(4):645–656.

Park, S. Y., Marsh, K. S., and Rhim, J. W. 2002. Characteristics of different molecular weight chitosan films affected by the type of organic solvents. *Journal of Food Science*, 67(1):194–197.

Patra, A., Verma, P. K., and Mitra, R. K. 2012. Slow relaxation dynamics of water in hydroxypropyl cellulose-water mixture traces its phase transition pathway: A spectroscopic investigation. *Journal of Physical and Chemistry B*, 116(5):1508–1516.

Pedroni, V. I., Schulz, P. C., Gschaider, M. E., and Andreucetti, N. 2003. Chitosan structure in aqueous solution. *Colloid and Polymer Science*, 282(1):100–102.

Pérez, S., Mazeau, K., and Hervé du Penhoat, C. 2000. The three-dimensional structures of the pectic polysaccharides. *Plant Physiology and Biochemistry*, 38:37–55.

Pérez, S., Rodríguez-Carvajal, M. A., and Doco, T. 2003. A complex plant cell wall polysaccharide: Rhamnogalacturonan II. A structure in quest of a function. *Biochimie*, 85:109–121.

Perez Espitia, P. J., Avena-Bustillos, R. J., and Du, W. X. 2014. Optimal antimicrobial formulation and physical-mechanical properties of edible films based on acai and pectin for food preservation. *Food Packaging and Shelf Life*, 2:38–49.

Petkowicz, C. L. O., Reicher, F., and Mazeau, K. 1998. Conformational analysis of galactomannans: From oligomeric segments to polymeric chains. *Carbohydrate Polymers*, 37:25–39.

Rao, M. S., Kanatt, S. R., Chawla, S. P., and Sharma, A. 2010. Chitosan and guar gum composite films: Preparation, physical, mechanical and antimicrobial properties. *India Carbohydrate Polymers*, 82(4):1243–1247.

Rao, M. S., Munoz, J., and Stevens, W. F. 2000. Critical factors in chitin production by fermentation of shrimp biowaste. *Applied Microbiology and Biotechnology*, 54:808–813.

Rao, V. S. R., Qasba, P. K., Balaji, P. V., and Chandrasekaran, R. 1998. *Conformation of Polysaccharides*. Australia, Canada, China, France, Germany, India, Japan, Luxembourg, Malaysia, The Netherlands, Russia, Singapore, Switzerland: Harwood Academic Publishers, 359p.

Reddy, K. S., Prabhakar, M. N., Babu, P. K., Venkatesulu, G., Rao, U. S. K., Rao, K. C., and Subha, M. C. S. 2012. Miscibility studies of hydroxypropylcellulose/polyethylene glycol in dilute solutions and solid state. *International Journal Carbohydrate Chemistry*, 2012:9p.

Reddy, K. S., Prabhakar, M. N., Rao, M., Suhasini, D. M., Reddy, V. N. M., Babu, P. K., Sudhakar, K., Babu, A. C., Subha, M. C. S., and Rao, K. C. 2013. Development and characterization of hydroxyl propyl cellulose/poly(viny) alcohol blends and their physico-chemical studies. *Indian Journal of Advances in Chemical Science*, 2(1):38–45.

Ren, H. 2007. Study of Properties and Application of Edible Soy Protein Isolate and Curdlan Composite Films. Master's thesis. GTID:2121360185987000.

Rhim, J. W., Wu, Y., Weller, C. L., and Schnepf, M. 1998. Physical characteristics of a composite film of soy protein isolate and propyleneglycol alginate. *Journal of Food Science*, 64(1):149–152.

Risbud, M., Hardikar, A., and Bhonde, R. 2000. Growth modulation of fibroblast by chitosan-polyvinyl pyrrolidone hydrogel: Implications for wound management. *Journal of Biosciences*, 25(1):147–159.

Robbins, S. R. J. 1988. Locust bean gum. In *A Review of Recent Trends in Selected Markets for Water-Soluble Gums*, ODNRI Bulletin No. 2. 108pp. London: Overseas Development Natural Resources Institute (now Natural Resources Institute, Chatham).

Roberts, G. A. F. 1997. Chitosan production routes and their role in determining the structure and properties of the product. In: Domard et al. (Eds.), *Advances in Chitin Sci.*, (vol. 2, pp. 22–31). National Taiwan Ocean University.

Rossman, J. M. 2009. Commercial manufacture of edible films. In: Embuscado, M. E., and Huber, K. (Eds.), *Edible Films and Coatings for Food Applications* (pp. 367–390). New York, USA: Springer.

Russo, R., Malinconico, M., and Santagata, F. 2007. Effect of cross-linking with calcium ions on the physical properties of alginate films. *Biomacromolecules*, 8:3193–3197.

Sabra, W., and Deckwer, W. D. 2005. Alginate—A polysaccharide of industrial interest and diverse biological functions. In Dumitriu, S. (Ed.), *Polysaccharides: Structural Diversity and Functional Versatility* (2nd ed., pp. 515–553). New York: Marcel Deckker.

Shekarabi, A. S., Oromiehie, A. R., Vaziri, A., Ardjmand, M., and Safekordi, A. A. 2014. Effect of glycerol concentration on physical properties of composite films prepared from plum gum and carboxymethyl cellulose. *Indian Journal of Fundamental and Applied Life Sciences*, 4(54):1241–1248.

Shimahara, H., Suzuki, H., Sugiyama, N., and Nishizawa, K. 1975a. Partial purification of ß-mannanases from the konjac tubers and their substrate specificity in relation to the structure of konjac glucomannan. *Agricultural and Biological Chemistry*, 39(2):301–312.

Shimahara, H., Suzuki, H., Sugiyama, N., and Nishizawa, K. 1975b. Isolation and characterization of oligosaccharides from an enzymic hydrolysate of konjac glucomannan. *Agricultural and Biological Chemistry*, 39(2):293–299.

Sousa, A. M. M., Sereno, A. M., Hilliou, L., and Goncalves, M. P. 2010. Biodegradable agar extracted from *Gracilaria vermiculophylla*: Film properties and application to edible coating. *Material Science Forum*, 636–637:739–744. Available at http://paginas.fe.up.pt/~sereno/publ/2010/Biodegradble_Agar_from_Gracilaria_Film_Prop_&_Applic_Coating.pdf.

Southwick, J. G., Jamieson, A. M., and Blackwell, J. 1982. Conformation of xanthan dissolved in aqueous urea and sodium chloride solutions. *Carbohydrate Research*, 99:117.

Sowa, W., Blackwood, A. C., and Adams, G. A. 1963. Neutral extracellular glucan of *Pullularia pullulans* (de Bary) Berkhout. *Canadian Journal of Chemistry*, 41:2314–2319.

Srinivasa, P., Baskaran, R., Ramesh, M., Harish Prashanth, K., and Tharanathan, R. 2002. Storage studies of mango packed using biodegradable chitosan film. *European Food Research and Technology*, 215(6):504–508.

Stone, B. A., and Clark, A. E. 1992. *Chemistry and Biology of (1,3)-β-Glucans*. Victoria, Australia: La Trobe University Press.

Sudo, S. 2011. Dielectric properties of the free water in hydroxypropyl cellulose. *Journal of Physical Chemistry B.*, 115(1):2–6.

Sugiyama, N., Shimahara, H., Andoh, T., Takemoto, M., and Kamata, T. 1972. Molecular weights of konjac mannans of various sources. *Agricultural and Biological Chemistry*, 36(8):1381–1387.

Sworn, G. 2000. Gellan gums. In Philips, G. O., and Williams, P. A. (Eds.), *Handbook of Hydrocolloids* (pp. 117–135). Boca Raton, FL: CRC Press/Woodhead Publishing.

Takigami, S. 2000. Konjac mannan. In Philips, G. O., and Williams, P. A. (Eds.), *Handbook of Hydrocolloids* (pp. 413–424). Boca Raton, FL: CRC Press/Woodhead Publishing.

Tapia, M. S., Rojas-Gratu, M. A., Rodriguez, F. J., Ramirez, J., Carmona, A., and Martin-Belloso, O. 2007. Alginate- and gellan-based edible films for probiotic coatings on fresh-cut fruits. *Journal of Food Science*, 72(4):190–196.

Tongdeesontorn, W., Mauer, L., Wongruong, S., Sriburi, P., and Rachtanapun, P. 2011. Effect of carboxymethyl cellulose concentration on physical properties of biodegradable cassava starch-based films. Purdue e-Pubs. 2-10-2011. http://docs.lib.purdue.edu/cgi/viewcontent.cgi?article=1009&context=foodscipubs.

Tseng, C. K. 1944. Agar—A valuable seaweed product. *Scientific Monthly, 58*:24–32.

Ueda, S., Fujita, K., Komatsu, K., and Nakashima, Z. 1963. Polysaccharide produced by the genus *Pullularia*. I. Production of polysaccharide by growing cells. *Applied Microbiology, 11*:211–215.

US Patent 5847109. Galactomannan product and composition containing same. Inventors: Garti, N., Aserin Abraham, A., and Sternheim, M.Z. 1998.

van de Velde, F., and de Ruiter, G. A. 2002. Carrageenan. In Vandamme, E. J., de Baets, S., and Steinbuchel, A. (Eds.), *Biopolymers—Polysaccharides II: Polysaccharides from Eukaryotes* (p. 245). Weinheim, Germany: Wiley-VCH. Available at http://www.wiley-vch.de/books/biopoly/pdf_v06/bpol6009_245_250.pdf.

Veiga-Santos, P., Oliveira, L. M., Ceredac, M. P., Alves, A. J., and Scamparinia, A. R. P. 2005. Mechanical properties, hydrophilicity and water activity of starch-gum films: Effect of additives and deacetylated xanthan gum. *Food Hydrocolloids, 19*(2):341–349.

Viebke, C. 2005. Order-Disorder conformational transition of xanthan gum. In Dumitriu, S. (Ed.), *Polysaccharides: Structural Diversity and Functional Versatility* (2nd ed., pp. 459–474). New York: Marcel Decker Inc.

Walkinshaw, M. D., and Arnott, S. 1981. Conformation and interactions of pectins. I. X-ray diffraction analyses of sodium pectate in neutral and acidified forms. *Journal of Molecular Biology, 153*(4):1055–1073.

Wallenfels, K., Bender, H., Keilich, G., and Bechtler, G. 1961. On pullulan, the glucan of the slime coat of *Pullularia pullulans*. *Angewandte Chemie, 73*:245–246.

Wallenfels, K., Keilich, G., Bechtler, G., and Freudenberger, D. 1965. Investigations on pullulan. IV. Resolution of structural problems using physical, chemical and enzymatic methods. *Biochem Z, 341*:433–450.

Weiss, J., Takhistov, P., and McClements, J. 2006. Functional materials in food nanotechnology. *Journal of Food Science, 71*(9):107–116.

Wielinga, W. C. 2000. Galactomannans. In: Philips, G. O., and Williams, P. A. (Eds.), *Handbook of Hydrocolloids* (pp. 413–423). Boca Raton, FL: CRC Press/Woodhead Publishing Ltd.

Win, N. N., Pengju, G., and Stevens, W. F. 2000. Deacetylation of chitin by fungal enzymes. In Uragani et al. (Eds.), *Advances in Chitin Science* (vol. 4, pp. 55–62). National Taiwan University.

Wu, C., Peng, S., Wen, C., Wang, X., Fan, L., Deng, R., and Pang, J. 2012. Structural characterization and properties of konjac glucomannan/curdlan blend films. *Carbohydrate Polymers, 89*(2):497–503.

Wu, Y. 1999. Development and application of multicomponent edible films. ETD collection for University of Nebraska, Lincoln. Paper AAI9952697. Available at http://digitalcommons.unl.edu/dissertations/AAI9952697.

Xiao, G., Zhu, Y., Wang, L., You, Q., Huo, P., and You, Y. 2011. Production and storage of edible film using gellan gum. *Procedia Environmental Sciences, 8*:756.

Yuen, S. 1974. Pullulan and its applications. *Process Biochemistry, 9*:7–9.

Yui, T., Ogawa, K., and Sarko, A. 1992. Molecular and crystal structure of konjac glucomannan in the mannan II polymorphic form. *Carbohydrate Research, 29*:41–55.

Zhang, H. T., Zhan, X. B., Zheng, Z. Y., Wu, J. R., English, N., Yu, X. B., and Lin, C. C. 2012. Improved curdlan fermentation process based on optimization of dissolved oxygen combined with pH control and metabolic characterization of *Agrobacterium* sp. ATCC 31749. *Applied Microbiology and Biotechnology, 93*(1):367–379.

Zimmer, R. C. 1984. Experiments on growth and maturation of fenugreek crop. *Canadian Plant Disease Survey, 64*:33–35.

3 Protein-Based Films and Coatings

Pedro Guerrero and Koro de la Caba

CONTENTS

Abstract 81
3.1 Introduction 82
3.2 Composition and Structure of Proteins 83
3.3 Modification of Protein-Based Materials 85
    3.3.1 Denaturation and Aggregation 86
    3.3.2 Plasticizing 88
    3.3.3 Chemical Crosslinking 90
3.4 Processing of Protein-Based Materials 92
    3.4.1 Processing in Solution 94
    3.4.2 Thermomechanical Processing 97
3.5 Applications of Protein-Based Materials 99
    3.5.1 Active Coatings 99
    3.5.2 Biodegradable Films 105
3.6 Conclusions 106
References 107

ABSTRACT

Environmental, social, and economic concerns related to the availability of petroleum-derived resources and plastic waste management have revived interest in renewable and biodegradable polymers. The durability of conventional plastics, which can be considered as an advantage for some applications, is actually leading to waste disposal problems since these materials are not biodegradable and are usually disposed after only one use, like in the case of food packaging materials. In addition to the disposal, the fabrication of industrial products should also consider sustainability concerns. In this regard, the use of raw materials from renewable resources or valorization of agro-industrial by-products or waste products seems to be the approach toward the manufacture of sustainable films and coatings for food packaging applications. Among those materials, proteins are available as by-products or waste from the food processing industry, showing high potential as raw material. Furthermore, proteins are well-known by their structural complexity, which provides high functional diversity. Therefore, the modification of proteins to obtain films and coating with enhanced functional and technological properties for specific applications, such as active coatings or biodegradable films, is receiving increasing interest for both scientists and industries. In addition to protein modification, both processing conditions and methods are the aims of the study, since these are key aspects not only to achieve the desirable properties, but also to be able to produce materials at industrial scale. In this context, this chapter is focused on the films and coatings based on animal and plant proteins, particularly on preparation, characterization, and application of active coatings and biodegradable films for food packaging. A global approach is considered to assess the whole life cycle of the protein-based materials from their origin to their disposal through the manufacture and use processes, taking into account scientific, technical, and environmental aspects involved in the production of these films and coatings.

3.1 INTRODUCTION

There has been an increasing interest in protein-based films and coatings in recent years. The use of renewable polymers as raw materials, as well as the environmental concerns related to nonbiodegradable materials, is probably the most important factor that has promoted this interest (Gandini and Belgacem, 2008; Hu et al., 2012; Luckachan and Pillai, 2011). Additionally, as long as food-grade proteins and additives are used, protein-based films and coatings are also edible (Krochta, 2002). In addition to renewable, biodegradable, and edible characteristics, availability also greatly affects the use of proteins as raw materials. The proteins used for the manufacture of materials are those found in greatest quantities, such as those of animal tissue structure (collagen and gelatin), cereal coproducts (gluten and zein), or reserve proteins of grains (soybean and sunflower). Edible films and coatings are prepared from proteins of plant (Martín-Closas and Pelacho, 2011; Mooney, 2009) and animal (Karim and Bhat, 2008) origin, abundantly available as by-products of the food processing industry (Figure 3.1).

Gelatin is obtained by partial hydrolysis of collagen, the main fibrous protein constituent in bones and skins, generated as waste during animal slaughtering (Gómez-Guillén et al., 2011). The most common gelatins used in the food industry are derived from bovine and porcine sources (Nur Hanani et al., 2012). Nevertheless, owing to bovine spongiform encephalopathy and religious constraints, fish gelatins, waste of fishery industries, have been given increasing attention as an alternative of

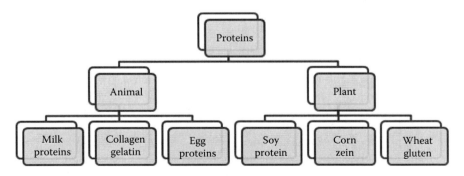

FIGURE 3.1 Types of proteins, considering their origin, and some examples of each category.

mammalian gelatins (Gómez-Guillén et al., 2009; Shyni et al., 2014). Milk proteins, such as casein and whey proteins, have also been formulated into coatings and films (Chen, 1995; Letendre et al., 2002). Whey protein is a by-product of cheese making (Ghanbarzadeh and Oromiehi, 2008), and caseinates are obtained by the acid precipitation of casein, the main protein in bovine milk (Audic and Chaufer, 2010). Egg products are also significant sources of proteins for the formation of films and coatings (Guérin-Dubiard and Audic, 2007). Egg yolk and egg white have a great number of applications in the food industry (Lim et al., 1998), leading to a surplus of egg albumen, a source of protein available as a by-product of dessert manufacturing industry (Fernández-Espada et al., 2013; Martín-Alfonso et al., 2014).

Besides animal proteins, plant proteins such as corn zein, wheat gluten, and soy protein are also used. Zein is the major storage protein of corn, containing 45%–50% of the protein fraction in corn (Shukla and Cheryan, 2001), and it is obtained as a by-product during starch production from corn by ethanolic extraction (Escamilla-García et al., 2013; Ghanbarzadeh et al., 2006). Gluten, the mixture of storage proteins of wheat, is also a by-product of the starch industry (Angellier-Cousy et al., 2011; Van Der Borght et al., 2005). Soy protein is also a by-product of food industry, obtained from soybeans through an extraction process to obtain soy oil (Russin et al., 2011; Singh et al., 2008). Although there are other vegetal sources of proteins, such as peanuts, peas, or rice, their use is more limited due to their lower availability (Vroman and Tighzert, 2009).

Owing to the diversity of sources for obtaining proteins and thus, the wide variety of properties that can be achieved when producing films and coatings from animal or vegetal proteins, these biopolymers have gained importance as potentially the most significant edible packaging materials (Janjarasskul and Krochta, 2010). In this context, the aim of this chapter is to review the work carried out related to the preparation and characterization of protein-based films and coatings for food packaging applications. First, essential aspects of proteins are shown; second, physical and chemical modifications of proteins to facilitate processing and/or improve properties; third, material processing is analyzed, both wet and dry processes; and finally, food packaging applications and environmental benefits are assessed.

3.2 COMPOSITION AND STRUCTURE OF PROTEINS

Proteins are heteropolymers built from up to 20 different monomers. Those monomers are amino acids joined together by peptide bonds, which are amide linkages formed by the condensation reaction of amino acids (Figure 3.2).

Amino acids differ from each other in the structure of side chains (R), which can be nonionized or ionized polar groups (basic or acid groups) and nonpolar groups

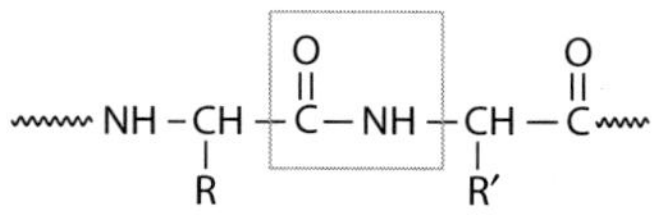

$$\sim\sim\sim NH-CH-\underset{R}{\overset{\overset{O}{\|}}{C}}-NH-CH-\underset{R'}{\overset{\overset{O}{\|}}{C}}\sim\sim\sim$$

FIGURE 3.2 Peptide bond in a protein macromolecule.

(McMurry, 2011). Proteins contain more than 100 amino acids, known as residues. The sequence of those amino acids constitutes the primary structure of proteins. The properties of proteins depend not only on the sequence of amino acids, but also on the way in which protein chains are folded in space (Zhang et al., 2014). Owing to their size, the orientation options of proteins could be enormous, but several factors limit the structural options, and it is possible to identify some common structures that appear repeatedly, such as α-helix, β-sheets, or unordered random coil structures, which are referred as secondary structures. Some factors that influence the conformational equilibrium of protein chains are the planarity of peptide bonds, hydrogen bonding of carbonyl groups to amino groups in peptide bonds, steric hindrance of neighboring groups, repulsion and attraction of charged groups, and the hydrophilic and hydrophobic characters of side groups. Most proteins do not adopt completely uniform conformations, and full descriptions of their preferred three-dimensional arrangements are defined as tertiary structures. In addition to the tertiary structure, the way in which polypeptide structures may aggregate constitutes the quaternary structure (Silva et al., 2014). This native structure of proteins may be altered by treatments that do not disrupt the primary structure but alter the secondary, tertiary, and quaternary structures.

According to their shape, proteins can be classified as fibrous or globular proteins, which show different behavior in aqueous solutions (Belitz et al., 2009). On the one hand, fibrous proteins have fiber-like structures and serve as the structural material in tissues. According to this structural function, fibrous proteins are relatively insoluble in water and unaffected by moderate changes in temperature and pH (Bourtoom, 2008). Fibrous proteins include collagens, which are the proteins of connective tissues, and keratins, which are major components of skin and hair. On the other hand, globular proteins, such as wheat gluten, corn zein, and soy protein, serve maintenance roles in living organisms and either dissolve or disperse in aqueous solutions (Ghanbarzadeh et al., 2007). Such proteins are generally more sensitive to temperature and pH changes than fibrous proteins (Yampolskaya and Platikanov, 2006).

Proteins from different sources have different characteristics, as they contain different amino acid contents, and show different structures. With regard to animal proteins, collagen is the primary protein component of animal connective tissues. All collagens comprise at least two different α-chains ($α_1$ and $α_2$), and their dimers (β-chain) and trimers (γ-chain). Collagen is composed of different polypeptides, which contain mostly glycine, proline, and hydroxyproline (Gelse et al., 2003; Wang et al., 2014). By denaturation of collagen, a high-molecular weight polypeptide is produced, called gelatin, which is water soluble (Cao et al., 2009; Gómez-Guillén et al., 2009). The amino acid composition of gelatin is very close to that of its parent collagen. Gelatin contains a large amount of proline, hydroxyproline, and lysine (Dangaran et al., 2009; Gómez-Guillén et al., 2002; Karim and Bhat, 2009). In relation to milk proteins, there are two major protein types: casein and whey protein (Creamer and MacGibbon, 1996). Casein is the main protein of milk, accounting for approximately 80% of the total protein content. It is a protein that can be separated into various fractions, such as α-, κ-, β-, and γ-casein, which differ in composition and molecular weight (Audic et al., 2003). Whey protein is the soluble fraction of milk protein after alkaline precipitation of caseinate. Various whey protein grades are commercially available, mainly whey protein concentrates (WPCs), containing 35–80wt% protein, and whey protein isolates (WPIs), with protein content above 90wt%. WPC and WPI also differ in the content of other components, such as carbohydrates and lipids and thus, the final properties of the films and coatings based on these whey proteins may also differ (Ramos et al., 2013). The two major constituents of whey protein are β-lactoglobulin and α-lactalbumin (Wang et al., 2013). In the case of egg proteins, egg white (albumen) is a complex protein, consisting of ovomucin fibers in an aqueous solution of numerous globular proteins (Gennadios et al., 1998), in which ovalbumin, a monomeric phosphoglycoprotein, represents about 54% of protein content (Fernández-Espada et al., 2013).

Regarding plant proteins, wheat gluten contains four main fractions of proteins: albumins (water soluble), globulins (soluble in dilute salt solution), but mainly gliadins (soluble in 70% ethanol), and high-molecular weight glutenins (partially soluble in dilute acids or alkali) (Chen et al., 2012; Lagrain et al., 2010). In relation to corn, it contains a prolamine protein, zein, exclusively found in corn (Anderson and Lamsal, 2011). Four different fractions of zein have been identified: α-, β-, γ-, and δ-zein, but α-zein is the predominant fraction present in commercial zein (Shi and Dumont, 2014). Zein comprises a large amount of hydrophobic amino acid residues, which are the basis for its hydrophobic character. In contrast, soy protein contains a high content of polar amino acid residues. Soy proteins are composed of a mixture of albumins and globulins, 90% of which are storage proteins with globular structure. Soy protein consists of four major fractions: 2S, 7S, 11S, and 15S, where S stands for Svedberg units, based on the rate of sedimentation. 7S (β-conglycinin) and 11S (glycinin) globulin fractions make up 70% of the total proteins in soybeans (Kinsella, 1979; Ning and Villota, 1994).

Proteins have potential properties for their application in the food packaging field due to their ability to form films and coatings with good barrier properties against oxygen (O_2), carbon dioxide (CO_2), aroma, and lipids (Falguera et al., 2011; Miller and Krochta, 1997). However, protein-based films and coatings have limited resistance to water vapor and thus, barrier and mechanical properties are compromised by moisture due to the inherent hydrophilicity of proteins. Therefore, formulations, preparation methods, and modification treatments must be optimized to achieve the desirable functional properties for specific applications.

3.3 MODIFICATION OF PROTEIN-BASED MATERIALS

Protein-based materials in the absence of secondary components possess inadequate physicochemical and mechanical properties for practical applications, mainly due to the inherent hydrophilicity of proteins and the strong molecular interactions among protein chains. Therefore, the use of unmodified proteins has been limited owing to their poor water and moisture resistance and brittleness. To overcome these problems, many attempts have been made to enhance the barrier and mechanical properties of protein-based films and coatings by plasticizing (Vieira et al., 2011) and/or crosslinking (Chabba and Netravali, 2005). The structural complexity of proteins provides a great functional diversity and thus, diverse methods to modify materials to obtain the desirable properties for food packaging purposes (Guilbert and Cuq, 2005). Modifying proteins can diminish the interactions among protein chains, improving mechanical properties. Furthermore, those modifications can also decrease the amount of free polar groups through physical or chemical interactions, enhancing barrier properties.

3.3.1 Denaturation and Aggregation

The structural unfolding of proteins, commonly known as denaturation (Figure 3.3), is often done deliberately in the course of modifying proteins (Hardy et al., 2008).

In the initial steps of the denaturation mechanism, native proteins undergo intramolecular transitions that differ from the native state only by minor conformational changes and then, protein unfolds, exposing hydrophobic amino acids buried in the native protein (Nishinari et al., 2014). These changes can be correlated with conformational changes found in protein secondary structures, which can be analyzed by self-deconvolution and curve fitting of Fourier transform infrared (FTIR) spectroscopy (Guerrero et al., 2013; Ramos et al., 2013). Afterwards, through noncovalent interactions, such as ionic, van der Waals, and hydrophobic interactions, or covalent bonds, such as disulfide bonds, denatured proteins take part in irreversible intermolecular interactions, which result in the formation of aggregates (Pereira et al., 2011). The extent of these reactions can be controlled by several factors, such as pH, temperature, application of electric fields, or treatment time, and the behavior of proteins during those treatments is essential to control their characteristics and the desired properties for specific applications.

Proteins incorporate both acid and basic functional groups and thus, the net charge of the protein in solution depends on solution pH. At acid pHs, both carboxylate and amine functions are protonated, so the protein has a positive net charge. At basic pHs, amino groups exist as neutral bases and carboxyl groups as their conjugate bases, so the protein has a negative net charge. At intermediate pHs, positively charged groups are balanced by negatively charged groups and the protein, on average, has no net charge. This characteristic pH is called the isoelectric point (Chou and Morr, 1979). At the isoelectric point, the proteins ability to interact with water is lower and thus, their solubility (Mojumdar et al., 2011). At extreme acidic and alkaline conditions, the strong repulsive forces or highly negative (pH > 12) or positive (pH < 1) charges are present along protein chains, preventing protein molecules from

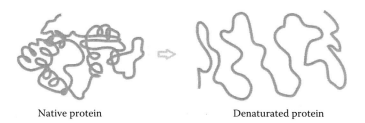

Native protein   Denaturated protein

FIGURE 3.3 Denaturation process of proteins.

associating and forming films (Flint and Johnson, 1981). Mauri and Añón (2006) investigated changes in the solubility of soy protein films prepared at pH 2, 8, and 11 and observed that protein networks were maintained by the same type of interactions (disulfide bonds, hydrophobic interactions, and hydrogen bonds) at the different pH values analyzed, but films prepared at extreme pHs had a denser microstructure, indicating the formation of aggregates during film formation. Therefore, proteins are a kind of polyelectrolytes, and pH values influence their association and dissociation behavior in aqueous solution (Song et al., 2011).

In addition to pH, the functional properties of the films and coatings depend on other preparation conditions, such as temperature and time (Denavi et al., 2009; Pérez-Gago and Krochta, 2000). Heating modifies the three-dimensional structure of proteins and the interactions in the native protein, exposing functional groups engaged in intramolecular bonding and thus, these groups become available for intermolecular interactions (Subirade et al., 1998; Wang and Damodaran, 1991). Pérez-Gago et al. (1999) found that transparent films from native and heat-denatured whey proteins can be formed; however, some functional properties of films, such as solubility, are different due to the differences in intermolecular bonding upon drying in both cases. Native whey proteins maintain their globular structure with most of the hydrophobic and sulfhydryl groups buried in the interior of the molecule and thus, cohesion relies mainly on hydrogen bonding, which leads to soluble films in water. In contrast, the intermolecular forces that promote cohesion in heat-denatured films involve hydrophobic bonds among the unfolded protein chains and covalent S–S bonding during drying, leading to insoluble films in water. Since some applications of protein-based films may require water insolubility to enhance product integrity, whereas others may require water solubility before consumption of the product, the degree of protein denaturation and unfolding can be controlled by heating time and temperature, depending on the desirable film properties.

Although temperature is one of the key parameters when analyzing denaturation and aggregation processes and thus, a particular emphasis has been given to its effect, more recently the use of emergent processing technologies, such as electric fields, has opened new perspectives for the development of innovative protein structures. By the use of electric fields, heating occurs by the transformation from electric to thermal energy, providing more uniform and rapid heating within the material in comparison with conventional technologies (Pereira et al., 2010). This approach avoids inducing an excessive denaturation of proteins. Xiang et al. (2011) analyzed the effect of pulsed electric fields (PEFs) on structural modification of WPI by using fluorescence spectroscopy. They found that PEF treatment increased the intrinsic tryptophan fluorescence of WPI, indicating changes in the polarity of tryptophan residues and resulting in a higher surface hydrophobicity. The effect of applying moderate electric fields (MEFs) has also been investigated on whey proteins. In comparison with PEF, MEF treatments use lower intensities. Pereira et al. (2011) found lower rates of WPI denaturation for MEF-treated samples compared with the samples treated by a conventional heating treatment. This behavior was kinetically supported by a lower rate constant (k) and denaturation reaction order (n). Furthermore, the thermodynamic parameters obtained in the study indicated that the rate-determining stage was protein unfolding over aggregation, as reflected by high activation energy (E_a) and enthalpy change (ΔH) together with a positive entropy change (ΔS). Although this treatment seems to be less aggressive for protein structure and shows a great potential for the manufacture of innovative protein-derived products with enhanced functional and technological properties (Rodrigues et al., 2015), further research is still needed for a better understanding of the effect of electric fields on conformational changes in native proteins and interactions between denatured proteins.

3.3.2 Plasticizing

Protein-based films and coatings without any additive have a brittle behavior, which makes processing difficult, but plasticization is often used for the modification of proteins to improve their processability and/or other properties demanded by food packaging materials (Aguirre et al., 2013; Lian et al., 1999). Mechanical properties are extremely important, since protein-based materials must show adequate resistance and deformability to maintain their integrity during processing, handling, transport, and storage (Paes et al., 2010; Vieira et al., 2011). For example, casein films produced without plasticizers are so brittle that it is not possible to determine tensile properties (Ghosh et al., 2009). However, the addition of 10wt% glycerol imparts flexibility due to the fact that unordered random coil structures, predominant in casein proteins, are transformed into helical structures, resulting in a more open molecular network (Siew et al., 1999).

Three theories have been proposed to explain the mechanism of the plasticizer effect (Sears and Darby, 1982). According to the lubricity theory, a plasticizer is considered as a lubricant to facilitate the movements of the macromolecules over each other; related to the gel theory, a plasticizer disrupts the polymer–polymer interactions, including hydrogen bonds, van der Waals, and ionic forces; and regarding the free volume theory, a plasticizer may depress the glass transition temperature by increasing polymer-free volume. The fundamental concept underlying these theories is that a plasticizer can interpose itself between the polymer chains and decrease the forces holding the chains together. Plasticizers exchange the intermolecular bonds among protein chains to bonds between the protein and the plasticizer, promoting conformational changes and resulting in higher deformability (Imre and Pukánszky, 2013).

Although water is a very efficient plasticizer, compounds with a higher boiling point are preferred because they lead to more stable properties (Shi and Dumont, 2014). The most commonly studied plasticizers are polyols (Bergo and Sobral, 2007; Gontard et al., 1993; McHugh and Krochta, 1994; Park and Chinnan, 1995), such as glycerol, sorbitol, ethylene glycol, propylene glycol, polyethylene glycols, and polypropylene glycol. The differences in composition, size, structure, and shape of plasticizers directly influence their ability to plasticize proteins (Orliac et al., 2003). At the same plasticizer concentration, Jongjareonrak et al. (2006) found that fish gelatin films plasticized with ethylene glycol showed the highest tensile strength, whereas glycerol-plasticized films showed the greatest elongation at break. Many studies have reported that plasticizing effect depends on the length of the plasticizer chain. Irissin-Mangata et al. (2001) observed that wheat gluten films plasticized with high-molecular weight (Mw = 1500 and 3400 Da) polyethylene glycols were opaque, indicating that these plasticizers were not miscible with protein and resulting in brittle films; however, low-molecular weight (Mw = 200 and 400 Da) polyethylene glycols resulted in continuous and homogeneous films, although not flexible enough.

Other hydrophilic compounds, such as diethanolamine or triethanolamine, which contain amino functional groups and not only hydroxyl groups, have also been tested to plasticize protein-based materials (Rahman and Brazel, 2004). However, glycerol is currently regarded as one of the most efficient plasticizers for proteins (Cao et al., 2009). Generally, smaller molecules are more easily incorporated into the protein matrix and exhibit a more efficient plasticizing effect (Sothornvit and Krochta, 2001). Plasticizers with different chemical structures produce different effects on film properties. The effect of different types and contents of plasticizers on mechanical properties of some vegetal and animal protein-based films are shown in Table 3.1.

As can be seen, the combination of different plasticizers can cause a further improvement of mechanical properties. To explain the synergistic effect of oleic acid and glycerol on zein films at a molecular level, Xu et al. (2012) examined the secondary structure of zein proteins by FTIR. Since the band corresponding to amide

TABLE 3.1
Mechanical Properties (Tensile Strength, TS, and Elongation at Break, EB) of Plasticized Protein-Based Films

Protein + Plasticizer	TS (MPa)	EB (%)	References
Casein + 30wt% glycerol	6.5 ± 1.0	58.1 ± 9.0	Cho et al. (2014)
Casein + 40wt% glycerol	1.6 ± 0.2	101.0 ± 13.7	
Bovine gelatin + 30wt% glycerol	17.0 ± 1.8	30.8 ± 4.1	Guerrero et al. (2014)
Bovine gelatin + 40wt% glycerol	7.8 ± 0.7	47.2 ± 5.1	
Soy protein + 30wt% glycerol	4.1 ± 0.4	105.4 ± 13.3	Guerrero et al. (2010)
Soy protein + 40wt% glycerol	1.6 ± 0.3	145.5 ± 22.6	
Soy protein + 50wt% glycerol	1.5 ± 0.2	170.2 ± 18.5	
Soy protein + 30wt% sorbitol	7.5 ± 1.0	29.7 ± 9.1	Garrido et al. (2014)
Soy protein + 40wt% sorbitol	5.3 ± 0.6	105.6 ± 5.2	
Soy protein + 50wt% sorbitol	2.9 ± 0.3	117.8 ± 22.9	
Wheat gluten + 25wt% glycerol	3.1 ± 0.1	346.0 ± 10.7	Rafieian et al. (2014)
Whey protein + 25wt% glycerol	3.1 ± 1.0	19.2 ± 2.8	Wagh et al. (2014)
Whey protein + 50wt% glycerol	1.6 ± 1.0	61.7 ± 1.6	
Whey protein + 30wt% sorbitol	12.1 ± 1.3	4.4 ± 0.5	Osés et al. (2009)
Zein	6.7 ± 0.4	2.0 ± 0.2	Shi et al. (2012)
Zein + 10wt% tributyl citrate	17.8 ± 4.3	4.5 ± 0.5	
Zein + 20wt% tributyl citrate	5.5 ± 0.6	3.0 ± 0.2	
Zein + 30wt% tributyl citrate	5.4 ± 1.3	3.3 ± 1.0	
Zein + 40wt% tributyl citrate	4.3 ± 0.2	2.4 ± 0.1	
Zein	5.7 ± 0.3	1.4 ± 0.0	Xu et al. (2012)
Zein + 10wt% glycerol	11.2 ± 0.9	1.7 ± 0.0	
Zein + 10wt% oleic acid	7.6 ± 0.2	2.1 ± 0.0	
Zein + 10wt% glycerol + 10wt% oleic acid	14.4 ± 0.9	2.4 ± 0.0	

I depends on the secondary structure of the protein backbone, it can be used for the analysis of different secondary structures; specifically, the bands at 1620 and 1680 cm^{-1} are assigned to β-sheets (Guerrero et al., 2014). Xu, Chai, and Zhang also observed a decrease in the intensity of those two bands with the incorporation of a glycerol and oleic acid blend, indicating the change of the conformation of the protein toward helical forms at the expense of β-sheets, as has also been shown in other works (Gillgren et al., 2009).

Wan et al. (2005) used scanning electron microscopy (SEM) to evaluate the cross sections of soy protein films plasticized with different combinations of plasticizers. They showed that the presence of large pores within the films resulted in elevated water vapor permeability (WVP). Therefore, plasticizers must be added at a certain amount to obtain films with improved flexibility without significant decrease in barrier properties (Sothornvit and Krochta, 2001). Recently, Shi et al. (2012) plasticized zein films with tributyl citrate to improve not only mechanical properties but also water resistance. They observed microstructural changes by atomic force microscopy (AFM), which indicated a significant decrease in surface roughness with the addition of tributyl citrate up to 20wt%; however, phase separation started at higher contents of plasticizer. Commonly, fatty acids are also incorporated into the formulations with the aim to improve not only mechanical properties but also WVP. In those cases, caution must be taken when measuring WVP values, since the measurements are dependent on temperature. For example, samples formulated with oleic acid show lower WVP at low temperatures due to the crystallization of the fatty acid (Shi and Dumont, 2014).

Regarding other barrier properties such as resistance against ultraviolet (UV) light, it is worth noting that films derived from proteins, such as casein, whey proteins, albumin, wheat gluten, and soy proteins, can resist UV light due to the presence of disulfide bonds within the network and the presence of amino acids such as tyrosine, tryptophan, and phenylalanine, well-known sensitive chromophores (Li et al., 2004). Guerrero et al. (2011) observed that soy protein films show better barrier properties to UV light than synthetic films such as low-density polyethylene (LDPE) or orientated polypropylene (OPP; Figure 3.4). Therefore, protein-based films could effectively prevent UV light transmission and product oxidation induced by UV light.

In addition to mechanical and barrier properties, the incorporation of phenolic acids as plasticizers can provide films with antioxidant and/or antimicrobial properties. Arcan and Yemenicioglu (2011) incorporated gallic acid into zein films and observed that this phenolic acid could increase the film flexibility, maintain film integrity following hydration, and show antimicrobial activity.

3.3.3 Chemical Crosslinking

Chemical modification of proteins is used to change protein–protein interactions and thus, improve specific properties. With this aim, proteins have been crosslinked with aldehydes and sugars, among others, to improve functional properties. In addition to free amino acids and relatively small number of functional groups on the terminal amino acids, only the functional groups on protein side chains are available for

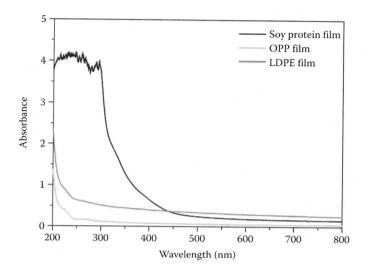

FIGURE 3.4 UV spectra of a soy protein film in comparison with orientated polypropylene (OPP) and low-density polyethylene (LDPE) films.

chemical reactions. Although there is no consensus regarding the exact crosslinking mechanism, it is believed that the reactive groups of the crosslinking agents interact mainly with the ε-amino group in lysine and the carboxyl group in aspartic and glutamic acids (Ghosh et al., 2009).

Various aldehydes have been used as crosslinkers of proteins, mainly formaldehyde and glutaraldehyde (Hernández-Muñoz et al., 2004). Ustunol and Mert (2004) observed that the solubility of whey protein films decreased and the tensile strength increased; however, WVP increased due to the additional polar groups provided by the incorporation of formaldehyde or glutaraldehyde. Regarding film solubility and tensile strength, a similar trend was also observed by Liu et al. (2004) when formaldehyde or glutaraldehyde was added to peanut protein films; in contrast, WVP decreased after the addition of aldehydes. Since the reaction between aldehydes and proteins proceeds via two steps (Gueguen et al., 1998): the first one corresponding to formation of methylol compounds, which is rapid and reversible, and the second one corresponding to the crosslinking step, which is much slower, it is difficult to compare results from different studies when different experimental conditions are employed in film preparation. Nevertheless, Gerrard et al. (2002, 2003) extensively investigated the crosslinking reactivity of formaldehyde and glutaraldehyde and observed that glutaraldehyde crosslinked protein faster, which was confirmed by the decrease in free lysine residues, whereas the extent of lysine loss was considerably lower with formaldehyde. With regard to the different reactivity of those two aldehydes, Ghosh et al. (2009) studied their effect on mechanical properties of casein films plasticized with 30wt% glycerol and observed that formaldehyde moderately increased tensile strength of films, while glutaraldehyde significantly improved both tensile strength and elongation at break. Formaldehyde has only one carbonyl group per molecule, while glutaraldehyde is a dicarbonyl compound and thus, results in a more tightly crosslinked network when reacts with amino groups in protein chains, as confirmed by SEM (Yue et al., 2012). In addition to formaldehyde and glutaraldehyde, Zárate-Ramírez et al. (2014) evaluated the incorporation of glyoxal into wheat gluten to control the microstructure and properties of gluten-based bioplastics. These authors observed that glyoxal produced bioplastics with better thermal and mechanical properties than formaldehyde and glutaraldehyde. This different behavior was related to the different reactivity and thus, the different kinetics of the crosslinking reactions induced by the three aldehydes mentioned above. Glyoxal-incorporated probes showed a more homogeneous and cohesive structure, a darker color and lower solubility values, in agreement with a higher degree of crosslinking. Other dialdehydes have also been used to improve water and mechanical resistance. Rhim et al. (2000) incorporated 10wt% dialdehyde starch into soy protein film-forming solutions and observed that film solubility decreased and tensile strength increased. More recently, Mu et al. (2012) prepared glycerol-plasticized gelatin films with dialdehyde carboxymethyl cellulose (DCMC) as crosslinking agent. The addition of DCMC caused an increase in tensile strength and thermal stability, although elongation at break decreased.

The crosslinking ability of formaldehyde and glutaraldehyde has been extensively studied; however, their substantial toxicity limits their use to nonedible packaging applications and thus, other compounds such as sugars have been investigated as crosslinkers (Dangaran et al., 2009). Maillard reaction of the amino group of lysine residues takes place in the presence of reducing sugars, such as lactose or glucose, although a nonreducing sugar like sucrose can also cause Maillard reaction, as shown by Guerrero et al. (2012). However, lactose modification led to more compact structures than sucrose modification, as observed by SEM. The early stage of the Maillard reaction involves the formation of the protein–sugar conjugates between the carbonyl group of carbohydrates and the amino group in proteins, leading to the Schiff base, which subsequently cyclizes to produce the Amadori compound (Figure 3.5). In the final stage, colored and insoluble polymeric compounds, referred as melanoidins, are formed (Yasir et al., 2007).

In recent years (Hauser et al., 2014), the antimicrobial activity of Maillard reaction products (MRPs) has been investigated. Several mechanisms have been proposed to explain the antimicrobial effect of MRPs, but it seems that melanoidins can interact with bacterial cells, since Mg^{2+}-chelating properties have been observed (Rufian-Henares and de la Cueva, 2009). This ability of melanoidins to chelate Mg^{2+} from the outer cell membrane would lead to bacterial death by disruption of the cell membrane. Furthermore, the ability of MRPs to generate H_2O_2, a very effective antimicrobial agent used in the food industry, would also contribute to their antimicrobial activity (Mueller et al., 2011).

3.4 PROCESSING OF PROTEIN-BASED MATERIALS

Protein-based films and coatings can be produced by using two techniques (Figure 3.6): wet and dry processes (Cuq et al., 1997). Regarding preparation of films, the wet process, known as solution casting, is based on the solubilization or dispersion

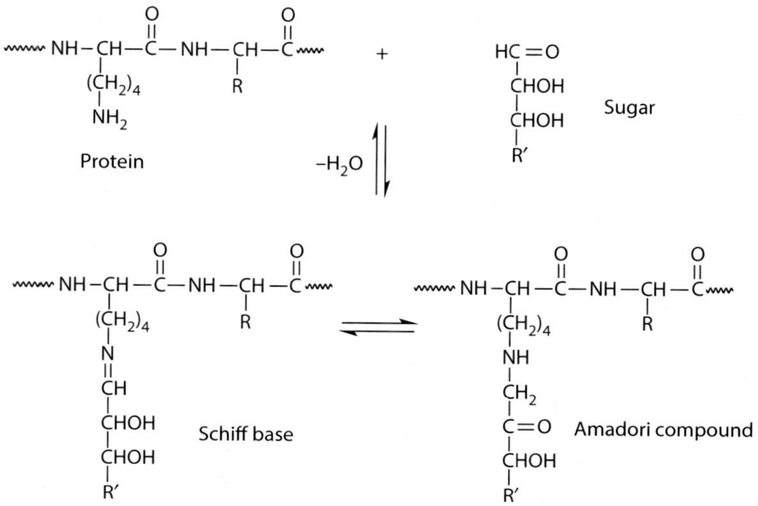

Protein-Based Films and Coatings

FIGURE 3.5 Maillard reaction between proteins and sugars.

of proteins in an aqueous medium and the subsequent evaporation of the solvent to obtain the film, whereas the dry process is based on the thermoplastic behavior of proteins in extrusion and compression molding (Hernandez-Izquierdo and Krochta, 2009). On the contrary, coatings involve the formation of films directly on the product surface. In the food industry, spraying and dipping are conventional techniques for applying coatings to food surfaces (Ribeiro et al., 2007; Zhong et al., 2014). For both coatings and films, the conditions used during the formation process, as well as the chemicals employed, have a significant effect on the final properties of the resulting coating or film (Sothornvit et al., 2007).

Preparation of packaging materials from proteins requires three steps: first, the rupture of low-energy intermolecular bonds that stabilize proteins in the native state; second, the arrangement and orientation of polymer chains; and third, the formation

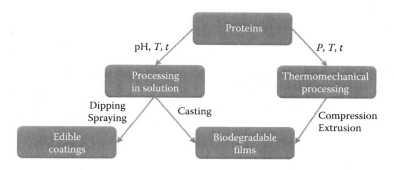

FIGURE 3.6 Processing methods and processing parameters, such as pH, temperature (T), time (t), and pressure (P), to prepare protein-based films and coatings.

of a three-dimensional network stabilized by new interactions and bonds (Jerez et al., 2007; Pommet et al., 2005). In both wet and dry processes, the adequate choice of processing conditions, such as pH, pressure, temperature, or time, is of great importance to achieve the desirable improvement of properties. Due to the complex nature of proteins, their processing into films and coatings is a challenging task.

3.4.1 Processing in Solution

Edible coatings can be applied by dipping or spraying after solubilization or dispersion of proteins. Typical methods for forming a coating include spraying and dipping. In the spraying operation, the coating is applied on food surface and then, a separate operation to coat the bottom surface of the product is required. Spraying is preferred for items possessing a large surface area. On the contrary, dipping is more convenient for irregularly shaped food objects. However, final coatings might be less uniform and multiple dipping might be necessary to ensure full coverage (Dangaran et al., 2009). In both cases, the coating operation time is usually fixed to control the coating thickness (Zhong et al., 2014).

Formation of films also involves the previous solubilization or dispersion of proteins. The solubility of proteins depends on the type of protein and its isoelectric point. Type A gelatin, produced from acid-treated collagen, has an isoelectric pH of 6–9, while the isoelectric point of type B gelatin is about 5 (Eysturskaro et al., 2009; Stainsby, 1987). Physicochemical properties of some animal and plant proteins are summarized in Table 3.2.

Proteins normally show lower solubility close to the isoelectric point (Massani et al., 2014). The isoelectric point of whey protein is around 5.2 and at pH lower than 4 and higher than 6, transparent whey protein films can be obtained, although only films prepared at basic pH are flexible (Moditsi et al., 2014). In relation to wheat gluten, its isoelectric point is 7.5 and thus, films do not form at pH 7–8, and dispersions at pH 5–6 are also very poor. Nevertheless, according to Gennadios et al. (1993), wheat gluten films can be obtained at basic or acidic conditions, although alkaline conditions result in stronger films (Kayserilioglu et al., 2001). Regarding

TABLE 3.2
Physicochemical Properties, Including Solvent Solubility and Isoelectric Point (pI), of Some Animal and Plant Proteins

Protein	Solvent	pI	References
Casein	Water	4.6	Creamer and MacGibbon (1996)
Egg albumen	Water	4.5	Mine (1995)
Gelatin	Water	5–9	Stainsby (1987)
Soy protein	Water	4.6	Kinsella (1979)
Wheat gluten	Alcohol–water	7.5	Wieser (2007)
Whey protein	Water	5.2	Kinsella and Whitehead (1989)
Zein	Alcohol–water	6.8	Avena-Bustillos and Krochta (1993)

soy proteins, films cannot be formed close to pH 4.5 (Sian and Ishak, 1990) and are mostly formed at alkaline conditions (Brandenburg et al., 1993; Jiang et al., 2012). These changes related to pH can be analyzed by FTIR spectroscopy.

The most characteristic peaks associated with peptide bonds are the band corresponding to the stretching of C = O (amide I) and the band corresponding to the stretching of C–N and to the vibration of N–H (amide II) (Lodha and Netravali, 2005; Subirade et al., 1998). As shown by Guerrero and de la Caba (2010) for soy protein films (Figure 3.7), the relative intensity between the band at 1630 cm^{-1} (amide I) and the band at 1530 cm^{-1} (amide II) can change depending on pH. The intensity of those two bands was similar at the isoelectric point, but the relative intensity between amide I and amide II bands became higher at pH 1.4 and lower at pH 10.0, indicating that the pH value affected protein structure and consequently, functional properties. Mauri and Añón (2008) showed that soy protein films formed at basic conditions (pH 8 and 11) exhibited 70% higher deformation than the films prepared at acidic conditions (pH 2) due to the presence of protein fractions in native state, allowing macromolecules to unfold during mechanical tests.

The isoelectric point of proteins and thus, their sensitivity to pH are associated with the content of ionized polar amino acids in proteins. In soy proteins, the high content of ionized polar amino acids (i.e., 25.4%) limits film formation at low pHs (Kim and Netravali, 2012). However, zein and keratin films form over a wide pH range due to their low content of ionized amino acids, 10.0% and 10.7%, respectively (Cabra et al., 2007; Cuq et al., 1998). Therefore, the structural diversity of proteins

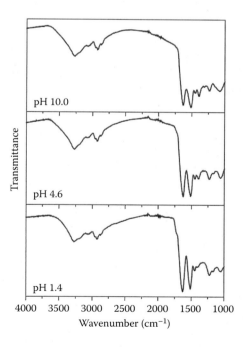

FIGURE 3.7 FTIR spectra of soy protein-based film-forming solutions at different pHs.

highly influences their solubility, processing conditions, and thus, the properties of the resulting films and coatings.

Although the thermal stability of proteins is dependent on their amino acid composition, temperature is a key factor in the solution processing of proteins since their conformation changes and thus, their degree of denaturation determines the type and proportion of covalent and noncovalent interactions between protein chains. Hoque et al. (2010) investigated the effect of temperature, from 40°C to 90°C, on the properties of glycerol-plasticized fish gelatin films. They found that thickness and transparency of the films were not affected by the heat treatment, while mechanical properties changed. The films prepared from solutions heated at 60–70°C showed higher tensile strength, whereas those films prepared by heating solutions at 80–90°C had higher elongation at break. These results were related to the melting transition temperature and enthalpy measured by differential scanning calorimetry (DSC). The higher melting transition temperature and enthalpy observed for the films heated at 60–70°C was related to the greater interchain interactions, mainly via hydrophobic interactions and hydrogen bonding, as observed by FTIR, causing stronger film network and higher tensile strength. In a similar way, Denavi et al. (2009) evaluated the influence of drying conditions, both temperature and relative humidity (RH), on mechanical properties, solubility, and barrier properties of soy protein films prepared by solution casting. They found that 70°C and 30% RH were the optimal drying conditions to achieve higher tensile strength and lower solubility and water vapor permeability (WVP). This behavior was attributed to a greater unfolding degree at these conditions, which allowed higher number of interactions between protein chains and led to a more compact structure, as observed by transmission electron microscopy (TEM).

To date, heat treatment is the most common method for protein processing in solution, but application of ultrasound technology is currently attracting much attention, since it can represent an alternative method to process proteins at lower temperatures using less time (Soria and Villamiel, 2010). Ultrasonic waves can be used to change specific surface area, rheological properties, and thus, physicochemical properties of proteins (Jambrak et al., 2009). Recently, Hu et al. (2013) examined the effect of low-frequency ultrasonication at different powers (200, 400, and 600 W) and times (15 or 30 min) on soy protein dispersions. The structural changes observed in this study were related to the decrease in noncovalent interactions after ultrasonic treatments. Changes in the secondary structure of soy proteins by ultrasonication were analyzed by circular dichroism. This analysis indicated that ultrasonic treatment resulted in an increase in α-helix component and a decrease in β-sheet component, in accordance with previous results obtained by other authors for whey proteins (Chandrapala et al., 2011). This effect was more pronounced at higher powers (400 and 600 W) and longer times (10 min). Therefore, controlling treatment conditions could lead to different structures and thus, to different functional properties.

Solvent casting has been the preferred method used by researches to prepare films due to the simplicity of the equipment required, which can consist of simple casting dishes. However, there are more advanced and sophisticated equipment that can produce larger films by mechanically spreading the solution to a fixed thickness. Kozempel and Tomasula (2004) developed a continuous process, where feed system and dryer parameters are the key factors to obtain films with good properties. Furthermore, the combination with emergent technologies in the food industry, such as high power ultrasound, is becoming more efficient for large-scale applications at lower temperatures (Patist and Bates, 2008).

3.4.2 Thermomechanical Processing

Although solution casting is a simple method to prepare protein-based films, it also has the major disadvantage of being difficult to scale-up for industrial purposes. For this reason, an increasing number of works (Balaguer et al., 2014; Janjarasskul et al., 2014; Reddy and Yang, 2013) produce protein-based films by dry methods in which the thermoplastic properties of the proteins are used to shape them under the combined effect of temperature and pressure. Thermomechanical processing by compression molding or extrusion can result in a highly efficient manufacturing method with commercial potential for large-scale production of films, mainly due to the short times used compared with solution casting. However, complexity and heterogeneity of intermolecular interactions within proteins are responsible for the lack of flow region, and fabrication of materials by these technologies could be expected only in the presence of agents that break the intermolecular bonds that stabilize native proteins (Hernández-Izquierdo and Krochta, 2008). Therefore, although thermoplastic processing is adapted to synthetic materials for which fabrication parameters are optimized, process parameters, such as temperature, plasticizer concentration, and residence time, must be optimized for proteins.

The combination of high temperatures and pressures, and short times in compression molding causes the transformation of protein–plasticizer mixtures into viscoelastic melts; then, films are formed upon cooling through hydrogen bonding, ionic and hydrophobic interactions, and covalent bonds (Pol et al., 2002). The addition of plasticizers facilitates deformation and processing without thermal degradation. The effect of plasticizers on thermal degradation can be studied by thermogravimetric analysis (TGA), as shown in Figure 3.8. The weight loss becomes significant above 200°C for soy proteins (Wang et al., 2007), zein (Nedi et al., 2012), and wheat gluten (Jerez et al., 2007). Since weight loss during film formation at elevated pressures can be different from that at ambient or low pressures, the values of weight loss from TGA cannot be used as absolute values in processing under pressure, but degradation temperatures obtained by TGA can be indicative for choosing the optimum processing temperature. A temperature of about 130°C was found to be suitable for wheat gluten (Zárate-Ramírez et al., 2011) and 150°C for soy protein (Guerrero et al., 2010).

For continuous shaping of thermoplastic materials, extrusion is a highly efficient method. This technology is generally defined as a process to mix, homogenize, and shape materials by forcing it through a specifically designed opening. The employment of extrusion is one of the most versatile and well-established processes used to produce polymeric materials. Although it is widely used in the plastic industry for most of the synthetic polymers, extrusion is still a complicated process that has to be mastered for proteins. A great number and variety of extruders are now in operation, and there is a trend toward developing and using high-temperature, short-time extrusion processing (Belitz et al., 2009). Extrusion conditions, such as high barrel

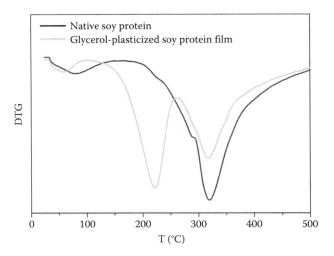

FIGURE 3.8 Derivative thermogravimetric (DTG) curves of soy protein-based materials.

temperatures, high pressures, and the use of low moisture during processing, bring about major conformational changes in the proteins during processing and thus, have important effects on the quality of the final extruded products. It is generally accepted that proteins are denatured during the extrusion process, so these unfolded protein chains can interact with each other and with added components, allowing the formation of a polymeric network. The heat and shear force produced by extrusion can restructure the protein and define the molecular network through dissociation and unfolding of chains, which can recombine and crosslink afterward (Shi and Dumont, 2014). Early formation of crosslinked structure reduces the mobility of protein chains and inhibits the elastic recovery without rupture after processing and thus, it should be avoided. Consequently, extrusion of proteins is only possible in a limited range of operating conditions to obtain continuous extrudates, instead of disrupted extrudates (Pommet et al., 2003).

The protein processing temperature in extrusion is narrow. The lower temperature limit is set by the denaturation temperature, while the upper limit is set by the increase in viscosity associated with extensive aggregation (Micard et al., 2001; Morel et al., 2002). Ullsten et al. (2006, 2009) found that extrusion temperature can be increased from 95°C to 135°C by incorporating only 1wt% salicylic acid in wheat gluten formulations, suggesting that the acid reduces protein aggregation as a consequence of the reorganization of the intramolecular disulfide bonds. Salicylic acid is a natural product that can act as a radical scavenger and delay crosslinking, permitting the increase in extrusion temperature (Sroka and Cisowski, 2003). Processing temperature also affects the specific mechanical energy (SME), the amount of mechanical energy dissipated as heat inside the material (Harper, 1989). High SME values can result in the extrudate rupture, but an increase in mixing temperature can be used to result in lower SME values. Nur Hanani et al. (2014) investigated the effect of extrusion temperature on mechanical and barrier properties of beef gelatin-based composite films incorporated with corn oil. They observed that increasing temperature from 90°C to 120°C improved tensile and puncture strength, as well as water vapor permeability. This behavior was related to a higher denaturation degree, causing a better distribution of oil within gelatin at higher temperatures, as evidenced by confocal laser scanning microscopy (CLSM). CLSM analysis showed an increase in the number of oil droplets and a decrease in size as temperature increased.

In addition to temperature, water plays an important role in extrusion processing due to its effect on heat transfer during extrusion (Chen et al., 2012). The addition of water results in accelerating the flow speed of extrudate coming from the extruder. Increasing water content results in increased heat transfer from the extruder barrel to the feed material and consequently, decreased viscosity, shear, and friction during extrusion. However, an excessive amount of water during protein extrusion could decrease melt viscosity in excess, leading to a low motor torque and low SME input and resulting in a low product temperature that could reduce the degree of interactions and thus, the protein transformation (Chen et al., 2010; Lin et al., 2000). Besides water, common plasticizers can also be used to facilitate processability of proteins by extrusion. Nur Hanani et al. (2013) analyzed the effect of glycerol content on functional properties of gelatin-based films obtained by extrusion and observed that the plasticizer content was a key factor to achieve the optimum equilibrium between mechanical and barrier properties. More recently, Janjarasskul et al. (2014) compared the barrier and tensile properties of whey protein films prepared by solution casting and by extrusion followed by compression. The elimination of pores in the matrix was observed due to the effect of compression after extrusion, which caused a significant improvement of the functional properties of whey protein films.

3.5 APPLICATIONS OF PROTEIN-BASED MATERIALS

In the specific context of food packaging materials, food packages must serve some important functions, including containment and protection of food. In addition to sensory and safety aspects relating to foods, the development and selection of materials also involve other issues such as logistics, marketing, financial, legislation, and environmental concerns (Haugaard et al., 2001; Müller et al., 2014). Consumer pressures in relation to health, food quality, convenience, and more proactive attitudes toward reducing the environmental impact of packaging wastes have increased the interest in active coatings and biodegradable films (Butler, 2009; Kuorwel et al., 2011). Some recent works focused on protein-based films and coatings are summarized in Table 3.3.

3.5.1 Active Coatings

Active materials are defined as materials that are intended to extend the shelf life or to maintain or improve the condition of packaged food (European Commission, 2009). Active coatings are applied directly on food surface and formed according to the same mechanisms associated with solvent casting of films. A dilute protein solution is applied to the surface of the food product and the coating forms upon evaporation of solvent.

TABLE 3.3
Some Recent Works Related to the Preparation and Characterization of Protein-Based Materials

Protein	Plasticizer	Processing	Characterization Method	References
Casein	Glycerol	Dipping	DSC, GAB, MC, RHE, SSD	Talens et al. (2012)
Casein	Water	Spraying	CIELAB, MC, MEC	Onwulata (2008)
Casein	Glycerol	Solution casting	AFM, G^{60}, OP, MEC, WVP	Fabra et al. (2011)
Egg albumen	Glycerol	Compression molding	AA, DMTA, TGA	Fernández-Espada et al. (2013)
Gelatin	Sorbitol	Solution casting	CIELAB, DSC, FTIR, MEC, RHE, SEM, TSM, UV, WVP	Núñez-Flores et al. (2013)
Gelatin	Sorbitol	Solution casting	CIELAB, FTIR, MC, MEC, SEM, UV, WCA, WVP, XRD	Kanmani and Rhim (2014)
Gelatin	Glycerol	Extrusion	FTIR, OP, SEM, TSM, WVP	Nur Hanani et al. (2012)
Soy protein	Glycerol	Solution casting	DSC, FTIR, TGA, MEC	Guerrero et al. (2010)
Soy protein	Glycerol	Compression molding	DSC, FTIR, TGA, MEC	Guerrero and de la Caba (2010)
Soy protein	Glycerol	Extrusion	DSC, FTIR, MC, SEM, SME, TSM, XRD	Guerrero et al. (2012)
Soy protein	Sorbitol	Solution	CIELAB, FTIR, G^{60}, MEC, XRD, WCA	Garrido et al. (2014)
Soy protein	Sorbitol	Compression	CIELAB, G^{60}, MEC, XRD, WCA	Garrido et al. (2013)
Wheat gluten	Glycerol	Solution	FTIR, SEM, WAXD, WVP	Guilherme et al. (2010)
Wheat gluten	Glycerol	Compression	CIELAB, DSC, MC, OP, SDS-PAGE, MEC, WVP	Balaguer et al. (2014)
Wheat gluten	Water	Extrusion	DSC, GAB, SW, SEM, TEM	Guillard et al. (2013)

(Continued)

TABLE 3.3 (*Continued*)
Some Recent Works Related to the Preparation and Characterization of Protein-Based Materials

Protein	Plasticizer	Processing	Characterization Method	References
Whey protein	Glycerol	Dipping	AA	Fernández-Pan et al. (2014)
Whey protein	Glycerol	Spraying	ADH, IA	Lin and Krochta (2006)
Whey protein	Glycerol	Solution casting	CIELAB, OP, SEM, SW, MEC, WCA, WVP	Kurek et al. (2014)
Whey protein	Glycerol	Compression	MEC, OP, SEM, WVP	Janjarasskul et al. (2014)
Whey protein	Water	Extrusion	CD, FS, HPLC, SDS-PAGE	Qi and Onwulata (2011)
Zein	Oleic acid	Solution casting	AFM, RHE, MEC, WCA	Wu et al. (2011)
Zein	Sugars	Solution casting	SEM, WCA, WVP	Ghanbarzadeh et al. (2007)
Zein	Glycerol	Compression molding	DSC, DMTA, MEC	Ghanbarzadeh and Oromiehi (2009)
Zein	Olive oil	Compression molding	DSC, DMTA	Ghanbarzadeh and Oromiehi (2008)

Note: AA = antimicrobial analysis, ADH = adhesion, AFM = atomic force microscopy, CD = circular dichroism, CIELAB = color measurements, DMTA = dynamic mechanical thermal analysis, DSC = differential scanning calorimetry, FS = fluorescence spectroscopy, FTIR = Fourier transform infrared spectroscopy, G^{60} = gloss at 60° incidence angle, GAB = Guggenheim-Anderson-Boer model for moisture sorption, HPLC = high-performance liquid chromatography, IA = image analysis, MC = moisture content, MEC = mechanical test, OP = oxygen permeability, RHE = rheology, SDS-PAGE = sodium dodecyl sulfate-polyacrylamide gel electrophoresis, SEM = scanning electron microscopy, SME = specific mechanical energy, SSD = surface solid density, SW = swelling, TEM = transmission electron microscopy, TGA = thermogravimetric analysis, TSM = total soluble matter, UV = ultraviolet spectroscopy, WCA = water contact angle, WVP = water vapor permeability, XRD = x-ray diffraction.

Two kinds of forces are involved in edible protein-based coatings: protein–protein (cohesion) and protein–food (adhesion). The degree of cohesion affects coating properties such as resistance, flexibility, and permeability, while the degree of adhesion affects the coating stability and its ability to enhance food quality. Those forces depend on protein composition and structure, the type and concentration of the additives incorporated into formulations, the processing methods and conditions employed, as well as the food on which they are applied (Falguera et al., 2011; Gennadios et al., 1997). Active coatings can be applied on fresh, frozen, and processed food, such as fruits, vegetables, meats, poultry, and seafood, improving food appearance and quality (Lin and Zhao, 2007; Maté and Krochta, 1996; Xu et al., 2001). Some examples of the application of edible coatings on food products are shown in Table 3.4.

Active coatings improve food quality by controlling mass transfer, moisture and oxygen diffusion, and flavor and aroma losses, in addition to maintaining physicochemical characteristics of food (Lacroix and Vu, 2014). These coatings can incorporate active compounds, such as antimicrobials and antioxidants, to protect food against microbial spoilage and quality loss, resulting in extension of shelf life (Guilbert et al., 1996). Furthermore, edible coatings are consumed along with food and thus, they can provide additional nutrients and enhance sensory characteristics (Debeaufort et al., 1998). In this regard, whey protein has an exceptional biological value that exceeds that of other edible proteins such as egg proteins (Smithers, 2008). Furthermore, edible proteins may be used as carriers of bioactive compounds, such as folic and ascorbic acids, omega-3 fatty acids, and riboflavin, among others (Maltais et al., 2010; Silva-Weiss et al., 2013). Due to the above reason, the composition of edible coatings must conform to the regulations concerning food products (European Commission, 2009). As any material used for direct food contact,

TABLE 3.4
Some Recent Works Related to Active Coatings Based on Proteins and Their Application on Food Products

Protein	Active Compound	Food Product	References
Casein	Chitosan	Carrots, cheese, salami	Moreira et al. (2011)
Bovine gelatin	Probiotic bacteria	Hake portions	López de Lacey et al. (2012)
Fish gelatin	Chitosan	Fish fillets	Nowzari et al. (2013)
Fish gelatin	Cinnamon oil	Fish fillets	Andevari and Rezaei (2011)
Fish gelatin	Potassium sorbate	White shrimp	Jiang et al. (2011)
Soy protein	Catechin	Walnuts	Kang et al. (2013)
Soy protein		Mozzarella cheese	Zhong et al. (2014)
Whey protein	Oregano essential oil	Chicken breast fillets	Fernández-Pan et al. (2014)
Whey protein	Transglutaminase	Salmon fillets	Rodriguez-Turienzo et al. (2013)
Whey protein		Chicken	Dragich and Krochta (2010)
Zein	Cinnamon oil, mustard oil	Cherry tomatoes	Yun et al. (2015)
Zein		Dates	Mehyar et al. (2014)

protein-based coatings used as carriers of compounds intended to migrate to food for preservative effects must meet regulations for food ingredients and must be generally recognized as safe (GRAS).

Edible coatings can prevent dripping, enhancing product presentation and making food attractive to consumers. Edible coatings can also be used to reduce flavor loss from food, as well as foreign odor pickup by food and to improve food nutritional value. Furthermore, proteins that contain cysteine, a potent free-radical target, can provide antioxidant properties, preventing degradation processes in food, such as lipid oxidation, microorganism growth, browning reactions, and vitamin loss (Bonilla et al., 2012). The antioxidant effect of protein-based coatings is also strongly linked to their low oxygen permeability, consequence of their ordered hydrogen-bonded network structure (Maté and Krochta, 1998). Nuts, rich in unsaturated fatty acids, are very prone to lipid oxidation. Maté and Krochta (1996) found that whey protein–coated nuts showed values above the acceptance limit during a 70-day storage period, while noncoated nuts only during a 20-day storage time. They also observed that the increase in coating thickness reduced lipid rancidity, indicating that the mechanism of protection against oxygen is related to oxygen permeability. The good oxygen barrier properties of films from collagen, wheat gluten, corn zein, soy protein, and whey protein have been largely documented in the literature (Gennadios et al., 1993; Hong and Krochta, 2006; Park and Chinnan, 1995). Protein-based coatings have been used to reduce lipid oxidation and brown coloration caused by myoglobin oxidation in meats. Specifically, beef, due to its high content of unsaturated fatty acids, oxidizes rapidly. Shon et al. (2010) highlighted the potential of using soy protein as edible coating to enhance the freshness of beef packaged aerobically during refrigerated storage. These authors found that soy protein coating reduced lipid oxidation, as measured by thiobarbituric acid–reactive substance (TBARS) values, and also moisture loss. Furthermore, color measurements were carried out and it was observed that the coating prevented the loss of beef lightness and maintained the characteristic reddish color of fresh beefs. Soy protein coatings have also been employed to delay lipid oxidation in other fatty foods such as salmon. Sathivel (2005) found that soy protein coatings significantly reduced TBARS values in salmon fillets during a 3-month frozen storage. Additionally, they observed that there was no effect on color values for cooked fillets. More recently, Rodriguez-Turienzo et al. (2011, 2012, 2013) analyzed the effects of whey protein coating formulations on salmon quality parameters. They concluded that glycerol-incorporated whey protein coating was effective for frozen salmon protection, since lipid oxidation was delayed with no modification of sensory properties of salmon fillets by using this coating.

Protein-based coatings have also been used to improve the quality of fresh fruits by reducing oxygen penetration into the fruit, thus reducing metabolic activity, controlling ripeness, and maintaining the desirable texture (Le Tien et al., 2001; Park, 1999). Zein coatings have been used to reduce moisture and firmness loss and improve sensory properties of fruits such as dates, which tend to mature quickly (Mehyar et al., 2014). Zein-coated dates, stored at 3°C for 21 days, showed retardation in fruit maturation from 7 to 14 days. Zein coatings have been also successfully used on other fruits such as apples (Bai et al., 2003). Zein coatings prepared with

10% zein and 10% propylene glycol were found to provide shine and extend shelf life compared with commercial wax coatings for apples. Apples have also been coated with gelatin at 5% (Moldao-Martins et al., 2003). These coatings reduced weight loss and therefore, maintained firmness, preserved freshness, and extended apple shelf life; in addition, the coating improved the fruit appearance. Furthermore, preventing weight loss by the combined effect of coating application and vacuum or modified atmosphere packaging can increase saleable weight of products and thus, have a significant economic impact.

Besides the barrier to oxygen provided by protein-based coatings, active coating formulations can also incorporate antioxidants. Sorbic, lactic, and acetic acids have a long history as GRAS food preservatives and have been extensively used as fungistatic and bacteriostatic agents for foods. In addition to antioxidant properties, the incorporation of organic acids, such as ascorbic and citric acids, has been proved to be effective in controlling bacterial proliferation and thus, extending food shelf life (Rojas-Graü et al., 2009). Guillard Issoupov et al. (2009) described the release kinetics of ascorbic acid from wheat gluten coatings. These authors developed a mathematical model to predict the release kinetics of additives from coatings and assess the efficacy of active coatings to optimize and minimize the concentration of antimicrobial agents incorporated into edible coatings. Nevertheless, antimicrobial packaging containing natural extracts, such as essential oils (EOs), are gaining interest (Coma, 2008; Sánchez-González et al., 2011; Teixeira et al., 2013). Whey protein coatings with oregano EO doubled the shelf life of chicken breast from 6 to 13 days during refrigerated storage keeping aerobic mesophilic bacteria, lactic acid bacteria, and *Pseudomonas* spp. below the recommended limits for consumption (Fernández-Pan et al., 2014). In the same manner, zein-based coatings containing cinnamon EO were investigated to enhance microbial safety of cherry tomatoes (Yun et al., 2015). These coatings reduced *Salmonella* population inoculated on tomatoes, prolonging the tomato shelf life. The antimicrobial mechanism of EOs is not well understood but it may be associated with the EO hydrophobicity, which damages cell membranes and disrupts cell structure and function (Turgis et al., 2009). Since active coatings are able to release antimicrobial compounds that inhibit or slow down bacterial growth during storage, particular attention must be focused on the formulations used. Therefore, due to the safety concern in the food packaging sector, those formulations based on natural compounds are receiving more and more interest (Mastromatteo et al., 2009).

Most of the active packaging materials currently used are broad spectrum antimicrobials that do not target specifically bacterial pathogenic species, which may be a small fraction of the total microbial load present in food systems (Payment and Locas, 2011). Furthermore, nonpathogenic microbes are necessary in the production of some dairy and fermented foods, highlighting the importance to target only pathogenic organisms while maintaining commensal bacteria (Vonaseka et al., 2014). Additionally, the commensal bacteria present may have benefits for human health or even control the growth of pathogenic bacteria (García et al., 2008). Therefore, there is a significant need to develop novel antimicrobial packaging materials with high specificity.

Protein-Based Films and Coatings

3.5.2 BIODEGRADABLE FILMS

The use of proteins for film manufacturing fits well within environmental sustainability strategy. First, the large amount of waste produced by the food industry can be converted into great valuable materials to be reused into food packaging systems, eliminating waste management problems from the economic and environmental point of view (Mirabella et al., 2014). Second, the development of protein films able to extend the food shelf life can also reduce the environmental impact due to the reduction of food losses (Barlow and Morgan, 2013; Williams and Wikström, 2011). Therefore, efforts should be carried out not only to reduce environmental impact from the packaging film itself, but also to enhance those functional properties of protein films that contribute to reduce food losses (Wikström et al., 2014). Finally, since governments have implemented legislation to reduce the amount of municipal waste sent to landfill, biodegradability/compostability is also one of the main focuses for choosing proteins as packaging materials (Shi et al., 2010).

Biodegradation provides the opportunity to degrade materials after useful life and enables to close the ideal life cycle, from cradle to cradle (Figure 3.9). Under appropriate conditions of moisture, temperature, and oxygen availability, biodegradation leads to disintegration of the film with no toxic or harmful residue (Rhim et al., 2013). The biodegradability of the film during composting occurs in three phases: mesophilic, thermophilic, and maturation phases (Verbeek et al., 2012). The mesophilic phase is an exothermic process in which easily oxidizable compounds are broken down by hydrolase enzymes and temperature rises, leading to the growth of thermophilic bacteria. During the thermophilic phase, proteins are biodegraded into amino acids and ammonia; in this phase, thermophilic bacteria and fungi take over biodegradation process, characterized by oxidation. The change from mesophilic to thermophilic phase must be accompanied with a pH change from acidic to alkaline to maintain high degradation rates in large-scale composting facilities

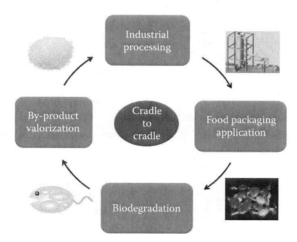

FIGURE 3.9 Ideal life cycle for protein-based films.

(Sundberg et al., 2004). After the active composting phases, compost requires a curing period to develop a valuable soil for plant growth (Kale et al., 2007).

To measure the environmental impact of products from the extraction of raw materials to ultimate disposal, life cycle assessment (LCA) is a practical key tool (Bier et al., 2011; Roy et al., 2009; Siracusa et al., 2014). The principles and framework of LCA are described in ISO 14040 (ISO, 2006); however, some limitations are related to data availability since industrial production and commercialization of biodegradable films are currently very competitive (Alvárez-Chávez et al., 2012; Iles and Martin, 2013). Furthermore, when dealing with novel products, data can be derived from experiments at laboratory scale, which could change when scaling to industrial production (Hospido et al., 2010).

Nevertheless, LCA can be used to identify the most pollutant phases of the life cycle for bio-based films, as a first step prior to the analysis of the changes needed during the design of products and processes to minimize negative environmental impacts. Although the environmental assessment of protein-based biodegradable films is largely unreported, some works have recently been published related to plant proteins. Deng et al. (2013) carried out LCA on wheat gluten materials. The LCA results exhibited that wheat gluten films produced by extrusion and incinerated to recover energy were favorable from an environmental perspective. Benefits in climate change and fossil depletion over LDPE films were measured. Although wheat gluten films suffer from common problems for bio-based materials, such as land occupation, the overall environmental performance indicated that wheat gluten can provide a promising source for bio-based polymer production. With regard to soy protein films, Leceta et al. (2014) showed that the extraction of raw materials was the stage with the highest environmental burden. Additionally, soybeans cultivation contributed to the environmental burden in land use category due to the use of glycerol, by-product from biodiesel production, as plasticizer. However, the end of life stage was the least pollutant phase for soy protein-based films due to their biodegradable nature, which allows composting as the end of life scenario and provides environmental benefits (Rudnik, 2008).

In spite of the fact that protein-based films seem to be more environmentally friendly materials when their origin and biodegradability are considered, the sustainability of biodegradable films must include all the stages of their life cycle, from cradle to cradle, as well as social and economic issues to assist in decision making (Philp et al., 2013). Therefore, recent developments were initiated to broaden LCA to life cycle sustainability analysis, an interdisciplinary framework for addressing knowledge from diverse scientific disciplines in such a way that the whole cause and effect can be assessed (Guinée et al., 2011).

3.6 CONCLUSIONS

Research and development of protein-based films and coatings in the food packaging field have been intensified, but their low presence in the market evidences that further studies must be carried out. The benefits of proteins can be numerous, but some drawbacks for potential applications need to be overcome. Protein-based films and coatings possess good barrier properties to oxygen, carbon dioxide, and lipids, but not to water vapor. Therefore, the challenge for successful commercial implementation of protein-based films and coatings is to improve and stabilize their functional properties during storage and use. In this context, extensive research is still needed on protein modification and processing methods.

Although protein-based materials are mostly processed in solution, this method is slow, requires large volumes of solvents, and thus, may be considered commercially less interesting than the dry process. Furthermore, protein-based materials developed by thermomechanical molding have better properties and are economically more efficient since they are less time consuming. With regard to the modification of proteins, this is still a long way from being a common method in food packaging, but it is increasingly being recognized as essential for two main reasons. Firstly, proteins fulfill multipurpose functions in materials science and these functions can be served better by modified than by native proteins. Secondly, environmental concerns promote the utilization of renewable and biodegradable materials. Furthermore, the use of by-products and waste streams from food processing industry for food packaging materials can positively impact the economics of food processes.

REFERENCES

Aguirre, A., Borneo, R., and León, A. E. 2013. Properties of triticale protein films and their relation to plasticizing–antiplasticizing effects of glycerol and sorbitol. *Industrial Crops and Products*, *50*, 297–303.

Alvárez-Chávez, C. R., Edwards, S., Moure-Eraso, R., and Geiser, K. 2012. Sustainability of bio-based plastics: General comparative analysis and recommendations for improvement. *Journal of Cleaner Production*, *73*, 47–56.

Anderson, T. J., and Lamsal, B. P. 2011. Zein extraction from corn, corn products, and coproducts and modifications for various application: A review. *Cereal Chemistry*, *88*, 159–173.

Andevari, G. T., and Rezaei, M. 2011. Effect of gelatin coating incorporated with cinnamon oil on the quality of fresh rainbow trout in cold storage. *International Journal of Food Science and Technology*, *46*, 2305–2311.

Angellier-Cousy, H., Gastaldi, E., Gontard, N., and Guillard, V. 2011. Influence of processing temperature on the water vapour transport properties of wheat gluten based agromaterials. *Industrial Crops and Products*, *33*, 457–461.

Arcan, I., and Yemenicioglu, A. 2011. Incorporating phenolic compounds opens a new perspective to use zein films as flexible bioactive packaging materials. *Food Research International*, *44*, 550–556.

Audic, J. L., and Chaufer, B. 2010. Caseinate based biodegradable films with improved water resistance. *Journal of Applied Polymer Science*, *117*, 1828–1836.

Audic, J. L., Chaufer, B., and Daufin, G. 2003. Non-food applications of milk components and dairy co-products: A review. *Lait*, *83*, 417–438.

Avena-Bustillos, R. J., and Krochta, J. M. 1993. Water vapor permeability of caseinate-based edible films as affected by pH, calcium crosslinking and lipid content. *Journal of Food Science*, *58*, 904–907.

Bai, J., Alleyne, V., Hagenmaier, R. D., Mattheis, J. P., and Baldwin, E. A. 2003. Formulation of zein coatings for apples (*Malus domestica* Borkh). *Postharvest Biology and Technology*, *28*, 259–268.

Balaguer, M. P., Gómez-Estaca, J., Cerisuelo, J. P., Gavara, R., and Hernández-Muñoz, P. 2014. Effect of thermo-pressing temperature on the functional properties of bioplastics made from a renewable wheat gliadin resin. *LWT-Food Science and Technology*, *56*, 161–167.

Barlow, C. Y., and Morgan, D. C. 2013. Polymer film packaging for food: An environmental assessment. *Resources, Conservation and Recycling, 78*, 74–80.

Belitz, H. D., Grosch, W., and Schieberle, P. 2009. *Food Chemistry* (4th ed.). Heidelberg: Springer (Chapter 1).

Bergo, P., and Sobral, P. J. A. 2007. Effects of plasticizer on physical properties of pigskin gelatin films. *Food Hydrocolloids, 21*, 1285–1289.

Bier, J. M., Verbeek, C. J. R., and Lay, M. C. 2011. Life cycle assessments of bioplastics: Applications and issues. *International Journal of Environmental, Cultural, Economic and Social Sustainability, 7*, 145–157.

Bonilla, J., Atarés, L., Vargas, M., and Chiralt, A. 2012. Edible films and coatings to prevent the detrimental effect of oxygen on food quality: Possibilities and limitations. *Journal of Food Engineering, 110*, 208–213.

Bourtoom, T. 2008. Edible films and coatings: Characteristics and properties. *International Food Research Journal, 15*, 237–248.

Brandenburg, A. H., Weller, C. L., and Testin, R. F. 1993. Edible films and coatings from soy protein. *Journal of Food Science, 58*, 1086–1089.

Butler, P. 2009. Consumer benefits and convenience aspects of smart packaging. In J. P. Kerry, and P. Butler (Eds.), *Smart Packaging Technologies* (pp. 233–245). Chichester: Wiley.

Cabra, V., Arreguin, R., Vazquez-Duhalt, R., and Farres, A. 2007. Effect of alkaline deamidation on the structure, surface, hydrophobicity, and emulsifying properties of the Z-19 α-zein. *Journal of Agricultural and Food Chemistry, 55*, 439–445.

Cao, N., Yang, X., and Fu, Y. 2009. Effects of various plasticizers on mechanical and water vapor barrier properties of gelatin films. *Food Hydrocolloids, 23*, 729–735.

Chabba, S., and Netravali, A. N. 2005. 'Green' composites. Part 1: Characterization of flax fabric and glutaraldehyde modified soy protein concentrate composites. *Journal of Materials Science, 40*, 6263–6273.

Chandrapala, J., Zisu, B., Palmer, M., Kentish, S., and Ashokkumar, M. 2011. Effects of ultrasound on the thermal and structural characteristics of proteins in reconstituted whey protein concentrate. *Ultrasonic Sonochemistry, 18*, 951–957.

Chen, H. 1995. Functional properties and applications of edible films made of milk proteins. *Journal of Dairy Science, 78*, 2563–2583.

Chen, L., Reddy, N., Wu, X., and Yang, Y. 2012. Thermoplastic films from wheat proteins. *Industrial Crops and Products, 35*, 70–76.

Chen, F. L., Wei, Y. M., and Ojokoh, A. O. 2010. System parameters and product properties response of soybean protein extruded at wide moisture range. *Journal of Food Engineering, 96*, 208–213.

Cho, S. W., Skrifvars, M., Hemanathan, K., Mahimaisenan, P., and Adekunte, K. 2014. Regenerated cellulose fibre reinforced casein films: Effect of plasticizer and fibres on the film properties. *Macromolecular Research, 22*, 701–709.

Chou, D. H., and Morr, C. V. 1979. Protein–water interactions and functional properties. *Journal of American Oil Chemists' Society, 56*, 53–62.

Coma, V. 2008. Bioactive packaging technologies for extended shelf life of meat-based products. *Meat Science, 78*, 90–103.

Creamer, L. K., and MacGibbon, A. K. H. 1996. Some recent advances in the basic chemistry of milk proteins and lipids. *International Dairy Journal, 6*, 539–568.

Cuq, B., Gontard, N., and Guilbert, S. 1997. Thermoplastic properties of fish myofibrillar proteins: Application to biopackaging fabrication. *Polymer, 38*, 4071–4078.

Cuq, B., Gontard, N., and Guilbert, S. 1998. Proteins as agricultural polymers for packaging production. *Cereal Chemistry, 75*, 1–9.

Dangaran, K., Tomasula, P. M., and Qi, P. 2009. Structure and function of protein-based edible films and coatings. In M. E. Embuscado, and K. C. Huber (Eds.), *Edible Films and Coatings for Food Applications* (pp. 25–56). New York, NY: Springer.

Debeaufort, F., Quezada-Gallo, J., and Voilley, A. 1998. Edible films and coatings: tomorrow's packaging: A review. *Critical Reviews in Food Science*, 39, 299–313.

Denavi, G., Tapia-Blácido, D. R., Añón, M. C., Sobral, P. J. A., Mauri, A. N., and Menegalli, F. C. 2009. Effects of drying conditions on some physical properties of soy protein films. *Journal of Food Engineering*, 90, 341–349.

Deng, Y., Achten, W. M. J., Van Acker, K., and Duflou, J. R. 2013. Life cycle assessment of wheat gluten powder and derived packaging film. *Biofuels, Bioproducts and Biorefining*, 7, 429–458.

Dragich, A. M., and Krochta, J. M. 2010. Whey protein solution coating for fat-uptake reduction in deep-fried chicken breast strips. *Journal of Food Science*, 75, S43–S47.

Escamilla-García, M., Calderón-Domínguez, G., Chanona-Pérez, J. J., Farrera-Rebolllo, R. R., Andraca-Adame, J. A., Arzate-Vázquez, I., Mendez-Mendez, J. V., and Moreno-Ruiz, L. A. 2013. Physical and structural characterisation of zein and chitosan edible films using nanotechnology tools. *International Journal of Biological Macromolecules*, 61, 196–203.

European Commission. 2009. Commission Regulation No 450/2009/EC on active and intelligent materials and articles intended to come into contact with food. *Official Journal of the European Union*, 30 May 2009, 3–11.

Eysturskaro, J., Haug, I. J., Ulset, A., and Draget, K. I. 2009. Mechanical properties of mammalian and fish gelatins based on their weight average molecular weight and molecular weight distribution. *Food Hydrocolloids*, 23, 2315–2321.

Fabra, M. J., Hambleton, A., Talens, P., Debeaufort, F., and Chiralt, A. 2011. Effect of ferulic acid and α-tocopherol antioxidants on properties of sodium caseinate edible films. *Food Hydrocolloids*, 25, 1441–1447.

Falguera, V., Quintero, J. P., Jiménez, A., Aldemar Muñoz, J., and Ibarz, A. 2011. Edible films and coatings: Structures, active functions and trends in their use. *Trends in Food Science and Technology*, 22, 292–303.

Fernández-Espada, L., Bengoechea, C., Cordobés, F., and Guerrero, A. 2013. Linear viscoelasticity characterization of egg albumen/glycerol blends with applications in material moulding processes. *Food and Bioproducts Processing*, 91, 319–326.

Fernández-Pan, I., Carrión-Granda, X., and Maté, J. I. 2014. Antimicrobial efficiency of edible coatings on the preservation of chicken breast fillets. *Food Control*, 36, 69–75.

Flint, F. O., and Johnson, R. F. P. 1981. A study of film formation by soy protein isolate. *Journal of Food Science*, 46, 1351–1353.

Gandini, A., and Belgacem, M. N. 2008. The state of the art. In M. N. Belgacem, and A. Gandini (Eds.), *Monomers, Polymers and Composites from Renewable Resources* (pp. 1–16). Oxford, UK: Elsevier.

García, P., Martínez, B., Obeso, J. M., and Rodríguez, A. 2008. Bacteriophages and their application in food safety. *Letters in Applied Microbiology*, 47, 479–485.

Garrido, T., Etxabide, A., Leceta, I., Cabezudo, S., de la Caba, K., and Guerrero, P. 2014. Valorization of soya by-products for sustainable packaging. *Journal of Cleaner Production*, 64, 228–233.

Garrido, T., Etxabide, A., Peñalba, M., de la Caba, K., and Guerrero, P. 2013. Preparation and characterization of soy protein thin films: Processing and properties correlation. *Materials Letters*, 105, 110–112.

Gelse, K., Poschi, E., and Aigner, T. 2003. Collagens—Structure, function and biosynthesis. *Advanced Drug Delivery Reviews*, 55, 1531–1546.

Gennadios, A., Brandenburg, A. H., Weller, C. L., and Testin, R. F. 1993. Effect of pH on properties of wheat gluten and soy protein isolate films. *Journal of Agricultural and Food Chemistry*, 41, 1835–1839.

Gennadios, A., Handa, A., Froning, G. W., Weller, C. L., and Hanna, M. A. 1998. Physical properties of egg white-dialdehyde starch films. *Journal of Agricultural and Food Chemistry*, 46, 1297–1302.

Gennadios, A., Hanna, M. A., and Kurth, L. B. 1997. Application of edible coatings on meats, poultry and seafoods: A review. *LWT-Food Science and Technology*, *30*, 337–350.

Gerrard, J. A., Brown, P. K., and Fayle, S. E. 2002. Maillard crosslinking of food proteins I: The reaction of glutaraldehyde, formaldehyde and glyceraldehyde with ribonuclease. *Food Chemistry*, *79*, 343–349.

Gerrard, J. A., Brown, P. K., and Fayle, S. E. 2003. Maillard crosslinking of food proteins II: The reactions of glutaraldehyde, formaldehyde and glyceraldehyde with wheat proteins *in vitro* and *in situ*. *Food Chemistry*, *80*, 35–43.

Ghanbarzadeh, B., Musavi, M., Oromiehi, A. R., Rezaky, K., Razmi Rad, E., and Milani, J. 2007. Effect of plasticizing sugars on water vapor permeability, surface energy and microstructure properties of zein films. *LWT-Food Science and Technology*, *40*, 1191–1197.

Ghanbarzadeh, B., and Oromiehi, A. R. 2008. Studies on glass transition temperature of mono and bilayer protein films plasticized by glycerol and olive oil. *Journal of Applied Polymer Science*, *109*, 2848–2854.

Ghanbarzadeh, B., and Oromiehi, A. R. 2009. Thermal and mechanical behavior of laminated protein films. *Journal of Food Engineering*, *90*, 517–524.

Ghanbarzadeh, B., Oromiehi, A. R., Musavi, M., D-Jomeh, Z. E., Rad, E. R., and Milani, J. 2006. Effect of plasticizing sugars on rheological and thermal properties of zein resins and mechanical properties of zein films. *Food Research International*, *39*, 882–890.

Ghosh, A., Ali, M. A., and Dias, G. J. 2009. Effect of cross-linking on microstructure and physical performance of casein protein. *Biomacromolecules*, *10*, 1681–1688.

Gillgren, T., Barker, S. A., Belton, P. S., Georget, D. M. R., and Stading, M. 2009. Plasticization of zein: A thermomechanical, FTIR, and dielectric study. *Biomacromolecules*, *10*, 1135–1139.

Gómez-Guillén, M. C., Giménez, B., López-Caballero, E., and Montero, P. 2011. Functional and bioactive properties of collagen and gelatin from alternative sources: A review. *Food Hydrocolloids*, *25*, 1813–1827.

Gómez-Guillén, M. C., Pérez-Mateos, M., Gómez-Estaca, J., López-Caballero, E., Giménez, B., and Montero, P. 2009. Fish gelatin: A renewable material for developing active biodegradable films. *Trends in Food Science and Technology*, *20*, 3–16.

Gómez-Guillén, M., Turnay, J., Fernández-Díaz, M., Ulmo, N., Lizarbe, M., and Montero, P. 2002. Structural and physical properties of gelatin extracted from different marine species: A comparative study. *Food Hydrocolloids*, *16*, 25–34.

Gontard, N., Guilbert, S., and Cuq, J. L. 1993. Water and glycerol as plasticizers affect mechanical and water vapor barrier properties of an edible wheat gluten film. *Journal of Food Science*, *58*, 206–211.

Gueguen, J., Viroben, G., Noireaux, P., and Subirade, M. 1998. Influence of plasticizers and treatments on the properties of films from pea proteins. *Industrial Crops and Products*, *7*, 149–157.

Guérin-Dubiard, C., and Audic, J. J. 2007. Egg protein-based films and coatings. In R. Huopalahti, R. López-Fandiño, M. Anton, and R. Schade (Eds.), *Bioactive Egg Compounds* (pp. 265–273). Heidelberg: Springer.

Guerrero, P., Beatty, E., Kerry, J. P., and de la Caba, K. 2012. Extrusion of soy protein with gelatin and sugars at low moisture content. *Journal of Food Engineering*, *110*, 53–59.

Guerrero, P., and de la Caba, K. 2010. Thermal and mechanical properties of soy protein films processed at different pH by compression. *Journal of Food Engineering*, *100*, 261–269.

Guerrero, P., Garrido, T., Leceta, I., and de la Caba, K. 2013. Films based on proteins and polysaccharides: Preparation and physical-chemical characterization. *European Polymer Journal*, *49*, 3713–3721.

Guerrero, P., Kerry, J. P., and de la Caba, K. 2014. FTIR characterization of protein-polysaccharide interactions in extruded blends. *Carbohydrate Polymers*, *111*, 598–605.

Guerrero, P., Nur Hanani, Z. A., and de la Caba, K. 2014. The effect of plasticizer content and disaccharide type on the mechanical, barrier and physical properties of bovine gelatin-based films. *Journal of Engineering Science and Technology*, 9, 364–373.

Guerrero, P., Retegi, A., Gabilondo, N., and de la Caba, K. 2010. Mechanical and thermal properties of soy protein films processed by casting and compression. *Journal of Food Engineering*, 100, 145–151.

Guerrero, P., Stefani, P. M., Ruseckaite, R. A., and de la Caba, K. 2011. Functional properties of films based on soy protein isolate and gelatin processed by compression molding. *Journal of Food Engineering*, 105, 65–72.

Guilbert, S., and Cuq, B. 2005. Material formed from proteins. In C. Bastioli (Ed.), *Handbook of Biodegradable Polymers* (pp. 339–384). Shropshire: Rapra Technology Limited.

Guilbert, S., Gontard, N., and Gorris, L. G. M. 1996. Prolongation of the shelf-life of perishable food products using biodegradable films and coatings. *LWT-Food Science and Technology*, 29, 10–17.

Guilherme, M. R., Mattoso, L. H. C., Gontard, N., Guilbert, S., and Gastaldi, E. 2010. Synthesis of nanocomposite films from wheat gluten matrix and MMT intercalated with different quaternary ammonium salts by way of hydroalcoholic solvent casting. *Composites Part A: Applied Science and Manufacturing*, 41, 375–382.

Guillard, V., Chevillard, A., Gastaldi, E., Gontard, N., and Angellier-Coussy, H. 2013. Water transport mechanisms in wheat gluten based (nano) composite materials. *European Polymer Journal*, 49, 1337–1346.

Guillard, V., Issoupov, V., Redl, A., and Gontard, N. 2009. Food preservative content reduction by controlling sorbic acid release from a superficial coating. *Innovative Food Science and Emerging Technologies*, 10, 108–115.

Guinée, J. B., Heijungs, R., Huppes, G., Zamagni, A., Masoni, P., Buonamici, R, Ekvall, T., and Rydberg, T. 2011. Life cycle assessment: Past, present, and future. *Environmental Science and Technology*, 45, 90–96.

Hardy, J. H., Römer, L. M., and Scheibel, T. R. 2008. Polymeric materials based on silk proteins. *Polymer*, 49, 4309–4327.

Harper, J. M. 1989. Food extruders and their applications. In C. Mercier, P. Linko, and J. M. Harper (Eds.), *Extrusion Cooking* (pp. 1–15). Minnesota: American Association of Cereal Chemists.

Haugaard, V. K., Udsen, A. M., Mortensen, G., Hoegh, L., Petersen, K., and Monahan, F. 2001. Potential food application of biobased materials. *Starch/Stärke*, 53, 189–200.

Hauser, C., Müller, U., Sauer, T., Augner, K., and Pischetsrieder, M. 2014. Maillard reaction products as antimicrobial components for packaging films. *Food Chemistry*, 145, 608–613.

Hernández-Izquierdo, V. M., and Krochta, J. M. 2009. Thermal transitions and heat-sealing of glycerol-plasticized whey protein films. *Packaging Technology and Science*, 22, 255–260.

Hernández-Izquierdo, V. M., and Krochta, J. M. 2008. Thermoplastic processing of proteins for film formation: A review. *Journal of Food Science*, 73, R30–R39.

Hernández-Muñoz, P., Villalobos, R., and Chiralt, A. 2004. Effect of cross-linking using aldehydes on properties of glutenin-rich films. *Food Hydrocolloids*, 18, 403–411.

Hong, S. I., and Krochta, J. M. 2006. Oxygen barrier performance of whey-protein-coated plastic films as affected by temperature, relative humidity, base film and protein type. *Journal of Food Engineering*, 77, 739–745.

Hoque, M. S., Benjakul, S., and Prodpran, T. 2010. Effect of heat treatment of film-forming solution on the properties of film from cuttlefish (*Sepia pharaonis*) skin gelatin. *Journal of Food Engineering*, 96, 66–73.

Hospido, A., Davis, J., Berlin, J., and Sonesson, U. 2010. A review of methodological issues affecting LCA of novel food products. *The International Journal of Life Cycle Analysis*, 15, 44–52.

Hu, X., Cebe, P., Weiss, A. S., Omenetto, F., and Kaplan, D. L. 2012. Protein-based composite materials. *Materials Today*, 15, 208–215.

Hu, H., Wu, J., Li-Chan, E. Y. C., Zhu, L., Zhang, F., Xu, X., and Pan, S. 2013. Effects of ultrasound on structural and physical properties of soy protein isolate (SPI) dispersions. *Food Hydrocolloids*, 30, 647–655.

Iles, A., and Martin, N. 2013. Expanding bioplastics production: Sustainable business innovation in the chemical industry. *Journal of Cleaner Production*, 45, 38–49.

Imre, B., and Pukánszky, B. 2013. Compatibilization in bio-based and biodegradable polymer blends. *European Polymer Journal*, 49, 1215–1233.

Irissin-Mangata, J., Bauduin, G., Boutevin, B., and Gontard, N. 2001. New plasticizers for wheat gluten films. *European Polymer Journal*, 37, 1533–1541.

ISO. 2006. ISO 14040: Environmental management—Life Cycle Assessment—Principles and framework. ISO, Geneva.

Jambrak, A. R., Lelas, V., Mason, T. J., Kresic, G., and Badanjak, M. 2009. Physical properties of ultrasound treated soy proteins. *Journal of Food Engineering*, 93, 386–393.

Janjarasskul, T., and Krochta, J. M. 2010. Edible packaging materials. *Annual Review of Food Science and Technology*, 1, 415–448.

Janjarasskul, T., Rauch, D. J., McCarthy, K. L., and Krochta, J. M. 2014. Barrier and tensile properties of whey protein-candelila wax film/sheet. *LWT-Food Science and Technology*, 56, 377–382.

Jerez, A., Partal, P., Martinez, I., Gallegos, C., and Guerrero, A. 2007. Protein-based bioplastics: Effect of thermo-mechanical processing. *Rheologica Acta*, 46, 711–720.

Jiang, M., Liu, S., and Wang, Y. 2011. Effects of antimicrobial coating from catfish skin gelatin on quality and shelf life of fresh white shrimp (*Penaeus vannamei*). *Journal of Food Science*, 76, M204–M209.

Jiang, J., Xiong, Y. L., Newman, M. C., and Rentfrow, G. K. 2012. Structure-modifying alkaline and acidic pH-shifting processes promote formation of soy proteins. *Food Chemistry*, 132, 1944–1950.

Jongjareonrak, A., Benjakul, S., Visessanguan, W., and Tanaka, M. 2006. Effects of plasticizers on the properties of edible films from skin gelatin of bigeye snapper and brownstripe red snapper. *European Food Research and Technology*, 222, 229–235.

Kale, G., Kijchavengkul, T., Auras, R., Rubino, M., Selke, S. E., and Singh, S. P. 2007. Compostability of bioplastic packaging materials: An overview. *Macromolecular Bioscience*, 7, 255–277.

Kang, H. J., Kim, S., You, Y. S., Lacroix, M., and Han, J. 2013. Inhibitory effect of soy protein coating formulations on walnut (*Juglans regia* L.) kernels against lipid oxidation. *LWT-Food Science and Technology*, 51, 393–396.

Kanmani, P., and Rhim, J. W. 2014. Physicochemical properties of gelatin/silver nanoparticle antimicrobial composite films. *Food Chemistry*, 148, 162–169.

Karim, A. A., and Bhat, R. 2008. Gelatin alternatives for the food industry: Recent development, challenges and prospects. *Trends in Food Science and Technology*, 19, 644–656.

Karim, A. A., and Bhat, R. 2009. Fish gelatin: Properties, challenges, and prospects as an alternative to mammalian gelatins. *Food Hydrocolloids*, 23, 563–576.

Kayserilioglu, B. S., Stevels, W. M., Mulder, W. J., and Akkas, N. 2001. Mechanical and biochemical characterisation of wheat gluten films as a function of pH and co-solvent. *Starch/Stärke*, 53, 381–386.

Kim, J. T., and Netravali, A. N. 2012. Physical properties of biodegradable films of soy protein concentrate/gelling agent blends. *Macromolecular Materials and Engineering*, 297, 176–183.

Kinsella, J. E. 1979. Functional properties of soy proteins. *Journal of the American Oil Chemists' Society*, 56, 242–258.

Kinsella, J. E., and Whitehead, D. M. 1989. Proteins in whey: Chemical, physical and functional properties. *Advances in Food and Nutrition Research, 33*, 343–438.

Kozempel, M., and Tomasula, P. M. 2004. Development of a continuous process to make casein films. *Journal of Agricultural and Food Chemistry, 52*, 1190–1195.

Krochta, J. M. 2002. Proteins as raw materials for films and coatings: Definitions, current status, and opportunities. In A. Gennadios (Ed.), *Protein-based Films and Coatings* (pp. 1–41). Boca Raton, FL: CRC Press.

Kuorwel, K. K., Cran, M. J., Sonneveld, K., Miltz, J., and Bigger, S. W. 2011. Antimicrobial activity of biodegradable polysaccharide and protein-based films containing active agents. *Food Science, 76*, R90–R102.

Kurek, M., Galus, S., and Debeaufort, F. 2014. Surface, mechanical and barrier properties of bio-based composite films based on chitosan and whey protein. *Food Packaging and Shelf Life, 1*, 56–67.

Lacroix, M., and Vu, K. D. 2014. Edible coating and film materials: Proteins. In J. H. Han (Ed.), *Innovations in Food Packaging* (pp. 277–304). Waltham, MA: Elsevier.

Lagrain, B., Goderis, B., Brijs, K., and Delcour, J. A. 2010. Molecular basis of processing wheat gluten toward biobased materials. *Biomacromolecules, 11*, 533–541.

Leceta, I., Etxabide, A., Cabezudo, S., de la Caba, K., and Guerrero, P. 2014. Bio-based films prepared with by-products and wastes: Environmental assessment. *Journal of Cleaner Production, 64*, 218–227.

Le Tien, C., Vachon, C., Mateescy, M. A., and Lacroix, M. 2001. Milk protein coatings prevent oxidative browning of apples and potatoes. *Journal of Food Science, 66*, 512–516.

Letendre, M., D'Aprano, G. D., Lacroix, M., Salmieri, S., and St-Gelais, D. 2002. Physicochemical properties and bacterial resistance of biodegradable milk protein films containing agar and pectin. *Journal of Agricultural and Food Chemistry, 50*, 6017–6022.

Li, H., Liu, B. L., Gao, L. Z., and Chen, H. L. 2004. Studies on bullfrog skin collagen. *Food Chemistry, 84*, 65–69.

Lian, F., Wang, Y., and Sun, X. S. 1999. Curing process and mechanical properties of protein-based polymers. *Journal of Polymer Engineering, 19*, 383–393.

Lim, L. T., Mine, Y., and Tung, M. A. 1998. Transglutaminase cross-linked egg white protein-films: Tensile properties and oxygen permeability. *Journal of Agricultural and Food Chemistry, 46*, 4022–4029.

Lin, S., Huff, H. E., and Hsieh, F. 2000. Texture and chemical characteristics of soy protein meat analogue extruded at high moisture. *Journal of Food Science, 65*, 264–269.

Lin, S. Y., and Krochta, J. M. 2006. Fluidized-bed system for whey protein film coating of peanuts. *Journal of Food Process Engineering, 29*, 532–546.

Lin, D., and Zhao, Y. 2007. Innovations in the development and application of edible coatings for fresh and minimally processed fruits and vegetables. *Comprehensive Reviews in Food Science and Food Safety, 6*, 60–75.

Liu, C. C., Tellez-Garay, A. M., and Castell-Perez, M. E. 2004. Physical and mechanical properties of peanut protein films. *LWT-Food Science and Technology, 3*, 731–738.

Lodha, P., and Netravali, A. N. 2005. Thermal and mechanical properties of environment-friendly "green" plastics from stearic acid modified-soy protein isolate. *Industrial Crops and Products, 21*, 49–64.

López de Lacey, A. M., López-Caballero, M. E., Gómez-Estaca, J., Gómez-Guillén, M. C., and Montero, P. 2012. Functionality of *Lactobacillus acidophilus* and *Bifidobacterium bifidum* incorporated to edible coatings and films. *Innovative Food Science and Emerging Technologies, 16*, 277–282.

Luckachan, G. E., and Pillai, C. K. S. 2011. Biodegradable polymers: A review on recent trends and emerging perspectives. *Journal of Polymers and the Environment, 19*, 637–676.

Maltais, A., Remondetto, G. E., and Subirade, M. 2010. Tabletted soy protein cold-set hydrogels as carriers of nutraceuticals substances. *Food Hydrocolloids*, *24*, 518–524.

Martín-Alfonso, J. E., Félix, M., Romero, A., and Guerrero, A. 2014. Development of new albumen based biocomposites formulations by injection moulding using chitosan as physicochemical modifier additive. *Composites Part B: Engineering*, *61*, 275–281.

Martín-Closas, L., and Pelacho, A. M. 2011. Agronomic potential of biopolymer films. In D. Plackett (Ed.), *Biopolymers: New Materials for Sustainable Films and Coatings* (pp. 277–299). Chichester: Wiley.

Massani, M. B., Botana, A., Eisenberg, P., and Vignolo, G. 2014. Development of an active wheat gluten film with *Lactobacillus curvatus* CRL705 bacteriocins and a study of its antimicrobial performance during ageing. *Food Additives and Contaminants: Part A*, *31*, 164–171.

Mastromatteo, M., Barbuzzi, G., Conte, A., and Del Nobile, M. A. 2009. Controlled release of thymol from zein based films. *Innovative Food Science and Emerging Technologies*, *10*, 222–227.

Maté, J. I., and Krochta, J. M. 1996. Whey protein coating effect on the oxygen uptake of dry roasted peanuts. *Journal of Food Science*, *61*, 1202–1207.

Maté, J. I., and Krochta, J. M. 1998. Oxygen uptake model for uncoated and coated peanuts. *Journal of Food Engineering*, *35*, 299–312.

Mauri, A. N., and Añón, M. C. 2006. Effect of solution pH on solubility and some structural properties of soybean protein isolate films. *Journal of the Science of Food and Agriculture*, *86*, 1064–1072.

Mauri, A. N., and Añón, M. C. 2008. Mechanical and physical properties of soy protein films with pH-modified microstructures. *Food Science and Technology International*, *14*, 119–125.

McHugh, T. H., and Krochta, J. M. 1994. Sorbitol vs glycerol-plasticized whey protein edible films: Integrate oxygen permeability and tensile property evaluation. *Journal of Agricultural and Food Chemistry*, *42*, 841–845.

McMurry, J. 2011. *Organic Chemistry* (8th ed.). Belmont: Brooks/Cole (Chapter 26).

Mehyar, G. F., El Assi, N. M., Alsmairat, N. G., and Holley, R. A. 2014. Effect of edible coatings on fruit maturity and fungal growth on Berhi dates. *International Journal of Food Science and Technology*, *49*, 2409–2417.

Micard, V., Morel, M. H., Bonicel, J., and Guilbert, S. 2001. Thermal properties of raw and processed wheat gluten in relation with protein aggregation. *Polymer*, *42*, 477–485.

Miller, K. S., and Krochta, J. M. 1997. Oxygen and aroma barrier properties of edible films: A review. *Trends in Food Science and Technology*, *8*, 228–237.

Mine, Y. 1995. Recent advances in the understanding of egg white protein functionality. *Trends in Food Science and Technology*, *6*, 225–232.

Mirabella, N., Castellani, V., and Sala, S. 2014. Current options for the valorization of food manufacturing waste: A review. *Journal of Cleaner Production*, *65*, 28–41.

Moditsi, M., Lazaridou, A., Moschakis, T., and Biliaderis, C. G. 2014. Modifying the physical properties of dairy protein film for controlled release of antifungal agents. *Food Chemistry*, *39*, 195–203.

Mojumdar, S. C., Moresoli, C., Simon, L. C., and Legge, R. L. 2011. Edible wheat gluten (WG) protein films. Preparation, thermal, mechanical and spectral properties. *Journal of Thermal Analysis and Calorimetry*, *104*, 929–936.

Moldao-Martins, M., Beirao-da-Costa, S. M., and Beirao-da-Costa, M. L. 2003. The effects of edible coatings on postharvest quality of the "Bravo de Esmolfe" apple. *European Food Research and Technology*, *217*, 325–328.

Mooney, B. P. 2009. The second green revolution. Production of plant-based biodegradable plastics. *Biochemical Journal*, *418*, 219–232.

Moreira, M. R., Pereda, M., Marcovich, N. E., and Roura, S. I. 2011. Antimicrobial effectiveness of bioactive packaging materials from edible chitosan and casein polymers: Assessment on carrot, cheese, and salami. *Journal of Food Science, 76*, M54–M63.

Morel, M. H., Redl, A., and Guilbert, S. 2002. Mechanism of heat and shear mediated aggregation of wheat gluten protein upon mixing. *Biomacromolecules, 3*, 48–497.

Mu, C., Guo, J., Lib, X., Lin, W., and Li, D. 2012. Preparation and properties of dialdehyde carboxymethyl cellulose crosslinked gelatin edible films. *Food Hydrocolloids, 27*, 22–29.

Mueller, U., Sauer, T., Weigel, I., Pichner, R., and Pischetsrieder, M. 2011. Identification of H_2O_2 as a major antimicrobial component in coffee. *Food and Function, 2*, 265–272.

Müller, G., Hanecker, E., Blasius, K., Seidemann, C., Tempel, L., Sadocco, P., Jamnicki, S., and Bobu, E. 2014. End-of-life solutions for fibre and bio-based packaging materials in Europe. *Packaging Technology and Science, 27*, 1–15.

Nedi, I., Di Maio, E., and Iannace, S. 2012. The role of protein-plasticizer-clay interactions on processing and properties of thermoplastic zein bionanocomposites. *Journal of Applied Polymer Science, 125*, E314–E323.

Ning, L., and Villota R. 1994. Influence of 7S and 11S globulins on the extrusion performance of soy protein concentrates. *Journal of Food Processing and Preservation, 18*, 421–436.

Nishinari, K., Fang, Y., Guo, S., and Phillips, G. O. 2014. Soy proteins: A review on composition, aggregation and emulsification. *Food Hydrocolloids, 39*, 301–318.

Nowzari, F., Shábanpour, B., and Ojagh, S. M. 2013. Comparison of chitosan-gelatin composite and bilayer coating and film effect on the quality of refrigerated rainbow trout. *Food Chemistry, 141*, 1667–1672.

Núñez-Flores, R., Giménez, B., Fernández-Martín, F., López-Caballero, M. E., Montero, P., and Gómez-Guillén, M. C. 2013. Physical and functional characterization of active fish gelatin films incorporated with lignin. *Food Hydrocolloids, 30*, 163–172.

Nur Hanani, Z. A., Beatty, E., Roos, Y. H., Morris, M. A., and Kerry, J. P. 2012. Manufacture and characterization of gelatin films derived from beef, pork and fish sources using twin screw extrusion. *Journal of Food Engineering, 113*, 606–614.

Nur Hanani, Z. A., McNamara, J., Roos, Y. H., Morris, M. A., and Kerry, J. P. 2013. Effect of plasticizer content on the functional properties of extruded gelatin-based composite films. *Food Hydrocolloids, 31*, 264–269.

Nur Hanani, Z. A., O'Mahony, J. A., Roos, Y. H., Oliveira, P. M., and Kerry, J. P. 2014. Extrusion of gelatin-based composite films: Effects of processing temperature and pH of film forming solution on mechanical and barrier properties of manufactured films. *Food Packaging and Shelf Life, 2*, 91–101.

Nur Hanani, Z. A., Roos, Y. H., and Kerry, J. P. 2012. Use of beef, pork and fish gelatin sources in the manufacture of films and assessment of their composition and mechanical properties. *Food Hydrocolloids, 29*, 144–151.

Onwulata, C. I. 2008. Baking properties of milk protein-coated wheat bran. *Journal of Food Processing and Preservation, 32*, 24–38.

Orliac, O., Rouilly, A., Silvestre, F., and Rigal, L. 2003. Effects of various plasticizers on the mechanical properties, water resistance and aging of thermo-moulded films made from sunflower proteins. *Industrial Crops and Products, 18*, 91–100.

Osés, J., Fabregat-Vázquez, M., Pedroza-Islas, R., Tomás, S. A., Cruz-Orea, A., and Maté, J. I. 2009. Development and characterization of composite edible films based on whey protein isolate and mesquite gum. *Journal of Food Engineering, 92*, 56–62.

Paes, S. S., Yakimets, I., Wellner, N., Hill, S. E., Wilson, R. H., and Mitchell, J. R. 2010. Fracture mechanisms in biopolymer films using coupling of mechanical analysis and high speed visualization technique. *European Polymer Journal, 46*, 2300–2309.

Park, H. J. 1999. Development of advanced edible coating for fruits. *Food Science and Technology*, 10, 254–260.
Park, H. J., and Chinnan, M. S. 1995. Gas and water vapor barrier properties of edible films from protein and cellulosic materials. *Journal of Food Engineering*, 25, 497–507.
Patist, A., and Bates, D. 2008. Ultrasonic innovations in the food industry: From the laboratory to commercial production. *Innovative Food Science and Emerging Technologies*, 9, 147–154.
Payment, P., and Locas, A. 2011. Pathogens in water: Value and limits of correlation with microbial indicators. *Ground Water*, 49, 4–11.
Pereira, R. N., Souza, B. W. S., Cerqueira, M. A., Teixeira, J. A., and Vicente, A. A. 2010. Effects of electric fields on protein unfolding and aggregation: Influence on edible films formation. *Biomacromolecules*, 11, 2912–2918.
Pereira, R. N., Teixeira, J. A., and Vicente, A. A. 2011. Exploring the denaturation of whey proteins upon application of moderate electric fields: A kinetic and thermodynamic study. *Journal of Agricultural and Food Chemistry*, 59, 11589–11597.
Pérez-Gago, M. B., and Krochta, J. M. 2000. Drying temperature effect on water vapor permeability and mechanical properties of whey protein-lipid emulsion films. *Journal of Agricultural and Food Chemistry*, 48, 2687–2692.
Pérez-Gago, M. B., Nadaud, P., and Krochta, J. M. 1999. Water vapor permeability, solubility and tensile properties of heat-denatured versus native whey protein films. *Journal of Food Science*, 64, 1034–1037.
Philp, J. C., Bartsev, A., Ritchie, R. J., Baucher, M. A., and Guy, K. 2013. Bioplastics science from a policy vintage point. *New Biotechnology*, 30, 635–646.
Pol, H., Dawson, P., Acton, J., and Ogale, A. 2002. Soy protein isolate/corn-zein laminated films: Transport and mechanical properties. *Journal of Food Science*, 67, 212–217.
Pommet, M., Redl, A., Guilbert, S., and Morel, M. H. 2005. Intrinsic influence of various plasticizers on functional properties and reactivity of wheat gluten thermoplastic materials. *Journal of Cereal Science*, 42, 81–91.
Pommet, M., Redl, A., Morel, M.H., Domenek, S., and Guilbert, S. 2003. Thermoplastic processing of protein-based bioplastics: Chemical engineering aspects of mixing, extrusion and hot molding. *Macromolecular Symposia*, 197, 207–217.
Qi, P. X., and Onwulata, C. I. 2011. Physical properties, molecular structures, and protein quality of texturized whey protein isolate: Effect of extrusion temperature. *Journal of Agricultural and Food Chemistry*, 59, 4668–4675.
Rafieian, F., Shahedi, M., Keramat, J., and Simonsen, J. 2014. Mechanical, thermal and barrier properties of nano-biocomposite based on gluten and carboxylated cellulose nanocrystals. *Industrial Crops and Products*, 53, 282–288.
Ramos, O. L., Pereira, R. N., Rodrigues, R., Teixeira, J. A., Vicente, A. A., and Malcata, F. X. 2014. Physical effects upon whey protein aggregation for nano-coating production. *Food Research International*, 66, 344–355.
Ramos, O. L., Reinas, I., Silva, S. I., Fernandes, J. C., Cerqueira, M. A., Pereira, R. N., Pintado, M. E., and Malcata, F. X. 2013. Effect of whey protein purity and glycerol content upon physical properties of edible films manufactured therefrom. *Food Hydrocolloids*, 30, 110–122.
Rahman, M., and Brazel, C. S. 2004. The plasticizer market: An assessment of traditional plasticizers and research trends to meet new challenges. *Progress in Polymer Science*, 29, 1223–1248.
Reddy, N., and Yang, Y. 2013. Thermoplastic films from plant proteins. *Journal of Applied Polymer Science*, 130, 729–736.
Rhim, J. W., Gennadios, A., Handa, A., Weller, C. L., and Hanna, M. A. 2000. Solubility, tensile, and color properties of modified soy protein isolate films. *Journal of Agricultural and Food Chemistry*, 48, 4937–4941.

Rhim, J. W., Park, H. M., and Ha, C. S. 2013. Bio-nanocomposites for food packaging applications. *Progress in Polymer Science*, *38*, 1629–1652.

Ribeiro, C., Vicente, A. A., Teixeira, J. A., and Miranda, C. 2007. Optimization of edible coating composition to retard strawberry fruit senescence. *Progress in Polymer Science*, *44*, 63–70.

Rodrigues, R. M., Martins, A. J., Ramos, O. L., Malcata, F. X., Teixeira, J. A., Vicente, A. A., and Pereira, R. N. 2015. Influence of moderate electric fields on gelation of whey protein isolate. *Food Hydrocolloids*, *43*, 329–339.

Rodriguez-Turienzo, L., Cobos, A., and Diaz, O. 2012. Effect of edible coatings based on ultrasound-treated whey proteins in quality attributes of frozen Atlantic salmon (*Salmo salar*). *Innovative Food Science and Emerging Technologies*, *14*, 92–98.

Rodriguez-Turienzo, L., Cobos, A., and Diaz, O. 2013. Effects of microbial transglutaminase added edible coatings based on heated or ultrasound-treated whey proteins in physical and chemical parameters of frozen Atlantic salmon (*Salmo salar*). *Journal of Food Engineering*, *119*, 433–438.

Rodriguez-Turienzo, L., Cobos, A., Moreno, V., Caride, A., Vieites, J. M., and Diaz, O. 2011. Whey protein-based coatings on frozen Atlantic salmon (*Salmo salar*): Influence of the plasticiser and the moment of coating on quality preservation. *Food Chemistry*, *128*, 187–194.

Rojas-Graü, M. A., Soliva-Fortuny, R., and Martín-Belloso, O. 2009. Edible coatings to incorporate active ingredients to fresh-cut fruits: A review. *Trends in Food Science and Technology*, *20*, 438–447.

Roy, P., Nei, D., Orikasa, T., Xu, Q., Okadome, H., Nakamura, N., and Shiina, T. 2009. A review of life cycle assessment (LCA) on some food products. *Journal of Food Engineering*, *90*, 1–10.

Rudnik, E. 2008. Properties and applications. In E. Rudnik (Ed.), *Compostable Polymer Materials* (pp. 38–69). Oxford, UK: Elsevier.

Rufian-Henares, J. A., and de la Cueva, S. P. 2009. Antimicrobial activity of coffee melanoidins: A study of their metal-chelating properties. *Journal of Agricultural and Food Chemistry*, *57*, 432–438.

Russin, T. A., Boye, J. I., Arcand, Y., and Rajamohamed, S. H. 2011. Alternative techniques for defatting soy: A practical review. *Food and Bioprocess Technology*, *4*, 200–223.

Sánchez-González, L., Vargas, M., González-Martínez, C., Chiralt, A., and Cháfer, M. 2011. Use of essential oils in bioactive edible coatings: A review. *Food Engineering Reviews*, *3*, 1–16.

Sathivel, S. 2005. Chitosan and protein coatings affect yield, moisture loss, and lipid oxidation of pink salmon (*Oncorhynchus gorbuscha*) fillets during frozen storage. *Journal of Food Science*, *70*, E455–E459.

Sears, J. K., and Darby J. R. 1982. Mechanism of plasticizer action. In J. K. Sears, and J. R. Darby (Eds.), *The Technology of Plasticizers* (pp. 35–77). New York: Wiley.

Shi, B., Bunyard, C., and Palfery, D. 2010. Plant polymer biodegradation in relation to global carbon management. *Carbohydrate Polymers*, *82*, 401–404.

Shi, W., and Dumont, M. J. 2014. Review: Bio-based films from zein, keratin, pea, and rapeseed protein feedstocks. *Journal of Materials Science*, *49*, 1915–1930.

Shi, K., Yu, H., Rao, S. L., and Lee, T. S. 2012. Improved mechanical property and water resistance of zein films by plasticization with tributyl citrate. *Journal of Agricultural and Food Chemistry*, *60*, 5988–5993.

Shon, J., Eo, J. H., and Eun, J. B. 2010. Effect of soy protein isolate coating on quality attributes of cut raw *Han-Woo* (Korean cow) beef, aerobically packaged and held refrigerated. *Food Control*, *33*, 42–60.

Shukla, R., and Cheryan, M. 2001. Zein: The industrial protein from corn. *Industrial Crops and Products*, *13*, 171–192.

Shyni, K., Hema, G. S., Ninan, G., Mathew, S., Joshy, C. G., and Lakshmanan, P. T. 2014. Isolation and characterization of gelatin from the skins of skipjack tuna (*Katsuwonus pelamis*), dog shark (*Scoliodon sorrakowah*), and rohu (*Labeo rohita*). *Food Hydrocolloids*, 39, 68–76.

Sian, N. K., and Ishak, S. 1990. Effect of pH on formation, proximate composition and rehydration capacity of winged bean and soybean protein-lipid film: A research note. *Journal of Food Science*, 55, 261–262.

Siew, D. C. W., Heilmann, C., Easteal, A. J., and Cooney, R. P. 1999. Solution and film properties of sodium caseinate/glycerol and sodium caseinate/polyethylene glycol edible coating systems. *Journal of Agricultural and Food Chemistry*, 47, 3432–3440.

Silva, N. H. C. S., Vilela, C., Marrucho, I. M., Freire, C. S. R., Neto, C. P., and Silvestre, A. J. D. 2014. Protein-based materials: From sources to innovative sustainable materials for biomedical applications. *Journal of Materials Chemistry B*, 2, 3715–3740.

Silva-Weiss, A., Ihl, M. Sobral, P. J. A., Gómez-Guillén, M. C., and Bifani, B. 2013. Natural additives in bioactive edible films and coatings: Functionality and applications in foods. *Food Engineering Reviews*, 5, 200–216.

Singh, P., Kumar, R., Sabapathy, S. N., and Bawa, A. S. 2008. Functional and edible uses of soy protein products. *Comprehensive Reviews in Food Science and Food Safety*, 7, 14–28.

Siracusa, V., Ingrao, C., Lo Giudice, A., Mbohwa, C., and Dalla Rosa, M. 2014. Environmental assessment of a multilayer polymer bag for food packaging and preservation: An LCA approach. *Food Research International*, 62, 151–161.

Smithers, G. W. 2008. Whey and whey proteins: From "gutter-to-gold." *International Dairy Journal*, 18, 695–704.

Soria, A. C., and Villamiel, M. 2010. Effect of ultrasound on the technological properties and bioactivity of food: A review. *Trends in Food Science and Technology*, 21, 323–331.

Song, F., Tang, D. L., Wang, X. L., and Wang, Y. Z. 2011. Biodegradable soy protein-isolate based materials: A review. *Biomacromolecules*, 12, 3369–3380.

Sothornvit, R., and Krochta, J. M. 2001. Plasticizer effect on mechanical properties of β-lactoglobulin films. *Journal of Food Engineering*, 50, 149–155.

Sothornvit, R., Olson, C. W., McHugh, T. H., and Krochta, J. M. 2007. Tensile properties of compression-molded whey protein sheets: Determination of molding condition and glycerol-content effects and comparison with solution-cast films. *Journal of Food Engineering*, 78, 855–860.

Sroka, Z., and Cisowski, W. 2003. Hydrogen peroxide scavenging, antioxidant and anti-radical activity of some phenolic acids. *Food and Chemical Toxicology*, 41, 753–758.

Stainsby, G. 1987. Gelatin gels. In A. M. Pearson, T. R. Dutson, and A. J. Bailey (Eds.), *Advances in Meat Research, Collagen as a Food* (pp. 209–222). New York: Van Nostrand Reinhold Company.

Subirade, M., Kelly, I., Guéguen, J., and Pézolet, M. 1998. Molecular basis of film formation form a soybean protein: Comparison between the conformation of glycinin in aqueous solution and in films. *International Journal of Biological Macromolecules*, 23, 241–249.

Sundberg, C., Smars, S., and Jönsson, H. 2004. Low pH as an inhibiting factor in the transition from mesophilic to thermophilic phase in composting. *Bioresource Technology*, 95, 145–150.

Talens, P., Pérez-Masía, R., Fabra, M. J., Vargas, M., and Chiralt, A. 2012. Application of edible coatings to partially dehydrated pineapple for use in fruit-cereal products. *Journal of Food Engineering*, 112, 86–93.

Teixeira, B., Marques, A., Ramos, C., Neng, N. R., Nogueira, J. M. F., Saraiva, J. A., and Nunes, M. L. 2013. Chemical composition and antibacterial and antioxidant properties of commercial essential oils. *Industrial Crops and Products*, 43, 587–595.

Turgis, M., Han, J., Caillet, S., and Lacroix, M. 2009. Antimicrobial activity of mustard essential oil against *Escherichia coli* O157:h7 and *Salmonella typhi*. *Food Control*, 20, 1073–1079.
Ullsten, N., Cho, S. W., Spencer, G., Gällstedt, M., Johansson, E., and Hedenqvist, M. S. 2009. Properties of extruded vital wheat gluten sheets with sodium hydroxide and salicylic acid. *Biomacromolecules*, 10, 479–488.
Ullsten, N., Gällstedt, M., Johansson, E., Gräslund, A., and Hedenqvist, M. S. 2006. Enlarged processing window of plasticized wheat gluten using salicylic acid. *Biomacromolecules*, 7, 771–776.
Ustunol, Z., and Mert, B. 2004. Water solubility, mechanical, barrier, and thermal properties of cross-linked whey protein isolate-based films. *Journal of Food Science*, 69, FEP129–FEP133.
Van Der Borght, A., Goesaert, H., Veraverbeke, W. S., and Delcour, J. A. 2005. Fractionation of wheat and wheat flour into starch and gluten: Overview of the main processes and the factors involved. *Journal of Cereal Science*, 41, 221–237.
Verbeek, C. J. R., Hicks, T., and Langdon, A. 2012. Biodegradation of bloodmeal-based thermoplastics in green-waste composting. *Journal of Polymers and the Environment*, 20, 53–62.
Vieira, M. G. A., Silva, M. A., Santos, L. O., and Beppu, M. M. 2011. Natural-based plasticizers and biopolymer films: A review. *European Polymer Journal*, 47, 254–263.
Vonaseka, E., Lea, P., and Nitin, N. 2014. Encapsulation of bacteriophages in whey protein films for extended storage and release. *Food Hydrocolloids*, 37, 7–13.
Vroman, I., and Tighzert, L. 2009. Biodegradable polymers. *Materials*, 2, 307–344.
Wagh, Y. R., Pushpadass, H. A., Emerald, F. M. E., and Nath, B. S. 2014. Preparation and characterization of milk protein films and their application for packaging of Cheddar cheese. *Journal of Food Science and Technology*, 51, 3767–3775.
Wan, V. C. H., Kim, M. S., and Lee, S. Y. 2005. Water vapor permeability and mechanical properties of soy protein isolate edible films composed of different plasticizer combinations. *Journal of Food Science*, 70, E387–E391.
Wang, C. H., and Damodaran, S. 1991. Thermal gelation of globular proteins: Influence of protein conformation on gel strength. *Journal of Agricultural and Food Chemistry*, 39, 433–438.
Wang, H., Jiang, J., and Fu, L. 2007. Properties of molded soy protein isolate plastics. *Journal of Applied Polymer Science*, 106, 3716–3720.
Wang, L., Liang, Q., Chen, T., Wang, Z., Xu, J., and Ma, H. 2014. Characterization of collagen from the skin of Amur sturgeon (*Acipenser schrenckii*). *Food Chemistry*, 38, 104–109.
Wang, Y., Xiong, Y. L., Rentfrow, G. K., and Newman, M. C. 2013. Oxidation promotes cross-linking but impairs film-forming properties of whey proteins. *Journal of Food Engineering*, 115, 11–19.
Wieser, H. 2007. Chemistry of gluten proteins. *Food Microbiology*, 24, 115–119.
Wu, L. Y., Wen, Q. B., Yang, X. Q., Xu, M. S., and Yin, S. W. 2011. Wettability, surface microstructure and mechanical properties of films based on phosphorus oxychloride-treated zein. *Journal of the Science of Food and Agriculture*, 91, 1222–1229.
Wikström, F., Williams, H., Verghese, K., and Clune, S. 2014. The influence of packaging attributes on consumer behaviour in food-packaging life cycle assessment studies: A neglected topic. *Journal of Cleaner Production*, 73, 100–108.
Williams, H., and Wikström, F. 2011. Environmental impact of packaging and food losses in a life cycle perspective: A comparative analysis of five food items. *Journal of Cleaner Production*, 19, 43–48.
Xiang, B. Y., Ngadi, M. O., Ochoa-Martinez, L. A., and Simpson, M. V. 2011. Pulsed electric field-induced structural modification of whey protein isolate. *Food and Bioprocess Technology*, 4, 1341–1348.

Xu, H., Chai, Y., and Zhang, G. 2012. Synergistic effect of oleic acid and glycerol on zein film plasticization. *Journal of Agricultural and Food Chemistry, 60,* 10075–10081.

Xu, S., Chen, X., and Sun, D. W. 2001. Preservation of kiwifruit coated with an edible film at ambient temperature. *Journal of Food Engineering, 50,* 211–216.

Yampolskaya, G., and Platikanov, D. 2006. Proteins at fluid interfaces: Adsorption layers and thin liquid films. *Advances in Colloid and Interface Science, 128–130,* 159–183.

Yasir, B. M., Sutton, K. H., Newberry, M. P., Andrews, N. R., and Gerrard, J. A. 2007. The impact of Maillard cross-linking on soy proteins and tofu texture. *Food Chemistry, 104,* 1502–1508.

Yue, H. B., Cui, Y. D., Shuttleworth, P. S., and Clark, J. H. 2012. Preparation and characterisation of bioplastics made from cottonseed protein. *Green Chemistry, 14,* 2009–2016.

Yun, J., Fan, X., Li, X., Jin, T. Z., Jia, X., and Mattheis, J. P. 2015. Natural surface coating to inactivate *Salmonella enterica* serovar *typhimurium* and maintain quality of cherry tomatoes. *International Journal of Food Microbiology, 193,* 59–67.

Zárate-Ramírez, L. S., Martínez, I., Romero, A., Partal, P., and Guerrero, A. 2011. Wheat gluten-based materials plasticised with glycerol and water by thermoplastic mixing and thermomoulding. *Journal of the Science of Food and Agriculture, 91,* 625–633.

Zárate-Ramírez, L. S., Romero, A., Martínez, I., Bengoechea, C., Partal, P., and Guerrero, A. 2014. Effect of aldehydes on thermomechanical properties of gluten-based bioplastics. *Food and Bioproducts Processing, 92,* 20–29.

Zhang, W. B., Yu, X., Wang, C. L., Sun, H. J., Hsieh, I. F., Li, Y., Van Horn, R., and Cheng, S. Z. D. 2014. Molecular nanoparticles are unique elements for macromolecular science: From "nanoatoms" to giant molecules. *Macromolecules, 47*(4), 1221–1239.

Zhong, Y., Cavender, G., and Zhao, Y. 2014. Investigation of different coating application methods on the performance of edible coatings on Mozzarella cheese. *LWT-Food Science and Technology, 56,* 1–8.

4 Edible Coatings and Films from Lipids, Waxes, and Resins

*Jorge A. Aguirre-Joya, Berenice Álvarez,
Janeth M. Ventura, Jesús O. García-Galindo,
Miguel A. De León-Zapata, Romeo Rojas,
Saúl Saucedo, and Cristóbal N. Aguilar*

CONTENTS

Abstract 122
4.1 Background of Edible Coatings and Films Based
 on Waxes, Lipids, and Resins 122
    4.1.1 Why Use Edible Coatings and Films? 123
4.2 Lipids, Waxes, and Resins Used as Raw Materials
 for Edible Coatings and Films 124
    4.2.1 Lipids 124
        4.2.1.1 Sunflower Oil 125
        4.2.1.2 Coconut Oil 125
        4.2.1.3 Palm Oil 126
        4.2.1.4 Cocoa Butter 126
    4.2.2 Waxes 126
        4.2.2.1 Beeswax 127
        4.2.2.2 Candelilla Wax 128
        4.2.2.3 Carnauba Wax 128
        4.2.2.4 Jojoba Oil 128
    4.2.3 Resins Classification 129
        4.2.3.1 Gum Arabic 129
        4.2.3.2 Gum Tragacanth 130
        4.2.3.3 Mesquite Gum 130
        4.2.3.4 Karaya Gum 130
        4.2.3.5 Ghatti Gum 130
        4.2.3.6 Vanilla Oleoresin 131
        4.2.3.7 Mastic Resin 131
4.3 Edible Coatings and Films Based on Lipids 131
    4.3.1 Lipids Used in Edible Coatings and Films 131
    4.3.2 Properties of Lipids in Edible Coatings and Films 132
    4.3.3 Water Vapor Barrier 133

4.3.4 Plasticizer Properties 134
4.3.5 Applications of Edible Coatings and Films Based on Lipids 134
4.4 Edible Coatings and Films Based on Natural Waxes 138
4.5 Concluding Remarks 138
References 142

ABSTRACT

Edible coatings and films are defined like thin layers of edible materials applied on food products that play an important role on their conservation, distribution, and marketing. The use of edible coatings and films improves appearance, being an effective barrier to transmission of gases, solving problems of migration of moisture, oxygen, carbon dioxide, and aromas, and reducing maturation processes, thus extending storage time and quality of vegetables or fruits. Usually edible coatings and films are classified according to their structural material. They can be composed of hydrocolloids, which consist of polysaccharides or proteins, or hydrophobic compounds (e.g., lipids, waxes, and resins), forming a simple coating or film, or by a mixture of these components. In this latter case, they are known as "composite films" with the aim of taking advantage of the properties of each compound and the synergy between them. These materials do not intend to fully replace traditional packaging; instead, their use allows providing further properties and adding benefits by incorporating natural additives to be used for food preservation or enrichment. A diverse quantity of natural waxes, lipids, and resins has been investigated to improve quality and time of shelf life and/or add colors and functional compounds, among others, to foods.

4.1 BACKGROUND OF EDIBLE COATINGS AND FILMS BASED ON WAXES, LIPIDS, AND RESINS

Edible coatings and films have been used for centuries and extensively studied for the last 20 years (Rossman, 2009). These materials do not pretend to fully replace traditional packaging; instead, the idea is to provide additional properties and add benefits by incorporating natural additives such as antioxidants, antimicrobials, and nutrients to be applied, for example, for food preservation or enrichment (Pavalath and Orts, 2009). These materials can be used to protect the product from mechanical damage, physical, chemical, and microbiological activity and also to prevent moisture loss, create a shiny surface, as well as control gas and lipid migration (Olivas and Barbosa-Cánovas, 2009). Edible coatings and films are mostly applied in highly perishable products such as fruits and vegetables, highly conditioned by the attainment of characteristics such as adequate cost, functional attributes, mechanical, and optical properties (Rojas-Grau et al., 2009). The term "edible" coating or film has been related to food applications. Edible coatings and films are defined like thin layers of edible materials applied on food products that play an important role on their conservation, distribution, and marketing (Falguera et al., 2011). Specifically, an edible coating is a thin layer of edible material formed as a coating on a food product that is applied in liquid form (structural matrix) on the food. On the contrary, an edible film is a preformed thin layer made of edible material that is first molded as solid sheets and can be placed on or between food components (Falguera et al., 2011; McHugh, 2000). All its components must be safe to eat or composed of generally recognized as safe (GRAS) materials and must have approval of regulatory agencies (e.g., the Food and Drug Administration [FDA], in the United States, or the European Food Safety Authority [EFSA], in the European Union). Characteristics of edible coatings and films depend on different parameters such as the kind of material implemented, the structural matrix, the conditions under which films are preformed and the type and concentration of additives (Guilbert et al., 1996).

Usually edible coatings and films are classified according to their structural material. They can be composed of hydrocolloids, which consist of polysaccharides or proteins, or hydrophobic compounds (e.g., lipids, waxes, and resins) or a mixture of these components. In this latter case, they are known as "composite films" with the aim of taking advantage of the properties of each compound and the synergy between them (Altenhofen et al., 2009; Falguera et al., 2011). All the components have a function into the system. The polymer or mixture of polymers to be used will form the base matrix of the film and is crucial to be compatible with all the ingredients to provide the required properties for the end use. Plasticizers are often included in the system and are added to the solution to enhance softness, flexibility, elongation, clarity, among others, and their choice will depend on the type of polymer used. Waxes, lipids, and resins can be included in the formulation to provide good moisture barrier properties because of their low affinity for water (Krochta, 1997). The properties that lipids confer to the matrix will depend on the characteristics of the lipid component, such as physical state, degree of saturation, and fatty acid content. The efficiency of an edible film as a barrier to moisture transfer cannot be simply improved with the addition of hydrophobic materials to the formulation, unless the formation of a homogeneous and continuous lipid layer inside the hydrocolloid matrix is achieved (Karbowiak et al., 2007). Several techniques are used for introducing lipids into edible films: (i) they can be the only constituent of the film, (ii) they can form a layer over a hydrocolloid layer, or (iii) they can form an emulsion with the hydrocolloid. The influence of lipids in edible coatings and films is particularly important when they are added to these systems. In water, lipid molecules form micelles, monolayers, bilayers, or vesicles due to the strong cohesive self-attraction of water molecules that repels the hydrocarbon chains. Variations in polarity can explain their different efficiency when integrated in edible films. Waxes belong to the class of nonpolar lipids, which means that they have no polar constituents or possess a hydrophilic part that in practice does not permit interaction with water; this character explains why waxes are the most efficient lipid barriers. The interactions with proteins and polysaccharides result from a subtle balance between forces of different nature, mainly electrostatic and hydrophobic, due to their amphiphilic character. The efficiency of lipids depends on several factors but mainly to the homogeneity of the film, structure, and physical state, among others (Callegarin, 1997).

4.1.1 WHY USE EDIBLE COATINGS AND FILMS?

Food preservation helps food to maintain its original properties/quality up to the point of consumption. It is accomplished by protecting the food product from surrounding environment conditions by controlling migration and mass transfer, including water loss. Indeed, product quality depends on the organoleptic, microbiological, nutritional, and functional properties (Lin and Zhao, 2007). These properties are subjected to changes during storage and distribution, mainly due to moisture migration either from the food to the environment or between different parts within a composite heterogeneous product (Debeaufort and Voilley, 2009).

The use of edible coatings and films improves appearance, being an effective barrier to transmission of gases, solving problems of migration of moisture, oxygen, carbon dioxide, and aromas (Campos et al., 2011; De León-Zapata et al., 2015), and reducing maturation processes, thus extending storage time and quality of vegetables or fruits (Beristain et al., 1999). Several lipids, for example, candelilla wax (De León-Zapata et al., 2015), carnauba wax (Chiumarelli and Hubinger, 2012; Jo et al., 2014), beeswax (Fagundes et al., 2014; Khanzadi et al., 2015); polysaccharides, for example, pectin (Otoni et al., 2014; Sánchez et al., 2015), chitosan (Arancibia et al., 2015), mesquite gum (Bosquez-Molina et al., 2003), tara gum (Antoniou et al., 2014), starch (Chen et al., 2010), galactomannans (Chen and Nussinovitch, 2000; Martínez-Ávila et al., 2014b), Arabic gum (Cruz et al., 2015), cellulose (Atef et al., 2015), and proteins (Hopkins et al., 2015; Khanzadi et al., 2015; Tong et al., 2015) have been used as base materials to formulate edible coatings and films. It is known that polysaccharide-based films have poor water vapor barrier properties, whereas most single hydrophobic coatings or films have high moisture resistance, although they form brittle films.

Moisture barrier properties of hydrophilic films can be improved by incorporating hydrophobic materials such as waxes or long-chain saturated fatty acids through emulsion or lamination technology (Bosquez-Molina et al., 2003). The presence of wax compounds can reduce water vapor permeability, but it can affect coatings transparence and mechanical properties, besides its possible impact in the aftertaste—which may impair the sensory characteristics of food (Bourlieu et al., 2009; Chiumarelli and Hubinger, 2014). Among the range of different hydrophobic substances, waxes provide the best barrier to moisture. Among these, beeswax, candelilla wax, and carnauba wax have GRAS status and are FDA approved for use as coatings in fruits and vegetables, or as additives in beverages and confectionery products (Kowalczyk and Baraniak, 2014).

4.2 LIPIDS, WAXES, AND RESINS USED AS RAW MATERIALS FOR EDIBLE COATINGS AND FILMS

4.2.1 Lipids

The use of lipids in edible films generally is an option to reduce the water vapor permeability due to its hydrophobic nature (Gurr and James, 1971). In addition to giving greater flexibility to the films, the lipid forms thicker and more brittle films (Vieira et al., 2011). In the case of proteins as an example of high polar polymer, shelf addition by diffusion is not important due to the minimal fixed order of the macromolecules, caused principally by the constitutional molecular forces holing up the polymer chains (Wittaya, 2012). In the case of cellulosic films and coatings, cellulose has a bone with a solid ring structure chain, where proteins are boned and tend to form helical chain structures (Banker, 1966).

4.2.1.1 Sunflower Oil

Sunflower oil is obtained from the sunflower seed. Sunflower seeds are processed by cold extraction to obtain the oils. Sunflower oil extraction involves cleaning the seeds, grinding, pressing, and extracting the crude oil and then refining. In terms of extraction by solvents, hexane is most commonly used. To consider the oil edible, it is necessary to remove the impurities present and then perform a refining operation to provide it greater stability through processes of degumming, neutralization, solvent purification, and bleaching. This oil contains a high concentration of vitamin E and low levels of saturated fat, which is advantageous for human consumption. The major constituents of sunflower oil are linoleic and oleic acids, as well as palmitic and stearic acids (Table 4.1) (Cindric et al., 2007; Guillen and Cabo, 1997, Souza et al., 2004; Tan and Man, 2000).

4.2.1.2 Coconut Oil

Coconut oil is a vegetable oil, also known as cocoa butter, extracted from the coconut pulp (*Cocos nucifera*). Fresh coconut meat is mechanically pressed to obtain

TABLE 4.1
Composition of Sunflower, Coconut, and Palm Oils and Cocoa Butter in Terms of Type and Content of Fatty Acids

Fatty Acid	Composition	Content (%)
Sunflower oil	Palmitic acid$_{C16}$	5.9–7.0
	Stearic acid$_{C18}$	4.6–6.0
	Oleic acid$_{C18:1}$	14.0–39.4
	Linoleic acid$_{C18:2}$	74.0–48.3
Coconut oil	Caproic acid$_{C6}$	0.01–0.24
	Caprylic acid$_{C8}$	5.0–9.0
	Capric acid$_{C10}$	6.0–10.0
	Lauric acid$_{C12}$	43–52
	Myristic acid$_{C14}$	13–19
	Palmitic acid$_{C16}$	7.7–10.5
	Stearic acid$_{C18}$	1.0–3.0
	Arachidic acid$_{C20}$	0.05–0.45
Palm oil	Myristic acid$_{C14}$	0.9–1.5
	Palmitic acid$_{C16}$	40.5–43.0
	Stearic acid$_{C18}$	2.7–4.0
	Oleic acid$_{C18:1}$	40.7–43.1
	Linoleic acid$_{C18:2}$	10.8–11.23
Cocoa butter	Palmitic acid$_{C16}$	25.7–25.8
	Oleic acid$_{C18:1\ cis}$	33.4–35.0
	Stearic acid$_{C18}$	34.1–34.9
	Linoleic acid$_{C18:2}$	3.2–4.0
	Arachidic acid$_{C20}$	0.5–1.5
	Palmitic acid$_{C16}$	25.7–25.8

coconut milk. Coconut milk is then chilled to 10°C to break the emulsion to help the separation of water and coconut butter separation. The resulting milk is transferred to a mixing vessel and heated to 45°C. Finally, the product is filtered to remove any suspended solids (Hamid et al., 2011). However, this method is expensive; therefore, centrifuges are used industrially to scale up the extraction process. In this case, the application of previous treatments is required which may include salts, acids, and high or low temperatures. The extraction may also be achieved using organic solvents, being hexane the most commonly used. The main constituent of coconut oil is lauric acid; however, it is also possible to find myristic, palmitic, capric, and caprylic acids in considerable quantities (Table 4.1) (Jitputti et al., 2006; Marina et al., 2009; Nevin and Rajamohan, 2004; Pehowich et al., 2000).

4.2.1.3 Palm Oil

Palm oil is obtained from the palm fruit (*Elaeis guineensis*), and it is the second most produced after soybean oil. Palm oil contains vitamins A and E in abundance. This oil is extracted through several steps: first, the fruit is removed with the aid of steam, which allows separating the fruit from the leaves and clusters; subsequently, the fruit is carried to digesters, where it is converted into pulp by heating; finally, pressing and centrifugation processes are carried out to separate the crude oil, which is then filtered and clarified to obtain a purified oil. In the early stages of fruit formation, the oil content is very low, but with fruit ripening the formation of oil increases rapidly. The major constituents of palm oil are palmitic and oleic acids; however, linoleic acid is also present in considerable quantity (Table 4.1) (Bell et al., 2002; Fitzherbert et al., 2008; Koh and Wilcove, 2008).

4.2.1.4 Cocoa Butter

Cocoa butter is a natural fat found in cocoa beans and is removed during the manufacturing process of chocolate, being separated from the cocoa mass by pressing. Cocoa beans are fermented, roasted, and then separated from their hulls, whereas cocoa liquor is pressed to extract cocoa butter, leaving a solid mass called cocoa cake. The major constituents of the resulting oil are oleic and stearic acids, followed by palmitic acid (Table 4.1) (Gutiérrez et al., 2014; Jahurul et al., 2013; Torres-Moreno et al., 2015).

4.2.2 WAXES

Waxes are used as barriers to gas and moisture transfer (applied on the skin of fresh fruits) and to improve the surface appearance of food products (e.g., the sheen on sweet). Waxes are even the most efficient edible compounds providing a barrier to humidity (Maftoonazad et al., 2013). The plant surfaces that are exposed to the atmosphere, such as leaves, fruits, petals, and stems, are covered with a hydrophobic, water-repellent substance, called wax. The outer surface of epidermis is covered with a substance called cutting, which is usually impregnated with wax; together they comprise the cuticle. The insoluble polymer cutting is composed of cross-linked hydroxy fatty acids (Kolattukudy, 1975). Generally, esters of fatty acids with alcohols of high molecular weight are highly insoluble in aqueous media. At room temperature, they are strong and tough (Martínez-Ávila et al., 2014a).

4.2.2.1 Beeswax

Wax is the material that bees use to build their hives. It is produced by young honey bees that secrete it as a liquid through their wax glands (Reybroeck et al., 2010). Wax is produced by all species of honey bees; however, when produced by different species of bees, it may display slightly different chemical and physical properties. It is a solid material at ambient temperature and has a light yellow color. Beeswax is extracted by beekeepers by different processes; it is subjected to hot water or steam to melt the wax, then leaks are made to increase its purity. It is a very stable material and highly appreciated by their properties (Table 4.2) (Hepburn et al., 2014; Maia and Nunes, 2013).

TABLE 4.2
Composition of Beeswax, Candelilla Wax, Carnauba Wax, and Jojoba Oil in Terms of Type and Content of Fatty Acids

Waxes	Properties	Values
Beeswax	Melting point (°C)	62–65
	Acid value (mg KOH g^{-1})	18–22
	Saponification value (mg KOH g^{-1})	14.0–39.4
	Solubility	Insoluble in water, soluble in alcohol
	Density 20°C (g cm^{-3})	0.95–0.97
Candelilla wax	Melting point (°C)	67–79
	Acid value (mg KOH g^{-1})	12–22
	Saponification value (mg KOH g^{-1})	35–87
	Solubility	Insoluble in water, soluble in ether, chloroform, and benzene
	Density 20°C (g cm^{-3})	–
Carnauba wax	Melting point (°C)	82.5–86
	Acid value (mg KOH g^{-1})	2–10
	Saponification value (mg KOH g^{-1})	78–88
	Solubility	Insoluble in water, soluble in ether, chloroform, and benzene
	Density 20°C (g cm^{-3})	0.996–0.998
Jojoba oil	Melting point (°C)	6–8
	Acid value (mg KOH g^{-1})	0.120–0.128
	Saponification value (mg KOH g^{-1})	88–96
	Solubility	Insoluble in water, soluble in ether, chloroform, and benzene
	Density 20°C (g cm^{-3})	0.863–0.873

Note: –, not found.

4.2.2.2 Candelilla Wax

Candelilla wax is a 100% natural substance, hard, brittle, and easy to be pulverized and exhibits a color that can be from light brown to yellow, depending on the refining and bleaching degree (Rojas-Molina et al., 2013). Its surface can reach high brightness levels, which is one of the most valued properties for various applications such as coating of chocolates and fruits, to stabilize emulsions and formulate lipsticks, body creams, among others (Cabello et al., 2013). Candelilla wax is recognized by the FDA as a GRAS substance for food industry applications. This wax can be extracted by immersing candelilla plant into acid solutions, being then heated to the boiling point, when the wax floats in the solution surface as foam; this foam is then collected and cooled to room temperature reaching the solid phase (Ochoa-Reyes et al., 2010). Finally, several processes such as filtration, decantation, and bleaching can be applied to refine it, providing the required characteristics (Rojas-Molina et al., 2011). Chemically, candelilla wax is composed of esters of fatty acids with high-molecular-weight alcohols. Wax particles are highly insoluble in aqueous media (Table 4.2) (Arato et al., 2013).

4.2.2.3 Carnauba Wax

Carnauba wax is obtained from the carnauba palm, a Brazilian tree (*Copernicia prunifera*). In hot and dry weather conditions, the plant secretes wax to protect the leaves from damage. Carnauba wax is used for a wide array of products, from cosmetics to food products and polishes (Wang et al., 2001). Some of those products are, for example, candies/sweets, chewing gums, chocolates, confectionary sugar, fruit coating, polishing wax (for car, leather, floor, or furniture), food packaging as coating, or plastic film among others.

Wax extraction is carried out in the dry months of the year and starts with the cutting of the carnauba palm leaves, when they are dried, allowing their separation in flake form. Subsequently, the wax is removed mechanically and is subjected to a refining process which involves filtering, decanting, and bleaching. The extent of this process will depend on the desired characteristics of the wax. Carnauba wax is bright yellow, has a high hardening capacity, and is extremely difficult to saponify (Table 4.2) (Barman et al., 2011, 2014; Dantas et al., 2013).

4.2.2.4 Jojoba Oil

Jojoba oil, also known as jojoba liquid wax, is extracted from mature seeds of the jojoba shrub *Simmondsia chinensis* and is manually harvested. Jojoba is produced in wild fields in the southeast of the United States and northeastern Mexico. The seed is rough and has an oil content ranging from 50% to 60%, whereas jojoba oil has a light color and a distinctive scent. Jojoba oil is obtained by cold pressing of the seeds, where the oil is separated, being then refined according to the final application. The oil can also be extracted through the use of solvents, hexane being the most used. This material has a mixture of triglycerides in its composition, while having the composition of a wax (Table 4.2), including long-chain (C18–C22) esters of unsaturated fatty acids (El-Boulifi et al., 2015; Palla et al., 2014; Sánchez et al., 2014).

4.2.3 Resins Classification

Generally the word "resin" is associated to the hydrocarbonic secretion or exudate of many plants, particularly of conifers, but the International Union of Pure and Applied Chemistry (IUPAC) defines resin as a soft solid or highly viscous substance, usually containing prepolymers with reactive groups, from vegetable sources (generally exudates) (IUPAC, 2006). Depending on their nature, resins can be defined as natural or synthetic, although both can be used for coating or film formation, or as component of these materials.

Resins are secretions produced by certain plant species and are appreciated for their properties as a structural material in food packaging, providing gas barrier, emulsifier, and adhesion properties in some applications. Resins can be classified into two major categories: natural and synthetic; these categories also have subcategories that will be described below.

Natural resins are a water-insoluble combination of compounds derived from trees, generally conifers. The profusion of compounds present in natural resins has a hydroaromatic structure that serves to protect the tree against pathogens (Jackrel and Wootton, 2015). Resins are found in the nature free of any solvent. These substances are often combined with other compounds, for example, isomeric carboxylic acids such as abietic and pimaric acids (U.S. Patent and Trademark Office, 2015).

There are many examples and uses of natural resins to improve some characteristics of films and or coatings, like those reported by Kim et al. (2015), who demonstrated that the incorporation of gums such as arabic, κ-carrageenan, xanthan, and gellan improves the water solubility and humidity of tapioca starch film. Particularly 0.2% of gellan was the most effective to improve mechanical properties and stability during storage conditions.

Other natural resin used as structural material for edible films and coatings is the mesquite gum with candelilla wax, also the functionality of mesquite gum to form edible coatings and films is even superior (in certain conditions) to the characteristics of arabic gum, used as guide mark as emulsifier, and film former (Bosques-Molina et al., 2010).

Synthetic resins are man-made resins that are more stable and homogenous than natural ones as a result of the polymerization process. Synthetic resins are used in plastics, food containers, paints, varnishes, and textiles. The difference between resins and waxes is that resins are a viscous substance consisting of prepolymers and functional groups, while waxes are firstly formed by fatty acids, mainly esters. The Food and Agriculture Organization (FAO) of the United Nations classifies resins in two groups: "hard resins" such as Copal, Damar, Mastic, and Dragon Blood and "soft resins and balsams" such as Benzoin, Stryrax, Peru, Tolu Balsam, Copaiba, Elemi, Asafoetida, and Galbanum (FAO, 1995).

4.2.3.1 Gum Arabic

Gum arabic, also known as acacia tree gum, is mainly obtained from *Acacia senegal* and is a resin exuded by the plant having the function of protecting the wounds and cracks in the bark. It is a polysaccharide of low protein content, soluble in water, composed of six different carbohydrates, that is, galactose, rhamnose, arabinopyranose, arabinofuranose, glucuronic, and 4-*o*-methyl glucuronic acids.

The concentration used varies considerably from 5% to 55% according to the intended application. Gum arabic dissolves quickly in cold water or in hot water, and it is the most soluble and less viscous hydrocolloid (Chranioti and Tzia, 2014; Desplanques et al., 2012; Nie et al., 2013).

4.2.3.2 Gum Tragacanth

Gum tragacanth is the exudate of the *Astragalus gummifer* tree. It contains a mixture of polysaccharides, that is, the tragacantic acid, which is insoluble in water and responsible for the water-absorbing properties of the gum, and arabinogalactan, which is a water-soluble polymer responsible for the gum solubility. Arabinogalactan has an acidic pH and a high viscosity at concentrations of about 1%, acting as emulsifier and stabilizer. Gum tragacanth has the highest viscosity among all hydrocolloids extracted from plants and produces viscous colloidal sols with texture similar to soft gels. It is soluble in cold water, stable to heat and acid pH (pH < 2) (Anderson and Bridgeman, 1985; Balaghi et al., 2010; Mohammadifar et al., 2006).

4.2.3.3 Mesquite Gum

Mesquite gum is a resin secreted by the mesquite tree (*Prosopis* spp.) under water and heat stress. It has a structure and properties very similar to gum arabic. It is highly soluble in water, particularly at neutral pH (between 6 and 7). Mesquite gum consists of high-molecular-weight proteoglycan (arabinogalactan protein), which is slightly acidic heterogeneous and polydisperse. It is composed mainly of L-arabinose, D-galactose followed by D-glucose, D-mannose, D-xylose, and glucuronic acid. It contains a protein fraction between 2% and 4.8% which confers emulsifying properties (Alftrén et al., 2012; Mirhosseini and Amid, 2012).

4.2.3.4 Karaya Gum

Karaya gum is a resin exuded by trees of the genus *Sterculia*, which is produced in northern and central India. This gum is an acetylated polysaccharide, mainly constituted by D-galacturonic acid, L-rhamnose, and D-galactose in addition to D-glucuronic acid. It has a low solubility in water, is very adhesive, and is currently used in low concentrations (ca. between 0.2% and 0.4%). When Karaya gum is prepared with cold water, it has a high viscosity; however, with increasing temperature, the water solubility decreases, as well as the viscosity, but if this gum is heated to boiling point, it loses these properties permanently. The viscosity is low with the presence of salts and extreme pH values. With an alkaline pH, it becomes a sticky paste (Khandelwal et al., 2012; Patil and Talele, 2014).

4.2.3.5 Ghatti Gum

Ghatti gum is a resin exuded from *Anogeissus latifolia* tree, which is produced mainly in India. It is a polysaccharide constituted by arabinose, galactose, mannose, xylose, and glucuronic acid. It is soluble in water but contains a water-insoluble fraction, forming a gel. Its solubility increases with increasing temperature and the best solubility is attained at a pH 6 (Deshmukh et al., 2012).

4.2.3.6 Vanilla Oleoresin

Vanilla is used in food as a natural flavor. This extract is obtained from the bean or pod of a tropical vanilla orchid (*Vanilla planifolia* Andrews [syn. *Vanilla fragrans* Salisb.]). Originally the Aztecs from Mexico cultivated vanilla and later in 1520, it was brought to Europe by the Spaniards (Walton et al., 2003). The Association of Official Analytical Chemists (AOAC) has two methods to quantify vanilla resin in vanilla extract: a qualitative (AOAC 960.36.192) and a quantitative (AOAC 926.09-1096) method. The vanilla oleoresin has phenolic compounds such as *p*-hydroxybenzoic, *p*-hydroxybenzaldehyde, and vanillic acid, all considered as antioxidants and nutraceutical.

4.2.3.7 Mastic Resin

The mastic resin, also known as mastic gum, is an exudate obtained after carving the trunk and branches of the tree *Pistacia lentiscus* L. var. *chia*, and this chia variety is exclusively grown in the southern region of the island of Chios in the Aegean Sea, Greece. The pharmaceutical properties of mastic resin crude and its essential oil have been reported by Assimopoulou et al. (2005) and include their antioxidant, anticancer, and antiulcer (gastric and duodenal) effect, and homeostatic, immunostimulant, and antimicrobial (against *Salmonella*, *Staphylococcus*, and *Helicobacter pylori*) activities. Those authors suggest the possible use in the formulation of cosmetic and food supplements. Mastic resin is a complex mix of terpenes (simple monoterpenes, oxygenated monoterpenes, and sesquiterpenes) and miscellaneous materials (Vourinen et al., 2015).

4.3 EDIBLE COATINGS AND FILMS BASED ON LIPIDS

4.3.1 Lipids Used in Edible Coatings and Films

The lipids that may be incorporated in the formulation of coatings and films are triglycerides, acetylated monoglycerides, fatty alcohols, fatty acid esters of sucrose, fatty acids (Martín-Polo et al., 1992b; Shellhammer and Krochta, 1997), and oils (vegetable, animal, and mineral) (Baldwin et al., 1995a,b).

The lipids most commonly used in edible coatings and films are fatty acids with a number of carbon atoms between 14 and 18, fatty alcohols, as well as stearyl alcohol, hydrogenated and nonhydrogenated vegetable oils (Milovmovic and Picuric-Jovanevic, 2001; Wong et al., 1992). Their regulatory status is different in the United States and Europe (Table 4.3).

An important topic included in the regulatory status is the presence of allergens. Many edible coatings and films are made with ingredients that could cause allergic reactions. Within these allergens, milk, soybeans, fish, peanuts, nuts, and wheat are the most relevant. Several edible coatings and films are formed from milk protein (i.e., whey and casein), wheat protein (i.e., gluten), soy protein, and peanut protein (Rojas-Grau et al., 2009). Therefore, the presence of a coating containing a known allergen on a food must also be clearly labeled.

TABLE 4.3
Lipid Materials Used in Edible Coatings and Films and Their Regulatory Status in the United States and Europe

Lipid	Uses/Regulatory Status (21 CRF)	Uses/European Union (Directive 95/2/EC)
Corn oil	Coating and emulsifying agent, texturizer (GRAS)	Included as vegetable oil
Lauric acid	Defoaming agent, lubricant (172.860)	Emulsifier, coating agent (E-570)
Mineral oil	Coating agent, confections (172.878)	Not permitted
Oleic acid	Emulsifier, binder, lubricant (172.862)	Emulsifier, coating agent (E-570)
Palm oil	Coating, emulsifier, texturizer (GRAS, 184.1585)	Included as vegetable oil
Esters glycerides of wood rosin	Coating component (175.300)	Not permitted

Note: 21 CRF: regulation numbers in title 21 of the U.S. Code of Federal Regulations are provided; GRAS: generally recognized as safe; Directive 95/2/EC: code numbers for food additives approved by the European Union (Directive 95/2/EC) are provided.

4.3.2 PROPERTIES OF LIPIDS IN EDIBLE COATINGS AND FILMS

Although a structural matrix should be built with components such as proteins and/or polysaccharides, once they provide the necessary flexibility, stability, and mechanical strength (Table 4.4) (Koelsch, 1994; Martin-Belloso et al., 2005). Lipids are capable to provide a moisture barrier to the coating and film, and when lipids are added to a protein or polysaccharide emulsions, they can improve the physical properties (Galus and Kadzińska, 2016).

TABLE 4.4
Properties of Lipids in Edible Coatings and Films

Properties	References
Reduce transpiration	Shellhammer and Krochta (1997), Figueroa et al. (2011)
Reduce dehydration	Shellhammer and Krochta (1997), Figueroa et al. (2011)
Reduce weight loss	Shellhammer and Krochta (1997), Figueroa et al. (2011)
Reduce abrasion	Figueroa et al. (2011)
Improve gloss of fruits and sugar confectionery products	Figueroa et al. (2011), Pérez-Gago et al. (2005)
Liposoluble additives support	Pérez-Gago et al. (2005)
Emulsifiers	Debeaufort and Voilley (1995)
Improve mechanical strength	Koelsch (1994)
Plasticizers	Koelsch (1994), Baldwin et al. (1995a,b)

4.3.3 Water Vapor Barrier

Lipids are hydrophobic compounds, so edible coatings and films made from these materials exhibit good barrier properties to water vapor due to their low polarity (Shellhammer and Krochta, 1997), as shown in Table 4.5.

Polarity of lipids depends on the distribution of chemical groups, length of aliphatic chains, and presence and degree of unsaturation (Morillon et al., 2002). Unsaturated fatty acids are less efficient to control moisture transfer because of their higher polarity (Hagenmaier and Baker, 1997) and significantly lower melting point in comparison with saturated fatty acids (Rhim and Shellhammer, 2005). Branching of acyl chain also results in increased water vapor permeability because of the increased mobility of hydrocarbon chains and less efficient lateral packing of acyl chains (Janjarasskul and Krochta, 2010).

Mono-, di-, and triglycerides (partial esters) can also be used as coating materials. Their functional properties, especially water vapor permeability, are dependent on their chemical structures and are used in the formulation of edible films and coatings as good emulsifiers, especially for stabilizing emulsified films and also for increasing adhesion between parts with different hydrophobicity, for example, between the film and the food or between the lipidic layer and the hydrocolloid layer in bilayer films (Debeaufort and Voilley, 1995). Long-chain triglycerides are insoluble in water, whereas short-chain molecules are partially water soluble (Bourlieu et al., 2008).

TABLE 4.5
Water Vapor Permeabilities of Edible Coatings and Films Based on Lipids

Film	Conditions	Water Vapor Permeability ($g\ mm\ m^{-2}\ d^{-1}\ kPa^{-1}$)	References
Fatty acids (myristic, palmitic, and stearic acids)	23°C, 12/56% RH	0.22–3.47	Martín-Polo et al. (1992a), Shellhammer and Krochta (1997), Milovmovic and Picuric-Jovanevic (2001)
Hydro peanut oil	25°C, 100/0% RH	3.3	Lovegren and Feuge (1954)
Acetylated monoglycerides	25°C, 100/0% RH	1.9–13	Lovegren and Feuge (1954)
Cocoa butter	25°C, 22/44% RH	3.6	Milovmovic and Picuric-Jovanevic (2001)
Peanut oil	25°C, 22/44% RH	13.8	Milovmovic and Picuric-Jovanevic (2001)
Hydrogenated cotton oil	27°C, 100/0% RH	0.13	Milovmovic and Picuric-Jovanevic (2001)
Tripalmitin	28°C, 100/0% RH	0.19	Shellhammer and Krochta (1997)

Note: RH = relative humidity.

Acetic acid esters of monoglycerides, called acetylated monoglycerides, have also been used as food coating materials. Their moisture barrier tends to improve with increasing degree of acetylation, possibly because of removal of hydrophilic (i.e., hydroxyl) groups (Bourlieu et al., 2008).

Lipids have also been added to edible films and coatings based on proteins and polysaccharides to improve the barrier properties to water vapor (Table 4.6).

Other lipids that have been used to improve the barrier properties to water vapor of edible coatings and films are oleic acid (Tagi et al., 2012), rapeseed oil (Kokoszka et al., 2010), lecithin, and polyglycerol polyricinoleate (Khan et al., 2013).

4.3.4 Plasticizer Properties

Lipids are compounds of low volatility and function as plasticizers, which are added to the coating (Kester and Fennema, 1986), considering two forces: one between the molecules of the film (cohesion) and other between the coating and the substrate (adhesion) (Christie, 1982; Guilbert, 1986). Lipids affect the mechanical and physical properties of the coating (Table 4.7) (elasticity, flexibility, wettability, permeability, strength, and shear strength) (Koelsch, 1994; Lazaridou and Biliaderis, 2002; Park and Chinnan, 1995) because they decrease intramolecular forces between polymer chains, thereby producing a decrease in the cohesive strength, tension, and glass transition temperature (Koelsch, 1994; Lazaridou and Biliaderis, 2002).

The main plasticizers are glycerol, fatty acids, and monoglycerides (Bósquez-Molina and Vernon-Carter, 2005; Fennema et al., 1994). Hydrophilic plasticizers such as glycerol are compatible with the polymeric material forming the film and increase the ability to absorb polar molecules such as water (Fennema et al., 1994). Currently, most coatings are added with glycerol (Raybaudi-Massilia et al., 2008; Rojas-Grau et al., 2007), using it as a plasticizer or simply to ensure greater barrier properties to water loss in the coating. The increased permeability with the content of plasticizer may be related to the hydrophilicity of the molecule of the plasticizer (Turhan and Sahbaz, 2004), because water vapor permeability increases with increasing plasticizer content; this increase is relatively modest for glycerol contents up to 30%, being more pronounced above that value (Bósquez-Molina and Vernon-Carter, 2005).

Proteins and polysaccharide-based coatings are very efficient barriers to O_2 and CO_2 and in some cases to water vapor (Koelsch, 1994); these barriers can be further improved by the addition of lipids (Table 4.7) as plasticizers (Koelsch, 1994).

Other lipids that have been used as plasticizers to improve the barrier properties of edible coatings and films are oleic acid (Tagi et al., 2012) and rapeseed oil (Kokoszka et al., 2010).

4.3.5 Applications of Edible Coatings and Films Based on Lipids

Moisture loss affects food quality and weight, thus causing economical losses during the marketing process (Avena-Bustillos et al., 1994a). Lipids are generally applied in thin layers or as composites with a polymeric matrix (Pérez-Gago et al., 2005). Lipids alone or in combination with other compounds are used as food edible coatings to

TABLE 4.6
Effect of Lipids in Water Vapor Permeability of Edible Coatings and Films Based on Proteins and Polysaccharides

Film	Conditions	WVP (g mm $m^{-2} d^{-1}$ kPa)	References
SA + PA/HPMC:PEG (37/9:1)	25°C, 85/0% RH	2.0	Kamper and Fennema (1984)
WG + Gly (2.4:1)	26°C, 50/100% RH	108	Aydt et al. (1991)
WG + Gly (5:1)	30°C, 100/0% RH	5.1	Gontard et al. (1992)
SPI + Gly (1.7:1)	25°C, 50/100% RH	154	Brandenburg et al. (1993)
CZ + PEG + Gly (5.9:1)	25°C, 50/100% RH	47	Butler and Vergano (1994)
CZ + PEG + Gly (2.6:1)	25°C, 50/100% RH	107	Butler and Vergano (1994)
WPI + Gly (4:1)	25°C, 0/77% RH	70	McHugh and Krochta (1994)
SPI + Gly (4:1)	28°C, 0/78% RH	39	Stuchell and Krochta (1994)
SC + Gly (2:1)	23°C, 55/72% RH	310	Banerjee and Chen (1995)
CC + Gly (2:1)	23°C, 55/77% RH	190	Banerjee and Chen (1995)
WPI + Gly (2:1)	23°C, 55/73% RH	291	Banerjee and Chen (1995)
WPC + Gly (2:1)	23°C, 55/74% RH	255	Banerjee and Chen (1995)
FMP + Gly (1.9:1)	20°C, 100/0% RH	6.1	Cuq et al. (1995)
CZ + Gly (4.9:1)	21°C, 85/0% RH	9.6	Park and Chinnan (1995)
EWP + Gly (3.3:1)	25°C, 50/72% RH	211	Gennadios et al. (1996)
EWP + Gly (2:1)	25°C, 50/70%–80% RH	256	Gennadios et al. (1996)
RC + Gly (1.4:1)	38°C, 0/90% RH	45	Chick and Ustunol (1998)
CZ + Glyc + MCTVO	25°C, 100/50% RH	8.28	Weller et al. (1998)
CZ + Glyc + SW + MCTVO	25°C, 100/50% RH	0.115–0.450	Weller et al. (1998)
CZ + Glyc + CW + MCTVO	25°C, 100/50% RH	0.115–0.195	Weller et al. (1998)
LAC + Gly (1.4:1)	38°C, 0/90% RH	55	Chick and Ustunol (1998)
SC + Gly	25°C, 50/70%–80% RH	4.3	Siew et al. (1999)
PPC + Gly (0.6:1)	38°C, 0/50% RH	37	Jangchud and Chinnan (1999)
AP + FA + FAL + BW + VO	ND	69–325	McHugh and Senesi (2000)
HPMC + SA	27°C, 0/97% RH	0.12	Milovmovic and Picuric-Jovanevic (2001)
MC + PEG + MA	223°C, 12/56% RH	3.5	Milovmovic and Picuric-Jovanevic (2001)
HPMC + PEG + AM	21°C, 0/85% RH	8.2	Milovmovic and Picuric-Jovanevic (2001)
AX + PA + Gly (16:90:10)	25°C, 22/100% RH	1.52	Peróval et al. (2002)
AX + SA + Gly (16:90:10)	25°C, 22/100% RH	1.19	Peróval et al. (2002)
AX + T + Gly (16:90:10)	25°C, 22/100% RH	1.18	Peróval et al. (2002)
AX + OK35 + Gly (16:90:10)	25°C, 22/100% RH	1.24	Peróval et al. (2002)
AX + Gly (16:15)	25°C, 22/84% RH	13.82	Phan The et al. (2002)
AX + HPKO + Gly (16:25:15)	25°C, 22/84% RH	9.31	Phan The et al. (2002)
MG + CW + Gly	20°C, 70/80% RH	8.50	Bósquez-Molina and Vernon-Carter (2005)

(Continued)

TABLE 4.6 (Continued)
Effect of Lipids in Water Vapor Permeability of Edible Coatings and Films Based on Proteins and Polysaccharides

Film	Conditions	WVP (g mm $m^{-2} d^{-1}$ kPa)	References
MG + CW + Gly + Ca	20°C, 70/80% RH	8.63	Bósquez-Molina and Vernon-Carter (2005)
MG + CWMO + Gly	20°C, 70/80% RH	6.89	Bósquez-Molina and Vernon-Carter (2005)
MG + CWMO + Gly + Ca	20°C, 70/80% RH	6.86	Bósquez-Molina and Vernon-Carter (2005)
WPI + SA + Gly (10:40:4)	105°C, 50/100% RH	2.8	Fernández et al. (2007)
SH/AG + Gly (1/6:1)	25°C, 22/99% RH	0.89	Phan The et al. (2008)
SH + PEG/AG + Gly (7:1/6:1)	25°C, 22/99% RH	1.83	Phan The et al. (2008)
SH/CAS + Gly (1/6:1)	25°C, 22/99% RH	ND	Phan The et al. (2008)
SH + PEG/CAS + Gly (7:1/6:1)	25°C, 22/99% RH	2.51	Phan The et al. (2008)
CH96 + Gly (1:20)	250°C, 100/0% RH	1.1	Ziani et al. (2008)
CH96 + Gly + T20 (1:20:5)	250°C, 100/0% RH	1.27	Ziani et al. (2008)
CH60.9 + Gly (1:20)	250°C, 100/0% RH	1	Ziani et al. (2008)
CH60.9 + Gly + T20 (1:20:5)	250°C, 100/0% RH	1.3	Ziani et al. (2008)
BW + SA + (HPMC-Gly) (1:0.2:(2:1))	23°C, 40/100% RH	4.8	Navarro-Tarazaga et al. (2008)
X + WMO + P20 + Gly (5:5:1:2)	23°C, 52/100% RH	1.5–2.4	Kim et al. (2012)
G + OO + Gly (5:5:0.4)	25°C, 100/0% RH	4.986	Ma et al. (2012)
G + OO + Gly (5:10:0.4)	25°C, 100/0% RH	5.372	Ma et al. (2012)
G + OO + Gly (5:15:0.4)	25°C, 100/0% RH	5.161	Ma et al. (2012)
G + OO + Gly (5:20:0.4)	25°C, 100/0% RH	4.194	Ma et al. (2012)
AS + Gly (7.5:20)	23°C, 85/0% RH	1.5	Jost et al. (2014)
AS + Gly (7.5:25)	23°C, 85/0% RH	1.9	Jost et al. (2014)
AS + Gly (7.5:30)	23°C, 85/0% RH	2.4	Jost et al. (2014)

Note: WVP = water vapor permeability; RH = relative humidity; MC = methylcellulose; HPMC = hydroxypropylmethylcellulose; PEG = polyethyleneglycol; AM = acetylated monoglycerides; SC = sodium caseinate; C = calcium caseinate; RC = rennet casein; LAC = lactic acid casein; WPI = whey protein isolate; WPC = whey protein concentrate; FMP = fish myofibrillar protein; EWP = egg white protein; SPI = soy protein isolate; PPC = peanut protein concéntrate; CZ = corn zein; WG = wheat gluten; SA = stearic acid; MA = myristic acid; AX = arabinoxylans; HPKO = hydrogenated palm kernel oil; PA = palmitic acid; SA = stearic acid; T = triolein; OK35 = hydrogenated pal oil; BW = beeswax; X = xanthan; WMO = white mustard oil; P20 = polysorbate 20; G = gelatin; OO = olive oil; Sor = sorbitol; SW = sunflower oil; FA = fatty acids; FAL = fatty alcohols; VO = vegetable oil; AP = apple puree; CH96 = chitosan (96% of deacetylation); CH60.9 = chitosan (60.9% of deacetylation); T20 = tween 20; CZ = corn zein; MCTVO = medium-chain triglyceride vegetable oil; CW = carnauba wax; SW = sorghum wax; Glyc = glycerin; AS = sodium alginate; MG, mesquite gum; CW = candelilla wax; CWMO = candelilla wax + mineral oil; Ca = calcium; ND = not determined; SH = shellac; AG = agar; CAS = cassava starch.

TABLE 4.7
Effect of Lipids as Plasticizers on the Properties of Edible Coatings and Films

Film Composition	Plasticizer	TS (MPa)	EM (MPa)	E (%)	References
WPI + Gly (2.3:1)	Gly	14	490	31	McHugh and Krochta (1994)
WPI + Gly (5.7:1)	Gly	29	1100	5	McHugh and Krochta (1994)
EWP + Gly (3.3:1)	Gly	4	ND	12	Gennadios et al. (1996)
EWP + Gly (2:1)	Gly	1	ND	32	Gennadios et al. (1996)
CZ + Glyc + MCTVO	Gly	1.54	ND	161.8	Weller et al. (1998)
AX + AP + Gly (16:90:10)	AP	7.8	59.2	1.7	Peróval et al. (2002)
AX + AS + Gly (16:90:10)	AS	7.1	51.96	2.5	Peróval et al. (2002)
AX + T + Gly (16:90:10)	T	8.8	25.84	10.8	Peróval et al. (2002)
AX + OK35 + Gly (16:90:10)	OK35	6.4	26.65	8.9	Peróval et al. (2002)
WPI + SA + Gly (10:30:4)	Gly	7.5	ND	10	Fernández et al. (2007)
BW + SA + (HPMC-Gly) (1:0.2:(2:1))	Gly	4.9	450	9.3	Navarro-Tarazaga et al. (2008)
CH96 + Gly (1:20)	Gly	47.1	ND	67.3	Ziani et al. (2008)
CH96 + Gly + T20 (1:20:5)	Gly	43.6	ND	52.5	Ziani et al. (2008)
CH60.9 + Gly (1:20)	Gly	46.2	ND	59	Ziani et al. (2008)
CH60.9 + Gly + T20 (1:20:5)	Gly	44.5	ND	56.7	Ziani et al. (2008)
X + AMB + P20 + Gly (5:5:1:2)	Gly	5.1–8.8	ND	2.9–5	Kim et al. (2012)
G + OO + Gly (5:5:0.4)	Gly	8.28	124.04	73.19	Ma et al. (2012)
G + OO + Gly (5:10:0.4)	Gly	6.72	102.15	63.96	Ma et al. (2012)
G + OO + Gly (5:15:0.4)	Gly	7.14	92.07	78.05	Ma et al. (2012)
G + OO + Gly (5:20:0.4)	Gly	10.89	146.73	84.91	Ma et al. (2012)
ACS + Gly (90:10)	Gly	31.6	1055.0	4.3	Muscat et al. (2013)
CS + CW + SA + Gly (3:0.2:0.8:1.5)	Gly	0.729	0.221	31.074	Chiumarelli and Hubinger (2014)
AS + Gly (7.5:20)	Gly	71	ND	7.3	Jost et al. (2014)
AS + Gly = 7.5:25	Gly	63.5	ND	7.8	Jost et al. (2014)
AS + Gly = 7.5:30	Gly	15.6	ND	29.1	Jost et al. (2014)
AS + Gly = 7.5:40	Gly	5.3	ND	34.9	Jost et al. (2014)

Note: WPI = whey protein isolate; EWP = egg white protein; SC = sodium caseinate; Gly = glycerol; ACS = amylose corn starch; SA = stearic acid; HPKO = hydrogenated palm kernel oil; PA = palmitic acid; HPMC = hydroxypropylmethylcellulose; T = triolein; OK35 = hydrogenated pal oil; BW = beeswax; X = xanthan; WMO = white mustard oil; P20 = polysorbate 20; G = gelatin; OO = olive oil; CH96 = chitosan (96% of deacetylation); CH60.9 = chitosan (60.9% of deacetylation); T20 = tween 20; CZ = corn zein; MCTVO = medium-chain triglyceride vegetable oil; CW = carnauba wax; SW = sorghum wax; Glyc = glycerin; AS = sodium alginate; ND = not determined; TS = tensile strength; EM = elastic modulus; E = elongation.

prevent the main mechanism of moisture loss, which is the diffusion of water vapor due to a pressure gradient between the inside and outside of food (Maftoonazad and Ramaswamy, 2005).

Several studies have demonstrated advantages of lipid coatings when applied to food products (Table 4.8), by preserving their quality and improving shelf life.

4.4 EDIBLE COATINGS AND FILMS BASED ON NATURAL WAXES

Several waxes were used to manufacture edible coatings and films such as grain sorghum wax (Weller et al., 1998), resin wax (Meighani et al., 2014), paraffin wax (Bisen et al., 2012; Sugimoto et al., 2012), mineral, microcrystalline, oxidized, or nonoxidized polyethylene wax (Debeaufort and Voilley, 2009). But their limited availability and high cost limits their use in the food industry. Among the waxes commercially available, beeswax, candelilla wax, and carnauba wax are the most used to manufacture edible coatings and films, and it may be due to their high availability in international markets (Njombolwana et al., 2013).

In the last years, these three waxes have been successfully used in the preservation of various food products such as pears (Cruz et al., 2015), apples (Chiumarelli and Hubinger, 2012; Kowalczyk and Baraniak, 2014; Ochoa et al., 2011; Saucedo-Pompa et al., 2007), strawberries (Velickova et al., 2013), avocado (Saucedo-Pompa et al., 2009), oranges (Hagenmaier, 2000), mandarins (Porat et al., 2005), cherry tomatoes (Fagundes et al., 2014), plums (Navarro-Tarazaga et al., 2011), mangoes (Baldwin et al., 1999), and green bell peppers (Ochoa-Reyes et al., 2013). In all reported cases, the incorporation of waxes extended the shelf life of fruits and vegetables (Table 4.9).

4.5 CONCLUDING REMARKS

The human need to produce, commercialize, offer, and buy safer food involves the work of different knowledge areas and researchers, since the good agricultural practices from the field to the use of new and current technologies such as modified atmospheres, cold storage, application of nonthermal technologies, like electric field and ultrasound.

In recent years, researchers have been focused on the application of edible coatings and films to prolong the shelf life and food preservation of diverse highly perishable foods mainly fruits and vegetables. Edible coatings and films are made of natural and safe products, especially from food wastes and subproducts such as pectin, gums, resins, waxes, and lipids. They are excellent vehicles for adding natural antioxidants, antimicrobials, vitamins, and/or other functional bioactives to minimize the negative effects of the maturity process and the spoilage caused by the microbial invasion and improve quality and nutritional value of the foods.

The application of edible coatings and films has demonstrated to be an excellent technology to prolong shelf life of fruits and vegetables with natural compounds and natural alternative antimicrobials and antioxidants helping to hold the organoleptic quality of these fruits and vegetables.

TABLE 4.8
Applications and Functions of Lipid-Based Edible Coatings and Films

Lipid	Application	Function	References
Sodium caseinate + stearic acid or acetylated monoglyceride	Carrots	Improve firmness and appearance	Avena-Bustillos et al. (1994b)
Polysaccharide/lipid bilayer	Apple	O_2/CO_2 barrier, gloss	Wong et al. (1994)
Whey proteins + acetylated monoglycerides + lactic acid	Smoked salmon	Antioxidant barrier, H_2O barrier, antimicrobial barrier	Sensidoni and Peressini (1997)
Sunflower oil + corn starch + glycerol	Carrots	H_2O barrier	García et al. (2000)
Corn starch + methylcellulose + cocoa butter or soybean oil	Bakery food	H_2O barrier	Bravin et al. (2002)
Lipid based	Green bell pepper	$O_2/CO_2/H_2O$ barrier	Conforti and Ball (2002)
Hydroxypropylmethylcellulose + lipid composite	Plum	$O_2/CO_2/H_2O$ barrier	Pérez-Gago et al. (2003a)
Hydroxypropylmethylcellulose + stearic acid + glycerol	"Fortune" mandarins	H_2O barrier	Pérez-Gago et al. (2003b)
Arabinoxylans + linseed oil	Stuffed biscuits	H_2O barrier	Peróval et al. (2003)
Oleic acid + carnauba wax + shellac + morpholine + ammoniac + polydimethylsiloxane antifoam	Apples	$O_2/CO_2/H_2O$ barrier improve firmness	Bai et al. (2003)
Ethanol + ammonium hydroxide + stearic or palmitic acid + glycerol + beeswax	Organic strawberries	Improved firmness and appearance H_2O barrier	Tanada-Palmu and Grosso (2005)
Mineral oil + candelilla wax + mesquite gum + sorbitol + calcium	Persian limes	H_2O barrier	Bósquez-Molina and Vernon-Carter (2005)
Corn starch + methylcellulose + glycerol + soybean oil	Crackers	H_2O barrier	Bravin et al. (2006)
Apple puree + fatty acids + fatty alcohols + vegetable oil	Apples minimally processed	H_2O barrier antioxidant barrier antimicrobial barrier	McHugh and Senesi (2000), Rojas-Grau et al. (2007)

(Continued)

TABLE 4.8 (Continued)
Applications and Functions of Lipid-Based Edible Coatings and Films

Lipid	Application	Function	References
Beeswax + stearic acid + hydroxypropylmethylcellulose + glycerol	Mandarins (Ortanique)	$O_2/CO_2/H_2O$ barrier	Navarro-Tarazaga et al. (2008)
High Amylose + glycerol	Roasted hazelnuts	Antioxidant barrier	Travaglia et al. (2009)
Cottonseed and safflower oil + paraffin wax	Green bell pepper	H_2O barrier	Beaulieu et al. (2009)
Galactomannans + collagen	Mangoes and apples	$O_2/CO_2/H_2O$ barrier	Lima et al. (2010)
Corn zein + oleic acid	Unshelled macadamia nuts	$O_2/CO_2/H_2O$ barrier, antioxidant barrier	Colzato et al. (2011)
Beeswax + stearic acid + hydroxypropylmethylcellulose + glycerol	"Angeleno" plums	H_2O barrier	Navarro-Tarazaga et al. (2011)
White mustard oil + polysorbate 20 + xanthan gum + glycerol	Smoked salmon	Antioxidant barrier, H_2O barrier	Kim et al. (2012)
Stearic acid + glycerol + carnauba wax + cassava starch	Apples minimally processed	$O_2/CO_2/H_2O$ barrier	Chiumarelli and Hubinger (2014)
Coconut oil + glycerol + starch + antioxidants of green tea	Tomatoes	H_2O barrier, improve firmness, antimicrobial barrier, antioxidant retention	Das et al. (2013)
Locust bean gum + oleic acid + morpholine + carnauba wax	Red bell peppers	Gloss	Marmur et al. (2013)
Maize zein + gallic acid or catequin + carnauba wax + soybean L-α-lecithin + glycerol + lysozyme	Fresh Kashar cheese	Antimicrobial barrier, antioxidant barrier, antioxidant retention	Unalan et al. (2013)

Note: O_2 = oxygen; CO_2 = carbon dioxide; H_2O = water.

TABLE 4.9
Applications and Functions of Wax-Based Edible Coatings and Films

Wax	Application	Function	References
Candelilla wax + polimeric resin + jojoba oil + gallic acid	Apples	H_2O barrier, antimicrobial	Ochoa et al. (2011)
Candelilla wax + pectin + glycerol + ellagic acid	Avocados	Antioxidant barrier, H_2O barrier antimicrobial barrier	Saucedo-Pompa et al. (2009)
Mineral oil + candelilla wax + mesquite gum + sorbitol + calcium	Persian limes	H_2O barrier	Bósquez-Molina and Vernon-Carter (2005)
Beeswax + stearic acid + hydroxypropylmethylcellulose + glycerol	Mandarins (ortanique)	$O_2/CO_2/H_2O$ barrier	Navarro-Tarazaga et al. (2008)
High Amylose + glycerol	Roasted hazelnuts	Antioxidant barrier	Travaglia et al. (2009)
Cottonseed and safflower oil + paraffin wax	Green bell pepper	H_2O barrier	Beaulieu et al. (2009)
Beeswax + stearic acid + hydroxypropylmethylcellulose + glycerol	"Angeleno" plums	H_2O barrier	Navarro-Tarazaga et al. (2011)
Stearic acid + glycerol + carnauba wax + cassava starch	Apples minimally processed	$O_2/CO_2/H_2O$ barrier	Chiumarelli and Hubinger (2014)
Locust bean gum + oleic acid + morpholine + carnauba wax	Red bell peppers	Gloss	Marmur et al. (2013)
Maize zein + gallic acid or catequin + carnauba wax + soybean L-α-lecithin + glycerol + lysozyme	Fresh Kashar cheese	Antimicrobial barrier, antioxidant barrier, antioxidant retention	Unalan et al. (2013)

Note: O_2 = oxygen; CO_2 = carbon dioxide; H_2O = water.

REFERENCES

Alftrén, J., Peñarrieta, J. M., Bergenståhl, B., and Nilsson, L. 2012. Comparison of molecular and emulsifying properties of gum arabic and mesquite gum using asymmetrical flow field-flow fractionation. *Food Hydrocolloids*, 26(1), 54–62.

Altenhofen, M., Krause, A. C., and Guenter, T. 2009. Alginate and pectin composite films crosslinked with Ca^{2+} ions: Effect of the plasticizer concentration. *Carbohydrate Polymers*, 77(1), 736–742.

Anderson, D. M. W., and Bridgeman, M. M. E. 1985. The composition of the proteinaceous polysaccharides exuded by *Astragalus microcephalus*, *Astragalus gummifer* and *Astragalus kurdicus*—The sources of Turkish Gum Tragacanth. *Phytochemistry*, 24(10), 2301–2304.

Antoniou, J., Liu, F., Majeed, H., Qazi, H. J., and Zhong, F. 2014. Physicochemical and thermomechanical characterization of tara gum edible films: Effect of polyols as plasticizers. *Carbohydrate Polymers*, 111, 359–365.

AOAC 926.09. 1990. Vanilla resins in vanilla extract. Paper chromatographic qualitative test. Official Methods of Analysis of the Association of Official Analytical Chemists, 15th Edition, 1990. Agricultural Chemicals; Contaminants; Drugs, Volume One.

Arancibia, M. Y., Alemán, A., López-Caballero, M. E., Gómez-Guillén, M. C., and Montero, P. 2015. Development of active films of chitosan isolated by mild extraction with added protein concentrate from shrimp waste. *Food Hydrocolloids*, 43(1), 91–99.

Arato Garza, M. J., Speelman, S., and Van Huylenbroeck, G. 2013. Integración de la inversión privada en el desarrollo de la cadena de valor de los productos forestales no-maderables Mexicanos: Oportunidad para el desarrollo rural sostenible de las comunidades del Desierto de Chihuahua. In Comercio agrícola y América Latina: Cuestiones, controversias y perspectivas. Buenos Aires, Argentina: FLACSO Argentina.

Assimopoulou, A. N., Sinakos, Z., and Papageorgiou, V. P. 2005. Radical scavenging activity of *Crocus sativus* L. extract and its bioactive constituents. *Phytotherapy Research*, 19, 997–1000.

Atef, M., Rezaei, M., and Behrooz, R. 2015. Characterization of physical, mechanical, and antibacterial properties of agar-cellulose bionanocomposite films incorporated with savory essential oil. *Food Hydrocolloids*, 45, 150–157.

Avena-Bustillos, R. J., Krochta, J. M., Salveit, M. E., Rojas-Villegas, R. J., and Sauceda-Perez, J.A. 1994a. Optimization of edible coating formulations on zucchini to reduce water loss. *Journal Food Engineering*, 21(1), 197–214.

Avena-Bustillos, R. J., Cisneros-Zevallos, L. A., Krochta, J. M., and Saltveit, J. M. E. 1994b. Application of casein-lipid edible film emulsions to reduce white blush on minimally processed carrots. *Postharvest Biology and Technology*, 4(4), 319–329.

Aydt, T. P., Weller, C. L., and Testin, R. F. 1991. Mechanical and barrier properties of edible corn and wheat protein films, *Transactions of the ASAE*, 34(1), 207–211.

Bai, J., Hagenmaier, R. D., and Baldwin, E. A. 2003. Coating selection for "delicious" and other apples. *Postharvest Biology and Technology*, 28(1), 381–390.

Balaghi, S., Mohammadifar, M. A., and Zargaraan, A. 2010. Physicochemical and rheological characterization of gum tragacanth exudates from six species of Iranian *Astragalus*. *Food Biophysics*, 5(1), 59–71.

Baldwin, E. A., Burns, J. K., Kazokas, W., Brecht, J. K., Hagenmaier, R. D., Bender, R. J., and Pesis, E. 1999. Effect of two edible coatings with different permeability characteristics on mango (*Mangifera indica* L.) ripening during storage. *Postharvest Biology and Technology*, 17(1), 215–226.

Baldwin, E. A., Nisperos, M., and Baker, R. 1995. Use of edible coatings to preserve quality of lightly (and slightly) processed products. *Critical Reviews in Food Science and Nutrition*, 35(1), 509–524.

Baldwin, E. A., Nisperos-Carriedo, M. O., and Baker, R. A. 1995. Edible coatings for lightly processed fruits and vegetables. *Horticultural Science*, *30*(1), 35–38.

Banerjee, R., and Chen, H. 1995. Functional properties of edible films using whey protein concentrate. *Journal of Dairy Science*, *78*(1), 1673–1683.

Banker, G. S. 1966. Film coating theory and practice. *Journal of Pharmaceutical Science*, *55*(1), 81–89.

Barman, K., Asrey, R., and Pal, R. K. 2011. Putrescine and carnauba wax pretreatments alleviate chilling injury, enhance shelf life and preserve pomegranate fruit quality during cold storage. *Scientia Horticulturae*, *130*(4), 795–800.

Barman, K., Asrey, R., Pal, R. K., Kaur, C., and Jha, S. K. 2014. Influence of putrescine and carnauba wax on functional and sensory quality of pomegranate (*Punica granatum* L.) fruits during storage. *Journal of Food Science and Technology*, *51*(1), 111–117.

Beaulieu, J. C., Park, H. S., Mims, A. G. B., and Kuk, M. S. 2009. Extension of green bell pepper shelf life using oilseed-derived lipid films from soapstock. *Industrial Crops and Products*, *30*(2), 271–275.

Bell, J. G., Henderson, R. J., Tocher, D. R., McGhee, F., Dick, J. R., Porter, A., and Sargent, J. R. 2002. Substituting fish oil with crude palm oil in the diet of Atlantic salmon (*Salmo salar*) affects muscle fatty acid composition and hepatic fatty acid metabolism. *The Journal of Nutrition*, *132*(2), 222–230.

Beristain, C. I., García, H. S., and Vernon-Carter, E. J. 1999. Mesquite gum (*Prosopis juliflora*) and maltodextrin blends as wall materials for spay-dried encapsulated orange peel oil. *Food Science and Technology International*, *5*, 353–356.

Bisen, A., Pandey, S., and Patel, N. 2012. Effect of skin coatings on prolonging shelf life of kagzi lime fruits (*Citrus aurantifolia* Swingle). *Journal of Food Science and Technology*, *49*(6), 753–759.

Bosquez-Molina, E., Guerrero-Legarreta, I., and Vernon-Carter, E. J. 2003. Moisture barrier properties and morphology of mesquite gum–candelilla wax based edible emulsion coatings. *Food Research International*, *36*, 885–893.

Bosques-Molina, E., Tomás, S. A., and Rodríguez-Huezo, M. E. 2010. Influence of $CaCl_2$ on the water vapor permeability and the surface morphology of mesquite gum based edible films. *LWT—Food Science and Technology*, *43*(23), 1419–1425.

Bósquez-Molina, E., and Vernon-Carter, E. J. 2005. Efecto de plastificantes y calcio en la permeabilidad al vapor de agua de películas a base de goma de mezquite y cera de candelilla. *Revista Mexicana de Ingeniería Química*, *4*(1), 157–162.

Bourlieu, C., Guillard, V., Vallès-Pamiès, B., and Gontard, N. 2008. Edible moisture barriers for food product stabilization. In J. M. Aguilera, & P. J. Lillford (Eds.), *Food Materials Science* (pp. 547–575). New York: Springer.

Bourlieu, C., Guillard, V., Vallès-Pamiès, B., Guilbert, S., and Gontard, N. 2009. Edible moisture barriers: How to assess of their potential and limits in food products shelf-life extension? *Critical Reviews in Food Science and Nutrition*, *49*(5), 474–499.

Brandenburg, A. H., Weller, C. L., and Testin, R. F. 1993. Edible films and coatings from soy protein. *Journal of Food Science*, *58*(1), 1086–1089.

Bravin, B., Peressini, D., and Sensidoni, A. 2002. Development and application of polysaccharide-lipid based edible film. *Industrie Alimentari*, *41*(419), 1177–1185.

Bravin, B., Peressini, D., and Sansidoni, A. 2006. Development and application of polysaccharide–lipid edible coating to extend shelf-life of dry bakery products. *Journal of Food Engineering*, *76*(1), 280–290.

Butler, B. L., and Vergano, P. J. 1994. Degradation of edible films in storage. ASAE Paper No. 946551, St. Joseph, MI: American Society of agricultural Engineers.

Cabello, J. C., Sáenz-Galindo, A., Barajas, L., Pérez, C., Ávila, C., and Valdés, J. A. 2013. Cera de Candelilla y sus aplicaciones [Candelilla wax and their applications]. *Avances en Química*, *8*(2), 105–110.

Callegarin, F., Quezada Gallo, J-A., Debeaufort, F., and Voiley, A. 1997. Lipids and biopackaging. *Journal of the American Oil Chemists' Society*, 74(10), 1183–1192.

Campos, C., Gerschenson, L., and Flores, S. 2011. Development of edible films and coatings with antimicrobial activity. *Food Bioprocess Technology*, 4(1), 849–875.

Chen, C.-H., Kuo, W.-S., and Lai, L.-S. 2010. Water barrier and physical properties of starch/decolorized hsian-tsao leaf gum films: Impact of surfactant lamination. *Food Hydrocolloids*, 24(2–3), 200–207.

Chen, S., and Nussinovitch, A. 2000. Galactomannans in disturbances of structured wax–hydrocolloid-based coatings of citrus fruit (easy-peelers). *Food Hydrocolloids*, 14(6), 561–568.

Chick, J., and Ustunol, Z. 1998. Mechanical and barrier properties of lactic acid and rennet precipitated casein-based edible films. *Journal of Food Science*, 63(1), 1024–1027.

Chiumarelli, M., and Hubinger, M. D. 2012. Stability, solubility, mechanical and barrier properties of cassava starch—Carnauba wax edible coatings to preserve fresh-cut apples. *Food Hydrocolloids*, 28(1), 59–67.

Chiumarelli, M., and Hubinger, M. D. 2014. Evaluation of edible films and coatings formulated with cassava starch, glycerol, carnauba wax and stearic acid. *Food Hydrocolloids*, 38, 20–27.

Chranioti, C., and Tzia, C. 2014. Arabic gum mixtures as encapsulating agents of freeze-dried fennel oleoresin products. *Food and Bioprocess Technology*, 7(4), 1057–1065.

Christie, W. W. 1982. *Lipid Analysis-Isolation, Separation, Identification and Structural Analysis of Lipids* (2nd ed.). Pergamon Press: New York.

Cindric, I. J., Zeiner, M., and Steffan, I. 2007. Trace elemental characterization of edible oils by ICP–AES and GFAAS. *Microchemical Journal*, 85(1), 136–139.

Colzato, M., Scramin, J. A., Forato, L. A., Colnago, L. A., and Assis, O. B. G. 2011. 1H NMR investigation of oil oxidation in macadamia nuts coated with zein-based films. *Journal of Food Processing and Preservation*, 35(6), 790–796.

Conforti F. D., and Ball J. A. 2002. A comparison of lipid and lipid/hydrocolloid based coatings to evaluate their effect on postharvest quality of green bell peppers. *Journal Food Quality*, 25(1), 107–116.

Cruz, V., Rojas, R., Saucedo-Pompa, S., Martínez, D. G., Aguilera-Carbó, A. F., Alvarez, O. B., Rodríguez, R., Ruiz, J., and Aguilar, C. N. 2015. Improvement of shelf life and sensory quality of pears using a specialized edible coating. *Journal of Chemistry*, 1, 1–6.

Cuq, B., Aymard, C., Cuq, J.-L., and Guilbert, S. 1995. Edible packaging films based on fish myofibrillar proteins: Formulation and functional properties. *Journal of Food Science*, 60(1), 1369–1374.

Dantas, A. N. D. S., Magalhães, T. A., Matos, W. O., Gouveia, S. T., and Lopes, G. S. 2013. Characterization of carnauba wax inorganic content. *Journal of the American Oil Chemists' Society*, 90(10), 1475–1483.

Das, D. K, Dutta, H., and Lata-Mahanta, C. 2013. Development of a rice starch-based coating with antioxidant and microbe-barrier properties and study of its effect on tomatoes stored at room temperature. *LWT—Food Science and Technology*, 50(1), 272–278.

De León-Zapata, M. A., Sáenz-Galindo, A., Rojas-Molina, R., Rodríguez-Herrera, R., Jasso-Cantú, D., and Aguilar, C. N. 2015. Edible candelilla wax coating with fermented extract of tarbush improves the shelf life and quality of apples. *Food Packaging and Shelf Life*, 3, 70–75.

Debeaufort, E, & Voilley, A. 1995. Effect of surfactants and drying rate on barrier properties of emulsified edible film. *International Journal Food Science and Technology*, 30(1), 183–190.

Debeaufort, F., and Voilley, A. 2009. Lipid-based edible films and coatings. In K. C. Huber, & M. E. Embuscado (Eds.), *Edible Films and Coatings for Food Applications* (pp. 135–168). New York: Springer.

Deshmukh, A. S., Setty, C. M., Badiger, A. M., and Muralikrishna, K. S. 2012. Gum ghatti: A promising polysaccharide for pharmaceutical applications. *Carbohydrate Polymers*, 87(2), 980–986.

Desplanques, S., Renou, F., Grisel, M., and Malhiac, C. 2012. Impact of chemical composition of xanthan and acacia gums on the emulsification and stability of oil-in-water emulsions. *Food Hydrocolloids*, 27(2), 401–410.

El-Boulifi, N., Sánchez, M., Martínez, M., and Aracil, J. 2015. Fatty acid alkyl esters and monounsaturated alcohols production from jojoba oil using short-chain alcohols for biorefinery concepts. *Industrial Crops and Products*, 69, 244–250.

Fagundes, C., Palou, L., Monteiro, A. R., and Pérez-Gago, M. B. 2014. Effect of antifungal hydroxypropyl methylcellulose-beeswax edible coatings on gray mold development and quality attributes of cold-stored cherry tomato fruit. *Postharvest Biology and Technology*, 92, 1–8.

Falguera, V., Quintero, J. P., Jímenez, A., Aldemar M. J., and Ibarz, A. 2011. Edible films and coatings: Structures, active functions and trends in their use. *Trends in Food Science & Technology*, 22, 292–303.

FAO. 1995. Food and Agriculture Administration. Available at http://www.fao.org/docrep/v9236e/v9236e00.htm. Accessed on January, 2015.

Fennema, O., Donhowe, I. G., and Kester, J. J. 1994. Lipid type and location of the relative humidity gradient influence on the barrier properties of lipids to water vapor. *Journal of Food Engineering*, 22(1), 225–239.

Fernández, L., Díaz de Apodaca, E., Cebrián, M., Villarán, M. C., and Maté, J. I. 2007. Effect of the unsaturation degree and concentration of fatty acids on the properties of WPI-based edible films. *European Food Research Technology*, 224(1), 415–420.

Figueroa, J., Salcedo, J., Aguas, Y., Olivero, R., and Narváez, G. 2011. Recubrimientos comestibles en la conservación del mango y aguacate, y perspectiva, al uso del propóleo en su formulación [Edible coatings conservation of mango and avocado, and perspective, the use of propolis in its formulation]. *Revista Colombiana Ciencia Animal*, 3(2), 386–400.

Fitzherbert, E. B., Struebig, M. J., Morel, A., Danielsen, F., Brühl, C. A., Donald, P. F., and Phalan, B. 2008. How will oil palm expansion affect biodiversity? *Trends in Ecology & Evolution*, 23(10), 538–545.

Galus, S., and Kadzińska, J. 2016. Whey protein edible films modified with almond and walnut oils. *Food Hydrocolloids*, 52, 78–86.

García, E. A., Martino, M. N., and Zaritzky, N. E. 2000. Lipid addition to improve barrier properties of edible starch-based films and coatings. *Journal of Food Science*, 65(6), 941–947.

Gennadios, A., Weller, C. L., Hanna, M. A., and Froning, G. W. 1996. Mechanical and barrier properties of egg albumen films. *Journal of Food Science*, 61, 585–589.

Gontard, N., Guilbert, S., and Cuq, J. L. 1992. Edible wheat gluten films: Influence of the main process variables on film properties using response surface methodology. *Journal of Food Science*, 57(199), 190–195.

Guilbert, S. 1986. Technology and application of edible protective films. In M. Mathlouthi (Ed.), *Food Packaging and Preservation. Theory and Practice* (pp. 371–394). New York: Elsevier Applied Science Publishers.

Guilbert, S., Gontard, N., and Gorris, L. G. M. 1996. Prolongation of the shelf-life of perishable food products using biodegradable films and coatings. *LWT—Food Science and Technology*, 29, 10–17.

Guillen, M. D., and Cabo, N. 1997. Characterization of edible oils and lard by Fourier transform infrared spectroscopy. Relationships between composition and frequency of concrete bands in the fingerprint region. *Journal of the American Oil Chemists' Society*, 74(10), 1281–1286.

Gurr, M. I., and James, A. T. 1971. *Lipid Biochemistry and Introduction*. Ithaca, New York: Cornell University Press.

Gutiérrez, R. H., Lares, M., Pérez, E. E., and Álvarez, C. 2014. Effect of roasting on the fatty acid profile of cocoa butter that was extracted by two methods, from Barlovento, Venezuela. *Acta Horticulturae*, *1016*, 119–124.

Hagenmaier, R. D. 2000. Evaluation of a polyethylene–candelilla coating for "Valencia" oranges. *Postharvest Biology and Technology*, *19*, 147–154.

Hagenmaier, R. D., and Baker, R. A. 1997. Edible coatings from morpholine-free wax microemulsions. *Journal of Agricultural and Food Chemistry*, *45*, 349–352.

Hamid, M. A., Sarmidi, M. R., Mokhtar, T. H., Sulaiman, W. R. W., Sulaiman, R. A., and Aziz, R. A. 2011. Innovative integrated wet process for virgin coconut oil production. *Journal of Applied Science*, *11*(13), 2467–2469.

Hepburn, H. R., Pirk, C. W. W., and Duangphakdee, O. 2014. The chemistry of beeswax. In H. R. Hepburn, C. W. W. Pirk, and O. Duangphakdee (Eds.), *Honeybee Nests* (pp. 319–339). Berlin Heidelberg: Springer.

Hopkins, E. J., Chang, C., Lam, R. S. H., and Nickerson, M. T. 2015. Effects of flaxseed oil concentration on the performance of a soy protein isolate-based emulsion-type film. *Food Research International*, *67*, 418–425.

IUPAC 2006. *Compendium of chemical terminology*, 2nd ed. (the "Gold Book"). Compiled by A. D. McNaught & A. Wilkinson. Blackwell Scientific Publications, Oxford (1997). XML on-line corrected version: http://goldbook.iupac.org (2006) created by M. Nic, J. Jirat, B. Kosata; updates compiled by A. Jenkins. ISBN 0-9678550-9-8. doi:10.1351/goldbook.http://goldbook.iupac.org/RT07166.html.

Jackrel, S. L., and Wootton J. T. 2015. Cascading effects of induced terrestrial plant defences on aquatic and terrestrial ecosystem function. *The Royal Society Publishing, Proceedings B*, *285*, 1805.

Jahurul, M. H. A., Zaidul, I. S. M., Norulaini, N. A. N., Sahena, F., Jinap, S., Azmir, J., and Omar, A. M. 2013. Cocoa butter fats and possibilities of substitution in food products concerning cocoa varieties, alternative sources, extraction methods, composition, and characteristics. *Journal of Food Engineering*, *117*(4), 467–476.

Jangchud, A., and Chinnan, M. S. 1999. Peanut protein film as affected by drying temperature and pH of film-forming solution. *Journal of Food Science*, *64*, 153–157.

Janjarasskul, T., and Krochta, J. M. 2010. Edible packaging materials. *Annual Review of Food Science and Technology*, *1*, 415–448.

Jitputti, J., Kitiyanan, B., Rangsunvigit, P., Bunyakiat, K., Attanatho, L., and Jenvanitpanjakul, P. 2006. Transesterification of crude palm kernel oil and crude coconut oil by different solid catalysts. *Chemical Engineering Journal*, *116*(1), 61–66.

Jo, W. S., Song, H. Y., Song, N. B., Lee, J. H., Min, S. C., and Song, K. B. 2014. Quality and microbial safety of "Fuji" apples coated with carnauba-shellac wax containing lemongrass oil. *LWT—Food Science and Technology*, *55*(2), 490–497.

Jost, V., Kobsik, K., Schmid, M., and Noller, K. 2014. Influence of plasticiser on the barrier, mechanical and greaseresistance properties of alginate cast films. *Carbohydrate Polymers*, *110*, 309–319.

Kamper, S. L., and Fennema, O. 1984. Water vapor permeability of edible bilayer films. *Journal of Food Science*, *49*, 1478–1481.

Karbowiak, T., Debeaufort, F., and Voilley, A. 2007. Influence of thermal process on structure and functional properties of emulsion-based edible films. *Food Hydrocolloids*, *21*(5–6), 879–888.

Kester, J., and Fennema, O. 1986. Edible films and coatings: A review. *Food Technology*, *40*, 47–59.

Khan, M. K. I., Maan, A. A., Schutyser, M., Schroen, K., and Boom, R. 2013. Electrospraying of water in oil emulsions for thin film coating. *Journal of Food Engineering*, *119*(4), 776–780.

Khandelwal, M., Ahlawat, A., and Singh, R. 2012. Polysaccharides and natural gums for colon drug delivery. *The Pharma Innovation*, *1*(1), 9–13.

Khanzadi, M., Jafari, S. M., Mirzaei, H., Chegini, F. K., Maghsoudlou, Y., and Dehnad, D. 2015. Physical and mechanical properties in biodegradable films of whey protein concentrate–pullulan by application of beeswax. *Carbohydrate Polymers, 118*, 24–29.

Kim, I. H., Yang, H. J., Noh, B. S., Chung, S. J., and Min, S. C. 2012. Development of a defatted mustard meal-based composite film and its application to smoked salmon to retard lipid oxidation. *Food Chemistry, 133*, 1501–1509.

Kim, R. S. B., Choia, Y. G., Kimb, J. Y., and Lima, S. G. 2015. Improvement of water solubility and humidity stability of tapioca starch film by incorporating various gums. *LWT—Food Science and Technology*. Available at http://dx.doi.org/10.1016/j.lwt.2015.05.009.

Koelsch, C. 1994. Edible water vapour barriers properties and promise. *Trends in Food Science and Technology, 51*, 76–81.

Koh, L. P., and Wilcove, D. S. 2008. Is oil palm agriculture really destroying tropical biodiversity? *Conservation Letters, 1*(2), 60–64.

Kokoszka, S., Debeaufort, F., Lenart, A., and Voilley, A. 2010. Liquid and vapour water transfer through whey protein/lipid emulsion films. *Journal of the Science of Food and Agriculture, 90*(10), 1673–1680.

Kolattukudy, P. E. 1975. Biochemistry of cutin, suberin and waxes on the lipid barriers of plants. In T. Galliard, & E. I. Mercer (Eds.), *Recent Advances in the Chemistry and Biochemistry of Plant Lipids* (p. 203). New York: Academic Press.

Kowalczyk, D., and Baraniak, B. 2014. Effect of candelilla wax on functional properties of biopolymer emulsion films—A comparative study. *Food Hydrocolloids, 41*, 195–209.

Krochta, J. M. 1997. Edible composite moisture-barrier films. In B. Blakistone (Ed.), *Packaging Yearbook* (pp. 38–54). Washington, DC: National Food Processing Association.

Lazaridou, A., and Biliaderis, C. G. 2002. Thermophysical properties of chitosan, chitosan starch and chitosan pullulan films near the glass transition. *Carbohydrate Polymers, 48*, 179–190.

Lima, A. M., Cerqueira, M. A., Souza, B. W. S. et al. 2010. New edible coatings composed of galactomannans and collagen blends to improve the postharvest quality of fruits—Influence on fruits gas transfer rate. *Journal of Food Engineering, 97*, 101–109.

Lin, D., and Zhao, Y. 2007. Innovations in the development and application of edible coatings for fresh and minimally processed fruits and vegetables. *Comprehensive Reviews in Food Science and Food Safety, 6*(3), 60–75.

Lovegren, N. V., and Feuge, R. O. 1954. Permeability of acetostearin products of water vapor. *Journal of Agricultural and Food Chemistry, 2*, 558–563.

Ma, W., Tang, C. H., Yin, S. W. et al. 2012. Characterization of gelatin-based edible films incorporated with olive oil. *Food Research International, 49*, 572–579.

Maftoonazad, N., Badii, F., and Shahamirian, M. 2013. Recent innovations in the area of edible films and coatings. *Recent Patents on Food, Nutrition & Agriculture, 5*(3), 201–213.

Maftoonazad, N., and Ramaswamy, H. 2005. Postharvest shelf-life extension of avocados using methyl cellulose-based coating. *LWT—Food Science and Technology, 38*(6), 617–624.

Maia, M., and Nunes, F. M. 2013. Authentication of beeswax (*Apis mellifera*) by high-temperature gas chromatography and chemometric analysis. *Food Chemistry, 136*(2), 961–968.

Marina, A. M., Man, Y. C., Nazimah, S. A. H., and Amin, I. 2009. Chemical properties of virgin coconut oil. *Journal of the American Oil Chemists' Society, 86*(4), 301–307.

Marmur, T., Elkind, Y., and Nussinovitch, A. 2013. Increase in gloss of coated red peppers by different brushing procedures. *Food Science and Technology, 51*, 531–536.

Martin-Belloso, O., Soliva-Fortuny, R., and Baldwin, A. 2005. Conservación mediante recubrimientos comestibles [New technologies for preservation of fresh cut vegetables]. In G. González-Aguilar, A. Gardea, & F. Cuamea-Navarro, (Eds.), *Nuevas tecnologías de conservación de productos vegetales frescos cortados* (p. 558). Sonora, México: Centro de Investigaciones en Alimentación y Desarrollo A.C. Hermosillo.

Martínez-Ávila, G. C. G., Aguilera, A. F., Saucedo, S., Rojas, R., Rodriguez, R., and Aguilar, C. N. 2014a. Fruit wastes fermentation for phenolic antioxidants production and their application in manufacture of edible coatings and films. *Critical Reviews in Food Science and Nutrition*, 54(3), 303–311.

Martínez-Ávila, G. C. G., Hernández-Almanza, A. Y., Sousa, F. D., Moreira, R., Gutierrez-Sanchez, G., and Aguilar, C. N. 2014b. Macromolecular and functional properties of galactomannan from mesquite seed (*Prosopis glandulosa*). *Carbohydrate Polymers*, 102, 928–931.

Martín-Polo, M., Maughuin, C., and Voilley, A. 1992a. Hydrophobic films and their efficiency against moisture transfer. Influence of the film preparation technique. *Journal of Agricultural and Food Chemistry*, 40(1), 407–412.

Martín-Polo, M., Voilley, A., Blond, G. et al. 1992b. Hydrophobic films and their efficiency against moisture transfer. Influence of the physical state. *Journal of Agricultural and Food Chemistry*, 40(2), 413–418.

McHugh, T. H. 2000. Protein–lipid interactions in edible films and coatings. *Nahrung*, 44, 148–151.

McHugh, T. H., and Krochta, J. M. 1994. Milk protein based edible films and coatings. *Food Technology*, 48(1), 97–103.

McHugh, T. H., and Senesi, E. 2000. Apple wraps: A novel method to improve the quality and extend the shelf life of fresh-cut apples. *Journal of Food Science*, 65(3), 480–490.

Meighani, H., Ghasemnezhad, M., and Bakhshi, D. 2014. Effect of different coatings on postharvest quality and bioactive compounds of pomegranate (*Punica granatum* L.) fruits. *Journal of Food Science and Technology*, 52, 4507–4514.

Milovmovic, M., and Picuric-Jovanevic, K. 2001. Lipids and biopackaging usage of lipids in edible films. *Journal of Agricultural Sciences*, 46(1), 79–87.

Mirhosseini, H., and Amid, B. T. 2012. A review study on chemical composition and molecular structure of newly plant gum exudates and seed gums. *Food Research International*, 46(1), 387–398.

Mohammadifar, M. A., Musavi, S. M., Kiumarsi, A., and Williams, P. A. 2006. Solution properties of targacanthin (water-soluble part of gum tragacanth exudate from *Astragalus gossypinus*). *International Journal of Biological Macromolecules*, 38(1), 31–39.

Morillon, V., Debeaufort, F., Blond, G., Capelle, M., and Voilley, A. 2002. Factors affecting the moisture permeability of lipid-based edible films: A review. *Critical Reviews in Food Science and Nutrition*, 42(1), 67–89.

Muscat, D., Adhikari, R., McKnight, S., Guo, Q., and Adhikari, B. 2013. The physicochemical characteristics and hydrophobicity of high amylose starch–glycerol films in the presence of three natural waxes. *Journal of Food Engineering*, 119, 205–219.

Navarro-Tarazaga, M. L., Del Rio, M. A., Krochta, J. M., and Pérez-Gago, M. B. 2008. Fatty acid effect on hydroxypropyl methylcellulose-beeswax edible film properties and postharvest quality of coated "Ortanique" Mandarins. *Journal of Agricultural Food Chemistry*, 56, 10689–10696.

Navarro-Tarazaga, M. L., Massa, A., and Pérez-Gago, M. B. 2011. Effect of beeswax content on hydroxypropyl methylcellulose-based edible film properties and postharvest quality of coated plums (cv. Angeleno). *LWT—Food Science and Technology*, 44(10), 2328–2334.

Nevin, K. G., and Rajamohan, T. 2004. Beneficial effects of virgin coconut oil on lipid parameters and *in vitro* LDL oxidation. *Clinical Biochemistry*, 37(9), 830–835.

Nie, S. P., Wang, C., Cui, S. W., Wang, Q., Xie, M. Y., and Phillips, G. O. 2013. The core carbohydrate structure of *Acacia seyal* var. *seyal* (*Gum arabic*). *Food Hydrocolloids*, 32(2), 221–227.

Njombolwana, N. S., Erasmus, A., van Zyl, J. G., du Plooy, W., Cronje, P. J. R., and Fourie, P. H. 2013. Effects of citrus wax coating and brush type on imazalil residue loading, green mould control and fruit quality retention of sweet oranges. *Postharvest Biology and Technology*, *86*, 362–371.

Ochoa, E., Saucedo-Pompa, S., Rojas-Molina, R., Heliodoro de la Garza, Charles-Rodríguez, A. V., and Aguilar, C. N. 2011. Evaluation of a candelilla wax-based edible coating to prolong the shelf-life quality and safety of apples. *American Journal of Agricultural and Biological Sciences*, *6*(1), 92–98.

Ochoa-Reyes, E., Martínez-Vazquez, G., Saucedo-Pompa, S., Montañez, J., Rojas-Molina, R., Leon-Zapata, M. A. d., Rodríguez-Herrera, R., and Aguilar, C. N. 2013. Improvement of shelf life quality of green bell peppers using edible coating formulations. *Journal of Microbiology, Biotechnology and Food Sciences*, *2*(6), 2448–2451.

Ochoa-Reyes, E., Saucedo-Pompa, S., Garza, H. D. L., Martínez, D. G., Rodríguez, R., and Aguilar-Gonzalez, C. N. 2010. Extracción Tradicional de Cera de *Euphorbia antysiphilitica*. *Acta Química Mexicana*, *2*(3), 1–13.

Olivas, G. I., and Barbosa-Cánovas, G. 2009. Edible films and coatings for fruits and vegetables. In G. I. Olivas, & G. Barbosa-Cánovas (Eds.), *Edible Films and Coatings for Food Applications* (pp. 211–244). New York: Springer.

Otoni, C. G., Moura, M. R. D., Aouada, F. A., Camilloto, G. P., Cruz, R. S., Lorevice, M. V., Soares, N. d. F. F., and Mattoso, L. H. C. 2014. Antimicrobial and physical-mechanical properties of pectin/papaya puree/cinnamaldehyde nanoemulsion edible composite films. *Food Hydrocolloids*, *41*, 188–194.

Palla, C., Hegel, P., Pereda, S., and Bottini, S. 2014. Extraction of jojoba oil with liquid CO_2 + propane solvent mixtures. *The Journal of Supercritical Fluids*, *91*, 37–45.

Park, H. J., and Chinnan, M. S. 1995. Gas and water vapor barrier properties of edible films from protein and cellulosic materials. *Journal of Food Energy*, *25*, 497–507.

Patil, S. H., and Talele, G. S. 2014. Natural gum as mucoadhesive controlled release carriers: Evaluation of cefpodoxime proxetil by D-optimal design technique. *Drug Delivery*, *21*(2), 118–129.

Pavalath, A, E., and Orts, W. 2009. Edible films and coatings: Why, what and how? In K. C. Huber, & M. E. Embuscado (Eds.), *Edible Films and Coatings for Food Applications* (pp. 1–23). New York: Springer.

Pehowich, D. J., Gomes, A. V., and Barnes, J. A. 2000. Fatty acid composition and possible health effects of coconut constituents. *The West Indian Medical Journal*, *49*(2), 128–133.

Pérez-Gago, M. B., Rojas, C., and Del Río, M. A. 2003a. Effect of hydroxypropyl methylcellulose-lipid edible composite coatings on plum (cv. Autumn giant) quality during storage. *Journal Food Science*, *68*, 879–883.

Pérez-Gago, M. B., Rojas, C., and Del Río, M. A. 2003b. Effect of hydroxypropyl methylcelulose-beeswax edible composite coating on postharvest quality of "Fortune" Mandarins. *Acta Horticulturae*, *599*, 583–587.

Pérez-Gago, M. B., Serra, M., Alonso, M., Mateos, M., and Del Río, M. A. 2005. Effect of whey proteinand hydroxypropyl methylcellulose-based edible composite coatings on color change of fresh-cut apples. *Postharvest Biology and Technology*, *36*, 77–85.

Peróval, C., Debeaufort, F., Desprea, D., and Voilley, A. 2002. Edible Arabinoxylan-based films. Effects of lipid type on water vapor permeability, film structure, and other physical characteristics. *Journal of Agricultural and Food Chemistry*, *50*, 3977–3983.

Peróval, C., Debeaufort, F., Seuvre, A. M., Chevet, B., Despré, D., and Voilley, A. 2003. Modified arabinoxylan-based films. Part B. Grafting of omega-3 fatty acids by oxygen plasma and electron beam irradiation. *Journal of Agricultural and Food Chemistry*. *51*(10), 3120–3126.

Phan The, D., Debeaufort, F., Luu, D., and Voilley, A. 2008. Moisture barrier, wetting and mechanical properties of shellac/agar or shellac/cassava starch bilayer bio-membrane for food applications. *Journal Membrane Science*, *325*, 277–283.

Phan The, D., Peróval, C., Debeaufort, F. et al. 2002. Arabinoxylan-lipids-based edible films and coatings. 2. Influence of sucroester nature on the emulsion structure and film properties. *Journal of Agricultural and Food Chemistry*, 50(2), 267–272.

Porat, R., Weiss, B., Cohen, L., Daus, A., and Biton, A. 2005. Effects of polyethylene wax content and composition on taste, quality, and emission of off-flavor volatiles in "Mor" mandarins. *Postharvest Biology and Technology*, 38(3), 262–268.

Raybaudi-Massilia, R. M., Mosqueda-Melgar, J., and Martín-Belloso, O. 2008. Edible alginate-based coating as carrier of antimicrobials to improve shelf-life and safety of fresh-cut melon. *International Journal of Food Microbiology*, 121(3), 313–327.

Reybroeck, W., Jacobs, F. J., De Brabander, H. F., and Daeseleire, E. 2010. Transfer of sulfamethazine from contaminated beeswax to honey. *Journal of Agricultural and Food Chemistry*, 58(12), 7258–7265.

Rhim, J. W., and Shellhammer, T. H. 2005. Lipid-based edible films and coatings. In J. H. Han (Ed.), *Innovations in Food Packaging* (pp. 362–383). London, UK: Academic Press.

Rojas-Grau, M., Raybaudi-Massilia, R., Soliva-Fortuny, R. et al. 2007. Apple puree-alginate edible coating as carrier of antimicrobial agents to prolong shelf-life of fresh-cut apples. *Postharvest Biology and Technology*, 45, 254–264.

Rojas-Grau, M., Soliva-Foruny, R., and Martín-Belloso, O. 2009. Edible coatings to incorporate active ingredients to fresh-cut fruits: A review. *Trends in Food Science and Technology*, 20, 438–447.

Rojas-Molina, R., León-Zapata, M. A., Saucedo-Pompa, S., Aguilar-González, M. Á., and Aguilar-González, C. N. 2013. Chemical and structural characterization of candelilla (*Euphorbia antisyphilitica* Zucc.). *Journal of Medicinal Plants Research*, 7(12), 702–705.

Rojas-Molina, R., Saucedo-Pompa, S., León-Zapata, M. Á. d., Jasso-Cantú, D., and Aguilar-González, C. N. 2011. Ensayo: Pasado, presente y futuro de la candelilla. *Revista Mexicana de Ciencias Forestales*, 2(6), 7–18.

Rossman, J. M. 2009. Commercial manufacture of edible films. In K. C. Huber, & M. E. Embuscado (Eds.), *Edible Films and Coatings for Food Applications* (pp. 367–390). New York: Springer.

Sánchez, A. D., Andrade-Ochoa, S., Aguilar, C. N., Contreras-Esquivel, J. C., and Nevárez-Moorillón, G. V. 2015. Antibacterial activity of pectic-based edible films incorporated with Mexican lime essential oil. *Food Control*, 50, 907–912.

Sánchez, M., Marchetti, J. M., Boulifi, N. E., Martínez, M., and Aracil, J. 2014. Jojoba oil biorefinery using a green catalyst. Part I: Simulation of the process. *Biofuels, Bioproducts and Biorefining*, 9(2), 129–138.

Saucedo-Pompa, S., Jasso-Cantú, D., Ventura-Sobrevilla, J. V., Sáenz-Galindo, A., Rodriguez-Herrera, R., and Aguilar, C. N. 2007. Effect of candelilla wax with natural antioxidants on the shelf life quality of fresh-cut fruits. *Journal of Food Quality*, 30, 823–836.

Saucedo-Pompa, S., Rojas-Molina, R., Aguilera-Carbó, A. F., Saenz-Galindo, A., Garza, H. d. L., Jasso-Cantú, D., and Aguilar, C. N. 2009. Edible film based on candelilla wax to improve the shelf life and quality of avocado. *Food Research International*, 42, 511–515.

Sensidoni, A., and Peressini, D. 1997. Edible films: Potential innovation for fish products. *Industrie Alimentari*, 36(356), 129–133.

Shellhammer, T. H., and Krochta, J. M. 1997. Whey protein emulsion film performance as affected by lipid type and amount. *Journal of Food Science*, 62, 390–394.

Siew, D. C. W., Heilmann, C., Esteal, A. J., and Cooney, R. P. 1999. Solution and film properties of sodium caseinate/glycerol and sodium caseinate/polyethylene glycol edible coating systems. *Journal of Agricultural and Food Chemistry*, 47(1), 3432–3440.

Souza, A., Santos, J. C., Conceição, M. M., Silva, M. C., and Prasad, S. 2004. A thermoanalytic and kinetic study of sunflower oil. *Brazilian Journal of Chemical Engineering*, *21*(2), 265–273.

Stuchell, Y. M., and Krochta, J. M. 1994. Enzymatic treatments and thermal effects on edible soy protein films. *Journal of Food Science*, *59*(1), 1332–1337.

Sugimoto, I., Nagai, A., and Okamoto, M. 2012. Simple low-vacuum coating of paraffin wax on carbonaceous gas sensing layers. *Vacuum*, *86*(12), 1905–1910.

Tagi, A., Askar, K.A., Nagy, K., Mutihac, L., and Stamatin, I. 2012. Effect of different concentrations of olive oil and oleic acid on the mechanical properties of albumen (egg white) edible films. *African Journal of Biotechnology*, *10*(60), 12963–12972.

Tan, C. P., and Man, Y. C. 2000. Differential scanning calorimetric analysis of edible oils: Comparison of thermal properties and chemical composition. *Journal of the American Oil Chemists' Society*, *77*(2), 143–155.

Tanada-Palmu, P. S., and Grosso, C. R. F. 2005. Effect of edible wheat gluten-based films and coatings on refrigerated strawberry (*Fragaria ananassa*) quality. *Postharvest Biology and Technology*, *36*(1), 199–208.

Tong, X., Luo, X., and Li, Y. 2015. Development of blend films from soy meal protein and crude glycerol-based waterborne polyurethane. *Industrial Crops and Products*, *67*, 11–17.

Torres-Moreno, M., Torrescasana, E., Salas-Salvadó, J., and Blanch, C. 2015. Nutritional composition and fatty acids profile in cocoa beans and chocolates with different geographical origin and processing conditions. *Food Chemistry*, *166*, 125–132.

Travaglia, E., Coisson, J. D., Bordiga, M., Martelli, A., and Arlorio, M. 2009. Improving the quality of roasted hazelnuts during their shelf-life using film coating starch-based. *Czech Journal of Food Sciences*, *27*(1), 346.

Turhan, K. N., and Sahbaz, F. 2004. Water vapor permeability, tensile properties and solubility of methylcellulose-based edible films. *Journal of Food Engineering*, *61*(1), 459–466.

Unalan, I. U., Arcan, I., Korel, F., and Yemenicioglu, A. 2013. Application of active zein-based films with controlled release properties to control *Listeria monocytogenes* growth and lipid oxidation in fresh Kashar cheese. *Innovative Food Science and Emerging Technologies*, *20*(1), 208–214.

U.S. Patent Trade Market Office. 2015. (p. 530–1) Available at http://www.uspto.gov/web/patents/classification/uspc530/defs530.pdf.

Velickova, E., Winkelhausen, E., Kuzmanova, S., Alves, V. D., and Moldão-Martins, M. 2013. Impact of chitosan–beeswax edible coatings on the quality of fresh strawberries (*Fragaria ananassa* cv. *Camarosa*) under commercial storage conditions. *LWT—Food Science and Technology*, *52*(2), 80–92.

Vieira, M. G. A., da Silva, M. A., dos Santos, L. O., and Beppu, M. M. 2011. Natural-based plasticizers and biopolymer films: A review. *European Polymer Journal*, *47*(3), 254–263.

Vourinen, A., Seibert, J., Papageorglou, V. P., Rollinger, J. M., Odermatt, A., Schuster, D., and Assimopoulou, A. N. 2015. *Pistacia lentiscus* oleoresin: Virtual screening and identification of masticadienonic and isomasticadienonic acids as inhibitors of 11β-hydroxysteroid dehydrogenase 1. *Planta Medica*, *81*(6), 1–8.

Walton, N. J., Mayer, M. J., and Narbad, A. 2003. Vanillin. *Phytochemistry*, *63*(5), 505–515.

Wang, L., Ando, S., Ishida, Y., Ohtani, H., Tsuge, S., and Nakayama, T. 2001. Quantitative and discriminative analysis of carnauba waxes by reactive pyrolysis-GC in the presence of organic alkali using a vertical microfurnace pyrolyzer. *Journal of Analytical and Applied Pyrolysis*, *58–59*, 525–537.

Weller, C. L., Gennadios, A., and Saraiva, R. A. 1998. Edible bilayer films from zein and grain sorghum wax or carnauba wax. *LWT—Food Science and Technology*, *31*(3), 279–285.

Wittaya, T. 2012. Protein-based edible films: Characteristics and improvement of properties. In A. A. Eissa (Ed.), *Structure and Function of Food Engineering* (pp. 43–70). Rijeka, Croatia: InTech.

Wong, D. W. S., Gastineau, F. A., Gregorski, K. S., Tillin, S. J., and Pavlath, A. E. 1992. Chitosan-lipid films, microstructure and surface energy. *Journal of Agricultural and Food Chemistry*, *40*(1), 540–544.

Wong, D. W. S., Tillin, S. J., Hudson, J. S., and Pavlath, A. E. 1994. Gas exchange in cut apples with bilayer coatings. *Journal of Agricultural and Food Chemistry*, *42*(1), 2278–2285.

Ziani, K., Oses, J., Coma, V., and Mate, J. I. 2008. Effect of the presence of glycerol and Tween 20 on the chemical and physical properties of films based on chitosan with different degree of deacetylation. *Food Science and Technology*, *41*(1), 2159–2165.

5 Production and Processing of Edible Packaging
Stability and Applications

Nuria Blanco-Pascual and Joaquín Gómez-Estaca

CONTENTS

Abstract	153
5.1 Introduction	154
5.2 Edible Packaging Applications	155
5.2.1 Potential Applications	156
5.2.2 Water Vapor Barrier	158
5.2.3 Oxygen Barrier	159
5.2.4 UV Light Barrier	159
5.2.5 Organic Vapor Barrier	160
5.2.6 Fat Barrier	160
5.2.7 Improvement of Mechanical Properties of Food	160
5.2.8 Improvement of Sensory Properties	161
5.2.9 Release Systems of Active Compounds	161
5.3 Edible Packaging Stability	162
5.3.1 Film Matrix Reorganizations	162
5.3.2 Chemical Degradation	163
5.3.3 Enzymatic Degradation	164
5.3.4 Microbial Growth	165
5.4 Methods for the Production of Packaging Materials	166
5.4.1 Wet Processing	168
5.4.1.1 Tape Casting	168
5.4.1.2 Edible Coatings	169
5.4.2 Dry Processing	170
5.4.2.1 Thermopressing/Thermoforming	171
5.4.2.2 Extrusion	171
5.5 Conclusions	172
References	172

ABSTRACT

The processability and versatility of synthetic plastic materials have led to their widespread use as food packaging materials, resulting in problems of accumulation of waste in the environment. For this reason, the ability of biopolymer-based packaging materials to create novel applications for the improvement of food quality, safety, and stability is undergoing continuous study. One of the most worrisome questions addressed by the possible industrial applications of edible packaging materials is the scale-up of the production methods along with the uncertainty of their potential applications and stability. Most of the research regarding edible packaging applications are related to their barrier properties (water vapor, oxygen, aroma compounds, light, fat, etc.) as well as to their mechanical versatility or the ability to be a vehicle for active substances. However, opposite to synthetic packaging materials, edible packaging materials are from natural origin and so they are susceptible to chemical, enzymatic or microbial modification, and/or degradation, leading to a greater instability that may limit their applications or even their shelf life. Regarding the production of edible packaging materials, wet (casting) or dry (thermopressing, extrusion) methods can be employed depending on the raw materials, the food products, or the desired applications. In this chapter, these important subjects are discussed as some of the most important issues in the development of edible packaging.

5.1 INTRODUCTION

The organoleptic, hygienic, and nutritional characteristics of food products evolve during storage and commercialization and are not only dependent on their intrinsic properties, but also on the exchanges and migration of components between the food and the media. In this regard, packaging is the last processing step responsible for food preservation. One important function of packaging, when regarded as a food preservation technology, is to retard food product deterioration and to maintain and increase the quality and safety of the packaged foods. Thus, the main purpose of food packaging is to protect the food from microbial and chemical contamination, oxygen, water vapor, and light (Lopez-Rubio et al., 2004). The processability and versatility of synthetic plastic materials have led to their widespread use as food packaging materials. This massive use of plastics has resulted in the problem of accumulation of waste plastic materials in the environment, as well as a high dependence on fossil fuels.

Research on biopolymers is gaining interest, thanks to the use of renewable raw materials for both their production and their biodegradability. The most widely researched thermoplastic sustainable biopolymers in monolayer packaging applications are starch, polyhydroxyalkanoates (PHAs), and polylactic acid (PLA). Of these, starch and PLA are the most interesting biodegradable materials, as they are commercially available, are produced on a large scale, and possess interesting balance of properties (Fabra et al., 2014). In spite of the renewability, biocompatibility, and biodegradability of such polymers, those extracted from biomass (polysaccharides, proteins, lipids) are the ones with real edible applications. Edible packaging has been recognized as a useful alternative or a complement to the conventional packaging systems utilized in the food industry. Different research groups have studied the ability of biopolymer-based packaging materials to create novel applications for the improvement of the food quality, safety, stability, and convenience for regular consumers.

Edible packaging is generally presented as a film or a coating and has been developed utilizing diverse biopolymers, such as proteins, polysaccharides, lipids, and other edible components (Tharanathan, 2003). These materials are intended to be an integral part of the final product and to be eaten as such, or to be part of the package, taking advantage of some important properties of biopolymers as food packaging materials. Therefore, an edible packaging needs to have an acceptable physicochemical and microbiological stability to be an efficient barrier, make possible the transport of different compounds, and improve the final product. The almost unlimited number of combinations among biopolymers, with different physicochemical properties, plasticizers, and other additives, makes edible packaging technology a very versatile alternative to improve food quality. However, there are three main drawbacks of edible food packaging materials, in contrast to traditional ones, that may limit their use and applications. Due to their hydrophilic nature, they are more dependent on environmental conditions (temperature, humidity, oxygen, etc.), which may cause the modification of their properties. Moreover, as their constituents are from natural origin, edible packaging materials are subjected to several chemical, enzymatic, and microbial modifications or degradation, leading to a greater instability that may also limit their applications or even their shelf life. Additionally, the difficulty with the scale-up of the production of edible food packaging materials can retard their potential industrial applications. In the present chapter, it is discussed not only the main applications and limitations derived from the stability of edible packaging, but also the different methods for its production.

5.2 EDIBLE PACKAGING APPLICATIONS

Edible packaging from proteins and polysaccharides possesses some properties that make them very attractive as food packaging materials. They present good optical properties (gloss and transparency), excellent fat barrier properties, high oxygen and organic vapor barriers at low/intermediate relative humidity (RH), selective permeability to gases (high CO_2/O_2 permeability relationship in comparison with other synthetic polymers), and moderate mechanical properties. However, they have poor water barrier properties due to their high hydrophilic character. On the contrary, lipid-based edible films (e.g., fatty acids, waxes, etc.) are hydrophobic; therefore, they have good water barrier but low mechanical properties, both dependent on the presence of polar groups, polymer chain length, and insaturation or acetylation degree (Debeaufort and Voilley, 2009). Mainly used as edible coating materials, lipids do not have a good stand-alone film-making capacity; hence, they are commonly supported on other polymer matrices (Campos et al., 2011) to improve their water barrier property (Talens and Krochta, 2005).

An important characteristic of films made from hydrocolloids is their great sensitivity to water. This may be a drawback, due to the fact that their mechanical properties, oxygen barrier, and in some cases even their integrity can be seriously compromised (Hernandez-Munoz et al., 2005). However, the hydrophilic nature of protein and polysaccharide materials can be used positively for the development of active packages, because the mass transport properties of these materials are also affected by the presence of water (Lopez-Rubio et al., 2004). Therefore, the release of functional compounds (antimicrobials, antioxidants, nutraceuticals, among others) deliberately introduced into the packaging material during its formulation is favored in humid environments, such as applied to high water activity food systems (Balaguer et al., 2013a).

In addition, other materials that are not water sensitive, such as polypropylene, have shown little or no ability to release active compounds (Wessling et al., 1999).

Regarding their possible industrial applications, edible packaging needs to be suitable for handling and hygienic in its application, and it has to fulfill some requirements as both packaging materials and food ingredients, such as an acceptable physicochemical and microbiological stability and lack of toxicity. As packaging components, they need to be as tasteless as possible, unless edible films presented specific sensory characteristics that would need to be compatible with the containing food product. Edible films need to be organoleptically and functionally compatible with the food they are intended for. In this latest scenario, edible films would have not only a packaging application but also a culinary one (Arboleya et al., 2008; Falguera et al., 2011).

Some of the most important properties to be evaluated in edible packaging are their permeability to water vapor and gases, mechanical properties, wettability, solubility, adhesion, cohesion, transparency, sensory properties, and microbial stability. Once these are well established, the composition and behavior can be predicted and optimized depending on their final food application (Falguera et al., 2011). In this regard, for a successful edible packaging food application, it is important both to understand the food product problematic and to clearly establish the edible packaging application over this food product. Therefore, it is necessary to take into account how the final food product can be affected by the film/coating-forming solution properties, the change over time suffered by edible packaging, the ways of interaction between the polymer and the food product, and the effect of the storage conditions on the edible packaging (Bourtoom, 2008).

5.2.1 Potential Applications

The commercial applications of edible packaging materials have been historically limited due to the problems related to their mechanical and barrier characteristics compared to synthetic polymers (Falguera et al., 2011). However, some studies have revealed how these films and coatings can complement or even be essential to improve the quality of some fresh, treated, or frozen products (Baldwin et al., 2011). Edible films can be located in the surface of the product or between different parts within it, such as alternated with slices or covering individual parts included in the final product. Edible films may improve the mechanical and barrier properties of the final food product; facilitate its integrity, handling, and carriage; or reduce the mass transport among food components. Historically, edible coatings have helped to protect and separate small pieces of food to enable the final product consumption, which was found to be a very interesting application for sticky ingredients or additives in the food industry. One of the most challenging targets in edible packaging is to achieve a selective mass transfer, allowing, for example, vegetables respiration but limiting their water losses during the storage time. At the moment, most of the works regarding edible packaging applications are related to their barrier properties such as the water vapor permeability (WVP), gas permeability, mechanical versatility, or the ability to be a vehicle of active substances. The following discussion on some of the most relevant potential applications of edible packaging describes the more important aspects that need to be taken into account. Furthermore, Table 5.1 summarizes

TABLE 5.1
Intended Applications of Edible Packaging Systems along with Their Main Advantages and Some Examples Found in the Literature

Intended Application	Advantage	Example of Packaging System
Avoid dehydration of fresh foods during refrigerated or frozen storage (fruits, vegetables, meat, fish, etc.)	Diminish weight loss, improve texture	Edible coating of casein–lipid on minimally processed carrots (Avena-Bustillos et al., 1994) Edible film of chitosan and essential oils on trout fillets (Albertos et al., 2015)
Avoid rehydration of dried foods (nuts, chips, crackers, freeze-dried fruits, etc.)	Maintain characteristic texture	Edible coating of starch–soybean oil on crackers (Bravin et al., 2006)
Reduce mechanical injury during storage and transportation (fruits and vegetables)	Improve mechanical properties, reduce damage	Edible coating of chitosan on strawberry (Hernández-Muñoz et al., 2006)
Reduce smashing of fragile products during manipulation and transportation (nuts, chocolates, etc.)	Improve mechanical properties, reduce cracking	Protein coatings on nuts or chocolates (Kokoszka and Lenart, 2007)
Retard oxygen transmission rate (fruits, vegetables, fish, meat, oils, nuts, etc.)	Reduce lipid oxidation, reduce fruits and vegetables respiration rate, reduce enzymatic browning, reduce oxidation of vitamins	Maize and cassava starch coating to reduce oxidation of provitamin A during pumpkin drying (Lago-Vanzela et al., 2013) Whey protein coating on peanuts to reduce lipid oxidation (Maté and Krochta, 1998) Wheat starch bags packaging pistachio kernels to reduce lipid oxidation (Javanmard et al., 2011) Chitosan coating to reduce the respiration rate of apples (Gemma and Du, 1998) Carrageenan, carboxymethylcellulose (CMC), and sodium alginate composite films to inhibit enzymatic browning of cut peach (Huimin et al., 2009)
Fat barrier (butter, fried bread products, etc.)	Reduce fat uptake upon frying, improve fat resistance of packaging materials	Edible coating of alginate, whey protein, etc., to reduce fat uptake upon frying (Mellema et al., 2003) Wheat gluten-coated paper to pack oily products (Guillaume et al., 2010)
UV light barrier	Reduce UV light-induced lipid oxidation	Gelatin coating on conventional packaging materials (Farris et al., 2009)

(Continued)

TABLE 5.1 (*Continued*)
Intended Applications of Edible Packaging Systems along with Their Main Advantages and Some Examples Found in the Literature

Intended Application	Advantage	Example of Packaging System
Release of antioxidant compounds (whichever product sensitive to oxidation)	Retard or inhibit lipid oxidation	Gelatin films incorporated with oregano or rosemary aqueous extracts to retard lipid oxidation of cold-smoked sardine (Gómez-Estaca et al., 2007)
Release of antimicrobial compounds (whichever product sensitive to microbial growth)	Retard or inhibit microbial growth	Gelatin–chitosan films incorporated with clove essential oil to retard microbial growth of fresh cod fillets (Gómez-Estaca et al., 2010) Gliadin films incorporated with cinnamaldehyde to retard mold growth on bread or spread cheese (Balaguer et al., 2013c)

the most relevant applications of edible packaging, along with some examples of practical applications found in the literature.

5.2.2 Water Vapor Barrier

The water vapor transmission is a main concern of hydrophilic edible packaging films, where water interacts with the polymer matrix and the permeation increases nonlinearly with the water vapor pressure (Cervera et al., 2004). According to Fabra et al. (2012), it is important to promote the use of edible food packaging materials in food preservation systems designed for low RH conditions to maximize their water vapor and gases barrier properties. Nevertheless, the water vapor barrier property may be improved by changing the formulation, such as: (i) designing composite films with lipid compounds (Fabra et al., 2012); (ii) using less hydrophilic plasticizers or surfactants (Andreuccetti et al., 2009, 2011; Hernandez-Munoz et al., 2004); and (iii) increasing the average polymer molecular weight via cross-linking by either physical (Pérez-Gago et al., 1999) or chemical (Balaguer et al., 2013b; Blanco-Pascual et al., 2014b) treatments during the development of the matrices.

For some products, such as frozen products, fresh vegetables, fruits, and meat, it is interesting to minimize their surface dehydration, as, for instance, it has been achieved with casein–lipid edible coatings in minimally processed carrots (Avena-Bustillos et al., 1994). Similar results have been found by applying chitosan- and wheat gluten–based coatings to decrease decay incidence and weight loss of refrigerated and frozen fruits, also reducing color change, drip loss, and in general, maintaining their textural quality (Han et al., 2004; Hernandez-Munoz et al., 2006; Tanada-Palmu and Grosso, 2005). Edible coatings have also retarded the moisture transfer on frozen pink salmon, reducing up to 50% the moisture loss after

3 months of storage, as well as the drip loss after thawing (Sathivel, 2005). Albertos et al. (2015) applied chitosan edible films incorporated with essential oils on the preservation of refrigerated trout fillets, showing lower weight loss than the control sample. Edible packaging may be applied not only to limit water losses of food products but also to reduce water uptake by dehydrated food during their processing and storage, therefore improving the final food product quality. This is the case, for example, of freeze-dried powders and bakery products. Thereby, Bravin et al. (2006) demonstrated the reduction of the hydration kinetic of crackers stored in an environment with high water vapor activity (a_w) using starch–soybean oil coatings, whereas Huang et al. (2011) achieved a quality enhancement of freeze-dried strawberries by applying whey protein edible coatings.

5.2.3 Oxygen Barrier

Oxygen is responsible for degradation processes such as senescence of fruits and vegetables, lipid oxidation, microbial growth, enzymatic browning, and loss of nutrients. Several edible films and coatings have been applied to prevent these oxygen negative effects in different food products. For instance, during pumpkin drying, provitamin A carotenoids oxidizes, which can be minimized by maize and cassava starch coating pretreatment (Lago-Vanzela et al., 2013). Chitosan coatings are effective on reducing lipid oxidation of both refrigerated and frozen fish (Jeon et al., 2002; Sathivel, 2005). The spontaneous oxidation of myoglobin in refrigerated ground beef patties was also lowered by application of chitosan films (Suman et al., 2010). Edible whey protein coating applied in peanuts resulted in lower oxygen uptake and lower rancidity, which was dependent on coating thickness and homogeneity (Maté and Krochta, 1998). Javanmard et al. (2011) developed bags made from whey starch that effectively reduced lipid oxidation of pistachio kernels.

In many fruits and vegetables, controlling gas exchange (O_2/CO_2) reduces respiration rate and improves shelf life and quality. However, it is well known that excessive restriction of gas exchange can lead to anaerobiosis and the development of off-flavors. As an example of this application, chitosan coating has been reported to modify the internal atmosphere of tomatoes (Elghaouth et al., 1992) and apples (Gemma and Du, 1998) by depletion of endogenous O_2 and a rise in CO_2 without achieving anaerobiosis.

5.2.4 UV Light Barrier

It is well known that UV light catalyzes oxidation reactions, including lipid oxidation and loss of functional or bioactive ingredients such as vitamins, carotenoids, polyphenols, etc. in foods (Cardenia et al., 2013; López-Rubio and Lagaron, 2011; Suwannateep et al., 2012). Edible packaging commonly presents UV barrier properties. For example, protein films usually absorb in the UV region due to the presence of aromatic amino acids such as tryptophan, tyrosine, and phenylalanine (Fang et al., 2002; Gomez-Guillen et al., 2007; Shiku et al., 2004). The UV barrier properties may also be enhanced by the addition of compounds absorbing in this region, as for example, polyphenols (Gomez-Estaca et al., 2007, 2009; Gomez-Guillen et al., 2007) and pigments (Blanco-Pascual et al., 2014a, 2014c), conferring the resulting edible film an improved UV barrier with interesting applications in food packaging in addition to their antioxidant effect.

5.2.5 Organic Vapor Barrier

The film organic vapor transfer is also important to be taken into account due to its role in the aroma compounds retention and the prevention of external odor or solvent penetration into the food product, especially during long-term storage conditions (Embuscado and Huber, 2009). Compared with WVP and gas permeability measurements, aroma film permeability studies are challenging and limited, but some methods have been proposed (Miller and Krochta, 1998). Protein coatings containing fat have been applied to preserve flavor and aroma of fresh and dried fruits (Debeaufort and Voilley, 1994; Kokoszka and Lenart, 2007). In this regard, beeswax emulsion composite edible films retained the volatile compounds responsible for the pineapple orange aroma during 2 days of storage at 21°C (Nisperos-Carriedo et al., 1990), while carnauba, semperfresh, and *Aloe vera* edible coatings have also shown good aroma preservation and inhibition of nondesirable odors development applied to mango fruit (Dang et al., 2008).

5.2.6 Fat Barrier

Edible packaging has been already applied to limit the penetration of oil during the frying, such as alginate, cellulose and its derivatives, whey protein, soy protein, albumin, gluten, corn, and pectin (Mellema, 2003). Promising results have been observed in edible coatings applied in vegetables (De Grandi Castro Freitas et al., 2009; Kilincceker and Hepsag, 2012), cereals (Williams and Mittal, 1999), fruits (Singthong and Thongkaew, 2009), or fishery products (Chen et al., 2008). However, the decrease in food fat intake normally comes together with a decrease in food moisture (Mallikarjunan et al., 1997), which can be minimized using different composite formulations, as Albert and Mittal (2002) showed with soy protein/methyl cellulose and soy/whey protein coatings applied in cereal products. Other application as fat barrier is the application of protein coatings on paper, increasing the resistance to oil in the material (Guillaume et al., 2010; Trezza and Vergano, 1994).

5.2.7 Improvement of Mechanical Properties of Food

As previously discussed, the reduction of water transfer among food and environment achieved through the application of edible films or coatings is directly related to food texture improvement, thanks to avoiding of dehydration of fresh foods (vegetables, meats, fish, etc.) of rehydration of dried ones (nuts, chips, crackers, etc.). French fries, for example, are frequently coated to ensure their protection during freezing storage and minimize water loses before frying or even minimize loss of crispness after frying (Pavlath and Orts, 2009). In other occasions, edible films or coatings may be applied to directly improve the mechanical resistance of foods, preventing smashing of products such as nuts or chocolates during manipulation, storage, and transportation (Kokoszka and Lenart, 2007). In this connection, medicine pills are normally coated to prevent crumbling. A similar application was reported by Xie et al. (2002), who applied different edible coating formulations to minimize eggshell breakage and postwash bacterial contamination, with excellent results. Edible films and coatings are also used to improve the mechanical resistance of fruits and vegetables, preventing mechanical injury produced during postharvest handling and processing (Hernandez-Munoz et al., 2006).

5.2.8 Improvement of Sensory Properties

Sensory characteristics such as taste, odor, transparency, color, shininess, softness, and stickiness can be both modified and manipulated with edible films and coatings. In this regard, whey protein coating showed the same gloss effect as shellac, zein, hydroxyl methylcellulose, or dextrin coatings, more stability at higher relative humidities with lack of flavor and color (Trezza and Krochta, 2000). Therefore, the use of whey protein coating is feasible in food products based on chocolates or nuts (Lee et al., 2002). On the contrary, when nutraceutical ingredients are incorporated, the taste, flavor, and appearance of edible coatings can be modified and lead to customer rejection of the product. It has been also observed that fresh strawberries with chitosan-based coatings containing vitamin E reduced the final product glossiness (Rojas-Grau et al., 2009). Carrageenan, carboxymethyl cellulose, and sodium alginate coating composite films can effectively inhibit the browning of fresh-cut peach fruits (Huimin et al., 2009).

5.2.9 Release Systems of Active Compounds

Edible films and coatings are generally in intimate contact with food; thereby, they are excellent vehicles for the incorporation of active compounds that are gradually released over time. These compounds can improve the food product appearance and sensory properties (e.g., incorporation of flavors and colors), improve the food nutritional value (e.g., minerals, vitamins, and antioxidants), and improve shelf life or add potential bioactivity to the final product (e.g., antioxidants and antimicrobials). Furthermore, the hydrophilic nature of edible films facilitates the release of these active compounds, as compared to nonhydrophilic synthetic polymers. However, the effectiveness of these compounds' activity is going to depend on the RH of the environment. In this regard, it has been observed how humid environments favor the release of the active compounds (Balaguer et al., 2013a). Some examples of such applications are the incorporation of antioxidant extracts from oregano or rosemary on gelatin films to improve the oxidative stability of cold-smoked sardine (Gomez-Estaca et al., 2007) or the incorporation of cinnamaldehyde in gliadin films as antifungal for bread and cheese spread (Balaguer et al., 2013c). Besides, these compounds have been micro- and nanoencapsulated to improve the sustained release properties (Rojas-Grau et al., 2009). In this regard, Rojas-Grau et al. (2007) developed sodium alginate and gellan gum coatings with antibrowning agents and vegetable oils enriched with essential fatty acids that were successfully applied in minimally processed fruits.

5.3 EDIBLE PACKAGING STABILITY

Edible films are unstable biological systems that, due to the nature of their components, reactions among them and external parameters can be expected, resulting in molecular changes and reorganization over time (Cuq et al., 1996). The nature and extent of such reorganization is going to be dependent on the type of polymer, type and concentration of plasticizers and other additives, the presence of other components such as pigments or enzymes, and ambient conditions (temperature, humidity, among others). The main events influencing the stability of edible packaging are the reorganization of the film matrix (plasticizer migration, protein aggregation, starch crystallization), the chemical (lipid oxidation, nonenzymatic browning, degradation of active compounds) or enzymatic (proteolysis) degradation of film components, and microbial growth, which will affect not only the physicochemical properties of the films (mechanical, solubility, barrier, optical), but also its functionality and even its safety. The stability of edible packaging over time is an issue that has been the object of many studies due to the fact that it can be significantly different depending on the matrix composition. The main causes of edible film modification during storage are discussed below.

5.3.1 FILM MATRIX REORGANIZATIONS

Protein networks tend to aggregate over time. After a first network stabilization with hydrogen bonds, in many cases, protein molecules tend to interact forming hydrophobic interactions, disulfide and nondisulfide covalent bonds during storage, which normally result in an increase in the tensile resistance of the films, a reduction in their elongation ability, and a decrease in the film WVP and solubility (Leerahawong et al., 2012; Tongnuanchan et al., 2013). These cross-links formations are strongly influenced by different chemical occurrences, such as lipid oxidation or nonenzymatic browning, as well as by plasticizer migration and loss, which promotes protein–protein interactions. Furthermore, the presence of residual endogenous proteinases in fish muscle protein films can lead to slow protein hydrolysis during aging, providing more functional groups susceptible to participate in cross-linking reactions (Blanco-Pascual et al., 2014b).

Starch edible films suffer physical aging by crystallization and retrogradation, which alter film matrix structure and hence film physicochemical properties. This phenomenon, which consists of the rearrangement of amylose and linear parts of amylopectin to form a more crystalline structure stabilized by hydrogen bonds, usually occurs together with plasticizer migrations to the surface and has been related to elongation capacity losses and embrittlement (Van Soest and Knooren, 1997). Film aging or its effects on film properties may be alleviated by employing different plasticizers or their mixtures, developing composite edible films made from different polymers, or tailoring the polymer properties by chemical or pH modifications (Blanco-Pascual et al., 2014b; Cervera et al., 2004; Guo et al., 2014; Hernandez-Munoz et al., 2004; Krogars et al., 2003).

Polymer structure modifications suffered over time can also result in a less efficient binding capacity, hence different compounds migrate within the matrix.

For example, plasticizing water, as well as other added plasticizers, can slowly migrate to the film surface and evaporate. Due to the low-molecular-weight compounds, plasticizers are potential migrants in packaging applications (Hernandez-Munoz et al., 2004), and consequently, their losses are going to affect films' properties. Plasticizer migration speed and extent are going to depend on the matrix structure, for instance, the functional groups availability and polarity, as well as on the molecular weight of the plasticizer. Thus, low-molecular-weight plasticizers such as glycerol and sorbitol are more prone to migrate to the film surface and evaporate (Hernandez-Munoz et al., 2004; Taghizadeh et al., 2013). The main limitation on the use of glycerol is its high hydrophilicity, low thermal stability, and its surface migration observed in different edible films over time, such as myofibrillar-based films (Blanco-Pascual et al., 2014b), gluten-based films (Hernandez-Munoz et al., 2004; Park et al., 1994), whey protein–based films (Osés et al., 2009), or starch-based films (Kuutti et al., 1998). Despite the higher thermal stability of sorbitol compared to glycerol, it also presents high surface migration and demixing along with its tendency to recrystallization over time (Li and Huneault, 2011). The loss of the plasticizer during storage will induce a change on the films' properties, including lower weight loss, higher mechanical strength, lower elongation, and lower WVP (Hernandez-Munoz et al., 2004). In this regard, higher molecular weight polyols such as diglycerol and polyglycerol with more complex conformational structure have shown to be a significant advantage to lower film moisture uptake, achieve higher thermal stability, and obtain interfacial modification of biodegradable blends with improved miscibility and overall higher stability over time (Taghizadeh et al., 2013).

5.3.2 Chemical Degradation

The presence of lipid in edible polymers makes them more prone to oxidation. Lipid oxidation is particularly relevant in the case of edible films from fish muscle proteins, in which depending on the washing/protein extraction procedures, as well as on the species of origin, a considerable fat content may be present in the final film. This is also true for edible film formulations in which a lipid is incorporated to a polysaccharide or protein matrix to improve water barrier properties. Lipid oxidation products generated during the storage time provide the carbonyl groups (aldehyde and ketone) that can interact with the protein amino groups, forming nondisulfide covalent bonds, as well as volatile oxidation products responsible of rancid odor. The intra- and intermolecular cross-linking gives rise to the protein polymerization, strengthening of the film network, and decreasing of the protein solubility, generally along with a reduction of the film solubility. The WVP and the elongation at break also tend to decrease under these conditions and the film tensile strength increases (Tongnuanchan et al., 2011). However, a decrease in film breaking strength and elongation at break has been observed by other authors (Perez-Mateos et al., 2009), being indicative of the complexity of the phenomena taking place. The film's sensory properties are also affected by lipid oxidation, due to the development of undesirable rancid flavors.

Some plasticizers or their degradation products may react with free amino groups of proteins via the so-called Maillard reaction, giving rise to browning/yellowing and/or protein cross-linking and aggregation, with the subsequent modification of the film's physicochemical properties. This is the case of reducing sugars such as glucose, galactose, or fructose, as well as some disaccharides such as lactose or saccharose, which hydrolyze to their monosaccharides under some conditions, for example, at acidic pH (Chinabhark et al., 2007). Leerahawong et al. (2012), for example, observed how glucose and fructose reacted with the amino groups of muscle proteins during the storage of edible films. This reaction caused an increase in yellowness, as well as protein aggregation, revealed by the reduction of protein solubility, which resulted in the increase in the film strength and the reduction in the film elongation. Somanathan et al. (1992) also found yellowing of casein films after storing, as well as Gennadios et al. (1998) in egg white protein/dialdehyde starch films. In this latter case, yellowness developed immediately just after film production, as no storage trials were performed.

Although the Maillard reaction and the resulting yellowness are not a problem specific to fish muscle protein films, these matrices are the ones in which it has been majorly referred. This could be due to the fact that a residual lipid content is usually present in such kind of films, the oxidation of which provides the carbonyl groups involved in the Maillard reaction (Artharn et al., 2009; Tongnuanchan et al., 2011). However, the development of yellowness did not necessarily affect some film optical properties, maintaining during aging both film transparency and UV barrier (Artharn et al., 2009; Leerahawong et al., 2012; Tongnuanchan et al., 2013). As Tongnuanchan et al. (2011) observed in red tilapia muscle protein films, Maillard reaction commonly starts during the film-forming solution casting and drying, and is specially induced by alkaline pH. However, Tongnuanchan et al. (2011) and Blanco-Pascual et al. (2013) described how after storage at room temperature (20 days and 4 months, respectively), film yellowness was more evident in acidic protein films, even with the lowest possible lipid content. The development of nonenzymatic browning, as well as the extent of it, is dependent on many factors, such as the nature of the polymer, plasticizer, film developing, and storage conditions. An example of this difference is given by Hernández-Muñoz et al. (2003), who did not find any color change in glutenin films (plasticized with glycerol, triethanolamine, or sorbitol) after 16 weeks at 23°C and 50% RH.

Although some film properties may be affected during storage, there are others that can remain stable. For example, chitosan films (plasticized with glycerol and polyethylenglycol) have proved to present excellent oxygen barrier properties and acceptable water vapor barrier and mechanical properties, all of them stable up to 9 weeks at 23°C and 50% RH (Butler et al., 1996; Caner et al., 1998), in spite of the change of the film mechanical properties; and myofibrillar protein films (plasticized with saccharose) showed stable WVP and mechanical properties despite the yellowing suffered (Cuq et al., 1996).

5.3.3 Enzymatic Degradation

The presence of residual proteolytic enzymes is a factor to be considered when developing edible films from muscle proteins (e.g., fish, cephalopods, crustaceans, etc.). Temperatures that are arisen during film dry processing will irreversibly inactivate proteolytic enzymes. However, edible films produced by the casting method under mild temperatures will suffer modifications in their structure during production and storage due to the action of proteolytic enzymes. For film production, proteins need to be previously solubilized at acidic or alkaline pH (ranging from 2 to 3 and 10 to 11, respectively). Depending on the muscle processing method (e.g., application of previous washing steps), muscle soluble proteins (including proteolytic enzymes) may form part of the final film (Hamaguchi et al., 2007). The maximum proteolytic activity of such enzymes may be at acidic or alkaline pH depending on the fish species; as for giant squid (*Dosidicus gigas*) muscle, the higher activity was found at acidic pH (Blanco-Pascual et al., 2014b), whereas for shrimp (*Litopenaeus vannamei*) and big-eye snapper (*Priacanthus tayenus*) muscle, it was higher at alkaline pH (Chinabhark et al., 2007; Gómez-Estaca et al., 2014). The main effect of proteolysis is the degradation of the myosin heavy chain, with the consequent reduction of the average polymer molecular weight. In consequence, film's physicochemical properties generally worsen, showing lower mechanical strength and higher film water solubility and lower transparency (Gómez-Estaca et al., 2014).

Discoloration or enzymatic browning, caused by polyphenol oxidase, is another frequent food issue that normally affects fruits and vegetables, wherein phenolic compounds react with atmosphere oxygen in the presence of cooper. In general, edible packaging is not much susceptible to enzymatic browning, but generates concern when active compounds such as polyphenols or plant extracts are included in the formulation and applied to the preservation of fruits and vegetables with high polyphenol oxidase activity, leading to color changes and degradation of the bioactive. To protect these compounds and preserve their stability in the edible packaging, several encapsulation techniques are being performed and tested (Munin and Edwards-Lévy, 2011).

5.3.4 Microbial Growth

Microbial growth is a very important factor to be considered when evaluating the stability of edible packages, especially when they are manufactured from very unstable raw materials, such as fish muscle. In other cases, such as edible films produced from vegetal proteins or chitosan, the extent of microbial growth is not significant or negligible and thus it is not usually studied. However, it is important to be considered that in most fresh processed food products, the highest intensity of microbial contamination is found in the surface; therefore, edible packaging would be a target for microbial growth.

To prevent microbial growth in edible films, it is important to take extra precautions with the initial microbial load of the raw materials, as well as to select the most appropriate manufacture conditions. For example, edible films obtained by dry processing will be more stable from a microbiological point of view, due to the bactericidal effect of the temperatures up to 250°C that are commonly applied (Balaguer et al., 2014b; Cuq et al., 1999). The processing of the raw materials when producing films by casting is also an important factor to be considered. Therefore, Blanco-Pascual et al. (2013) developed edible film from giant squid muscle that had been solubilized by various methods, namely water, salt, acidic, or alkaline solubilization and found that films prepared from water- and salt-solubilized proteins were unstable from a microbiological point of view. Gómez-Estaca et al. (2014) found films from shrimp muscle protein cast at pH 2 to be moderately stable after 30 days of storage at 17°C and 58% RH, recording microbial counts of ~3 CFU/g, which was mainly attributed to the presence of lactic acid bacteria. In any case, once the edible packaging is manufactured, its microbiological stability is going to be affected not only by the film composition but also by the external conditions, mainly temperature, humidity, food product characteristics, and environmental contamination.

To improve microbial stability, antimicrobial compounds can be added to the formulations. This is the case of the work by Gómez-Estaca et al. (2014), in which cinnamaldehyde was added to the film-forming solution, obtaining microbial counts (aerobic plate counts, lactic acid bacteria, Enterobacteriaceae) under the detection limit all over the film storage period (30 days at 17°C, with 58% RH). There are also a considerable number of works dealing with the development of edible films with added antimicrobial compounds for food preservation (Balaguer et al., 2014a; Gómez-Estaca et al., 2010), which also exert their antimicrobial activity in the edible film.

The a_w is critical not only for microbial growth but also for chemical and enzymatic spoilage, being an important factor to be taken into account for edible packaging development. Reducing the water activity of edible films during its storage before the final application would prolong their shelf life. Nevertheless, due to the low microbial stability of films developed from animal proteins, as compared to those from polysaccharides or many vegetable proteins, it would be advisable to give them their intended use immediately after film production. The short storage life of such edible films should not be a problem if they are applied to food products with an even shorter shelf life. It is worth noting that edible packaging is usually applied in combination with other preservation methods as an additional hurdle to microbial growth, such as refrigeration (López de Lacey et al., 2014), high pressure (Gomez-Estaca et al., 2007), modified atmosphere packaging (Conte et al., 2009), or irradiation (Kang et al., 2007), thus improving not only food shelf life but also edible packaging microbial stability.

5.4 METHODS FOR THE PRODUCTION OF PACKAGING MATERIALS

Edible films may be produced by wet (casting) or dry (thermopressing, extrusion) methods. The solvent casting method is the most used technique at a laboratory scale; however, there are also industrial-scale continuous casting equipment. Furthermore, the dry thermoplastic process is gradually becoming a feasible commercial alternative to the edible packaging manufacturing process (Janjarasskul and Krochta, 2010; Pascall and Lin, 2013). A discussion on wet and dry processing methods to obtain edible films and/or coatings is provided in the following sections; Figure 5.1 summarizes the main methods for producing edible food packaging materials.

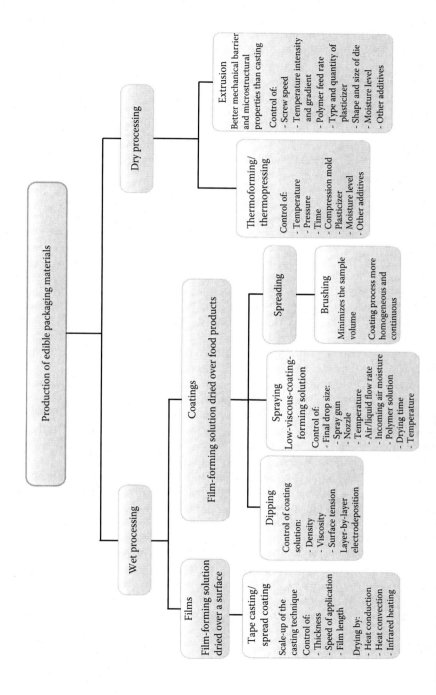

FIGURE 5.1 Schematic representation of some of the methods of production of edible food packaging materials.

5.4.1 Wet Processing

Wet processing uses the method of spreading and solvent evaporation or "casting." It consists in dissolving the biopolymers in suitable solvents to make a film-forming solution where the desired additives, functional compounds, or fillers are added (e.g., plasticizers, cross-linkers, lipids, antimicrobials, micro-/nanoparticles), followed by its spreading over a surface and the consequent evaporation of the solvent to obtain a film. When the film-forming solution, in which the food product has been dipped in, is dried over the food surface, an edible coating is obtained. This casting procedure, which is widely used in the development of laboratory-scale materials, has been employed to test the film-forming ability of a wide range of biopolymers and their mixtures, as well as the effect of different processing conditions and additives on their physicochemical properties. However, the classic casting method has two main drawbacks that make it impracticable at industrial scale. On the one hand, it is difficult to scale-up film production to larger size than 25–30 cm, and on the other hand, a long drying time (10–24 h) is needed.

5.4.1.1 Tape Casting

The tape casting technique (also known as spread casting or knife coating), which is commonly employed in paper, plastic, ceramic, and paint manufacturing industries, has been successfully employed to scale-up the production of edible films (De Moraes et al., 2013). In the tape casting process, a suspension is placed in a reservoir with a blade, whose height can be adjusted with micrometric screws. The suspension is cast as a thin layer on a support (tape), due to the movement of the carrier tape (continuous process) or the movement of the doctor blade (batch process). In this method, thickness, application speed, and film length can be controlled and different suitable supports can be chosen for each biopolymer solution. The film-forming solution spread over the support is dried by heat conduction, heat convection, or infrared heating (De Moraes et al., 2013).

The casting method may be applied not only to obtain stand-alone edible films, but also to manufacture conventional packaging materials coated with a layer of biopolymer material (Farris et al., 2009). These kind of applications allow obtaining packaging materials with improved physical properties (oxygen or UV light transmission barriers), or with the ability to release functional ingredients under medium or high RH conditions, maintaining the mechanical properties, such as heat sealability and printability, for example, of conventional plastic material. In this regard, Farris et al. (2009, 2010) developed a gelatin-based coating on various kinds of conventional plastic materials. Although the resulting materials were less transparent, they presented a greater barrier to transmission of oxygen and UV radiation, maintaining the heat sealability. Similar results have been obtained by Hong et al. (2004) and Hong and Krochta (2003), which developed a whey protein isolate coating over poly(vinyl chloride) or polypropylene with good adherence, transparency, and improved oxygen barrier properties. Apart from synthetic materials, paper has been extensively studied as substrate to develop protein-coated materials, due to the fact that it is manufactured from natural and renewable raw materials and offers the advantage of being both recyclable and biodegradable (Gastaldi et al., 2007).

Furthermore, due to the hydrophilic nature of biopolymers, they generally present good compatibility with paper. Several protein sources, such as caseinates, whey protein isolate, wheat gluten, soy protein isolate, or corn zein, have been employed to coat paper. In the case of whey protein isolate coating, oil resistance, printability of water-based ink, and water barrier properties were improved without compromising the optical and mechanical properties (Han and Krochta, 1999, 2001). Soy protein isolate coating made the coated paper/paperboard more resistant to water and grease but it decreased the tensile strength (Park et al., 2000; Rhim et al., 2006). Wheat gluten–coated papers showed better water and oil resistance, as well as lower permeability to oxygen, carbon dioxide, and water vapor (Guillaume et al., 2010). Similar results were also obtained for corn zein–coated paper (Trezza and Vergano, 1994; Trezza et al., 1998).

5.4.1.2 Edible Coatings

Edible coatings can be processed by dipping the food product in the coating solution, brushing and spraying the coating solution over the food product, among others (e.g., dripping, foaming, fluidized-bed coating, planning, or electrostatic coating). The method to be employed is going to depend on the characteristics of the coating materials and the food product, the intended effect of the coating and the costs.

Coating material can be applied by dipping food products in solutions and allowing the excess to drain while it dries and solidifies. Edible coating dipping is normally designed according to the coating solution density, viscosity, surface tension, and the withdrawal speed of the coating solution, as Cisneros-Zevallos and Krochta (2003) showed in hydroxypropyl methylcellulose coating of Fuji apples by dipping. This technique is also the most common method for applying coatings to preserve the quality of either fresh, whole or minimally processed fruits, vegetables, and meat products, by generally dipping food into the coating-forming solution around 5–30 s (Vargas et al., 2008). However, it has been observed that hydrophobic surfaces do not allow a good adhesion of the coating. Due to this fact, the dipping process normally consists in applying a multilayered edible coating by using a layer-by-layer electrodeposition with several dipping and washing steps, where different materials are physically or chemically bonded to each other (Ayranci and Tunç, 1997; Vargas et al., 2008).

Spraying is the conventional method generally used to apply low-viscous-coating-forming solutions in the food industry, existing programmable spray systems for its automation. Indeed, highly viscous solutions cannot be easily sprayed, being normally dipped and giving high thickness to the coating. The quality of the coating depends on the final drop size, which is going to rely on the spray gun, the nozzle, temperature, air and liquid flow rate, incoming air moisture, and the polymer solution itself. Classic spraying systems can produce a size-drop distribution up to 20 µm in a fine spray. Not only is drop size the main factor, but there are also others like the drying time, temperature, and method. Spraying is normally used as protection against decay of fruits (Meng et al., 2009), although Bravin et al. (2006) showed how spraying coating conferred worse water vapor barrier and mechanical properties than spread coating. Therefore, spray coating is less effective in controlling moisture transference in moisture-sensitive foods such as dry bakery products.

The brushing method, an alternative of spreading, has been also applied in different food products such as fresh vegetables, fruits, cheese, and eggs, presenting the advantages of minimizing the sample volume utilized and resulting in more homogeneous and continuous coating processes compared to dipping and spraying (Kim et al., 2008). Brushing has resulted for instance in less moisture loss for beans and strawberries than cellulose-based dipped coatings and films (Ayranci and Tunç, 1997), as well as better protective barrier and preservation of the internal eggs quality than both dipping and spraying chitosan-coating solutions (Kim et al., 2008).

5.4.2 Dry Processing

The dry processing of conventional polymers is commonly employed in the industry for the production of different kinds of food packages. As many biopolymers also behave as thermoplastic materials, it is possible to adapt this technology for the production of edible packaging. Thus, in the presence of plasticizers, at low moisture levels and high temperatures and with pressure or shear forces, many biopolymers acquire a viscoelastic behavior that allows them to be shaped for the production of all kinds of materials, including edible food packaging materials such as films, disposable pots, cups, trays, and lids. Up to the moment, the most widely employed biopolymers at an industrial scale are PHAs, hydroxypropyl cellulose, PLA and highly modified starch derivatives; some of them, although biocompatible and biodegradable, are not edible. The ability of other biopolymers to produce packaging materials by dry processing has been tested at a laboratory and pilot plant scale with good results, being shown that polymers extracted from biomass are the most indicated for edible applications. Among them, proteins stand up because they are denatured during heating and change their conformation, exposing functional groups that were originally hidden, permitting the establishment of new interactions and links that may produce improvements in the properties of the materials (Hernandez-Izquierdo and Krochta, 2008). Specifically, wheat gluten, whey proteins, gelatin, fish myofibrillar proteins, sunflower protein, corn zein, soy protein, and keratin have shown the ability to be processed by extrusion and/or thermopressing (Barone et al., 2006; Hernandez-Izquierdo and Krochta, 2008; Krishna et al., 2012; Rouilly et al., 2006).

There are two main methods of obtaining edible packaging materials by dry processing: thermopressing/thermoforming and extrusion. Although it is possible to obtain materials by the use of both technologies independently, sometimes they are combined, using, for example, extrusion for the mixing and partial modification of the components, and then obtaining the materials by thermopressing/thermoforming. Besides, it is also possible to use other processes typically employed in making conventional plastic materials, such as blowing and injection. In this regard, different types of extrusion systems can be utilized such as flat-die, cast-on-carrier, or blown-film extrusion. Belyamani et al. (2014), for instance, showed the possibility of developing sodium caseinate films, plasticized with glycerol, by plastic techniques such as injection, compression molding, and blowing extrusion.

5.4.2.1 Thermopressing/Thermoforming

This technology is commonly employed in the industry for the production of films or containers such as cups, pots, trays, and lids. Thermoforming consists in heating a plastic sheet to a pliable forming temperature, giving it a shape in a mold and trimming it to create the final package or container. Thermopressing is a type of thermoforming in which high temperature and pressure are applied to a mixture of plasticized polymer resin, which acquires viscoelastic properties, forming a film when cooled. These films are stabilized by hydrophobic and ionic interactions, hydrogen bonds, and/or covalent links (Hernandez-Izquierdo and Krochta, 2008). Thermopressing technology has been widely used at pilot plant scale to study the thermoplastic behavior of biopolymers, as well as the properties of the resulting films under different processing conditions or the addition of additives. The equipment for thermopressing consists of a pair of thermostated plates that act as a press, between which the plasticized biopolymer mixture to be processed is placed. The parameters susceptible to be regulated are temperature, pressure, time, type and content of plasticizer, moisture level, and other additives, among others. By thermopressing, it is also possible to manufacture closely adhering multilayer materials that may be of great interest in various applications for food packaging. An example of this kind of multilayer material development is a laminate made by thermopressing from wheat gluten and PLA, which combines the good oxygen transmission barrier properties of the former and the good mechanical properties of the latter (Cho et al., 2010). Pol et al. (2002) also developed laminated composed materials of soy and zein protein by thermopressing, which maintained the inherent good oxygen barrier properties of soy protein and a good water barrier, which is attributed to the higher hydrophobicity of zein. Among polysaccharides, both cellulose and starch have been processed by extrusion and compression molding (Avérous et al., 2001; Huneault and Li, 2012; Simon et al., 1998).

5.4.2.2 Extrusion

Extrusion is probably the most widely used polymer processing method. In consequence, the adaptation of this process to the production of films from biopolymers will help to extend their use for the production of edible packaging materials. Films developed by extrusion often result in better mechanical, barrier, and microstructural properties than those obtained with solution casting (Hernandez-Izquierdo et al., 2008).

An extruder basically consists of an endless screw inside a barrel with a double casing that permits control of temperature. The polymer is fed from a hopper and is pushed by the screw toward a die. During its residence in the extruder, the polymer is subjected to shear forces and high-temperature compression that permit the formation of a viscoelastic mass that takes on the desired form after passing through the die and cooling. Some of the parameters that influence the process and have to be controlled are screw speed, temperature intensity and gradient, polymer feed rate, type and quantity of plasticizer, the presence of other additives, moisture level, and the shape and size of the die. These parameters determine the extent of the changes that affect conformation, aggregation, and cross-linking of the biopolymer. Cross-linking reactions can give rise to an increase in the glass transition temperature and greater melt viscosity, which require the addition of plasticizers to increase the free volume and mobility of the molecules (Hernandez-Izquierdo and Krochta, 2008). As the temperature rises above glass transition, the plasticized biopolymer turns into a soft, rubbery material that can be shaped as desired, and when it cools, it retains the shape that it has been given. A sufficiently high extrusion temperature is required to achieve a high degree of denaturation and aggregation, needed to allow the processing of the polymer and consequently achieve a homogeneous film (Ullsten et al., 2009). In the case of wheat gluten films plasticized with glycerol, the lowest temperature at which they can be extruded is 90°C, and it is determined by the beginning of sulfhydryl/disulfide interchange reactions, whereas the highest temperature is conditioned by the increase in viscosity produced as a result of protein aggregation. Ullsten et al. (2006) improved the processability of wheat gluten by extrusion by adding salicylic acid, which acted as radical scavenger retarding the cross-linking temperature, thus increasing the maximum working temperature. Under these conditions, continuous films without irregularities can be obtained. Other proteins that have been successfully processed by extrusion are corn zein, soy protein, sunflower protein, whey protein, gelatin, and keratin (Barone et al., 2006; Hernandez-Izquierdo et al., 2008; Krishna et al., 2012; Rouilly et al., 2006; Wang and Padua, 2003; Zhang et al., 2001). Among polysaccharides, starch is the most employed one to obtain edible packaging by extrusion, but also others such as modified cellulose, pectin, and xanthan gum have been used (Arvanitoyannis and Biliaderis, 1999; Fishman et al., 2000; Flores et al., 2010; Li et al., 2011).

5.5 CONCLUSIONS

It is clear that biopolymers are here to stay. Although edible films have shown several potential applications, just a few industrial applications have been developed; anyway, as the technology continues to improve, the range of potential applications will expand as well.

At the moment, edible packaging is not meant to entirely replace the conventional packaging systems, but their unique versatility makes them envisioned to be used as primary edible packaging together with nonedible parts or as secondary package responsible for additional protection. Therefore, a feasible application of edible packaging can be the reduction of the complexity of overall traditional packaging systems without comprising their functions. To be a competitive product, edible packaging technology needs to fulfill some requirements like its presentation as a new, easy, and eco-friendly technology, developed through a low cost process by the exploitation of by-products and waste from agricultural, livestock, and food industries.

REFERENCES

Albert, S., and Mittal, G. S. 2002. Comparative evaluation of edible coatings to reduce fat uptake in a deep-fried cereal product. *Food Research International,* 35(5), 445–458.

Albertos, I., Rico, D., Diez, A. M., González-Arnáiz, L., García-Casas, M. J., and Jaime I. 2015. Effect of edible chitosan/clove oil films and high-pressure processing on the microbiological shelf life of trout fillets. *Journal of the Science of Food and Agriculture,* 95(14), 2858–2865.

Andreuccetti, C., Carvalho, R. A., Galicia-García, T., Martínez-Bustos, F., and Grosso, C. R. F. 2011. Effect of surfactants on the functional properties of gelatin-based edible films. *Journal of Food Engineering, 103*(2), 129–136.

Andreuccetti, C., Carvalho, R. A., and Grosso, C. R. F. 2009. Effect of hydrophobic plasticizers on functional properties of gelatin-based films. *Food Research International, 42*(8), 1113–1121.

Arboleya, J.-C., Olabarrieta, I., Luis-Aduriz, A., Lasa, D., Vergara, J., Sanmartín, E., and Martínez de Marañón, I. 2008. From the chef's mind to the dish: How scientific approaches facilitate the creative process. *Food Biophysics, 3*(2), 261–268.

Artharn, A., Prodpran, T., and Benjakul, S. 2009. Round scad protein-based film: Storage stability and its effectiveness for shelf-life extension of dried fish powder. *LWT—Food Science and Technology, 42*(7), 1238–1244.

Arvanitoyannis, I., and Biliaderis, C. G. 1999. Physical properties of polyol-plasticized edible blends made of methyl cellulose and soluble starch. *Carbohydrate Polymers, 38*(1), 47–58.

Avena-Bustillos, R. J., Cisneros-Zevallos, L. A., Krochta, J. M., and Saltveit Jr., M. E. 1994. Application of casein-lipid edible film emulsions to reduce white blush on minimally processed carrots. *Postharvest Biology and Technology, 4*(4), 319–329.

Avérous, L., Fringant, C., and Moro, L. 2001. Starch-based biodegradable materials suitable for thermoforming packaging. *Starch/Staerke, 53*(8), 368–371.

Ayranci, E., and Tunç, S. 1997. Cellulose-based edible films and their effects on fresh beans and strawberries. *Zeitschrift fur Lebensmittel—Untersuchung und Forschung, 205*(6), 470–473.

Balaguer, M. P., Borne, M., Chalier, P., Gontard, N., Morel, M. H., Peyron, S., and Hernandez-Munoz, P. 2013a. Retention and release of cinnamaldehyde from wheat protein matrices. *Biomacromolecules, 14*(5), 1493–1502.

Balaguer, M. P., Cerisuelo, J. P., Gavara, R., and Hernandez-Muñoz, P. 2013b. Mass transport properties of gliadin films: Effect of cross-linking degree, relative humidity, and temperature. *Journal of Membrane Science, 428*, 380–392.

Balaguer, M. P., Fajardo, P., Gartner, H., Gomez-Estaca, J., Gavara, R., Almenar, E., and Hernandez-Munoz, P. 2014a. Functional properties and antifungal activity of films based on gliadins containing cinnamaldehyde and natamycin. *International Journal of Food Microbiology, 173*, 62–71.

Balaguer, M. P., Gomez-Estaca, J., Cerisuelo, J. P., Gavara, R., and Hernandez-Munoz, P. 2014b. Effect of thermo-pressing temperature on the functional properties of bioplastics made from a renewable wheat gliadin resin. *LWT—Food Science and Technology, 56*(1), 161–167.

Balaguer, M. P., Lopez-Carballo, G., Catala, R., Gavara, R., and Hernandez-Munoz, P. 2013c. Antifungal properties of gliadin films incorporating cinnamaldehyde and application in active food packaging of bread and cheese spread foodstuffs. *International Journal of Food Microbiology, 166*(3), 369–377.

Baldwin, E. A., Hagenmaier, R., and Bai, J. (Eds.) 2011. *Edible Coatings and Films to Improve Food Quality*: Boca Raton, Florida: CRC Press.

Barone, J. R., Schmidt, W. F., and Gregoire, N. T. 2006. Extrusion of feather keratin. *Journal of Applied Polymer Science, 100*(2), 1432–1442.

Belyamani, I., Prochazka, F., and Assezat, G. 2014. Production and characterization of sodium caseinate edible films made by blown-film extrusion. *Journal of Food Engineering, 121*(1), 39–47.

Blanco-Pascual, N., Alemán, A., Gómez-Guillén, M. C., and Montero, M. P. 2014a. Enzyme-assisted extraction of κ/ι-hybrid carrageenan from *Mastocarpus stellatus* for obtaining bioactive ingredients and their application for edible active film development. *Food and Function, 5*(2), 319–329.

Blanco-Pascual, N., Fernández-Martín, F., and Montero, M. P. 2013. Effect of different protein extracts from *Dosidicus gigas* muscle co-products on edible films development. *Food Hydrocolloids, 33*(1), 118–131.

Blanco-Pascual, N., Fernandez-Martin, F., and Montero, P. 2014b. Jumbo squid (*Dosidicus gigas*) myofibrillar protein concentrate for edible packaging films and storage stability. *LWT—Food Science and Technology, 55*(2), 543–550.

Blanco-Pascual, N., Montero, M. P., and Gómez-Guillén, M. C. 2014c. Antioxidant film development from unrefined extracts of brown seaweeds *Laminaria digitata* and *Ascophyllum nodosum*. *Food Hydrocolloids, 37,* 100–110.

Bourtoom, T. 2008. Edible films and coatings: Characteristics and properties. *International Food Research Journal, 15*(3), 237–248.

Bravin, B., Peressini, D., and Sensidoni, A. 2006. Development and application of polysaccharide–lipid edible coating to extend shelf-life of dry bakery products. *Journal of Food Engineering, 76*(3), 280–290.

Butler, B. L., Vergano, P. J., Testin, R. F., Bunn, J. M., and Wiles, J. L. 1996. Mechanical and barrier properties of edible chitosan films as affected by composition and storage. *Journal of Food Science, 61*(5), 953–956.

Campos, C. A., Gerschenson, L. N., and Flores, S. K. 2011. Development of edible films and coatings with antimicrobial activity. *Food and Bioprocess Technology, 4*(6), 849–875.

Caner, C., Vergano, P. J., and Wiles, J. L. 1998. Chitosan film mechanical and permeation properties as affected by acid, plasticizer, and storage. *Journal of Food Science, 63*(6), 1049–1053.

Cardenia, V., Rodriguez-Estrada, M. T., Baldacci, E., and Lercker, G. 2013. Health-related lipids components of sardine muscle as affected by photooxidation. *Food and Chemical Toxicology, 57,* 32–38.

Cervera, M. F., Karjalainen, M., Airaksinen, S., Rantanen, J., Krogars, K., Heinämäki, J., and Yliruusi, J. 2004. Physical stability and moisture sorption of aqueous chitosan–amylose starch films plasticized with polyols. *European Journal of Pharmaceutics and Biopharmaceutics, 58*(1), 69–76.

Cisneros-Zevallos, L., and Krochta, J. M. 2003. Dependence of coating thickness on viscosity of coating solution applied to fruits and vegetables by dipping method. *Journal of Food Science, 68*(2), 503–510.

Conte, A., Gammariello, D., Di Giulio, S., Attanasio, M., and Del Nobile, M. A. 2009. Active coating and modified-atmosphere packaging to extend the shelf life of Fior di Latte cheese. *Journal of Dairy Science, 92*(3), 887–894.

Chen, C.-L., Li, P.-Y., Hu, W.-H., Lan, M.-H., Chen, M.-J., and Chen, H.-H. 2008. Using HPMC to improve crust crispness in microwave-reheated battered mackerel nuggets: Water barrier effect of HPMC. *Food Hydrocolloids, 22*(7), 1337–1344.

Chinabhark, K., Benjakul, S., and Prodpran, T. 2007. Effect of pH on the properties of protein-based film from bigeye snapper (*Priacanthus tayenus*) surimi. *Bioresource Technology, 98*(1), 221–225.

Cho, S. W., Gallstedt, M., and Hedenqvist, M. S. 2010. Properties of wheat gluten/poly(lactic acid) laminates. *Journal of Agricultural and Food Chemistry, 58*(12), 7344–7350.

Cuq, B., Gontard, N., Cuq, J. L., and Guilbert, S. 1996. Stability of myofibrillar protein-based biopackagings during storage. *Food Science and Technology-Lebensmittel-Wissenschaft & Technologie, 29*(4), 344–348.

Cuq, B., Gontard, N., and Guilbert, S. 1997. Thermoplastic properties of fish myofibrillar proteins: Application to biopackaging fabrication. *Polymer, 38*(16), 4071–4078.

Cuq, B., Gontard, N., and Guilbert, S. 1999. Effects of thermoulding process conditions on the properties of agro-materials based on fish myofibrillar proteins. *Food Science and Technology-Lebensmittel-Wissenschaft & Technologie, 32*(2), 107–113.

Dang, K. T. H., Singh, Z., and Swinny, E. E. 2008. Edible coatings influence fruit ripening, quality, and aroma biosynthesis in mango fruit. *Journal of Agricultural and Food Chemistry, 56*(4), 1361–1370.

Debeaufort, F., and Voilley, A. 1994. Aroma compound and water vapor permeability of edible films and polymeric packagings. *Journal of Agricultural and Food Chemistry, 42*(12), 2871–2875.

Debeaufort, F., and Voilley, A. 2009. Lipid-based edible films and coatings. In K. C. Huber and E. Embuscado (Eds.), *Edible Films and Coatings for Food Applications* (pp. 135–168). New York: Springer.

De Grandi Castro Freitas, D., Berbari, S. A. G., Prati, P., Fakhouri, F. M., Collares Queiroz, F. P., and Vicente, E. 2009. Reducing fat uptake in cassava product during deep-fat frying. *Journal of Food Engineering, 94*(3–4), 390–394.

De Moraes, J. O., Scheibe, A. S., Sereno, A., and Laurindo, J. B. 2013. Scale-up of the production of cassava starch based films using tape-casting. *Journal of Food Engineering, 119*(4), 800–808.

Elghaouth, A., Ponnampalam, R., Castaigne, F., and Arul, J. 1992. Chitosan coating to extend the storage life of tomatoes. *HortScience, 27*(9), 1016–1018.

Embuscado, M. E., and Huber, K. C. 2009. *Edible Films and Coatings for Food Applications* (pp. 245–250). Dordrecht, the Netherlands: Springer.

Fabra, M. J., López-Rubio, A., and Lagaron, J. M. 2014. Biopolymers for food packaging applications. In M. R. Aguilar and J. San Román (Eds.), *Smart Polymers and their Applications*, (pp. 476–509). Cambridge, UK: Woodhead Publishing.

Fabra, M. J., Talens, P., Gavara, R., and Chiralt, A. 2012. Barrier properties of sodium caseinate films as affected by lipid composition and moisture content. *Journal of Food Engineering, 109*(3), 372–379.

Falguera, V., Quintero, J. P., Jiménez, A., Muñoz, J. A., and Ibarz, A. 2011. Edible films and coatings: Structures, active functions and trends in their use. *Trends in Food Science & Technology, 22*(6), 292–303.

Fang, Y., Tung, M. A., Britt, I. J., Yada, S., and Dalgleish, D. G. 2002. Tensile and barrier properties of edible films made from whey proteins. *Journal of Food Science, 67*(1), 188–193.

Farris, S., Cozzolino, C. A., Introzzi, L., and Piergiovanni, L. 2010. Development and characterization of a gelatin-based coating with unique sealing properties. *Journal of Applied Polymer Science, 118*(5), 2969–2975.

Farris, S., Introzzi, L., and Piergiovanni, L. 2009. Evaluation of a bio-coating as a solution to improve barrier, friction and optical properties of plastic films. *Packaging Technology and Science, 22*(2), 69–83.

Fishman, M. L., Coffin, D. R., Konstance, R. P., and Onwulata, C. I. 2000. Extrusion of pectin/starch blends plasticized with glycerol. *Carbohydrate Polymers, 41*(4), 317–325.

Flores, S. K., Costa, D., Yamashita, F., Gerschenson, L. N., and Grossmann, M. V. 2010. Mixture design for evaluation of potassium sorbate and xanthan gum effect on properties of tapioca starch films obtained by extrusion. *Materials Science and Engineering: C, 30*(1), 196–202.

Gastaldi, E., Chalier, P., Guillemin, A., and Gontard, N. 2007. Microstructure of protein-coated paper as affected by physico-chemical properties of coating solutions. *Colloids and Surfaces A: Physicochemical and Engineering Aspects, 301*(1–3), 301–310.

Gemma, H., and Du, J. M. 1998. Effect of application of various chitosans with different molecular weights on the storability of "Jonagold" apple. In R. Bieleski, W. A. Laing and C. J. Clark (Eds.), *Postharvest '96—Proceedings of the International Postharvest Science Conference* (Vol. 464). International Society for Horticultural Science (ISHS), Leuven, Belgium.

Gennadios, A., Handa, A., Froning, G. W., Weller, C. L., and Hanna, M. A. 1998. Physical properties of egg white – dialdehyde starch films. *Journal of Agricultural and Food Chemistry, 46*(4), 1297–1302.

Gomez-Estaca, J., Bravo, L., Gomez-Guillen, M. C., Aleman, A., and Montero, P. 2009. Antioxidant properties of tuna-skin and bovine-hide gelatin films induced by the addition of oregano and rosemary extracts. *Food Chemistry, 112*(1), 18–25.

Gómez-Estaca, J., López de Lacey, A., López-Caballero, M. E., Gómez-Guillén, M. C., and Montero, P. 2010. Biodegradable gelatin-chitosan films incorporated with essential oils as antimicrobial agents for fish preservation. *Food Microbiology, 27*(7), 889–896.

Gomez-Estaca, J., Montero, P., Gimenez, B., and Gomez-Guillen, M. C. 2007. Effect of functional edible films and high pressure processing on microbial and oxidative spoilage in cold-smoked sardine (*Sardina pilchardus*). *Food Chemistry, 105*(2), 511–520.

Gómez-Estaca, J., Montero, P., and Gómez-Guillén, M. C. 2014. Shrimp (*Litopenaeus vannamei*) muscle proteins as source to develop edible films. *Food Hydrocolloids, 41*, 86–94.

Gomez-Guillen, M. C., Ihl, M., Bifani, V., Silva, A., and Montero, P. 2007. Edible films made from tuna-fish gelatin with antioxidant extracts of two different murta ecotypes leaves (*Ugni molinae* Turcz). *Food Hydrocolloids, 21*(7), 1133–1143.

Guillaume, C., Pinte, J., Gontard, N., and Gastaldi, E. 2010. Wheat gluten-coated papers for bio-based food packaging: Structure, surface and transfer properties. *Food Research International, 43*(5), 1395–1401.

Guo, J., Ge, L., Li, X., Mu, C., and Li, D. 2014. Periodate oxidation of xanthan gum and its crosslinking effects on gelatin-based edible films. *Food Hydrocolloids, 39*, 243–250.

Hamaguchi, P. Y., Weng, W., Kobayashi, T., Runglertkreingkrai, J., and Tanaka, M. 2007. Effect of fish meat quality on the properties of biodegradable protein films. *Food Science and Technology Research, 13*(3), 200–204.

Han, C., Zhao, Y., Leonard, S. W., and Traber, M. G. 2004. Edible coatings to improve storability and enhance nutritional value of fresh and frozen strawberries (*Fragaria ananassa*) and raspberries (*Rubus ideaus*). *Postharvest Biology and Technology, 33*(1), 67–78.

Han, J. H., and Krochta, J. M. 1999. Wetting properties and water vapor permeability of whey-protein-coated paper. *Transactions of the American Society of Agricultural Engineers, 42*(5), 1375–1382.

Han, J. H., and Krochta, J. M. 2001. Physical properties and oil absorption of whey-protein-coated paper. *Journal of Food Science, 66*(2), 294–299.

Hernandez-Izquierdo, V. M., and Krochta, J. M. 2008. Thermoplastic processing of proteins for film formation—A review. *Journal of Food Science, 73*(2), R30–R39.

Hernandez-Izquierdo, V. M., Reid, D. S., McHugh, T. H., Berrios, J. D. J., and Krochta, J. M. 2008. Thermal transitions and extrusion of glycerol-plasticized whey protein mixtures. *Journal of Food Science, 73*(4), E169–E175.

Hernandez-Munoz, P., Almenar, E., Ocio, M. J., and Gavara, R. 2006. Effect of calcium dips and chitosan coatings on postharvest life of strawberries (*Fragaria ananassa*). *Postharvest Biology and Technology, 39*(3), 247–253.

Hernandez-Munoz, P., Kanavouras, A., Lagaron, J. M., and Gavara, R. 2005. Development and characterization of films based on chemically cross-linked gliadins. *Journal of Agricultural and Food Chemistry, 53*(21), 8216–8223.

Hernández-Muñoz, P., López-Rubio, A., del-Valle, V., Almenar, E., and Gavara, R. 2003. Mechanical and water barrier properties of glutenin films influenced by storage time. *Journal of Agricultural and Food Chemistry, 52*(1), 79–83.

Hernandez-Munoz, P., Lopez-Rubio, A., Del-Valle, V., Almenar, E., and Gavara, R. 2004. Mechanical and water barrier properties of glutenin films influenced by storage time. *Journal of Agricultural and Food Chemistry, 52*(1), 79–83.

Hong, S. I., Han, J. H., and Krochta, J. M. 2004. Optical and surface properties of whey protein isolate coatings on plastic films as influenced by substrate, protein concentration, and plasticizer type. *Journal of Applied Polymer Science, 92*(1), 335–343.

Hong, S. I., and Krochta, J. M. 2003. Oxygen barrier properties of whey protein isolate coatings on polypropylene films. *Journal of Food Science, 68*(1), 224–228.

Huang, L., Zhang, M., Yan, W., Mujumdar, A. S., and Sun, D. 2011. Rehydration characteristics of freeze-dried strawberry pieces as affected by whey protein edible coatings. *International Journal of Food Science & Technology, 46*(4), 671–677.

Huimin, J., Tao, H., Liping, L., and Haiying, Z. 2009. Effects of edible coatings on browning of fresh-cut peach fruits. *Transactions of the Chinese Society of Agricultural Engineering, 25*(3), 282–286.

Huneault, M. A., and Li, H. 2012. Preparation and properties of extruded thermoplastic starch/polymer blends. *Journal of Applied Polymer Science, 126*(1), E96–E108.

Janjarasskul, T., and Krochta, J. M. 2010. Edible packaging materials. *Annual Review of Food Science and Technology, 1*(1), 415–448.

Javanmard, M., Ahangari, R., and Tavakolipour, H. 2011. *Journal of Food Process Engineering, 34*(4), 1156–1171.

Jeon, Y.-J., Kamil, J. Y. V. A., and Shahidi, F. 2002. Chitosan as an edible invisible film for quality preservation of herring and Atlantic cod. *Journal of Agricultural and Food Chemistry, 50*(18), 5167–5178.

Kang, H. J., Jo, C., Kwon, J. H., Kim, J. H., Chung, H. J., and Byun, M. W. 2007. Effect of a pectin-based edible coating containing green tea powder on the quality of irradiated pork patty. *Food Control, 18*(5), 430–435.

Kilincceker, O., and Hepsag, F. 2012. Edible coating effects on fried potato balls. *Food and Bioprocess Technology, 5*(4), 1349–1354.

Kim, S. H., No, H. K., and Prinyawiwatkul, W. 2008. Plasticizer types and coating methods affect quality and shelf life of eggs coated with chitosan. *Journal of Food Science, 73*(3), S111–S117.

Kokoszka, S., and Lenart, A. 2007. Edible coatings—formation, characteristics and use—A review. *Polish Journal of Food and Nutrition Sciences, 57*(4), 399–404.

Krishna, M., Nindo, C. I., and Min, S. C. 2012. Development of fish gelatin edible films using extrusion and compression molding. *Journal of Food Engineering, 108*(2), 337–344.

Krogars, K., Heinämäki, J., Karjalainen, M., Niskanen, A., Leskelä, M., and Yliruusi, J. 2003. Enhanced stability of rubbery amylose-rich maize starch films plasticized with a combination of sorbitol and glycerol. *International Journal of Pharmaceutics, 251*(1–2), 205–208.

Kuutti, L., Peltonen, J., Myllärinen, P., Teleman, O., and Forssell, P. 1998. AFM in studies of thermoplastic starches during ageing. *Carbohydrate Polymers, 37*(1), 7–12.

Lago-Vanzela, E. S., do Nascimento, P., Fontes, E. A. F., Mauro, M. A., and Kimura, M. 2013. Edible coatings from native and modified starches retain carotenoids in pumpkin during drying. *LWT—Food Science and Technology, 50*(2), 420–425.

Lee, S. Y., Dangaran, K. L., Guinard, J. X., and Krochta, J. M. 2002. Consumer acceptance of whey-protein-coated as compared with shellac-coated chocolate. *Journal of Food Science, 67*(7), 2764–2769.

Leerahawong, A., Tanaka, M., Okazaki, E., and Osako, K. 2012. Stability of the physical properties of plasticized edible films from squid (*Todarodes pacificus*) mantle muscle during storage. *Journal of Food Science, 77*(6), E159–E165.

Li, H., and Huneault, M. A. 2011. Comparison of sorbitol and glycerol as plasticizers for thermoplastic starch in TPS/PLA blends. *Journal of Applied Polymer Science, 119*(4), 2439–2448.

Li, M., Liu, P., Zou, W., Yu, L., Xie, F., Pu, H., and Chen, L. 2011. Extrusion processing and characterization of edible starch films with different amylose contents. *Journal of Food Engineering, 106*(1), 95–101.

López de Lacey, A. M., López-Caballero, M. E., and Montero, P. 2014. Agar films containing green tea extract and probiotic bacteria for extending fish shelf-life. *LWT—Food Science and Technology, 55*(2), 559–564.

Lopez-Rubio, A., Almenar, E., Hernandez-Munoz, P., Lagaron, J. M., Catala, R., and Gavara, R. 2004. Overview of active polymer-based packaging technologies for food applications. *Food Reviews International, 20*(4), 357–387.

López-Rubio, A., and Lagaron, J. M. 2011. Improved incorporation and stabilisation of β-carotene in hydrocolloids using glycerol. *Food Chemistry, 125*(3), 997–1004.

Mallikarjunan, P., Chinnan, M. S., Balasubramaniam, V. M., and Phillips, R. D. 1997. Edible coatings for deep-fat frying of starchy products. *Food Science and Technology-Lebensmittel-Wissenschaft & Technologie, 30*(7), 709–714.

Maté, J. I., and Krochta, J. M. 1998. Oxygen uptake model for uncoated and coated peanuts. *Journal of Food Engineering, 35*(3), 299–312.

Mellema, M. 2003. Mechanism and reduction of fat uptake in deep-fat fried foods. *Trends in Food Science & Technology, 14*(9), 364–373.

Meng, F. Z., Jiang, Y., Sun, Z. H., Yin, Y. Z., and Li, Y. Y. 2009. Electrohydrodynamic liquid atomization of biodegradable polymer microparticles: Effect of electrohydrodynamic liquid atomization variables on microparticles. *Journal of Applied Polymer Science, 113*(1), 526–534.

Miller, K., and Krochta, J. 1998. Measuring aroma transport in polymer films. *Transactions of the ASAE, 41*(2), 427–433.

Munin, A., and Edwards-Lévy, F. 2011. Encapsulation of natural polyphenolic compounds: A review. *Pharmaceutics, 3*(4), 793–829.

Nisperos-Carriedo, M. O., Shaw, P. E., and Baldwin, E. A. 1990. Changes in volatile flavor components of pineapple orange juice as influenced by the application of lipid and composite films. *Journal of Agricultural and Food Chemistry, 38*(6), 1382–1387.

Osés, J., Fernández-Pan, I., Mendoza, M., and Maté, J. I. 2009. Stability of the mechanical properties of edible films based on whey protein isolate during storage at different relative humidity. *Food Hydrocolloids, 23*(1), 125–131.

Park, H. J., Bunn, J. M., Weller, C. L., Vergano, P. J., and Testin, R. F. 1994. Water vapor permeability and mechanical properties of grain protein based films as affected by mixtures of polyethylene glycol and glycerin plasticizers. *Transactions of the ASAE, 37*(4), 1281–1285.

Park, H. J., Kim, S. H., Lim, S. T., Shin, D. H., Choi, S. Y., and Hwang, K. T. 2000. Grease resistance and mechanical properties of isolated soy protein-coated paper. *Journal of the American Oil Chemists' Society, 77*(3), 269–273.

Pascall, M. A., and Lin, S. J. 2013. The application of edible polymeric films and coatings in the food industry. *Food Processing and Technology, 4*(2), e116–e117.

Pavlath, A. E., and Orts, W. 2009. Edible films and coatings: Why, what, and how? In M. E. Embuscado and K. C. Huber (Eds). *Edible Films and Coatings for Food Applications* (pp. 1–23). New York: Springer.

Pérez-Gago, M. B., Nadaud, P., and Krochta, J. M. 1999. Water vapor permeability, solubility, and tensile properties of heat-denatured versus native whey protein films. *Journal of Food Science, 64*(6), 1034–1037.

Perez-Mateos, M., Montero, P., and Gomez-Guillen, M. C. 2009. Formulation and stability of biodegradable films made from cod gelatin and sunflower oil blends. *Food Hydrocolloids, 23*(1), 53–61.

Pol, H., Dawson, P., Acton, J., and Ogale, A. 2002. Soy protein isolate/corn-zein laminated films: Transport and mechanical properties. *Journal of Food Science, 67*(1), 212–217.

Rhim, J. W., Lee, J. H., and Hong, S. I. 2006. Water resistance and mechanical properties of biopolymer (alginate and soy protein) coated paperboards. *LWT—Food Science and Technology, 39*(7), 806–813.

Rojas-Grau, M. A., Soliva-Fortuny, R., and Martin-Belloso, O. 2009. Edible coatings to incorporate active ingredients to fresh-cut fruits: A review. *Trends in Food Science & Technology, 20*(10), 438–447.

Rojas-Grau, M. A., Tapia, M. S., Rodriguez, F. J., Carmona, A. J., and Martin-Belloso, O. 2007. Alginate and gellan-based edible coatings as carriers of antibrowning agents applied on fresh-cut Fuji apples. *Food Hydrocolloids, 21*(1), 118–127.

Rouilly, A., Meriaux, A., Geneau, C., Silvestre, F., and Rigal, L. 2006. Film extrusion of sunflower protein isolate. *Polymer Engineering and Science, 46*(11), 1635–1640.

Sathivel, S. 2005. Chitosan and protein coatings affect yield, moisture loss, and lipid oxidation of pink salmon (*Oncorhynchus gorbuscha*) fillets during frozen storage. *Journal of Food Science, 70*(8), E455–E459.

Shiku, Y., Hamaguchi, P. Y., Benjakul, S., Visessanguan, W., and Tanaka, M. 2004. Effect of surimi quality on properties of edible films based on Alaska pollack. *Food Chemistry, 86*(4), 493–499.

Simon, J., Müller, H. P., Koch, R., and Müller, V. 1998. Thermoplastic and biodegradable polymers of cellulose. *Polymer Degradation and Stability, 59*(1–3), 107–115.

Singthong, J., and Thongkaew, C. 2009. Using hydrocolloids to decrease oil absorption in banana chips. *LWT—Food Science and Technology, 42*(7), 1199–1203.

Somanathan, N., Naresh, M., Arumugam, V., Ranganathan, T., and Sanjeevi, R. 1992. Mechanical properties of alkali treated casein films. *Polymer Journal, 24*(7), 603–611.

Suman, S. P., Mancini, R. A., Joseph, P., Ramanathan, R., Konda, M. K. R., Dady, G., and Yin, S. 2010. Packaging-specific influence of chitosan on color stability and lipid oxidation in refrigerated ground beef. *Meat Science, 86*(4), 994–998.

Suwannateep, N., Wanichwecharungruang, S., Haag, S. F., Devahastin, S., Groth, N., Fluhr, J. W., and Meinke, M. C. 2012. Encapsulated curcumin results in prolonged curcumin activity *in vitro* and radical scavenging activity ex vivo on skin after UVB-irradiation. *European Journal of Pharmaceutics and Biopharmaceutics, 82*(3), 485–490.

Taghizadeh, A., Sarazin, P., and Favis, B. D. 2013. High molecular weight plasticizers in thermoplastic starch/polyethylene blends. *Journal of Materials Science, 48*(4), 1799–1811.

Talens, P., and Krochta, J. M. 2005. Plasticizing effects of beeswax and carnauba wax on tensile and water vapor permeability properties of whey protein films. *Journal of Food Science, 70*(3), E239–E243.

Tanada-Palmu, P. S., and Grosso, C. R. F. 2005. Effect of edible wheat gluten-based films and coatings on refrigerated strawberry (*Fragaria ananassa*) quality. *Postharvest Biology and Technology, 36*(2), 199–208.

Tharanathan, R. N. 2003. Biodegradable films and composite coatings: Past, present and future. *Trends in Food Science & Technology, 14*(3), 71–78.

Tongnuanchan, P., Benjakul, S., and Prodpran, T. 2011. Roles of lipid oxidation and pH on properties and yellow discolouration during storage of film from red tilapia (*Oreochromis niloticus*) muscle protein. *Food Hydrocolloids, 25*(3), 426–433.

Tongnuanchan, P., Benjakul, S., Prodpran, T., and Songtipya, P. 2013. Properties and stability of protein-based films from red tilapia (*Oreochromis niloticus*) protein isolate incorporated with antioxidant during storage. *Food and Bioprocess Technology, 6*(5), 1113–1126.

Trezza, T. A., and Krochta, J. M. 2000. Color stability of edible coatings during prolonged storage. *Journal of Food Science, 65*(7), 1166–1169.

Trezza, T. A., and Vergano, P. J. 1994. Grease resistance of corn zein coated paper. *Journal of Food Science, 59*(4), 912–915.

Trezza, T. A., Wiles, J. L., and Vergano, P. J. 1998. Water vapor and oxygen barrier properties of corn zein coated paper. *Tappi Journal, 81*(8), 171–176.

Ullsten, N. H., Cho, S. W., Spencer, G., Gallstedt, M., Johansson, E., and Hedenqvist, M. S. 2009. Properties of extruded vital wheat gluten sheets with sodium hydroxide and salicylic acid. *Biomacromolecules, 10*(3), 479–488.

Ullsten, N. H., Gallstedt, M., Johansson, E., Graslund, A., and Hedenqvist, M. S. 2006. Enlarged processing window of plasticized wheat gluten using salicylic acid. *Biomacromolecules, 7*(3), 771–776.

Van Soest, J. J. G., and Knooren, N. 1997. Influence of glycerol and water content on the structure and properties of extruded starch plastic sheets during aging. *Journal of Applied Polymer Science, 64*(7), 1411–1422.

Vargas, M., Pastor, C., Chiralt, A., McClements, D. J., and González-Martínez, C. 2008. Recent advances in edible coatings for fresh and minimally processed fruits. *Critical Reviews in Food Science and Nutrition, 48*(6), 496–511.

Wang, Y., and Padua, G. W. 2003. Tensile properties of extruded zein sheets and extrusion blown films. *Macromolecular Materials and Engineering, 288*(11), 886–893.

Wessling, C., Nielsen, T., Leufvén, A., and Jägerstad, M. 1999. Retention of α-tocopherol in low-density polyethylene (LDPE) and polypropylene (PP) in contact with foodstuffs and food-simulating liquids. *Journal of the Science of Food and Agriculture, 79*(12), 1635–1641.

Williams, R., and Mittal, G. S. 1999. Water and fat transfer properties of polysaccharide films on fried pastry mix. *LWT—Food Science and Technology, 32*(7), 440–445.

Xie, L., Hettiarachchy, N. S., Ju, Z. Y., Meullenet, J., Wang, H., Slavik, M. F., and Hanes, M. E. 2002. Edible film coating to minimize eggshell breakage and reduce post-wash bacterial contamination measured by dye penetration in eggs. *Journal of Food Science, 67*(1), 280–284.

Zhang, J., Mungara, P., and Jane, J. 2001. Mechanical and thermal properties of extruded soy protein sheets. *Polymer, 42*(6), 2569–2578.

6 Mass Transfer Measurement and Modeling for Designing Protective Edible Films

Valérie Guillard, Carole Guillaume, Mia Kurek, and Nathalie Gontard

CONTENTS

Abstract	182
6.1 Introduction	182
6.2 Main Functions of Edible Films and Coatings as Regulating Agent of Mass Transfer	183
6.2.1 Mechanisms of Food Spoilage and Degradation: Importance of the Surrounding Atmosphere	183
6.2.2 Edible Films and Coatings: Actor to Control Mass Transfer Phenomena	184
6.2.3 Which Material for Which Targeted Mass Transfer?	185
6.2.4 A Case Study: Regulation of Moisture Transfer in Composite Food Product	186
6.2.5 Assessment of Edible Films and Coating Performance and Compatibility	186
6.3 Characterization of Mass Transfer Phenomena	189
6.3.1 Basics on Mass Transfer	189
6.3.2 Measurement of Mass Transfer Parameters	190
6.3.2.1 Permeation	190
6.3.2.2 Diffusivity	191
6.3.2.3 Detection System	191
6.3.2.4 Experimental Setup and Numerical Treatment of Data for Effective Diffusivity Determination	192
6.3.2.5 Sorption	193
6.3.3 Factors Affecting Mass Transfer Parameters	194
6.3.3.1 Films Composition and Structure	194
6.3.3.2 Influence of the Diffusing Molecule	197
6.3.3.3 Influence of External Parameters	198

6.4 Edible Films Dimensioning Based on Mathematical Modeling 199
　　6.4.1 Case A: Simplified Approach Considering Only Permeation
　　　　　　in the Coating/Film 202
　　6.4.2 Case B: Permeation Coupled with Reaction..............................204
　　6.4.3 Case C: Unsteady-State Transfer from the Film into the Food 205
　　6.4.4 Case D: Coupling Diffusion and Reaction 205
6.5 Conclusion 206
References...207

ABSTRACT

A wide range of film-forming compounds are available today and facilitate the design and tailoring of edible films and coatings with optimized functional properties. The main targeted functional properties are mass transfer properties, that is, solubility, diffusivity, and permeability compatible with the targeted applications, for example, moisture or O_2 barrier, controlled release of an active agent, O_2 scavenging property, among others. The dimensioning of an edible film or coating relies thus essentially on the knowledge of its targeted mass transfer property. Despite recent huge efforts made in the development of integrated approach for building the specifications of edible film and coating, a realistic assessment of the film or coating efficacy is still critical. Mathematical modeling of mass transfer phenomena involved in the coated food is the base of a better realistic assessment of the performance of protective edible films. The aim of this chapter is to make a progress report on the measurement of mass transfer properties and on the modeling of mass transfer into the coating/food system. The last section illustrates how mathematical tools could be used to adequately dimension edible films or coatings.

6.1 INTRODUCTION

Edible films or coatings have been generally defined as thin layers of material which are eaten by the consumer and permit to master mass transfer (moisture, gas, solute) between the food and the environment or between different components of the food itself (e.g., case of composite food). Fruit waxing was first used in the twelfth century in China and further in England for meat larding during the sixteenth century (Kester and Fennema, 1986). Even today, this technique still remains an important tool to maintain the quality of slightly modified fruits, fruit pieces (Dhall, 2013; Olivas and Barbosa-Cánovas, 2005; Valencia-Chamorro et al., 2011; Vargas et al., 2008), or processed products (Debeaufort et al., 2000; Gennadios et al., 1997; Guillard et al., 2003a; Koelsch, 1994; Roca et al., 2008a, 2008b).

The targeted mass transfer when edible films and coatings are used mainly concern gas and vapor permeation between the surrounding atmosphere and the product (in that case, the targeted molecules are O_2, CO_2, water vapor, and organic vapors such as aroma compounds) and in a less extent solutes diffusion between two media in contact (such as colorings and salts). In both cases, the main objective is to avoid the unwanted diffusion of a molecule in food compartment, where it should not exist. For instance, edible coatings are very useful to decrease O_2 influx into coated fruit to slow down its respiration metabolism and thus, increase its shelf life. They are also a powerful technique to prevent water transfer between two components of contrasted water activity of a same food product called composite food (i.e., ready-to-eat foods such as sandwiches, confectionaries, mixed salads, among others) and thus maintain food sensory properties (mainly crispness of the cereal-based phase).

Edible films and coatings are also used for controlling diffusion of an active molecule such as antioxidant or antimicrobial. In that case, the active compound is voluntarily added in the edible formulation. The edible film or coating raw material is chosen for its diffusion properties of the active molecule: very slow diffusion is usually wanted when the active compound must be kept in the edible coating—that is, on the food surface—while intermediate to high diffusion rates are preferred when release of the compound into food in contact is targeted.

Whatever the aim of the transfer involved (limiting the unwanted diffusion or favoring the desired one), the most important properties to know when defining the specifications of an edible film or coating for a targeted application is the window of optimal mass transfer properties required, such as permeability, solubility, and diffusivity. This window of optimal mass transfer properties will condition the choice of the raw material, the structure, and processing conditions used for making the edible film or coating. The definition of the window of optimal mass transfer properties is a tricky task when an empirical experimental approach is used. In that case, the formulations of edible solution are chosen *a priori*, then tested in real conditions of use (in contact with the product and in real storage conditions) until the right formulation is found. This is time and cost consuming with a high risk of failure. To circumvent this "trial-and-error" approach, more integrated approaches based on the use of mathematical models of mass transfer have been proposed. They permit to calculate the targeted mass transfer properties of the coating prior testing and then to reason its formulation on the basis of these specifications.

This chapter aims at describing (1) the methodologies used for measuring permeability, solubilities, and diffusivities of thin films focusing on the specificities of edible films and coatings and (2) the modeling approaches used for dimensioning edible films and coatings, that is, defining *a priori* the optimal mass transfer properties for a target application. The first part of this chapter will present the main functions of edible films and coatings as regulating agent of mass transfer, then the second part will recall the basics on mass transfer and how measuring mass transfer properties. The last part will be dedicated to the modeling aspects.

6.2 MAIN FUNCTIONS OF EDIBLE FILMS AND COATINGS AS REGULATING AGENT OF MASS TRANSFER

6.2.1 Mechanisms of Food Spoilage and Degradation: Importance of the Surrounding Atmosphere

Food spoilage is related to a loss in food quality due to microbial contamination, nonoptimal sensory attributes (i.e., loss of texture, flavor and/or odor, color, and aspect), and to a minor extent to a loss in nutritional value. Mass transfer phenomena between the surrounding atmosphere and the food product are driving these reactions of degradation. Food quality of inert or living food evolves with time of storage due to physical, chemical, or biological reactions mainly driven by oxygen and water vapor transfer from the surrounding. For instance, oxygen mass transfer strongly impacts the oxidation rate of O_2-sensitive compounds, such as vitamins or fatty acids, leading to depreciation of nutritional properties and enhancement of aerobes growth which, in turn, could significantly affect the composition of oxygen and carbon dioxide through their respiring metabolism.

Another example of food degradation is related to moisture transfer. Loss in dry biscuit texture is mainly related to remoistening of the product via the external atmosphere or from another component in contact displaying a higher water activity (a_w) value. This last type of phenomenon occurs mainly in composite food product where components having different texture and thus, a_w, are voluntarily associated (e.g., ready-to-eat food such as cereal-based mix for breakfast, sandwiches, cereal-based composite in contact with savory and wet filling) (Bourlieu et al., 2009).

In all the aforementioned examples, mass transfers are involved between the product and its surrounding atmosphere or between the different phases of food itself (Angellier-Coussy et al., 2013; Bourlieu et al., 2007, 2009c).

6.2.2 Edible Films and Coatings: Actor to Control Mass Transfer Phenomena

In summary, the main role of edible films and coatings is to regulate mass transfer between the product and its surrounding atmosphere and between the different parts of the product itself. In the first case, mainly with coating materials that are formed directly on the solid food product (e.g., waxing of fruits and vegetables) are considered. In the second case, an edible film is placed at the interface between two food components to prevent mass exchange between these two phases (e.g., yoghurt-containing crispy cereal-based phase, biscuit filled with a moist, savory cream, etc.) (Figure 6.1). The molecule of interest diffusing in the medium depends on the

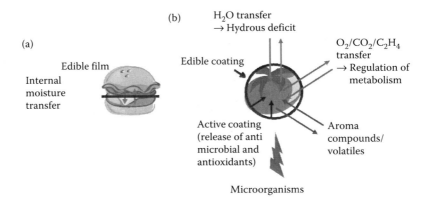

FIGURE 6.1 Main mass transfers occurring in the coating/food system. (a) Composite food. (b) Fresh produce.

targeted application. When a coating is used as barrier between the product and its external atmosphere, O_2 and/or CO_2 are usually targeted and the objective is thus to regulate O_2/CO_2 absorption by a respiring product. This superficial layer could be ingested by the consumer or not (if, e.g., the fruit is peeled before its consumption) (Dhall, 2013; Olivas and Barbosa-Cánovas, 2005). When an internal edible layer is placed between two phases, the main objective is to prevent moisture transfer between the component of high a_w toward the component of lower a_w (Bourlieu et al., 2009c; Debeaufort, 1998; Debeaufort et al., 2000). Then, the targeted products are generally a cereal-based component associated to a moist savory filling. Internal layer could also be used, in a less extent, to master flavorings or colorings migration from a colored phase in a white or pale phase (e.g., dry berries dispersed in a dairy product) (Debeaufort et al., 2003). In this last case, the edible layer is obligatorily ingested by the consumer.

Beyond their barrier properties, edible coatings can be used as vector of preservatives such as antimicrobial, antioxidant, and anti-browning additives (e.g., ascorbic and citric acids, resveratrol, or tocopherols). These compounds help preserving foods by concentrating on the food surface, that is, where protection is needed, the required concentration of the active molecule to prevent microorganism growth or oxidation, supplement the diet, and/or protect the sensory and nutritive quality of the food itself (Guillard et al., 2009; Valencia-Chamorro et al., 2011). Recently, edible coatings have also been used to deliver and maintain desirable concentrations of color, flavor, spiciness, sweetness, saltiness, nutraceuticals, and probiotic organisms. That means only very small amount of additives is required per kilogram of food compared to the usual case when preservatives are added to the overall food product, at the quantity required to control the degradation phenomena and regulated by food additives legislation (EC 95/2, 1995). Numerous papers could be found in the scientific literature dealing with the development of antimicrobial edible coatings that we will not detail here (e.g., Cagri et al., 2004; Campos et al., 2011; Flores et al., 2007; Gutiérrez et al., 2009; López et al., 2007; Pelissari et al., 2009; Rojas-Graü et al., 2007).

In all the aforementioned applications, diffusion mechanisms are involved in the coating and also most of the times, in the food itself. The efficiency of an edible film is thus strongly dependent on its ability to let a migrant diffuse and the rate of this diffusion. As barrier properties are frequently required (moisture, aroma, flavoring barrier, among others), the term of edible barrier is often employed.

6.2.3 Which Material for Which Targeted Mass Transfer?

Materials used to produce edible films and coatings belong to three main classes: proteins, polysaccharides, and lipids and derivatives (Table 6.1). Properties of each class of materials as well as their processing conditions were well described in previous chapters. Depending on their mechanical resistance, lipid-based materials are usually good moisture barriers while proteins and polysaccharides generally provide good barrier to O_2 and aroma but are very sensitive to moisture.

TABLE 6.1
Which Raw Material to Use According to the Targeted Properties?

Properties	Proteins	Polysaccharides	Lipids	Plastics
Sensory	+ +	+ +	– –	Ø
Mechanical	+ +	+ + +	– – –	+ + +
Water barrier	– –	– –	+ + +	+ + +
Gas barrier (O_2 and CO_2)	+ + +	+ +	–	+
Aroma barrier	+ +	+ +	+ –	+ –

Source: Adapted from Bourlieu, C. et al. 2007. Edible moisture barriers: Materials, shaping techniques and promises in food product stabilization. In J. M. Aguilera and P. J. Lillford (Eds.), *Food Materials Science* (pp. 547–577). New York, NY: Springer.

Note: O_2 = oxygen; CO_2 = carbon dioxide; +++ = highly suitable; ++ = suitable; + = moderately suitable; – = unsatisfactory; – – = not recommended; – – – = strongly not recommended; Ø = not applicable.

6.2.4 A Case Study: Regulation of Moisture Transfer in Composite Food Product

Limitation of moisture transfer has always been, and still is, an important issue for the markets in developed societies in which modern ways of living have favored "composite" or "heterogeneous" ready-to-eat food product sales. These food products are based on components characterized by contrasted moisture content, chemical composition, and texture. They gather a wide range of food products (e.g., sandwiches, savory, sweet pies, and filled biscuits) but generally include a dry or intermediate a_w cereal-based component associated with a wet, high a_w filling. The contrasted organoleptic profiles resulting from the above-mentioned associations, along with composite food product practicality, can explain the high consumer demand for such products. However, moisture transfer triggered by a_w difference between components results in deleterious changes of the organoleptic properties of the product (such as loss of crispness, softening). Stabilization of such products remains a critical point, which limits possibilities of new products development, for example, using associations of textures. The solutions consisting in adding solutes and/or lipids, to reduce a_w difference between components, advocated in the 1980s are limited by nutritional concerns. On the contrary, current trends consist to structure, formulate, and diversify ready-to-eat food products so that they present optimized organoleptic and nutritional profiles and long shelf life.

In this context, edible coatings appear as a promising and competitive tool from economic, nutritional, and technologic points of view (Labuza, 1998).

6.2.5 Assessment of Edible Films and Coating Performance and Compatibility

Today, assessment of edible films and coatings performance and compatibility with the product to protect constitute major hurdles that hamper their general application in food products.

For example, when edible films and coatings are used as moisture barriers, determination of their barrier potentials to extend shelf life of food products represents a major scientific and economic challenge. This challenge has favored research in this field and given birth to plethoric reviews and scientific articles over the last 30 years. Among the main reviews on this topic, we can cite works of Bourlieu et al. (2007, 2009c), Debeaufort (1998), Debeaufort et al. (2000), and Koelsch (1994). These studies consist, most of the times, in characterizing moisture transfer through self-supported edible moisture barriers mainly using water vapor permeability measurements. Methodologies permitting the determination of performance of edible barrier placed *in situ*, that is, applied at the food surface or at the interface of a composite food, were scarcely presented.

Bourlieu et al. (2009c) proposed an overview of generalized methodologies permitting evaluation of edible films performance in real conditions of use. Contrary to the classical "trial-and-error" approach, these generalized methodologies rely in the use of mathematical modeling of mass transfer (Figure 6.2). By simply running one or two simulations, an estimation of the product food shelf life could be done, or, when used in a reverse manner, identification of the optimal window of water vapor permeability for the edible film could be achieved.

Such modeling approaches were successfully used several times to assess *in situ* moisture barrier performance of lipid-based films in cereal-based composite foods (Bourlieu et al., 2008; Guillard et al., 2003a, 2003b, 2004a, 2004b; Roca et al., 2008a, 2008b). Figure 6.3 lists the main groups of researchers in the last 30 years dealing with assessment of moisture barrier performance of edible films, mainly focusing on moisture barrier properties. Only the most recent studies dealt with the development of integrated approaches modeling mass transfer into the film and also into the associated components. The effect of an air gap, that is, imperfect contact

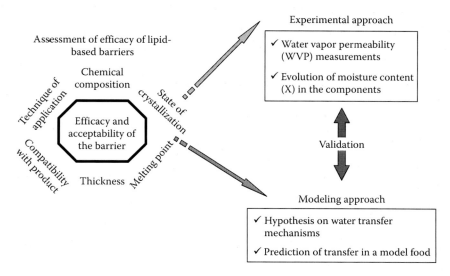

FIGURE 6.2 Two main routes of assessment of edible films efficacy: Example of moisture barrier edible films.

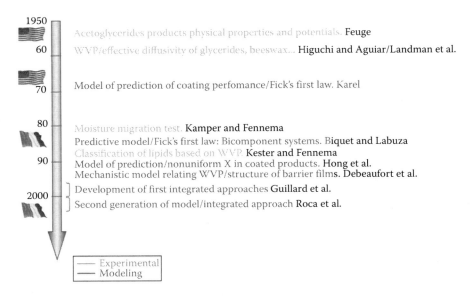

FIGURE 6.3 List the main group of researchers conducted in the last 30 years dealing with assessment of moisture barrier performance of edible films, mainly focusing on moisture barrier properties. (Adapted from Biquet, B. and Labuza, T. P. 1988a. *Journal of Food Science 53*(4), 989–998; Biquet, B. and Labuza, T. P. 1988b. *Journal of Food Processing and Preservation, 12*(2), 151–161; Debeaufort, F., Quezada-Gallo, J., and Voilley, A. 2000. *Food Packaging: Testing Methods and Applications*, 753, ACS Symposium Series, pp. 9–16. Feuge, R. O. 1954. *Food Technology, 8*, S20–S20; Guillard, V. et al. 2003a. *Journal of Food Science, 68*(3), 958–966; Higushi, W. I. 1958. *American Journal of Physical Chemistry, 62*(6), 649–653; Hong, Y. C., Bakshi, A. S., and Labuza, T. P. 1986. *Journal of Food Science, 51*(3), 554–558; Kamper, S. L. and Fennema, O. 1985. *Journal of Food Science, 50*(2), 382–384; Karel, M. and Labuza, T. P. 1969. *Aerospace Medical School*; Kester, J. J. and Fennema, O. R. 1989. *Journal of Food Science, 54*(6), 1383–1389; Kester, J. and Fennema, O. 1986. *Food Technology*, 40–47; Landmann, W., Lovegren, N. V., and Feuge, R. O. 1960. *Journal of the American Oil Chemists' Society, 37*(1), 1–4; Roca, E. et al. 2008a. *International Journal of Food Engineering, 4*(4), 1556–3758; Roca, E. et al. 2008b. *Journal of Food Engineering, 86*(1), 74–83.)

between the components, was also considered by Roca et al. (2008a, 2008b). These integrated approaches will be detailed in Section 6.4.

Such modeling approach was also used for dimensioning antimicrobial coatings, for example, in the study of Guillard et al. (2009), who used a dedicated model to assess the performance in real conditions of use of edible coatings containing sorbic acid as antimicrobial agent. This modeling approach couples mass transfer models (sorbic diffusion in the coated food) with a microorganism growth equation. For example, it allowed concluding that a wheat gluten layer, because of its high sorbic acid diffusivity, was not suitable for such application whereas a beeswax layer was. Beyond the choice of the edible material, such modeling tool could help calculating the optimal initial preservative quantity to add into the coating or the optimal coating thickness. In other words, these modeling tools are really decision support tools for helping to dimension the coating layer/food system.

6.3 CHARACTERIZATION OF MASS TRANSFER PHENOMENA

6.3.1 Basics on Mass Transfer

Mass transfer through edible films or coatings is usually characterized by Fick's laws which imply the knowledge of three main parameters: sorption, diffusion, and permeability.

Solubility describes the maximum quantity of a matter that could sorb in a given medium. For gas or vapor, this quantity depends on the surrounding partial pressure of the given gas or vapor. Concentration and partial pressure are related through Henry's law at a given temperature and a total pressure. Henry's law is usually validated for dilute systems and for gases at a surrounding atmosphere pressure of up to 1 atm. Solubility is the inverse of Henry's coefficient (Equation 6.1). Both are thermodynamic parameters. For the specific case of water vapor, Henry's law does not apply. The relationship between moisture content and water vapor partial pressure is provided by the water sorption isotherm curve.

$$p = K_H \times C = \frac{1}{S} \times C \tag{6.1}$$

where p is the pressure (Pa), C is the concentration (kg m^{-3}), K_H is the Henry's coefficient (Pa m^3 kg^{-1}), and S is the solubility (kg m^{-3} Pa^{-1}).

Diffusion coefficient or diffusivity is a nonthermodynamic kinetic parameter which indicates molecule mobility in a given medium. This transfer is due to the presence of a gradient of concentration of the considered molecule. It is usually described by the well-known Fick's second law (Fick, 1855), which connects local changes of concentration (kg m^{-3}), as function of time (s), and x is the distance between the interface and measurement point (m) with D, the diffusion coefficient (m^2 s^{-1}), according to

$$\frac{\partial C}{\partial t} = D \frac{\partial^2 C}{\partial x} \tag{6.2}$$

In steady state and for a monodirectional mass transfer, Fick's first law is applied as follows:

$$\frac{dC}{dt} = DA \frac{\Delta C}{\Delta x} \tag{6.3}$$

By combining Fick's first law and Henry's law, a new parameter appears in Fick's law, the permeability or coefficient of permeability (P).

$$\frac{dC}{dt} = DS \frac{\Delta p}{\Delta x} \tag{6.4}$$

where permeability is thus the product of diffusivity by solubility.

$$P = DS \tag{6.5}$$

The determination of permeability gives a rough idea of the barrier performance of an edible film or coating. But this measurement is made in fixed conditions that are scarcely extrapolable to real conditions of use. If more consequent modeling approaches are preferred, diffusivity and solubility parameters are thus needed.

6.3.2 Measurement of Mass Transfer Parameters

Methodologies used to measure mass transfer parameters are the same as those used for thin membranes such as packaging materials. Specificities of edible films are related to their mechanical properties: they are usually more fragile and then must be handled more carefully than a plastic film. Another specificity is their high sensitivity to moisture and temperature and therefore, special attention must be given to temperature and relative humidity (RH) control. Permeability, solubility, and diffusivity vary according to Arrhenius law and most of edible raw materials (except lipid compounds) are sensitive to moisture. These specificities will be described in the next section.

All the methodologies described below are developed for self-supported films. That means the evaluation of mass transfer parameters in coatings must be done on self-supporting films with the same formulation than the coating material. We know that the method of application of the coating material could strongly impact its barrier performance and change its value of diffusivity for instance because holes are often created in the structure when spraying is used (Guillard et al., 2004a, 2004b, 2004c). To evaluate the "true" mass transfer properties of the coating, indirect methods such as those described in Section 6.4 based on mathematical modeling and identification methodologies should be preferred.

6.3.2.1 Permeation

Permeation-based methods were primarily developed for the evaluation of permeability of polymer membranes. A thin sheet of material is placed between two constant concentration sources of the diffusant (e.g., between two compartments which are maintained at different isothermal RH by means of suitable buffer solutions or at given partial pressure values, in gases). After a time period, the surfaces of the sheet come into equilibrium with the diffusant sources, thus developing a constant gradient of surface concentrations thus leading to steady-state conditions of diffusion. Permeability or diffusivity can be estimated by measuring the flux of the diffusant, dC/dt, with known surface concentrations and thickness of the material sheet. This can be done experimentally by successive weighing of the diffusion cells at different time intervals for water vapor for instance or by using gas chromatography to detect O_2 or CO_2 flux in a flux of the vector gas. Diffusivity is thus calculated using Equation 6.3 and permeability using Equation 6.4, when partial pressures of the diffusing molecule are known on both sides of the membrane. These permeation measurement methodologies are essentially used for determining permeability values and seldom diffusivities. Different detection methods (gravimetric, manometric, colorimetric) can be used, depending on the diffusing molecule to be detected and different standards (American, German, Japanese) and are presented in Hagenmaier (2011). As regards the specific case of aroma compounds, laboratories have often developed their own systems for measuring permeability. An extensive review of different methods is given by Debeaufort et al. (2002), describing static and dynamic approaches. Newer approaches include combination of dynamic vapor sorption with purge-and-trap/fast gas chromatography (Zhou et al., 2004).

6.3.2.2 Diffusivity

Diffusivity could not be determined simply by a single experimental assessment. This property is obtained by fitting an adequate mathematical model or equation to experimental mass transfer kinetic data measured in transient state and by using an identification procedure. This measure implies generating a concentration gradient of the diffusing substance through a portion of the thin film or coating of known thickness and to monitor the resulting mass transfer into or through the film using an adequate detection system. Numerical treatment of experimental data using an appropriate equation or systems of equations implies considering some hypotheses on the system's conditions. These equations almost always derive from Fick's laws, although the available samples most often do not obey the underlying theory. Therefore, the identified diffusivity is an apparent or effective one, noted D_{eff}, representative of the overall mass transfer without considering the mass transfer mechanisms prevailing in the sample. Current technologies developed for determining diffusivity in foods and membranes could be applied regardless the characteristics of the sample used for such determinations.

6.3.2.3 Detection System

Gravimetry is the simplest methodology to detect gain or loss of matter by a piece of material, providing that the sensitivity of the balance is low enough to detect low weight variations. In practice, gravimetry is widely used for water transfer measurement and in a less extent for organic compounds such as aroma compounds and gases (CO_2 and O_2). Most often, giving the low amounts of O_2/CO_2 that could be dissolved in food and nonfood products, gravimetry is not suitable to detect gas mass change kinetics, and alternative methodologies must be selected. Pénicaud et al. (2012) and Chaix et al. (2014) recently reviewed the methods used for molecular quantification of, respectively, O_2 and CO_2 in the medium. Among all methodologies used for O_2 detection, luminescence-based sensors seem to be the most promising technology to detect O_2. Miniaturized sensors of this type were used in experimental setups dedicated to determining O_2 diffusion kinetics in liquid (Pénicaud et al., 2010a) and solid systems (Chaix, 2014). For CO_2, chemical titration remains the easiest methodology to implement at lab scale to determine the diffusivity of CO_2 in liquid or solid media (Chaix et al., 2014). Methodologies for detection of small molecules, mainly aroma compounds were reviewed by Karbowiak et al. (2009). Fluorescence recovery after photobleaching technique was shown to be a valuable tool for studying diffusion of small molecules in polymer films and may contribute to obtain information about their molecular organization and their potential functionality as barrier

(Karbowiak et al. 2006). Nuclear magnetic resonance spectroscopy investigation contributes to a better understanding and identification of the interactions between the polymer and the diffusant. Fourier transform InfraRed–attenuated total reflectance (FTIR-ATR) analysis enables to obtain the diffusion coefficient of the liquid water in the film (Karbowiak et al., 2009).

6.3.2.4 Experimental Setup and Numerical Treatment of Data for Effective Diffusivity Determination

Once the detection system is selected, the second step is the experimental setup required to create the gradient in the sample. This experimental setup conditions the complexity of the boundary conditions applied to the sample and the corresponding mathematical solution of Fick's laws used to identify D_{eff}.

The simplest experimental setup used is based on the permeation method previously described. This method is based on the time period prior to the establishment of the steady-state diffusion. If by some convenient means, one of the sheet surfaces is maintained at C_1 concentration, while the other at zero concentration, after a theoretically infinite time period, a steady-state condition of diffusion will be achieved. Assuming that the diffusivity is constant, the sheet is initially completely free of diffusant and that the diffusant is continually removed from the low concentration side, the amount of diffusant which will permeate through the sheet, when $t \to +\infty$ is given as a linear function of time by

$$W_t = \frac{D_{eff} C_l}{L}\left(t - \frac{L^2}{6 D_{eff}}\right) \quad (6.6)$$

where W_t is the amount of diffusant (kg m^{-2}).

By plotting W_t against time, after a relatively large time interval, a straight line results intercepting the t-axis at the quantity ($L^2/6D_{eff}$). From this intercept, the effective diffusivity, D_{eff}, may be calculated. Besides the simplicity of the experimental setup and the numerical treatment, this method has several drawbacks such as difficulties to produce a thin material sheet of constant thickness and homogeneous structure, an essential prerequisite for accurate D_{eff} determination. In addition, swelling of the membrane material under experimental conditions and erroneous measurement of the flow rate of diffusion could impact the determined value.

Transient mass transfer into food and nonfood products and more especially into edible films and coatings could be generated using desorption or adsorption measurements under controlled conditions. This is widely used for volatiles such as water vapor or aroma compounds. Adsorption or desorption evolution with time is monitored up to equilibrium. Time before reaching equilibrium varies as a function of sample geometries. The thinner the sample, the more rapidly the equilibrium will be achieved. Lab-made solutions are usually set up for controlling atmosphere around the sample and master the initial and boundary conditions. When gravimetry could be used to monitor mass change, integrated systems with microbalance allow performing automated, quick measures due to the small sample size and weight used (15–30 mg). Such systems exist for water vapor (Cahn microbalance, Dynamical

Vapour Sorption system, SMS, London) and are extensively used to determine water vapor diffusivity in edible films (Angellier-Coussy et al., 2011; Bourlieu et al., 2006; Guillard et al., 2003b, 2013). Dedicated microbalances were also used to determine diffusivity of aroma compounds into edible protein-based films (Chalier et al., 2009; Dury-Brun et al., 2008).

When the diffusing substance is a nonvolatile one, the absorption/desorption from or into controlled atmosphere is not possible. Adsorption of solutes such as NaCl or additives into a sample of material may be done by direct contact with another material enriched in the diffusing substance and considered as a source. This material could be of the same nature of the material investigated such as in the work of Pénicaud et al. (2010b) that determined ascorbic acid diffusivity into agar gel by this methodology. The authors put two cylindrical samples of the same radius but one spiked with ascorbic acid and the other free of this molecule. At time $t = 0$, the two cylinders are put into direct contact with each other and after a specified time interval, the diffusant concentration profile along the axis can be determined by slicing and assaying ascorbic acid content of each slice. Diffusivity was then identified from resulting experimental concentration distribution profiles using a dedicated numerical solution of Fick's second law with appropriate boundary conditions. The same approach was used by Warin et al. (1997), Motarjemi (1988), Karathanos and Kostaropoulos (1995) who have successfully used this method for measuring sucrose diffusivity within agar gels, moisture distribution profiles in meat, and moisture distribution in dough/raisin systems. More recently, this approach was also used by Guillard et al. (2003a) to identify moisture diffusivity in a sponge cake. This methodology based on slicing of the sample is difficult to adapt to thin layers such as edible films and coatings but the same principle of putting a source in contact may be used. In that specific case, a global determination of the concentration of the diffusing substance in the whole film could be done. The local content in the compound of interest could be assayed in one precise point (usually film interface) with techniques such as FTIR-ATR (Mauricio-Iglesias et al., 2009) or O_2-sensors that allow local quantification (Pénicaud et al., 2010a). Distribution profiles within thin layers of materials may be obtained using RAMAN spectroscopy, provided that the spectral signature of the compound to detect could be distinguished by RAMAN (Martinez-Lopez et al., 2014; Mauricio-Iglesias et al., 2009, 2011).

6.3.2.5 Sorption

Maximal solubility of a molecule into a given material is obtained at equilibrium of diffusion. Usually, the same set of data could be used to simultaneously determine diffusivity and calculate solubility. For the latter, and contrary to diffusivity (that could be determined using increases or decreases in the relative concentration, i.e., no calibration needed), solubility determinations need quantification of the number of molecules sorbed in the matter. Techniques to measure sorption or solubility of a given molecule into a material rely therefore on the availability and the accuracy of methodologies of quantification of this molecule. As mentioned earlier, if gravimetry (e.g., using a quartz crystal microbalance, McBain quartz spring balance, microbalance, Cahn electrobalance, etc.) is one of the simplest methodologies to detect sorption of a given substance, it could be used only if sorption induced significant weight changes in the sample which, in practice, is not always the case (see, e.g., the case of O_2). Gravimetry is widely used to determine water vapor sorption isotherms in various food and nonfood products including edible films and coatings. Specific methodologies and mathematical modeling of these water sorption curves can be found in detail in Guillard et al. (2013).

When gravimetry could not be used, indirect methods based on the measurement of the targeted molecule in the phase in contact and in equilibrium with the material (gas or liquid phase) are preferred. For instance, solubility of O_2 into solid products could be deduced from quantification, by gas chromatography, of gaseous O_2 in the atmosphere of known volume in equilibrium with the product (Chaix et al., 2014).

6.3.3 Factors Affecting Mass Transfer Parameters

Mass transfer properties of small molecules into edible films and coatings are principally influenced by the composition and structure of the polymer matrix, the properties of the diffusing molecule and its interaction with the polymer and finally by the environmental conditions (extrinsic characteristics). Today a plethora of publications and research can be found trying to characterize and explain variations of mass transfer properties as a function of intrinsic and extrinsic characteristics of the system. In the last 10 years, numerous books and more than 20,000 research articles described different edible films formulation and their properties; among them, we can cite as example the works of Baldwin et al. (2011), Bourtoom (2008), Debeaufort (1998), Dhall (2013), Embuscado and Huber (2009), Guilbert and Gontard (2005), Guilbert et al. (1995, 1996), Han (2014), Miller and Krochta (1997), among others. Recently, a special emphasis has been put on the relationship between structure and functional properties especially at the nanoscale. Modeling of this relationship was also tempted. In the following sections, general trends regarding intrinsic and extrinsic characteristics influencing mass transfer properties in edible films and coatings will be briefly recalled, keeping in mind that all the plethoric published information on that topic could not be extensively cited and described here.

6.3.3.1 Films Composition and Structure
6.3.3.1.1 Chemistry of the Raw Material
The chemical structure of the polymer and the nature of its functional groups have a dominant effect on mass transfer. For instance, proteins and polysaccharides are, by nature, relatively impermeable to gases and aromas while fats are usually good moisture barriers. The good gas barrier properties of carbohydrate polymers, at least at low RH, could be related to large degrees of hydrogen bonding. Unluckily, this specificity leading to good gas barrier properties often results in a high sensitivity to polar migrants such as water vapor, resulting in poor moisture barrier (Guilbert, 1986; Khwaldia et al., 2004; Nisperos-Carriedo, 1994). A well-known possibility to improve the moisture barrier properties of protein- and polysaccharide-based materials relies on their combination with hydrophobic compounds such fats in isotropic structures (intimate blends resulting in micro/nanoemulsions) or anisotropic

ones (bilayer structures). The effect of fat incorporation strongly depends on several aspects: the resulting structure (emulsions being usually less efficient than bilayer structures are regards moisture resistance); physical state of fat (high solid fat content is usually preferable to low solid fat content); fat chain length (longer chains being positively correlated to higher barrier properties); saturation degree of fatty acids; hydrogenation degree of oils; and crystalline form of the fats (Morillon et al., 2002). One must keep in mind that moisture barrier properties of fats and lipid compounds are usually inversely correlated to their mechanical properties. Indeed, if a high solid fat content, longer fatty acid chains, and high melting point are theoretically in favor of high moisture barrier properties, the poor mechanical properties of the resulting films and coatings provoke the unavoidable formation of cracks and channels in the structure under real conditions of use, leading to a decrease in their performance as moisture barrier (Bourlieu et al., 2009b).

6.3.3.1.2 Plasticizers

To avoid film brittleness caused by extensive intermolecular forces, addition of plasticizers is generally required in polysaccharide- and protein-based material. The incorporation of plasticizers has a relatively complex influence on film permeability (Stannett, 1968). Positioned between the macromolecular chains, these additives decrease the intermolecular forces, increase free space and chain mobility, and lower the glass transition temperature. Thereby, by modifying the local segmental motions, they have an indirect effect on transport parameters, that is, gas permeability normally increases with the plasticizer content (Arvanitoyannis and Biliaderis, 1998; Butler et al., 1996). Concomitantly to gas permeability increase, which could be desired in some cases, the addition of plasticizers usually even increases the sensitivity of biomacromolecules to moisture. This triggered moisture sensitivity is not only due to a simple increase in free volume and chain mobility, that generally favor water diffusion in the polymer network, but could also be due to the affinity for water of the plasticizer itself. For example, in polyol plasticized biopolymers, the gain in mobility was demonstrated to be due to the double effect of the plasticizer that simultaneously increased chain mobility and the amount of sorbed water compared to nonplasticized polymer solutions (Karbowiak et al., 2006).

6.3.3.1.3 Crystallinity

Klopffer and Flaconnèche (2001) extensively described how crystallinity affects mass transport in polymers. Polymer crystallinity can vary depending on processing conditions. In a pure crystalline medium, diffusion is supposed to be null. Therefore, in semicrystalline polymers, the crystalline zones act as excluded volumes for the sorption process and are impermeable barriers for the diffusion process. Diffusion can only occur in amorphous domains or through structural imperfections. Permeability and ability of the amorphous phase to let matter diffuse through it depends on its state: rubbery above glass transition or glassy below it. High degrees of crystallinity usually provide certain impermeability (Weinkauf and Paul, 1990). Crystalline zones, on one hand, increase the effective path length of diffusion but, on the other, seem to reduce polymer chains mobility in the amorphous phase (Klopffer and Flaconnèche, 2001). Therefore, the permeability of films below their glass transition tends to be extremely low.

6.3.3.1.4 Solvents

Solvents primarily used for edible films production comprise water, ethanol, or a combination of both. Solvent type and its polarity, acidity, and molecular weight influence films' structure and consequently their properties. For example, Caner et al. (1998) studied the effects of different types of acids (acetic, formic, lactic, and propionic), their concentrations, plasticizer concentrations, storage time on water vapor, and oxygen permeability of chitosan films. The oxygen and the water vapor permeabilities were in the following order: lactic < acetic < propionic < formic and acetic < propionic < formic < lactic. Thus, transfer properties can be controlled by proper design and optimization of the composition of edible films.

6.3.3.1.5 Nanostructure Voluntarily Created

Substances like cyclodextrine or nanoclays may be added to film composition aiming at improving films' functionality. The case of nanoclays is especially scarce for edible films and coatings because of consumer safety issues. However, polysaccharides such as cyclodextrins that create a nano- or microstructure could be voluntarily added in the formulation. Cyclodextrins are molecules that are able to form host–guest complexes with volatile guest compounds and display a very interesting RH-dependent volatile release because of the plasticizing effect of water on the molecule (Del Valle, 2004). This allows triggering the release when necessary, that is to say when the coating material is in contact with food. Thereby, by their active role entrapping active compounds, these substances play an important role in mass transfer. In addition to their active role, molecules such as cyclodextrins may have a passive effect on mass transfer by creating tortuosity in the structure, extending the diffusion pathway. This passive role on the diffusion and permeation properties of a film could be predicted using tortuosity-based models. Some of them were described by Bourlieu et al. (2009) for the specific case of lipid-based edible coatings. A general review on that topic has been made by Choudalakis and Gotsis (2009). We note that crystalline zones in a polymer create a given amount of tortuosity, the impact of which on mass transfer properties could be predicted using the same type of equations than those used for nanocomposites (Namin, 2011).

6.3.3.1.6 Structural Defects

To be effective, an edible film or coating must be uniform and fairly free from defects, pinholes, microscopic cracks, and curling ends. Even small irregularities can exponentially increase diffusion rate and that is not taken into account in the Fick's law (Pavlath and Orts, 2009). Principal preparation technologies used are solvent casting and extrusion for films, and spraying and dipping for coatings. During processing, drying rate, cooling, and tempering must be well mastered. Varying the cooling rate may result in films with completely different characteristics (Bourlieu et al., 2009b). The use of spraying technologies on foods that usually have some type of roughness on their surface favors the creation of imperfect structures, which display higher diffusivity values than the corresponding dense, self-supported structures used to mimic them during diffusion rate determinations (Guillard et al., 2004a, 2004b, 2004c). Rapid drying of cast polymers leads to premature packing of polymer chains that limits the development of optimal intermolecular associations and optimal structure, since the mobility of the polymer chains is restricted when the solvent concentration decreases and as such it can affect barrier performance (Guilbert, 1986, 1988). Although not all the polymers are extrudable at industrial scale, extrusion is however much more attractive, since it does not require solvent addition and a time of evaporation.

6.3.3.1.7 Posttreatments

Different posttreatments were studied to improve mass transfer properties of edible films and coatings by acting on their structure. For example, cross-linking obtained by means of chemical or physical posttreatments of the film is known to enhance barrier properties, restricting the movement of molecules by the formation of a tighter structure. For example, it was demonstrated that enzymatic treatment with transglutaminase was efficient in lowering the water vapor permeability of gelatine films (de Carvalho and Grosso, 2004). Different chemical treatments may be employed to achieve enhanced covalent bonding (e.g., treatment with formaldehyde, addition of tannic acid, glutaraldehyde, among others). However, such chemical reactions can create by-products and induce toxicity and must be first approved by legal authorities before being used in edible materials (Hernández-Muñoz et al., 2004). For this reason, alternatives to those controversial chemical reactions were proposed based on physical methods such as thermal, UV, and electron beam irradiation. These have been shown to cross-link proteins and polysaccharides and to improve their barrier properties (Jo et al., 2005; Ryzhkova et al., 2011; Wihodo and Moraru, 2013). Recently, it was also demonstrated that irradiation induced linkage between polymers of different origin, for example, gelatine and chitosan, and as such improved water vapor and oxygen barrier performance of resulting blends of these two materials (BenBettaïeb et al., 2015; Lacroix et al., 2002).

6.3.3.2 Influence of the Diffusing Molecule

Besides the composition and the structure of the polymer network, the nature of the diffusing molecule also plays an important role on mass transfer properties. Molecular weight, length, size, and shape have a strong influence on diffusivity, solubility, and permeability values.

The solubility of a molecule depends largely on its compatibility with the matrix. Generally, more easily condensed gases are more soluble in a polymer. For example, CO_2 is more condensable than O_2 and N_2, and it has a much higher solubility in water than O_2 and N_2 and more generally in all food and nonfood materials (Chaix et al., 2014). The solubility of O_2 is higher than that of N_2. The same observation was done for diffusivity of gases into matter: diffusivity of CO_2 is higher than that of O_2 which is in turn higher than that of N_2. Once permeability is directly related to the product of D by S (Equation 6.4), then the permeability of gases in polymer systems usually follows the order $CO_2 > O_2 > N_2$ (Salamone, 1996).

It is generally admitted that diffusivity is inversely proportional to the square root of the molecular weight of the diffusing molecule. This precept, validated several times in the field of food contact materials, was used as a basis for predicting diffusivity values of additives in plastic materials (Baner et al., 1996; Hinrichs and Piringer,

2002; Martinez-Lopez et al., 2014; Mercea, 2008; Welle, 2013). Therefore, diffusivities of small molecules such as gases (e.g., oxygen, carbon dioxide), water vapor, or volatiles such as ethylene are much more important than that of molecules of higher molecular weights such as flavors and oils. For molecules of similar molecular weight, globular molecules generally diffuse faster than elongated ones. Polar molecules diffuse faster than nonpolar ones, especially through polar films (Baldwin et al., 2011).

The diffusing molecule could also impact mass transfer properties by acting on the polymer structure itself through a plasticization effect. This effect, well known when water is the diffusing molecule (see next section dealing with the impact of external parameters such as RH), could also happen with other diffusing molecules such as organic volatile molecules. When considering aroma transfer through an edible film, some plasticization of the polymer network could occur, leading to a concentration-dependent diffusivity or permeability (Quezada Gallo et al., 1999).

6.3.3.3 Influence of External Parameters

6.3.3.3.1 Temperature

Temperature affects kinetic and thermodynamic phenomena, increasing mass transfer rates. Diffusion is the most temperature-sensitive transport parameter, in comparison to solubility and permeability. Given that the movement of gas molecules through a film is considered as a thermally activated process, temperature increase has always a positive effect on the diffusion coefficient (Rogers, 1985). It is expressed in terms of an Arrhenius-type relationship. Effect of temperature on solubility could also be represented by an Arrhenius law but most often with negative values of the activation energy. That means that increasing temperatures will decrease solubility values, as it is systematically the case for gases (O_2/CO_2) (Chaix et al., 2014). This is due to the higher condensability of gases at low temperatures, which favors their solubilization. For other molecules such as aroma compounds, hydrophobic interactions and hydrogen bonding established between the film's network and the molecule lead to either an increase or a decrease in solubility with increasing temperature, which must be evaluated case by case (Debeaufort et al., 2002). Considering the impact of temperature on both D and P, its effect on permeability is difficult to predict. Depending on the governing phenomenon, temperature could increase or decrease the value of P. Both may also balance out each other and in that case, P will remain constant. Generally, experimental observations showed that permeability increases with increasing temperature, indicating that the controlling mechanism is diffusion. However, it is difficult to generalize this, since there are always some exceptions. For example, for lipid-based films, temperature modifies the solid fat content and therefore affects the structure and barrier performance of the film. It results in unexpected, nonlinear impact of increasing temperature on D and P.

6.3.3.3.2 Relative Humidity

Being hydrophilic by nature, edible films and coatings made of proteins or polysaccharides are very sensitive to moisture. They display impressive gas barrier properties in dry conditions, and while RH increases, there is a drastic decrease in the gas barrier (McHugh and Krochta, 1994; MujicaPaz and Gontard, 1997). In real conditions of use (i.e., under food contact), edible films and coatings are generally in contact with high humidity or a_w (close to 1). It is therefore crucial to carefully control RH when measuring mass transfer properties of hydrophilic bio-based films and coatings. During hydration, water uptake softens the polymer network through plasticization, and polymer chains are consequently more mobile, resulting in an increase in water vapor transfer and also of other compounds such as gases and other volatiles. For example, when RH increased from 0% to 96%, permeability of oxygen in chitosan film increased 13 times (Despond et al., 2001). Oxygen permeability in starch films was increased by 6.5 and 48.1 times when the RH increased from 60% to 80% and from 60% to 90%, respectively (Gaudin et al., 2000). In wheat gluten films, O_2 and CO_2 permeabilities were increased by 25 and 632 times when RH increased from 0% to 100% (MujicaPaz and Gontard, 1997). Numerous other examples can be found in the scientific literature.

6.4 EDIBLE FILMS DIMENSIONING BASED ON MATHEMATICAL MODELING

Mathematical modeling tools have been developed to simplify dimensioning and design of edible films and coatings by minimizing the number of experimental trials required to achieve optimal systems. Generally, selection of films and coatings for a targeted application involves an empirical approach, the so-called trial-and-error approach, which implies numerous experimental trials. Besides the significant economic expenses of shelf-life tests, high risks of error based on a wrong selection of the raw materials or bad dimensioning of the coating thickness complicate the approach, making it most often not particularly relevant (Robertson, 2006).

To overcome the issue of the aforementioned empirical approach, mathematical modeling tools based on the prediction of mass transfer phenomena in the food/coating (or edible film) system have been proposed. Most of them are equal to those used for the food/packaging system. All mathematical approaches are based on the use of Fick's laws to describe mass transfer: diffusion within the food and the coating itself described by Fick's second law (Figure 6.4) or permeation through the film by using Fick's first law. These mass transfer models are combined with models predicting food shelf life, as follows: first or second order reaction for describing oxidation of a vitamin (Bacigalupi et al., 2013; Pénicaud et al., 2009, 2011); Michaelis–Menten equation to describe enzymatic reactions such as respiration (Cagnon et al., 2013; Guillard et al., 2012, 2015); and models of predictive microbiology to simulate microbial growth (Chaix et al., 2015). Mechanistic models coupling mass transfer and reaction models are the key for more relevant dimensioning of packaging and/or edible coating mass transfer parameters as a function of the targeted shelf life. Shelf life must be translated into the limiting element for food preservation, which could be a minimal level of a given vitamin, a maximum value of a microbial account or of moisture content (e.g., to allow maintaining crispness for instance of a cereal-based product).

Depending on the food/coating or film studied, boundary conditions vary and therefore the possible mathematical solution (analytical or numerical) for the differential equation represented by Fick's second law also changes. Crank (1980) proposed several analytical solutions for particular (often ideal) geometries and boundary conditions with constant diffusivity properties. When analytical solutions

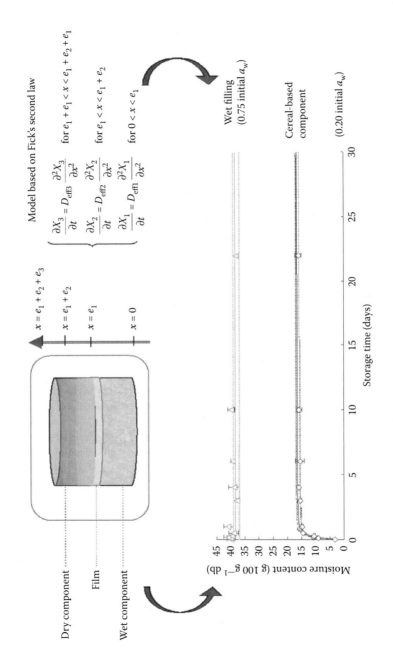

FIGURE 6.4 Example of mathematical model developed for predicting evolution of moisture content in cereal-based composite food with a moisture barrier film placed at the interface. (Adapted from Bourlieu, C. et al. 2006. *European Journal of Lipid Science and Technology*, 108(12), 1007–1020.)

are not available, especially for problems involving concentration-dependent diffusivity, complex geometries are at stake such as multicompartment systems, or when reaction models are coupled with mass transfer models, numerical solutions must be developed. They are based on the finite differences method: Fick's second law is discretized using one of the available discretization schemes (e.g., Crank–Nicolson).

In the mathematical modeling of mass transfer phenomena involved in a food/coating system, different approaches can be distinguished according to the number of simplified assumptions made and the complexity of the resulting mathematical treatment. Figure 6.5 presents a classification of the modeling approach starting from the simplest and finishing with the most complex ones. In the following sections, the four modeling cases proposed in Figure 6.5 will be briefly described with an emphasis on their main advantages and drawbacks.

(a)
1. Case A : Transfer only in the film, Fick's first law

In the film: Steady state

$$J = -D \frac{\partial c}{\partial x}$$

At the film/food interface: Not considered
In the food: Not considered
Application: Water transfer through edible film

$$\frac{dC_w}{dt} = WVP \cdot \frac{A \cdot (pe - e)}{\Delta x}$$

$\frac{dC_w}{dt}$ —The rate of moisture transferred per second
WVP—Water vapor permeability (gs^{-1} m^{-2} Pa)
A—Effective area of diffusion (m^2)
p, pe—The water vapor partial pressure in the coating and in the environment (Pa)
Δx—Film thickness (m)

2. Case B : Coupling permeation and reaction

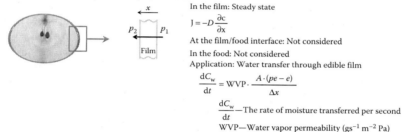

Application: Tailorpack MAP modeling tool (www.tailorpack.com)

O_2/CO_2 partial pressure inside = O_2/CO_2 transfer through the coating (Fick's first law) + O_2/CO_2 consumed/produced by respiratory activity (Michaelis-Menten equation)

Reaction → Respiration
→ Other biological reactions

$$P_{O_2} = \frac{Pe_{O_2} \cdot S}{e} - (P^{ext}_{O_2} - P^{food}_{O_2}) \cdot RR_{O_2} \cdot m$$

$$P_{CO_2} = \frac{Pe_{CO_2} \cdot S}{e} - (P^{ext}_{CO_2} - P^{food}_{CO_2}) \cdot RR_{O_2} \cdot m \cdot RQ$$

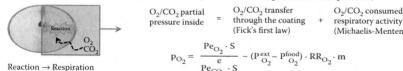

Pe_{O_2}, Pe_{CO_2}—O_2 and CO_2 permeability, respectively (mol m^{-1} s^{-1} Pa^{-1})
S—Coating surface (m^2)
e—Coating thickness (m)
RR_{O_2}—O_2 respiration rate (mmol kg^{-1} h^{-1})
m—Mass of the food (kg)
RQ—Respiration quotient (–)
Pj_i—Partial pressure of i in j (Pa)

FIGURE 6.5 Classification of the different mathematical modeling approaches used for prediction mass transfer in the coated/food system and for designing edible films and coatings. (a) Cases based on the use of Fick's first law. *(Continued)*

(b)

3. Case C: Unsteady-state transfer from the film to food, Fick's second law

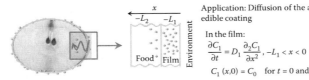

Application: Diffusion of the active compound from an edible coating

In the film:
$$\frac{\partial C_1}{\partial t} = D_1 \frac{\partial^2 C_1}{\partial x^2}, -L_1 < x < 0$$

$$\left. \begin{array}{l} C_1(x,0) = C_0 \quad \text{for } t = 0 \text{ and } -L_1 < x < 0 \\ \frac{\partial C1}{\partial x} = 0 \quad \text{for } x = -L_1 \end{array} \right\}$$

At the film/food interface:

$$\left. \begin{array}{l} D_1 \frac{\partial C_1}{\partial t} = D_2 \frac{\partial C_2}{\partial x} \quad \text{for } x = 0 \\ C_2 = kC_1 \quad \text{for } x = 0 \end{array} \right\}$$
with D_1, D_2 the diffusion coefficients
C_1, C_2 the solute contents
k the partition coefficient

In the food:
$$\frac{\partial C_2}{\partial t} = D_2 \frac{\partial_2 C_2}{\partial_2 x}, 0 < x < L_2$$

$C_2(x,0) = 0 \quad \text{for } t = 0 \text{ and } 0 < x < L_2$

for $x = L_2$

4. Case D: Coupling diffusion and reaction

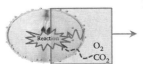

Application: Diffusion of active compound from edible coating coupled with the production rate of that compound

$$\frac{\partial C_A}{\partial t} = D \frac{\partial^2 C_A}{\partial x^2} + R$$

$$\frac{\partial C_A}{\partial t} = -k \cdot C_A$$

A—Diffusing compound (ex: active compound, O_2, CO_2)
R—Consumption or production rate of A
k—Reaction rate of release

Reaction → Release activated by another parameter
→ Chemical reaction
→ Respiration

If production of A influences an another reaction, for example, microbial growth, then further consumptions must be also taken into account:

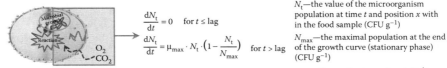

$$\frac{dN_t}{dt} = 0 \quad \text{for } t \leq \text{lag}$$

$$\frac{dN_t}{dt} = \mu_{max} \cdot N_t \cdot \left(1 - \frac{N_t}{N_{max}}\right) \quad \text{for } t > \text{lag}$$

N_t—the value of the microorganism population at time t and position x with in the food sample (CFU g^{-1})

N_{max}—the maximal population at the end of the growth curve (stationary phase) (CFU g^{-1})

μ_{max}—the maximal growth rate (s^{-1})

lag—the lag time for the microorganism concerned (s)

(Rosso et al., 1995)

FIGURE 6.5 (*Continued*) Classification of the different mathematical modeling approaches used for prediction mass transfer in the coated/food system and for designing edible films and coatings. (b) Cases based on the use of Fick's second law.

6.4.1 CASE A: SIMPLIFIED APPROACH CONSIDERING ONLY PERMEATION IN THE COATING/FILM

This approach is the simplest one involving no more than three simple calculations based on Fick's first law for dimensioning an edible film, that is, calculating the adequate permeability of the coating for a targeted application. An example of such dimensioning is given in Box 6.1. Such simplified calculations could be also used to predict the shelf life of the coated food with a given coating material of known permeability.

BOX 6.1 EXAMPLE OF COATING DIMENSIONING CALCULATION

An edible film is used to preserve a biscuit that contains 7% of water (g H_2O/dry basis). Inside the package, the biscuit generates a relative humidity of 53%. The storage conditions are 25°C and 93% RH. The packed amount of a biscuit is 125 g for a total area of 420 cm². The targeted thickness for the edible film is l = 50 µm). What is the acceptable water vapor permeability of the edible film to achieve a shelf life of 7 days, knowing that the maximal water content of the biscuit is 18% dry basis? Water vapor partial pressure at 25°C is 3167 Pa.

Calculation:

1. Maximal water intake (x) authorized for the biscuit

$$\left(\frac{7}{100+7}\right) \times 125 + x = \left(\frac{18}{100+18}\right) \times (125 + x)$$

$$x = 12.81 \text{ g of } H_2O$$

2. Maximal flow rate through the coating considering shelf life of 7 days and the calculated maximal water uptake

$$J = \frac{x}{t}$$

$$J = \frac{12.81}{7 \times 24 \times 3600}$$

$$J = 2.12 \times 10^{-5} \text{ g/s} \rightarrow J = \frac{2.12}{18} \times 10^{-5} = 1.18 \times 10^{-6} \text{ mol/s}$$

3. Calculation of the maximal water vapor permeability for the edible film

$$J = \frac{Pe \cdot A \cdot \Delta p}{l}$$

$$P = \frac{1.18 \times 10^{-6} \times 50 \times 10^{-6}}{420 \times 10^{-4} \times (0.93 - 0.53) \times 3167}$$

$$P = 3.88 \times 10^{-11} \text{ mol/m} \cdot \text{s} \cdot \text{Pa}$$

> Using this permeability as a limit, the user could retrieve in dedicated databases the appropriate coating material that could preserve its biscuit from remoistening during 7 days.
> ! Shelf-life calculation is slightly underestimated because it is estimated from the initial water vapor flow through the edible film.

The main drawback of such calculation based on the initial mass flow through the coating is that the shelf life and consequently the calculated permeability are underestimated; in reality, slightly higher permeability values would be also suitable for maintaining food quality throughout the targeted shelf life. To overcome this drawback, some authors introduced modifications in the model of Box 6.1 to incorporate in Fick's first law the time-dependence concentration in one side of the film. This modification is used essentially for water vapor transfer. The gain or loss of water by the food as a function of time is taken into account to recalculate the water partial pressure at the food/coating interface. This is made by using water sorption isotherm combined with Fick's first law. The resulting equation is given as follows:

$$\ln\left(\frac{m_e - m_i}{m_e - m_t}\right) = \frac{PAp_0 t}{eMb_s} \qquad (6.7)$$

where P is the permeability of the coating, A its surface, p_0 the saturated partial pressure of water at the studied temperature, e the thickness of the coating, M the food mass, b_s the slope of the linear part of the water sorption isotherm curve between the initial moisture content (m_i) and a critical moisture content (m_c), m_e the moisture content of the food corresponding to the water activity of the surrounding atmosphere, and m_t the moisture content at time t. This model was used, for example, by Biquet and Labuza (1988a, 1988b) to predict moisture gain in a model food covered by an edible coating (chocolate layer).

6.4.2 CASE B: PERMEATION COUPLED WITH REACTION

Food shelf life could be in some cases simplified by the means of one major reaction. Mass transfer phenomena in the coated system are thus strongly related to this reaction, and the use of a coating material is often dedicated to the mastering of this reaction. It is the case, for example, of respiration of fresh fruits and vegetables. The coating material is used to maintain a low level of O_2 and a moderate concentration of CO_2 in the close surrounding of the produce to slow down respiration rate and thus degradation rate. The passive modified atmosphere created around the product relies on the interplay of two phenomena: (1) gas permeation through the coating and (2) respiration by the product. The optimal atmosphere for a given product is reached providing that the coating has the suitable gas permeabilities. To identify these O_2/CO_2 permeabilities, a mathematical model could be used coupling Fick's first law and a respiration model (represented by Michaelis–Menten equation). An example of such tool can be obtained using a free web application (www.tailorpack.com). Starting from the optimal atmosphere for the product, such tool allows identifying the window of optimal gas permeabilities. The nature of the suitable coating (or packaging material) could be then inferred from these values of permeabilities by querying dedicated databases (Destercke et al., 2011; Guillard et al., 2015).

6.4.3 CASE C: UNSTEADY-STATE TRANSFER FROM THE FILM INTO THE FOOD

Case C is dedicated to bilayer systems when steady state in the film or coating could not be assumed. The distribution profile in the food itself cannot be neglected and influences in turn that in the coating. Consequently, at least two differential equations for Fick's law are considered, one for food and one for the coating. The same approach is used for films placed in sandwich between two phases (Figure 6.2). This approach was used by Rumsey and Krochta (1994), who studied various formulations of edible coatings and modeled moisture transfer in resulting coated foods. Guillard and coworkers used also this approach several times to predict the shelf life of various composite food products involving a cereal-based compartment associated with a wet phase of higher water activity (Guillard et al., 2003a, 2004a, 2004b, 2004c; Roca et al., 2008a, 2008b). They used this model to classify the *in situ* moisture barrier performance of various edible films (Bourlieu et al., 2006, 2008; Guillard et al., 2003b). The impact of fat composition, cooling rate (Bourlieu et al., 2009a, 2009b), and processing states (Guillard et al., 2004a, 2004b, 2004c) on the barrier performance of lipid-based films was investigated through simulations carried out by using this model. Comparison between various technologies of stabilization of composite foods including the use of edible films placed at the interface between the two phases of contrasted a_w was also carried out using a modeling study (Roca et al., 2008a, 2008b). More complex cases were also considered such as variations of volume due to shrinkage and swelling as in the study of Boudhrioua et al. (2003) or in the study of Fajardo et al. (2010), who developed a mathematical model for controlled release of natamycin from chitosan matrix applied on cheese considering the swelling of the matrix.

6.4.4 CASE D: COUPLING DIFFUSION AND REACTION

Case D is the most complex case, where unsteady-state diffusion occurs in all compartments (food and coating) and when reactions occur in the food and/or the coating itself. These reactions may consume or produce matter and thus the need of adding a term representing the net production rate of matter in Fick's second law equation (Figure 6.4). Applications of this model are, for example: O_2 diffusion coupled with oxidation (Pénicaud et al., 2009), gas diffusion coupled to microbial growth (Chaix, 2014), or antimicrobial diffusion coupled to microbial growth (Guillard et al., 2009). Such type of model opens up the field of possibilities for dimensioning controlled release systems such as active films and coatings with O_2 scavenging, antioxidant, or antimicrobial properties.

For example, formulation of edible films and coatings with inclusion of natural plant extracts as antimicrobial compounds has been and still is a major concern in the field of food preservation. The main outcome for these antimicrobial coatings is the maintenance of a high concentration in active substance, where it is required, that is, on the coated food surface. This local concentration is strongly dependent on the diffusion properties of the matrix used to make the controlled release system. Prediction of the distribution profile of the active compound in the coated food and the time during which the agent remains above the critical effective concentration is a real challenge to an efficient dimensioning of the coating (Ouattara et al., 2000). The interest of such modeling approach was well illustrated in the work of Guillard et al. (2009), who modeled the impact of sorbic acid diffusion in model food coated with active coatings of different nature on the growth rate of *Saccharomyces cerevisiae*. As the distribution profile of the active compound in the coated food is not always accessible by the experiment, alternatives have been proposed. Indeed, most often the release of an active agent from the film into the food or a food-simulating liquid is quantified by assaying the global content in active compound in the food or liquid. Modeling of this release is then performed. Several approaches are reported in literature about the modeling of active agents' release (Flores et al., 2007; Peppas et al., 2000; Siepmann and Siepmann, 2008) with or without coupling of the release with reaction.

6.5 CONCLUSION

Nowadays, food protection along the supply chain to master food quality and shelf life has been raised on the primary level in the food industry. Food waste and losses present an ethical and economical issue that could be partially solved by optimizing food protection during postharvest steps. Primary packaging and edible films and coatings, by their role in regulating mass transfer phenomena between food and external atmosphere, play a key role in prolonging food shelf life and reducing food waste and losses. Moreover, consumers are more concerned about what they eat and about the safety of packaging materials. Thus, novel, natural solutions have been developed over the last four decades. Among these natural solutions, edible films used as regulating agent or support of mass transfer present an alternative to chemical preservatives that are traditionally used in prolongation of food shelf life. However, each food product is characterized by its own properties and displays specific requirements as regard mass transfer phenomena that complicate the requisite tailoring of edible food films for a target product.

In this chapter, different approaches for measuring the crucial parameters that are mass transfer parameters (solubility, diffusivity, and permeability) have been presented. All these methodologies were derived from the characterization methods used of flexible packaging material. We particularly insisted on the unavoidable characterization of the *in situ* performance of edible films and coatings, that is, once used in contact with the food product. Up to date, there are several methodologies permitting evaluation of the performance of edible films in real conditions of use. The most efficient one relies on the use of mathematical modeling to predict in advance mass transfer phenomena in the coated/food system and therefore, the food shelf life. When used in a reverse manner, such type of numerical approaches permit to calculate the optimal window of permeabilities toward a given migrant suitable for the product or to optimize the edible film formulation, for instance, the initial quantity of antimicrobial agent to add in the formulation. Of course, these mathematical models are useless without input parameters of mass transfer, especially diffusivity data. Therefore, future research should bring an effort on (1) the development of fast methods for diffusivity and solubility determination (such as FTIR-ATR- or RAMAN spectroscopy–based methods) and (2) the mathematical modeling of food shelf life and its coupling with mass transfer mathematical models. This coupling is indeed absolutely indispensible to propose efficient numerical tool for designing efficient active edible films and coatings.

REFERENCES

Angellier-Coussy, H., Gastaldi, E., Gontard, N., and Guillard, V. 2011. Influence of processing temperature on the water vapour transport properties of wheat gluten based agromaterials. *Industrial Crops and Products*, 33(2), 457–461.

Angellier-Coussy, H., Guillard, V., Guillaume, C., and Gontard, N. 2013. Role of packaging in the smorgasbord of action for sustainable food consumption. *Agro Food Industry Hi-Tech*, 23, 15–19.

Arvanitoyannis, I. and Biliaderis, C. G. 1998. Physical properties of polyol-plasticized edible films made from sodium caseinate and soluble starch blends. *Food Chemistry*, 62(3), 333–342.

Bacigalupi, C., Lemaistre, Marie Hélène Peyron, S., Guillard, V., and Chalier, P. 2013. Changes in nutritional and sensory properties of orange juice packed in PET bottles: An experimental and modelling approach. *Food Chemistry*, 141(4), 3827–3836.

Baldwin, E.A., Hagenmaier, R., and Bai, J. 2011. *Edible Coatings and Films to Improve Food Quality* (2nd ed.). Boca Raton, FL: CRC Press.

Baner, A., Brandsch, J., Franz, R., and Piringer, O. 1996. The application of a predictive migration model for evaluating the compliance of plastic materials with European food regulations. *Food Additives & Contaminants*, 13(5), 587–601.

BenBettaïeb, N., Karbowiak, T., Bornaz, S., and Debeaufort, F. 2015. Spectroscopic analyses of the influence of electron beam irradiation doses on mechanical, transport roperties and microstructure of chitosan-fish gelatin blend films. *Food Hydrocolloid*, 46, 37–51.

Biquet, B. and Labuza, T. P. 1988a. Evaluation of the moisture permeability characteristics of chocolate films as an edible moisture barrier. *Journal of Food Science*, 53(4), 989–998.

Biquet, B. and Labuza, T. P. 1988b. New model gel system for studying water activity of foods. *Journal of Food Processing and Preservation*, 12(2), 151–161.

Boudhrioua, N., Bonazzi, C., and Daudin, J. D. 2003. Estimation of moisture diffusivity in gelatin-starch gels using time-dependent concentration distance curves at constant temperature. *Food Chemistry*, 82(1), 139–149.

Bourlieu, C., Ferreira, M., Barea, B., Guillard, V., Villeneuve, P., Guilbert, S., and Gontard, N. 2009a. Moisture barrier and physical properties of acetylated derivatives with increasing acetylation degree. *European Journal of Lipid Science and Technology*, 111(5), 489–498.

Bourlieu, C., Guillard, V., Ferreira, M., Powell, H., Vallès-Pàmies, B., Guilbert, S., and Gontard, N. 2009b. Effect of cooling rate on the structural and moisture barrier properties of high and low melting point fats. *Journal of the American Oil Chemists' Society*, 87(2), 133–145.

Bourlieu, C., Guillard, V., Powell, H., Vallès-Pàmies, B., Guilbert, S., and Gontard, N. 2006. Performance of lipid-based moisture barriers in food products with intermediate water activity. *European Journal of Lipid Science and Technology*, 108(12), 1007–1020.

Bourlieu, C., Guillard, V., Powell, H., Vallès-Pàmies, B., Guilbert, S., and Gontard, N. 2008. Modelling and control of moisture transfers in high, intermediate and low a_w composite food. *Food Chemistry*, 106(4), 1350–1358.

Bourlieu, C., Guillard, V., Vallès-Pamiès, B., and Gontard, N. 2007. Edible moisture barriers: Materials, shaping techniques and promises in food product stabilization. In: J. M. Aguilera and P. J. Lillford (Eds.), *Food Materials Science* (pp. 547–577). New York, NY: Springer.

Bourlieu, C., Guillard, V., Vallès-Pamiès, B., Guilbert, S., and Gontard, N. 2009c. Edible moisture barriers: How to assess of their potential and limits in food products shelf-life extension? *Critical Reviews in Food Science and Nutrition*, 49(5), 474–499.

Bourtoom, T. 2008. Review article: Edible films and coatings: Characteristics and properties. *Food Research International*, 15(3), 237–248.

Butler, B. L., Vergano, P. L., Testin, R. F., Bunn, J. M., and Wiles, J. L. 1996. Mechanical and barrier properties of edible chitosan film as affected by composition and storage. *Journal of Food Science*, 61(5), 953–961.

Cagnon, T., Méry, A., Chalier, P., Guillaume, C., and Gontard, N. 2013. Fresh food packaging design: A requirement driven approach applied to strawberries and agro-based materials. *Innovative Food Science and Emerging Technologies*, 20, 288–298.

Cagri, A., Ustunol, Z., and Ryser, E. T. 2004. Antimicrobial edible films and coatings. *Journal of Food Protection*, 67(4), 833–848.

Campos, C. A., Gerschenson, L. N., and Flores, S. K. 2011. Development of edible films and coatings with antimicrobial activity. *Food and Bioprocess Technology*, 4(6), 849–875.

Caner, C., Vergano, P. J., and Wiles, J. L. 1998. Chitosan film mechanical and permeation properties as affected by acid, plasticizer, and storage. *Journal of Food Science*, 63(6), 1049–1053.

de Carvalho, R. A. and Grosso, C. R. F. 2004. Characterization of gelatin based films modified with transglutaminase, glyoxal and formaldehyde. *Food Hydrocolloids*, 18(5), 717–726.

Chaix, E. 2014. Characterization and modelling of gas (O_2/CO_2) transfer in the food/packaging system in relationship with microbial growth (predictive microbiology). University of Montpellier 2, Montpellier, France.

Chaix, E., Couvert, O., Guillaume, C., Gontard, N., and Guillard, V. 2015. Predictive microbiology coupled with Gas (O_2/CO_2) transfer in food/packaging systems: How to develop an efficient decision support tool for food packaging. *Comprehensive Reviews in Food Science and Food Safety*, 14(1), 1–21.

Chaix, E., Guillaume, C., and Guillard, V. 2014. Oxygen and carbon dioxide solubility and diffusivity in solid food matrices: A review of past and current knowledge. *Comprehensive Reviews in Food Science and Food Safety*, 13(3), 261–286.

Chalier, P., Ben Arfa, A., Guillard, V., and Gontard, N. 2009. Moisture and temperature triggered release of a volatile active agent from soy protein coated paper: Effect of glass transition phenomena on carvacrol diffusion coefficient. *Journal of Agricultural and Food Chemistry*, 57(2), 658–665.

Choudalakis, G. and Gotsis, A. D. 2009. Permeability of polymer/clay nanocomposites: A review. *European Polymer Journal*, 45(4), 967–984.

Crank, J. 1980. *The Mathematics of Diffusion*. 2nd ed., Oxford Science Publications.

Debeaufort, F. 1998. Edible films and coatings: Tomorrow's packagings: A review. *Critical Reviews in Food Science and Nutrition*, 38(4), 299–313.

Debeaufort, F., Quezada-Gallo, J., and Voilley, A. 2000. Edible barriers: A solution to control water migration in foods. *Food Packaging: Testing Methods and Applications*, 753, ACS Symposium Series, pp. 9–16.

Debeaufort, F., Quezada-Gallo, J.-A., and Voilley, A. 2002. Edible films and coatings as aroma barriers. In A. Gennadios (Ed.), *Protein-Based Edible Films and Coatings* (pp. 579–600). Lancaster, PA: CRC Press.

Debeaufort, F., Quezada-Gallo, J., and Voilley, A. 2003. Flavour retention by edible barrier films. In J. LeQuere and P. Etievant (Eds.), *Flavour Research at the Dawn of the Twenty-First Century* (pp. 71–74). Paris, France: Lavoisier, Tec & Doc.

Del Valle, E. M. M. 2004. Cyclodextrins and their uses: A review. *Process Biochemistry*, 39(9), 1033–1046.
Despond, S., Espuche, E., and Domard, A. 2001. Water sorption and permeation in chitosan films: Relation between gas permeability and relative humidity. *Journal of Polymer Science Part B: Polymer Physics*, 39(24), 3114–3127.
Destercke, S., Buche, P., and Guillard, V. 2011. A flexible bipolar querying approach with imprecise data and guaranteed results. *Fuzzy Sets and Systems*, 169(1), 51–64.
Dhall, R. K. 2013. Advances in edible coatings for fresh fruits and vegetables: A review. *Critical Reviews in Food Science and Nutrition*, 53(5), 435–450.
Dury-Brun, C., Hirata, Y., Guillard, V., Ducruet, V., Chalier, P., and Voilley, A. 2008. Ethyl hexanoate transfer in paper and plastic food packaging by sorption and permeation experiments. *Journal of Food Engineering*, 89(2), 217–226.
Embuscado, M. and Huber, K. C. 2009. *Edible Films and Coatings for Food Applications*. New York: Springer.
Fajardo, P., Martins, J. T., Fuciños, C., Pastrana, L., Teixeira, J. A., and Vicente, A. A. 2010. Evaluation of a chitosan-based edible film as carrier of natamycin to improve the storability of Saloio cheese. *Journal of Food Engineering*, 101(4), 349–356.
Feuge, R. O. 1954. Acetoglycerides—New fat products of potential value for the food industry. *Food Technology*, 8, S20–S20.
Fick, A. 1855. On liquid diffusion. *Journal of Membrane Science*, 100, 33–38.
Flores, S., Conte, A., Campos, C., Gerschenson, L., and Del Nobile, M. 2007. Mass transport properties of tapioca-based active edible films. *Journal of Food Engineering*, 81(3), 580–586.
Flores, S., Haedo, A. S., Campos, C., and Gerschenson, L. 2007. Antimicrobial performance of potassium sorbate supported in tapioca starch edible films. *European Food Research and Technology*, 225, 375–384.
Gaudin, S., Lourdin, D., Forssell, P.M., and Colonna, P. 2000. Antiplasticisation and oxygen permeability of starch-sorbitol films. *Carbohydrate Polymers*, 43, 33–37.
Gennadios, A., Hanna, M. A., and Kurth, L. B. 1997. Application of edible coatings on meats, poultry and seafoods: A review. *LWT—Food Science and Technology 350*(4), 337–350.
Guilbert, S. 1986. Technology and application of edible protective films. In M. Mathlouthi (Ed.) *Food Packaging and Preservation. Theory and Practice* (pp. 371–394). London, UK: Elsevier Applied Science Publishing.
Guilbert, S. 1988. Use of superficial edible layer to protect intermediate moisture foods: Application to the protection of tropical fruit dehydrated by osmosis. In C. C. Seow (Ed.), *Food Preservation by Moisture Control* (pp. 119–219). London, UK: Elsevier Applied Science Publishers.
Guilbert, S. and Gontard, N. 2005. Agro-polymers for edible and biodegradable films: Review of agricultural polymeric materials, physical and mechanical characteristics. In J. H. Han (Ed.), *Innovations in Food Packaging* (pp. 263–276). London: Academic Press.
Guilbert, S., Gontard, N., and Cuq, B. 1995. Technology and applications of edible protective films. *Packaging Technology and Science*, 8, 339–346.
Guilbert, S., Gontard, N., and Gorris L. G. M. 1996. Prolongation of the shelf-life of perishable food products using biodegradable films and coatings. *LWT—Food Science and Technology*, 29(1–2), 10–17.
Guillard, V., Bourlieu, C., and Gontard, N. 2013. *Food Structure and Moisture Transfer A Modeling Approach*. London, UK: Springer.
Guillard, V., Broyart, B., Bonazzi, C., Guilbert, S., and Gontard, N. 2003a. Evolution of moisture distribution during storage in a composite food modelling and simulation. *Journal of Food Science*, 68(3), 958–966.
Guillard, V., Broyart, B., Bonazzi, C., Guilbert, S., and Gontard, N. 2003b. Preventing moisture transfer in a composite food using edible films: Experimental and mathematical study. *Journal of Food Science*, 68(7), 2267–2277.

Guillard, V., Broyart, B., Bonazzi, C., Guilbert, S., and Gontard, N. 2003c. Moisture diffusivity in sponge cake as related to porous structure evaluation and moisture content. *Journal of Food Science, 68*(2), 555–562.

Guillard, V., Broyart, B., Bonazzi, C., Guilbert, S., and Gontard, N. 2004a. Effect of temperature on moisture barrier efficiency of monoglyceride edible films in cereal-based composite foods. *Cereal Chemistry, 81*(6), 767–771.

Guillard, V., Broyart, B., Guilbert, S., Bonazzi, C., and Gontard, N. 2004b. Moisture diffusivity and transfer modelling in dry biscuit. *Journal of Food Engineering, 64*(1), 81–87.

Guillard, V., Buche, P., Destercke, S., Tamani, N., Croitoru, M., Menut, L., Guillaume, C., and Gontard, N. 2015. A decision support system to design modified atmosphere packaging for fresh produce based on a bipolar flexible querying approach. *Computers and Electronics in Agriculture, 111,* 131–139.

Guillard, V., Chevillard, A., Gastaldi, E., Gontard, N., and Angellier-Coussy, H. 2013. Water transport mechanisms in wheat gluten based (nano)composite materials. *European Polymer Journal, 49*(6), 1337–1346.

Guillard, V., Guilbert, S., Bonazzi, C., and Gontard, N. 2004c. Edible acetylated monoglyceride films: Effect of film-forming technique on moisture barrier properties. *Journal of the American Oil Chemists' Society, 81*(11), 1053–1058.

Guillard, V., Guillaume, C., and Destercke, S. 2012. Parameter uncertainties and error propagation in modified atmosphere packaging modelling. *Postharvest Biology and Biotechnology, 67*(1), 154–166.

Guillard, V., Issoupov, V., Redl, A., and Gontard, N. 2009. Food preservative content reduction by controlling sorbic acid release from a superficial coating. *Innovative Food Science and Emerging Technologies, 10*(1), 108–115.

Gutiérrez, L., Sánchez, C., Batlle, R., López, P., and Nerín, C. 2009. New antimicrobial active package for bakery products. *Trends in Food Science and Technology, 20*(2), 92–99.

Hagenmaier, R. D. 2011. Gas-exchange properties of edible films and coatings. In E. A. Baldwin, R. Hagenmaier and J. Bai (Eds.), *Edible Coatings and Films to Improve Food Quality* (2nd ed., pp. 137–155). Boca Raton, FL: CRC Press.

Han, J. H. 2014. *Innovations in Food Packaging* (2nd ed.). San Diego: Academic Press.

Hernández-Muñoz, P., López-Rubio, A., Del-Valle, V., Almenar, E., and Gavara, R. 2004. Mechanical and water barrier properties of glutenin films influenced by storage time. *Journal of Agricultural and Food Chemistry, 52*(1), 79–83.

Higushi, W. I. 1958. A new relationship for the dielectric properties of two- phase mixtures. *American Journal of Physical Chemistry, 62*(6), 649–653.

Hinrichs, K. and Piringer, O. (Eds.) 2002. Evaluation of migration models to use under directive 90/128/EEC. Final report contract SMT4-CT98-7513, European Commission, Directorate General for Research, EUR 20604 EN, Brussels.

Hong, Y. C., Bakshi, A. S., and Labuza, T. P. 1986. Finite element modeling of moisture transfer during storage of mixed multicomponent dried foods. *Journal of Food Science, 51*(3), 554–558.

Jo, C., Kang, H., Lee, N. Y., Kwon, J. H., and Byun, M. W. 2005. Pectin- and gelatin-based film: Effect of gamma irradiation on the mechanical properties and biodegradation. *Radiation and Physics Chemistry, 72*(6), 745–750.

Kamper, S. L. and Fennema, O. 1985. Use of an edible film to maintain water vapour gradients in foods. *Journal of Food Science, 50*(2), 382–384.

Karathanos, V. T. and Kostaropoulos, A. 1995. Diffusion and equilibrium of water in dough/raisin mixtures. *Journal of Food Engineering, 25*(1), 113–121.

Karbowiak, T., Debeaufort, F., and Voilley, A. 2006. Importance of surface tension characterization for food, pharmaceutical and packaging products: A review. *Critical Reviews in Food Science and Nutrition, 46*(5), 1–17.

Karbowiak, T., Debeaufort, F., Voilley, A., and Trystram, G. 2009. From macroscopic to molecular scale investigations of mass transfer of small molecules through edible packaging applied at interfaces of multiphase food products. *Innovative Food Science and Emerging Technologies*, *10*(1), 116–127.

Karel, M. and Labuza, T. P. 1969. Optimization of protective packaging of space food. Force contract F-43-609-68-C-0015, Aerospace Medical School.

Kester, J. and Fennema, O. 1986. Edible films and coatings—A review. *Food Technology*, 40:47–49.

Kester, J. J. and Fennema, O. R. 1989. An edible film of lipids and cellulose ethers: Barrier properties to moisture vapor transmission and structural evaluation. *Journal of Food Science*, *54*(6), 1383–1389.

Khwaldia, K., Banon, S., Desobry, S., and Hardy, J. 2004. Mechanical and barrier properties of sodium caseinate–anhydrous milk fat edible films. *International Journal of Food Science and Technology*, *39*(4), 403–411.

Klopffer, M. H. and Flaconnèche, B. 2001. Transport properties of gases in polymers: Bibliographic review. *Oil and Gas Science and Technology*, *56*(3), 223–244.

Koelsch, C. 1994. Edible water vapor barriers: Properties and promise. *Trends in Food Science and Technology*, *5*(3), 76–81.

Labuza, T. P. 1998. Moisture migration and control in multi-domain foods. *Trends food Science and Technology*, *9*(2), 47–55.

Lacroix, M., Le, T. C., Ouattara, B., Yu, H., Letendre, M., Sabato, S. F., Mateescu, M. A., and Patterson, G. 2002. Use of gamma irradiation to produce films from whey, casein and soy proteins: Structure and functional characteristics. *Radiation Physics and Chemistry*, *63*, 827–832.

Landmann, W., Lovegren, N. V., and Feuge, R. O. 1960. Permeability of some fat products to moisture. *Journal of the American Oil Chemists' Society*, *37*(1), 1–4.

López, P., Sánchez, C., Batlle, R., and Nerín, C. 2007. Development of flexible antimicrobial films using essential oils as active agents. *Journal of Agricultural and Food Chemistry*, *55*(21), 8814–8824.

Martinez-Lopez, B., Chalier, P., Guillard, V., Gontard, N., and Peyron, S. 2014. Determination of mass transport properties in food/packaging systems by local measurement with Raman microspectroscopy. *Journal of Applied Polymer Science*, *131*, 40958.

Mauricio-Iglesias, M., Guillard, V., Gontard, N., and Peyron, S. 2009. Application of FTIR and Raman microspectroscopy to the study of food/packaging interactions. *Food Additives & Contaminants. Part A, Chemistry, Analysis, Control, Exposure & Risk Assessment*, *26*, 1515–1523.

Mauricio-Iglesias, M., Guillard, V., Gontard, N., and Peyron, S. 2011. Raman depth-profiling characterization of a migrant diffusion in a polymer. *Journal of Membrane Science*, *375*(1–2), 165–171.

McHugh, T. H. and Krochta, J. M. 1994. Sorbitol- vs glycerol-plasticized whey protein edible films: Integrated oxygen permeability and tensile property evaluation. *Journal of Agricultural and Food Chemistry*, *42*, 841–845.

Mercea, P. 2008. Models for diffusion in polymers. In: O.-G. Piringer, and A. L. Baner (Eds.), *Plastic Packaging Materials for Food—Barrier Function, Mass Transport, Quality Assurance and Legislation*. Weinhein, Germany: Wiley-VCH Verlag GmbH.

Miller, K.S. and Krochta, J.M. 1997. Oxygen and aroma barrier properties of edible films: A review. *Trends in Food Science and Technology*, *8*(7), 228–237.

Morillon, V., Debeaufort, F., Blond, G., Capelle, M., and Voilley, A. 2002. Factors affecting the moisture permeability of lipid-based edible films: A review. *Critical Reviews in Food Science and Nutrition*, *42*(1), 67–89.

Motarjemi, Y. 1988. A study of some physical properties of water in foodstuffs. Water activity, water binding and water diffusivity in minced meat products. PhD Thesis, Lund University, Lund, Sweden.

MujicaPaz, H., and Gontard, N. 1997. Oxygen and carbon dioxide permeability of wheat gluten film: Effect of relative humidity and temperature. *Journal of Agricultural and Food Chemistry*, 45(10), 4101–4105.

Namin, P. M. 2011. Perméation des gaz dans les polymères semi-cristallins par modélisation moléculaire. PhD Thesis, Université Paris-Sud 11—UFR Scientifique d'Orsay, Paris, France.

Nisperos-Carriedo M. 1994. Edible coatings and films based on polysaccharides. In J. Krochta, E. Baldwin, and M. Nisperos-Carriedo (Eds.), *Edible Coatings and Films to Improve Food Quality* (pp. 305–335). Landcaster, PA: Technomic Publishing Company.

Olivas, G. I., and Barbosa-Cánovas, G. V. 2005. Edible coatings for fresh-cut fruits. *Critical Reviews in Food Science and Nutrition*, 45(7–8), 657–670.

Ouattara, B., Simard, R. E., Piette, G., Begin, A., and Holley, R. A. 2000. Diffusion of acetic and propionic acids from chitosan-based antimicrobial packging films. *Food and Chemical Toxicology*, 65(5), 768–773.

Pavlath, A. E. and Orts, W. 2009. Edible films and coatings: Why, what, and how? In M. Embuscado and K. C. Huber (Eds.), *Edible Films and Coatings for Food Applications* (pp. 1–25). New York, NY: Springer.

Pelissari, F. M., Grossmann, M. V. E., Yamashita, F., and Pineda, E. A. G. 2009. Antimicrobial, mechanical, and barrier properties of cassava starch-chitosan films incorporated with oregano essential oil. *Journal of Agricultural and Food Chemistry*, 57(16), 7499–7504.

Pénicaud, C., Broyart, B., Peyron, S., Gontard, N., and Guillard, V. 2011. Mechanistic model to couple oxygen transfer with ascorbic acid oxidation kinetics in model solid food. *Journal of Food Engineering*, 104(1), 96–104.

Pénicaud, C., Guillard, V., Peyron, S., and Gontard, N. 2009. Oxygen transfer coupled to oxidation reactions: Numerical tool for optimising nutritional quality of food. *Czech Journal of Food Sciences*, 27, S28–S28.

Pénicaud, C., Guilbert, S., Peyron, S., Gontard, N., and Guillard, V. 2010a. Oxygen transfer in foods using oxygen luminescence sensors: Influence of oxygen partial pressure and food nature and composition. *Food Chemistry*, 123(4), 127–1281.

Pénicaud, C., Peyron, S., Bohuon, P., Gontard, N., and Guillard, V. 2010b. Ascorbic acid in food: Development of a rapid analysis technique and application to diffusivity determination. *Food Research International*, 43(3), 838–847.

Pénicaud, C., Peyron, S., Gontard, N., and Guillard, V. 2012. Oxygen quantification methods and application to the determination of oxygen diffusion and solubility coefficients in food. *Food Reviews International*, 28(2), 113–145.

Peppas, N. A., Bures, P., Leobandung, W., and Ichikawa, H. 2000. Hydrogels in pharmaceutical formulations. *European Journal of Pharmaceutics and Biopharmaceutics*, 50(1), 27–46.

Quezada Gallo, J. A., Debeaufort, F., and Voilley, A. 1999. Interactions between aroma and edible films. 1. Permeability of methylcellulose and low-density polyethylene films to methyl ketones. *Journal of Agricultural and Food Chemistry*, 47(1), 108–113.

Robertson, G. L. 2006. *Food Packaging Principles and Practice*. Boca Raton, FL: CRC Press.

Roca, E., Adeline, D., Guillard, V., Guilbert, S., and Gontard, N. 2008a. Shelf life and moisture transfer predictions in a composite food product: Impact of preservation techniques. *International Journal of Food Engineering*, 4(4), 1556–3758.

Roca, E., Broyart, B., Guillard, V., Guilbert, S., and Gontard, N. 2008b. Predicting moisture transfer and shelf-life of multidomain food products. *Journal of Food Engineering*, 86(1), 74–83.

Rogers, C. E. 1985. Permeation of gases and vapours in polymers. In J. Comyn (Ed.), *Polymer Permeability* (Vol. 2, pp. 11–73). New York, NY: Elsevier Applied Science Publishers.

Rojas-Graü, M., Raybaudi-Massilia, R. M., Soliva-Fortuny, R. S., Avena-Bustillos, R. J., McHugh, T. H., and Martın-Belloso, O. 2007. Apple puree-alginate edible coating as carrier of antimicrobial agents to prolong shelf-life of fresh-cut apples. *Postharvest Biology and Technology, 45*(2), 254–264.

Rosso, L. et al. 1995. Convenient model to describe the combined effects of temperature and pH on microbial growth. *Applied and Environmental Microbiology, 61*(2), 610–616.

Rumsey, T. and Krochta, J. 1994. Mathematical modelling of moisture transfer in food systems with edible coatings. In J. Krochta, E. Baldwin, and M. Nisperos-Carriedo (Eds.), *Edible Coatings and Films to Improve Food Quality* (pp. 337–356), Landcaster, PA: Technomic Publishing Company.

Ryzhkova, A., Jarzak, U., Schafer, A., Baumer, M., and Swiderek, P. 2011. Modification of surface properties of thin polysaccharide films by low-energy electron exposure. *Carbohydrate Polymers, 83*(2), 608–615.

Salamone J. C. 1996. *Polymeric Material Encyclopedia.* Boca Raton, FL: CRC Press.

Siepmann, J. and Siepmann, F. 2008. Mathematical modeling of drug delivery. *International Journal of Pharmaceutics, 364*(2), 328–343.

Stannett, V. 1968. Simple gases. In J. Crank and G.S. Park (Eds.), *Diffusion in Polymers* (pp. 41–73). London, UK: Academic Press.

Valencia-Chamorro, S. A., Palou, L., del Río, M. A., and Pérez-Gago, M. B. 2011. Antimicrobial edible films and coatings for fresh and minimally processed fruits and vegetables: A review. *Critical Reviews in Food Science and Nutrition, 51*(9), 872–900.

Vargas, M., Pastor, C., Chiralt, A., McClements, D. J., and González-Martínez, C. 2008. Recent advances in edible coatings for fresh and minimally processed fruits. *Critical Reviews in Food Science and Nutrition, 48*(6), 496–511.

Warin, F., Gekas, V., Voirin, A., and Dejmek, P. 1997. Sugar diffusivity in agar gel/milk bilayer systems. *Journal of Food Science, 62*(3), 454–456.

Weinkauf, D. H. and Paul, D. R. 1990. Effects of structure order on barrier properties. In W. J. Koros (Ed.), *Barrier Polymers and Structures* (pp. 60–91). Washington, DC: American Chemical Society.

Welle, F. 2013. A new method for the prediction of diffusion coefficients in poly(ethylene terephthalate). *Journal of Applied Polymer Science, 129*(4), 1845–1851.

Wihodo, M. and Moraru, C. I. 2013. Physical and chemical methods used to enhance the structure and mechanical properties of protein films. *Journal of Food Engineering, 114*(3), 292–302.

Zhou, Q., Guthrie, B., and Cadwallader, K. R. 2004. Development of a system for measurement of permeability of aroma compounds through multilayer polymer films by coupling dynamic vapour sorption with purge-and-trap/fast gas chromatography. *Packaging Technology and Science, 17*(4), 175–185.

7 Edible Packaging
A Vehicle for Functional Compounds

Joana O. Pereira and M. Manuela Pintado

CONTENTS

Abstract 215
7.1 Introduction 216
7.2 Incorporating Functional Compounds: Why? 218
7.3 Advantages and Limitations of Functional Compounds 219
    7.3.1 Antioxidants and Antimicrobials 220
    7.3.2 Flavors 221
    7.3.3 Probiotics 221
7.4 Structural Matrix of Edible Packaging for Functional Compounds 222
7.5 Functionalities and Applications of Functional Edible Packaging 225
    7.5.1 Antioxidants and Antimicrobials 225
    7.5.2 Flavors 227
    7.5.3 Probiotics 229
    7.5.4 Other Functional Compounds 230
7.6 Conclusions 231
7.7 Future Trends 233
References 234

ABSTRACT

Nowadays, innovations constantly appear in food packaging, which lead to a demand for new foods, always aiming at creating a more efficient quality preservation system. Edible coatings and films are a good attempt to increase the storability of foods, controlling gas exchange, moisture, solutes migration, and oxidative reaction rates. In addition to these gains, edible coating/films can be used as carriers of bioactive compounds to improve the quality and enhance the nutritional value of food products, such as antimicrobials, antioxidants, flavors, probiotics, and other components, such as nutraceuticals or basic nutrients. Therefore, these approaches, in addition to being used to prolong shelf life, also provide functionality to food products. When the functional compounds are incorporated into the edible coating/film to perform their functions, they became a benefit for food as well as for consumers. Furthermore, the functional compounds are protected from the external factors and controlled release is allowed. In this sense, the aim of this chapter is to analyze the potential use of edible coatings and films incorporated with functional compounds, solve the disadvantages of direct application, and find the correct combination between the food product and the edible coating/film, which will ensure the success of the technology.

7.1 INTRODUCTION

The latest developments in new packaging concepts, which include packaging systems comprising bioactive components, are mentioned in this chapter. Although active coatings have been developed to play an active role in food preservation, the bioactive coating is a new concept of technology to assist in the production of functional foods, whose bioactive principles are designed to be contained within coatings or coating materials.

Advances in this area imply that diet and/or its components should contribute to reduce the risk of diseases, thus improving well-being and quality of life. These new concepts have led to the introduction of a new category of health-promoting food or functional food (Korhonen, 2002).

The term "functional foods," was first used in 1984 in Japan; however, there is no unique definition of what a functional food actually is (Siró et al., 2008). To date, a number of national authorities, academic bodies, and the industries have proposed definitions for functional food (Roberfroid, 2002). Thus, several definitions for functional foods can be checked (Bigliardi and Galati, 2013), including one found in literature: "Whole foods and fortified, enriched, or enhanced foods that have a potentially beneficial effect on health when consumed as part of a varied diet on a regular basis, at effective levels" (Hasler and Brown, 2009, p. 735).

The main difference between the known technology of edible packaging itself and bioactive coatings/films is that while edible packaging primarily deals with maintaining or increasing quality and safety of packaged foods, extending their shelf life, the bioactive packaging has a direct impact on consumer health by generating healthier packaged foods by a specific bioactive attribute (Han, 2005). Therefore, there is a new conceptual approach toward the development of functional foods using a new packaging technology in which to a food package or coating is given a distinctive role in increasing the impact of food on the health of the consumer.

In general, edible packaging can provide several functions that do not exist in conventional packaging systems. Edible coatings and films, directly consumed with the food, were initially developed to increase the shelf life of food, thereby controlling food spoilage reactions and microbial contamination, through the control of some relevant properties such as moisture, gas exchange, and oxidative reaction rate (Cuq et al., 1995; Han, 2000; Kester and Fennema, 1986; Naushad and Stading, 2007). Currently, edible coatings/films exhibit most interesting features. Bioactive edible coating and film is defined as a protective coating/film applied to the surface of a food and furthermore may possess other purposes, for example, adding high value to food products through the addition of functional compounds such as antioxidants, colors, flavors, nutraceuticals, nutrients, probiotics, prebiotics, and antimicrobials that increase the functionality of the coating and add extra functions to food products, as represent in Figure 7.1 (Min et al., 2005; Pranoto et al., 2005; Salmieri and Lacroix, 2006).

Edible Packaging

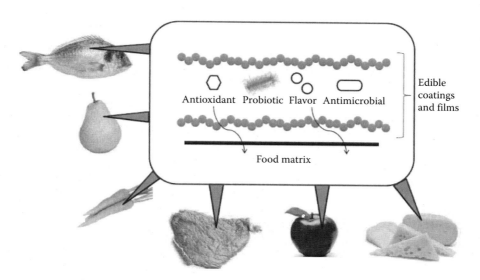

FIGURE 7.1 Representation of edible bioactive coatings and films, including the most used bioactive compounds (e.g., antioxidants, antimicrobials, probiotics, and flavors) and more frequent applications into food matrix (e.g., fruits, fish, meat, vegetables, and cheeses).

Edible coatings and films may be used as carriers for functional compounds, allowing their protection and controlled release strategy by which functional compounds are provided in the desired site and at desired time and rate (Pothakamury and Barbosa-Cánovas, 1995).

These functional compounds can be incorporated to produce new functional foods, increasing its useful life and enhancing the nutritional quality and consumer acceptance.

In general, active food packaging can provide several functions that do not exist in conventional packaging systems. The selection of functional compounds for incorporation is restricted to edible coatings/films, because they have to be consumed with the coating layers and foods together with guarantee of being safe for the consumer.

Functional compounds are components "extra nutrient" found in plants, animals, microorganisms, and marine organisms that normally occur in food in small quantities and can be obtained by extraction and biotechnological methods (Kris-Etherton et al., 2002).

The components that enable the functionality of the functional foods may be present naturally in products, but need appropriate strategies for maintaining bioactivity during application, processing, and storage of the formulated product and control the release of the functional component to the desired target (Lopez-Rubio et al., 2006; Ozimek et al., 2010).

A direct surface application of functional compounds has limited benefits, because the active substances can be neutralized on contact or diffused rapidly into the bulk of food (Min and Krochta, 2005; Siragusa and Dickson, 1992; Torres et al., 1985).

For development of functional foods, the incorporation of functional compounds into edible coating and films provides some advantages, like conveying substances that bring some benefits not only for food itself but also for the consumer (Falguera et al., 2011).

Additionally, as functional compounds may have certain disadvantages such as off flavors and an early loss of functionality (Silva-Weiss et al., 2013), the utilization of edible coatings/films as a carrier is a promising technique that can help solving these drawbacks.

Depending on the nature of the functional compound and coating materials, different techniques are used to coat the food. The techniques used for edible coatings include classic methodology of coating: spray fluidization, falling and pan coatings, spraying, dipping, and brushing. These processes are usually followed by drying in the case of aqueous products or by cooling for coatings based on lipids (Debeaufort et al., 1998). These processes are discussed in previous chapters.

For edible films, the used techniques are similar to those for manufacturing flexible plastic films: extrusion (or coextrusion of multilayer films), lamination, molding, and roll drying to remove solvent (Debeaufort et al., 1998; Guilbert et al., 1996; Hernandez-Izquierdo and Krochta, 2008, 2009). These techniques are more efficient than a direct application of functional agents on the food surface, because edible coatings/films delay the migration of the agents away from the surface, helping to maintain a high concentration of functional compounds where they are needed.

7.2 INCORPORATING FUNCTIONAL COMPOUNDS: WHY?

There are several functional compounds appropriate and necessary to our health and well-being. The most commonly used compounds in edible packaging are phytochemicals, flavors, and probiotics. All these compounds have various functions that make them much requested and appreciated.

Phytochemicals are chemicals of nonnutritious plants that contain protective compounds (antioxidants and antimicrobials) that prevent microbial growth and certain diseases. Several examples of these compounds are given below. They are mainly associated with the prevention and/or control of certain chronic diseases, such as cancer, diabetes, cardiovascular disease, and hypertension (Traka and Mithen, 2011). These compounds help to prevent cell damage, replication of malignant cells, and reduce cholesterol. In addition, several of these compounds are phenolics with antioxidant capacity for direct activity in elimination of free radicals and an indirect effect due to chelation of ions of pro-oxidant metals (Flora, 2009; Rahman, 2007; Shahidi, 2000; Wettasinghe and Shahidi, 1999). These compounds can be found in plants/seeds, but during certain processing steps, they are removed or lose their activity, failing in promoting health, or preventing disease (Mattila-Sandholm et al., 2002).

As edible coatings and films, in general, are consumed with product, therefore, the incorporation of these should not adversely affect consumer acceptance (Rojas-Graü et al., 2009). Hence, the flavors are usually incorporated with the main aim of increasing customer satisfaction and promote consumption and consumer acceptance.

Probiotics are considered "live microorganisms which when administered in adequate amounts, not be less than 10^6 CFU/g, confer a health benefit on the host" (FAO/WHO, 2002, p. 8). Among other effects attributed to the probiotics, the increased digestibility, the nutritional contribution for the production of conjugated linoleic acid, vitamins and short chain fatty acids, as well as antagonistic activities against enteric pathogens and modulation of intestinal flora are the most prominents (Gomes and Malcata, 1999; Reid, 2008; Vasiljevic and Shah, 2008).

The probiotics have key functionalities such as relief from lactose maldigestion, increased resistance to intestinal invasion by pathogenic species of bacteria, stimulation of the immune system, and protection against colon cancer (Chen and Walker, 2005; Daoud and Hani, 2013; Shida and Nanno, 2008; Zubillaga et al., 2001).

7.3 ADVANTAGES AND LIMITATIONS OF FUNCTIONAL COMPOUNDS

There are several functional compounds that can be added in our diet, through coated food, such as phenolic acids, carotenoids, vitamins, dietary fibers, and some specific molecules such as ascorbic acid (Ajila et al., 2010; Gonzalez-Aguilar et al., 2008; Gorinstein et al., 2011; Lanciotti et al., 2004; Soong and Barlow, 2004).

Nowadays, there is a growing demand for these food additives by consumers due to the attributes that demonstrated several benefits to human health.

The use of functional compounds has previously been directed to pharmaceutical or food products. In the latter area, the compounds are developed to create functional foods with some features such as antioxidants, antimicrobials, probiotics, and flavors (Min et al., 2005; Pranoto et al., 2005; Salmieri and Lacroix, 2006).

However, functional compounds may have certain constrains such as off flavors, an early loss of functionality, or interaction with other components of the food matrix causing the loss of quality of functional food products (Silva-Weiss et al., 2013). These adverse effects occur, since the manufacture of these foods implies a series of limitations and difficulties, namely

1. The compound introduced is not compatible with the food matrix (e.g., lipophilic compounds in foods with extensive aqueous phase) interacting with food compounds leading to loss of functionality and/or quality of functional food (McClements et al., 2009).
2. The need to adapt the production line for incorporation of bioactive compounds in food. Usually, this process involves considerable expenses that are accessible only to large companies (Homayouni et al., 2012).
3. Due to their sensitivity during the treatment, several bioactive compounds, as lipids, vitamins, peptides, fatty acids, antioxidants, minerals, and also living cells, such as probiotics, can be modified or even inactivated; some relevant treatment factors include temperature, pH, pressure, or stirring (de Vos et al., 2010).
4. The loss of functionality and quality of the product during storage time (Fogliano and Vitaglione, 2005), since changes resulting from interactions between food and bioactive compounds or modifications of specific compounds during storage conditions may result in significant losses.
5. The induction of undesirable flavors and aromas in foods through reactions of oxidation, Maillard among others, which if extensive may generate toxic compounds (Sun Pan et al., 2006).

Hence, the utilization of edible coatings/films as a carrier is a promising technique to overcome some of the limitations of functional compounds when applied as food additives. Specific advantages and limitations for the main bioactive compound groups used in edible coatings/films are presented below.

7.3.1 ANTIOXIDANTS AND ANTIMICROBIALS

The harmful effects of free radicals and other reactive oxygen compounds can be neutralized by antioxidants that promote human health and have beneficial effects on functional food technology. Tissues are under constant oxidative stress from free radicals, reactive oxygen species, and pro-oxidants generated both exogenous (heat and light) and endogenously (transition metals and H_2O_2). For this reason, the human organism developed antioxidant systems to control free radicals, lipid oxidation catalysts, oxidation intermediates, and secondary degradation products (Agati et al., 2007; Brown and Kelly, 2007; Chen, 2008; Iacopini et al., 2008; Nakatani, 2003).

Potential sources of antioxidants can be plants, fruits, vegetables, cereals, and herbs. These antioxidants include flavonoids, phenolic acids, carotenoids, tocopherols, among others and can inhibit oxidation, scavenge free radicals, or act as reducers (Khanduja and Bhardwaj, 2003; Ozsoy et al., 2009).

However, the use of these compounds may involve certain drawbacks, such as vulnerability to high temperature and light, high volatility, limited solubility, and unpleasant taste; these characteristics result in a loss of functionality, which limits its application (Fang and Bhandari, 2010). Some flavonoids (e.g., isoflavones) and phenols have limited solubility in lipophilic systems (Viskupicova et al., 2010). Many polyphenols (quercetin, kaempferol, taxifolin, procyanidines, salicin, thymol, and eugenol), terpenes, and carotenoids are astringent and have an unpleasant taste. Ferulic acid is easily volatilized and, therefore cannot inhibit the oxidation at elevated temperatures for extended periods of time (Nyström et al., 2007). Furthermore, various vitamins (e.g., C, E, and K) and some phytochemicals (e.g., phenol and flavonoids) are sensitive to UV-B and UV-C light radiation, limiting its use in food when they are exposed to such conditions (Durand et al., 2010).

Antimicrobials are compounds that perform the function of protection against microorganisms (Rauha et al., 2000). For this reason, it makes sense to describe these compounds as functional additives in foods, but this will be done briefly, because it will be discussed with more detail in the Chapter 8.

Natural antimicrobials most commonly used to increase food safety and lifetime are enzymes (e.g., lysozyme and lactoperoxidase), polysaccharides (e.g., chitosan), bacteriocins (e.g., nisin), and more recently herbs and spices (e.g., oregano, thyme, and cinnamon), and essential oils including terpenes, alcohols, ketones, phenols, acids, and aldehydes esters (Tajkarimi et al., 2010).

For their antioxidant and antimicrobial activities, essential oils have been widely studied. These are recognized as safe for health and have shown effectiveness against a wide range of bacteria and molds (Fisher and Phillips, 2008). Despite the advantages of essential oils derived from plant extracts, their high volatility, reactivity, and unpleasant aroma may limit its potential use (Del Toro-Sánchez et al., 2010). Moreover, these oils due to its hydrophobic character can represent a problem due

Edible Packaging

to the possible interactions between them and the food, essentially if the foods have high water content (Holley and Patel, 2005). Therefore, some packaging technologies can effectively reduce these constrains.

7.3.2 Flavors

Natural aromatic compounds are small molecules responsible for flavor and aroma in foods.

Among the most important sensory attributes on food, flavor and aroma are included and are often key indicators of lifetime, from the point of view of consumers and a determining factor in the purchase decision (Beaulieu and Lea, 2003).

In the universe of flavor compounds, the most important are the essential oils. These aromatic compounds are mostly constituted of short hydrocarbon chains, complemented with oxygen, nitrogen, and sulfur atoms attached to various points of the chain (Braca et al., 2008).

The loss of quality of food products can be related to the loss of aroma compounds, causing a reduction of flavor intensity and change in the typical food flavor.

These metabolic changes in the flavor are the result of synthesis or catabolism of either flavor compounds or compounds responsible for off flavors (Beaulieu and Lancaster, 2007).

There is a difficulty in keeping these compounds in foods, as they are generally sensible aroma, volatile, and hydrophobic, which is a disadvantage (Reineccius, 2009).

However, such deleterious reactions can be decreased with various preserving technologies, such as cooling and/or freezing, modified atmosphere, and edible coatings/films, with the aim of accomplishing consumer demands of a high-quality produce.

7.3.3 Probiotics

Probiotics are microorganisms intended to modify intestinal microbiota, as defined earlier. There is an increasing evidence that the maintenance of healthy intestinal microbiota may provide protection against gastrointestinal disorders, including gastrointestinal infections and intestinal diseases (Reid, 2008; Vasiljevic and Shah, 2008). For this reason, the incorporation of probiotics in food products has been increasing, to assure safe and healthy products.

Though a wide variety of genera and species of microorganisms are considered as potential probiotics (Holzapfel et al., 1998; Shah and Ravla, 2004), the most common bacteria commercially available are from the genera *Lactobacillus* and *Bifidobacterium*.

Some works have studied the effect of the incorporation of probiotics in food and have found some opportunities. Probiotic food products made out of fermentation of milk, cereals, fruits, vegetables, and meat are receiving great attention from the scientific world, as well as from general consumers (Gupta and Abu-Ghannam, 2012; Kołożyn-Krajewska and Dolatowski, 2012; Rößle et al., 2010; Rouhi et al., 2013).

The main challenge to the incorporation of multifunctional bacteria in food matrices is ensuring their viability, which cannot be less than 10^6 CFU mL^{-1}

(Madureira et al., 2011). The survival of probiotics during gastrointestinal transit is affected by the physical and chemical properties of food carriers, which may represent advantages or disadvantages, depending on the food matrices.

Low pH foods such as fruit juices, salads, and condiments present a problem for probiotic survival, as well as foods prepared or stored at high temperatures (Rodgers, 2007). Thus, the viability and growth of beneficial bacteria can be enhanced by proteins or polysaccharides—in particular, prebiotics that are film-forming biopolymers, such as whey proteins, or polymers that may be embedded in the films matrix, such as inulin or fructooligosaccharides.

7.4 STRUCTURAL MATRIX OF EDIBLE PACKAGING FOR FUNCTIONAL COMPOUNDS

Edible coatings and films offer some advantages such as edibility, biocompatibility, an esthetic appearance (color, transparent, low opacity, and brightness), barrier properties to solute or gas, moisture, being nontoxic, and nonpolluting (Baldwin et al., 2012; Dhall, 2013; Han, 2000; Kester and Fennema, 1986). In addition, edible coatings and films, by themselves acting as carriers of functional compounds, have been particularly considered in food preservation because of their ability to extend the shelf life, reduce the risk of pathogen growth on food surfaces, as well as provide a functional product with health benefits to the consumer. The incorporation of functional compounds into edible coatings and films is a way to protect these additives against severe environmental factors assuring that they can exert the desired effect on the expected target (Donhowe and Fennema, 1994; Franssen et al., 2003, 2004; Guilbert and Biquet, 1996; Oliveira et al., 2012; Rojas-Graü et al., 2009).

For the selection of a film-forming biopolymer for a specific functional compound, it must be considered the efficiency of both and possible interactions between them and other food components, since these interactions can modify their activity, as well as the characteristics of the film (Campos et al., 2010).

The evaluation of the sensory attributes of foods with edible packaging is usually performed by means of descriptive analysis (Eswaranandam et al., 2006) or consumer and free-choice profiling panels (Han et al., 2005). In some cases, especially when lipids are incorporated into coatings, consumers may reject the samples because of their artificial color and waxy appearance (Han et al., 2005; Tanada-Palmu and Grosso, 2005).

On the contrary, the compatibility of the various constituents is very important, especially to withstand stress factors caused by, for example, high pressures, electric fields, ultrasound, microwave radiation, gamma radiation, temperature, and light or to improve solubility and controlled release of bioactive compounds (Campos et al., 2010).

In edible coating/film formulation to incorporated functional compounds, different structural materials (Al-Hassan and Norziah, 2012) have been used such as proteins, lipids, polysaccharides, and composites. In the last years, several reviews have focused on such edible coatings and films based on lipids (Hambleton et al., 2009), protein (Ramos et al., 2012), and polysaccharides (Jiménez et al., 2013; Jridi et al., 2014), regarding the development and application of edible coatings/films thereof.

The production of biodegradable coatings/films by combining various polysaccharides, lipids, and proteins is considered with the aim of taking advantage of the properties of each compound and the synergy between them. The mechanical and barrier properties of these films not only depend on the compounds used in their formulation but also on their compatibility (Falguera et al., 2011).

The optimization of edible coatings/films composition is of great importance in this research field, since they must be formulated according to the properties of the raw material to which they have to be applied (Falguera et al., 2011).

Polysaccharides and proteins are the most widely investigated biopolymers and are great materials for the formation of edible packaging, as they demonstrate excellent mechanical and structural properties, but they have generally a poor barrier capacity against moisture transfer (Falguera et al., 2011; Iwata et al., 2000). This problem is not generally found in lipids due to their hydrophobic properties. To overcome the poor mechanical strength of lipids compounds, they can be employed in combination with hydrophilic materials by means of the formation of an emulsion or through a lamination with a hydrocolloid film lipid layer (Falguera et al., 2011). The efficiency of an edible coating/film against moisture transfer cannot be simply improved with the addition of hydrophobic materials in the formulation, unless the formation of a homogeneous and continuous lipid layer inside the hydrocolloid matrix is achieved (Falguera et al., 2011).

Polysaccharides used for edible coatings or films include cellulose, starch derivatives, pectin derivatives, seaweed extracts, exudate gums, microbial fermentation gums, and chitosan (Krochta and De Mulder-Johnston, 1997). Several studies investigated polysaccharide-based films and its derivatives regarding their physical, chemical, and biological properties (Arnon et al., 2015; Brasil et al., 2012; Bravin et al., 2006; Carneiro-da-Cunha et al., 2009; Oms-Oliu et al., 2008; Rojas-Graü et al., 2008).

Among polysaccharides, chitosan, and its derivatives showed a great number of applications focused on active coating systems (Arnon et al., 2014; Azevedo et al., 2014; Gol et al., 2013; Suseno et al., 2014; Vásconez et al., 2009). Chitosan has a vast potential that can be applied in the food industry because of its particular physicochemical properties such as biodegradability, biocompatibility with human tissues, null toxicity, and specificity, as well as its antimicrobial and antifungal activity (Aider, 2010). Chitosan is obtained by deacetylation of chitin, which is extracted from the exoskeleton of crustaceans and fungal cell walls. It has been extensively used in films and coatings due to its ability to inhibit the growth of various bacterial and fungal pathogens (Romanazzi et al., 2002). Chitosan has also been studied in combination with other biopolymers. New research and recent reviews on the use of chitosan gather some information on the effect of the deacetylation degree on its antimicrobial activity, use in active coating and its interaction with other components of food matrices (Aider, 2010; Devlieghere et al., 2004; Martínez-Camacho et al., 2010; No et al., 2002).

Protein-based edible coatings/films have received considerable attention in recent years because of their advantages, including their use as edible packaging materials, over the synthetic films. In addition, those films can supplement the nutritional value of the food (Gennadios and Weller, 1990). The mechanical properties of protein-based films are generally better than polysaccharide or lipid-based films, since proteins have a distinctive structure which confers a wider range of functional properties, mainly a high intermolecular binding potential (Cuq et al., 1995; Khwaldia et al., 2004). Nevertheless, the poor water vapor resistance of protein films and lower mechanical strength in comparison with synthetic polymers limit their application in food packaging. Many approaches exist to improve the barrier properties of edible protein coatings/films, such as by modifying properties of protein by chemical and enzymatic methods, combining with hydrophobic material or some polymers, or by using a physical method. Several proteins, including collagen, wheat gluten, corn zein, soy protein, whey protein, and bean protein, have been investigated for their film properties (Bourtoom, 2009). Protein edible coatings and films are usually produced from solutions of the protein by evaporation of solvent. The solvent is normally limited to water, ethanol, or ethanol–water mixtures. Generally, proteins must be denatured by heat, acid, bases, and/or solvents to form extended structures that are required for film formation. Once extended, protein chains can associate through hydrogen, ionic, hydrophobic, and covalent bonding. The chain-to-chain interaction that produces cohesive films is affected by the degree of chain extension and the nature and sequence of amino acid residues. The uniform distribution of polar, hydrophobic, and/or thiol groups along the polymer chain increase the likelihood of the respective interactions. The promotion of polymer chain-to-chain interaction results in films that are stronger but less flexible and permeable to gases, liquids, and vapors. Polymers containing groups that can be associated through hydrogen or ionic bonding result in films that are excellent oxygen barriers but susceptible to moisture. Thus, protein films are expected to be good oxygen barriers at low relative humidity (Wittaya, 2012).

Protein-based edible coatings and films can find application for the individual packaging of small portions of food, applied inside heterogeneous foods at the interfaces between different layers of components; they can be tailored to prevent the deterioration of intercomponent moisture and solute migration in foods. Moreover, protein-based edible coatings/films can function as carriers for antimicrobial and antioxidant agents, and they also can be used at the surface of food to control the diffusion rate of preservative substances from the surface to the interior of the food. Another possible application could be their use in multilayer food packaging materials together with nonedible coatings and films (Wittaya, 2012).

Edible barriers based on hydrophobic substances such as lipids were developed specifically for limiting moisture migration within foods. These hydrophobic substances are effective barriers against moisture migration because of their apolar nature (Morillon et al., 2002).

Lipids commonly used in coating formulations to preserve minimally processed products are stearic and palmitic acids and some vegetable oils, such as soybean and sunflower (Colla et al., 2006; Garcia et al., 2000; Martin-Belloso et al., 2005). Natural and synthetic waxes are also used, showing good gas barrier and better moisture barrier properties than coatings containing only fatty acids (Rhim and Shellhammer, 2005; Rojas-Argudo et al., 2009; Talens and Krochta, 2005). The emulsion stability and lipid particle size affect barrier properties of emulsified coatings, which make the study of physical stability important.

The main function of a lipid coating is to block the passage of moisture due to their relative low polarity. In contrast, the hydrophobic characteristic of lipid forms thicker and more brittle films. Consequently, they must be associated with film-forming agents such as proteins or cellulose derivatives (Debeaufort et al., 1993). Generally, water vapor permeability (WVP) decreases when the concentration of the hydrophobic phase increases. Lipid-based films are often supported on a polymer structure matrix, usually a polysaccharide, to provide mechanical strength (Bourtoom, 2008).

7.5 FUNCTIONALITIES AND APPLICATIONS OF FUNCTIONAL EDIBLE PACKAGING

The concept of functional edible packaging has been designed to respond to the current limitations in the production of functional foods and, in some cases, to improve quality or extend the shelf life of the food product through the addition of bioactive compounds.

The development of food packaging systems will provide a more efficient alternative, and in some cases, the unique means to produce healthy foods at industrial level (Lagaron, 2005).

7.5.1 Antioxidants and Antimicrobials

The antioxidant capacity is often combined with antimicrobial property, so in this context some studies including both properties are described, but the emphasis is given to the antioxidant capacity, since the antimicrobial effect is deeply described in Chapter 8.

The incorporation of antioxidants in edible packaging may be primarily a mean for carrying antioxidants with bioactive properties with benefits for the consumer, and also an interesting alternative to food preservation, since oxidation is one of the major problems affecting food quality. Antioxidants can be added into the edible packaging to protect against oxidative rancidity, degradation, and discoloration of certain foods (Baldwin et al., 1995; Vargas et al., 2008). Thus, extensive research has been conducted to adopt some natural antioxidants in alternative to synthetic ones used in functional packaging (Moore et al., 2003; Wu et al., 2001). Similar to what happens with biopreservatives and edible packaging materials, the natural antioxidants are also readily accepted by the consumers and they are not considered as chemicals.

There are some studies (summarized in Table 7.1) that demonstrate the antioxidant capacity of various compounds when incorporated into edible packaging.

Güçbilmez et al. (2007) reported that the incorporation of partially purified lysozyme into zein films in combination with chickpea albumin extract (CPAE), bovine serum albumin, and disodium EDTA. The retained antioxidant activity in the zein film with 2019 U cm^{-2} of lysozyme and 530 μg cm^{-2} of CPAE was almost 84% (6.8 nmol vit. C cm^{-2}) higher than that retained at the control film surfaces (12.5 nmol vit. C cm^{-2}). This study clearly showed the benefits of using CPAE to improve antioxidant activity in zein films.

Oussalah et al. (2004) developed films based on milk proteins containing mixtures of oregano, pimento, or oregano–pimento essential oils that were applied on

TABLE 7.1
Functionality of Edible Coatings and Films with Antioxidants and Antimicrobials as Bioactive Compounds

Antioxidants and Antimicrobials	Edible Packaging Material	Assigned Feature	Application	References
Ascorbic, citric, and oxalic acids	Carrageenan and whey protein coatings	Maintain color and prolong the shelf life	Apple slices	Lee et al. (2003)
Ascorbic acid and sorbic acid	Methylcellulose-based edible coatings	Retard browning and enhance texture	Cut-pear wedges	Olivas et al. (2003)
Gluconal CAL and DL-α-tocopheryl acetate	Chitosan-based coating	Delay the color change	Fresh and frozen strawberries	Han et al. (2004)
Oregano, pimento, and oregano–pimento essencial oils	Milk protein–based films	Antioxidant and antimicrobial (against *Escherichia coli* O157:H7 and *Pseudomonas* spp.) activities	—	Oussalah et al. (2004)
Stearic, citric, and ascorbic acids	Methylcellulose-based and polyethylene glycol edible coatings	Control and reduce oxygen, water permeability and vitamin C losses	Apricots and green peppers	Ayranci and Tunc (2004)
Cysteine or glutathione	Alginate and gellan edible coatings	Antioxidant activity	Fresh-cut apples	Rojas-Graü et al. (2007)
Purified lysozyme in combination with chickpea albumin extract (CPAE), bovine serum albumin, and disodium EDTA	Zein films	Improve antioxidant (lysosyme and CPAE—6.8 nmol vit. C cm^{-2}) and antimicrobial (against *E. coli* and *Bacillus subtilis*) activities	—	Güçbilmez et al. (2007)
N-acetylcysteine	Alginate and gellan edible coatings	Prevent browning	Apple wedges	Rojas-Graü et al. (2008)
Vitamin C and tea polyphenols	Alginate-based edible coating	Increase the sensory quality, retard chemical spoilage, and water loss	Fish	Song et al. (2011)
Catechin (CAT), gallic acid (GA), *p*-hydroxy benzoic, and ferulic acids	Zein films	Antioxidant (21.0 for CAT and 86.2 μmol trolox cm^{-2} for GA) and antimicrobial potential	—	Arcan and Yemenicioğlu (2011)
Ascorbic and citric acids	Alginate-based edible coating	Preserve color and increase the antioxidant potential	Fresh-cut mangoes	Robles-Sánchez et al. (2013)

Note: —, not found.

beef muscle slices. This oregano-based film stabilized lipid oxidation and was the most effective against *Escherichia coli* O157:H7 and *Pseudomonas* spp. in beef muscle samples, whereas pimento-based films presented the highest antioxidant activity.

A recent study reported that incorporation of specific phenolic compounds such as catechin (CAT), gallic acid (GA), *p*-hydroxy benzoic acid, and ferulic acid in zein films possesses antioxidant and antimicrobial potential. The trolox equivalent antioxidant capacity (TEAC) of different phenolic compounds was determined by area under the curve (AUC) method using ABTS free radical, and the TEACs for the values most significant antioxidant were 21.0 for CAT and 86.2 µmol trolox/cm^2 for GA (Arcan and Yemenicioğlu, 2011).

Other authors have shown that the some antioxidants such as ascorbic, citric, and oxalic acids incorporated into the carrageenan and whey protein coatings exhibited antioxidant activity, maintained color during 2 weeks, and effectively prolonged the shelf life of apple slices (Lee et al., 2003).

Similarly, the incorporation of ascorbic and sorbic acids into methylcellulose-based edible coatings was able to retard browning and to enhance texture of cut-pear wedges (Olivas et al., 2003). Ayranci and Tunc (2004) developed methylcellulose-based edible coatings, but with polyethylene glycol and incorporating citric, ascorbic, and stearic acids to control and reduce oxygen, water permeability, and vitamin C losses. They observed that coatings with stearic acid were effective in reducing water loss of fresh apricots and green peppers, whereas the coatings containing citric and ascorbic acids reduced the vitamin C loss of these fresh foods.

Han et al. (2004) observed that the incorporation of Gluconal® CAL and DL-α-tocopheryl acetate as antioxidants into chitosan-based coating, significantly delayed the color change of fresh and frozen strawberries.

Rojas-Graü et al. (2007) proved that antioxidant agents such as cysteine or glutathione incorporated into edible coatings based on alginate and gellan can be used to protect the surface of fresh-cut apples. The same group observed that alginate and gellan edible coatings containing *N*-acetylcysteine, as antioxidant compound, prevented apple wedges from browning (Rojas-Graü et al., 2008).

7.5.2 Flavors

The addition of flavors in edible packaging represents an effective method for adding flavors to foods, which allows control flavor loss and release.

Surface flavoring by edible coating/film permits to reduce flavor compounds in the food to be flavored assuring the improvement of taste. The flavors are concentrated at the surface in a very thin layer; it can be released very fast in the mouth and then provide a higher sensory impact. This can also be used for masking bitterness. Some studies have reported successful results incorporating flavors in edible coatings and films, which are summarized in Table 7.2.

Kaushik and Roos (2007) observed that the food matrix to incorporate the flavor should have good solubility and emulsifying properties, as well as low viscosity at high solid concentration. The efficiency of the film or coating to retain volatile compounds also depends on the nature of the medium matrix in contact.

TABLE 7.2
Functionality of Edible Coatings and Films with Flavors as Bioactive Compounds

Flavors	Edible Packaging Material	Assigned Feature	Application	References
Pandan leaf extract	Sorbitol-plasticized rice starch	Produce rice with aroma compounds. Flavor holding after 6 months of storage	Rice	Laohakunjit and Kerdchoechuen (2007)
n-hexanal	Alginate emulsion–based films	Protect aroma compounds	–	Hambleton et al. (2009)
Ethyl acetate, ethyl butyrate, ethyl iso-butyrate, ethyl hexanoate, ethyl octanoate, 2-pentanone, 2-heptanone, 2-octanone, 2-nonanone, 1-hexanol	Iota-carrageenan emulsion–based edible film	Controlled release of aroma compounds and maintain aroma and taste	–	Marcuzzo et al. (2010)
n-hexanal	Soy protein–based films	Capable of retaining and maintaining the n-hexanal with beeswax	–	Monedero et al. (2010)
n-hexanal and D-limonene	Iota-carrageenan films (with and without lipid)	Quickly release of aroma compounds at higher temperatures (25 and 37°C)	–	Fabra et al. (2012)

Note: –, not found.

Marcuzzo et al. (2010) demonstrated the incorporation of 10 different compounds in carrageenan films. The films prepared with carrageenan may release aroma compounds and thereby maintain the sensory characteristics such as aroma and taste for certain periods of time.

Hambleton et al. (2009) proved that the aroma compound n-hexanal incorporated into packaging made of polysaccharides such as alginate are well protected due to its low oxygen permeability. Fabra et al. (2012) analyzed the release of n-hexanal and D-limonene from edible coatings. Aroma compounds were released easily in water medium, being D-limonene released quickly at higher temperatures. Therefore, these coating technologies represent a promising approach for improving food aroma or flavor.

Monedero et al. (2010) observed that it was necessary to add beeswax to improve the capacity of soy protein–based films to retain n-hexanal, owing to the affinity of *n*-hexanal with the nonpolar lipids as beeswax, which contribute to link the aroma compound and also the hydrophilic properties of soy protein.

Coated nonaromatic milled rice was reported by Laohakunjit and Kerdchoechuen (2007) with sorbitol-plasticized rice starch containing natural pandan leaf extract (viz., *Pandanus amaryllifolius* Roxb.). The rice starch coating containing this extract produced rice with aroma compounds similar to that of aromatic rice.

Thus, these studies demonstrate the ability of edible packaging as carrier of flavorings, to enhance the sensory properties of the food, making it most appealing and acceptable by the consumers.

7.5.3 Probiotics

The probiotics are important for the well-being of humans but are sensible to different environmental factors and for that reason, it is important to protect and incorporate them in food. This is a hot topic, but very recent, so only few studies exist and are summarized in Table 7.3.

The first study describes the incorporation of probiotics, *Bifidobacterium lactis* Bb-12, in alginate–gellan coatings and films for coating fresh-cut apple and papaya. The authors maintained for 10 days of storage values higher than 10^6 CFU g^{-1}, thus demonstrating the viability of probiotics in films applied to fresh fruit (Tapia et al., 2007).

The incorporation of *Lactobacillus acidophilus* and *Bifidobacterium bifidum* into gelatin edible coatings applied to fish, and the assessment of its effect during chilled storage was performed by López de Lacey et al. (2012). During the storage, the bacteria remained viable and the H_2S-producing microorganisms were reduced in 2 log cycles. In a further experiment from the same group, a high pressure treatment was applied to fish coated with gelatin films incorporating bifidobacteria; the results showed a total reduction of total viable counts over 13 days of storage. The application of gelatin edible packaging incorporated with bacteria can be promising for fish preservation, especially when combined with other technologies such as a high pressure (López de Lacey et al., 2012).

Kanmani and Lim (2013) reported that the viability of probiotic strains, viz. *Lactobacillus reuteri* ATCC 55730, *Lactobacillus plantarum* GG ATCC 53103, and *L. acidophilus* DSM 20079 was maintained in starch–pullulan-based edible films. However, the viability of bacteria was influenced by the pullulan to starch ratio and storage temperature. Pure pullulan and pullulan/potato starch films presented the highest relative cell viabilities, followed by pullulan/tapioca starch films, while pure starch films exhibited lower cell viabilities after 30 days of storage at 4°C.

Recently, an experiment conducted by Soukoulis et al. (2014) demonstrated the development of probiotic pan bread by the application of edible coatings based in sodium alginate or sodium alginate/whey protein concentrate incorporated with *Lactobacillus rhamnosus* GG, followed by different drying steps. The presence of whey protein concentrate improved significantly the viability of *L. rhamnosus* GG throughout air drying at room temperature storage and the use of film-based exclusive on sodium alginate improved the viability throughout the simulated gastrointestinal conditions (Soukoulis et al., 2014). These studies represent promising advances

TABLE 7.3
Functionality of Edible Coatings and Films with Probiotics as Bioactive Compounds

Probiotics	Edible Packaging Material	Assigned Feature	Application	References
Bifidobacterium lactis Bb-12	Alginate–gellan coatings and films	Maintain the viability of probiotics in films	Fresh-cut apple and papaya	Tapia et al. (2007)
Lactobacillus acidophilus and *Bifidobacterium bifidum*	Gelatin edible coatings	Maintain the viability of bacteria and decrease the H_2S producing microorganisms. Preservation, especially when combined with high pressure	Fish	López de Lacey et al. (2012)
Lactobacillus reuteri ATCC 55730, *Lactobacillus plantarum* GG ATCC 53103, and *L. acidophilus* DSM 20079	Starch- and pullulan-based edible films	Effective delivery and carrier systems for probiotics; maintaining the viability during the storage	–	Kanmani and Lim (2013)
Lactobacillus rhamnosus GG	Sodium alginate or sodium alginate/whey protein concentrate–based edible coatings	Improved significantly the viability of bacteria throughout the simulated gastrointestinal conditions and throughout air drying and room temperature storage	Pan bread	Soukoulis et al. (2014)

Note: –, not found.

in the search for new applications of edible coatings and films as carriers of diverse probiotics, and open new possibilities for the development of novel food products with probiotics.

7.5.4 OTHER FUNCTIONAL COMPOUNDS

Edible packaging is an excellent carrier to enhance the nutritional value of foods by carrying basic nutrients and/or nutraceuticals that lacking or are present in only low amounts in foods (Bourbon et al., 2011; Mei et al., 2002; Park and Zhao, 2004).

Edible Packaging

Nutrients and nutraceuticals can be incorporated into the formulation of edible coatings and films, providing an alternative way to fortify unprocessed foods, such as fresh fruits, thus encouraging their consumption. The edible coatings and films promote the delivery of these compounds to foods mainly by preventing their interaction with other food components, for example, iron bioavailability is commonly affected by interactions with food ingredients (Lynch, 1997; Sandström, 2001; Thankachan et al., 2008).

Some studies have reported the effect of the addition of active compounds in the functionality of edible coatings and films and are summarized in Table 7.4.

Mei and Zhao (2003) indicated that a successful development of edible packaging containing high concentration of nutraceuticals strongly depends on the type of coating materials and concentration of nutraceuticals incorporated into the coating formulations.

Mei et al. (2002) developed xanthan gum–based coating incorporated with high concentration of calcium and vitamin E not only preventing moisture loss and surface whitening, but also significantly increasing the calcium and vitamin E contents of peeled baby carrots.

Park and Zhao (2004) reported the development of chitosan coatings containing high concentrations of calcium, zinc, or vitamin E providing alternative ways to fortify fresh fruits and vegetables. This application has been successfully demonstrated on fresh and frozen strawberries and red raspberries (Han et al., 2004).

Similarly, Hernández-Muñoz et al. (2006) observed that chitosan-coated strawberries retained more calcium gluconate than strawberries dipped into calcium solutions.

Souza et al. (2011) incorporated different lipid fractions: fish and vegetable oils, and stearic and oleic acids into chitosan films. Results showed that incorporation decreased the WVP (1.3–1.8 g mm m^{-2} day^{-1} kPa^{-1}) as compared with pure chitosan film (3.8 g mm m^{-2} day^{-1} kPa^{-1}). A higher reduction in WVP (65%) was found with the addition of refined fish oil to the continuous matrix of the films than with the addition of refined rice oil, oleic, or stearic acid (50%–60%).

A similar study performed by Jiménez et al. (2013) incorporated sodium caseinate, glycerol, and lipids (oleic acid and/or α-tocopherol) into films based on starch. After storage, films containing lipid were more stretchable. Lipid addition did not induce a notable decrease in WVP, but oxygen permeability was highly increased when they contain oleic acid. The incorporation of α-tocopherol greatly increased the antioxidant capacity of the films which affected oxygen permeability (Jiménez et al., 2013).

Bourbon et al. (2011) produced packaging films based on chitosan containing different functional compounds, for example, a peptide fraction from whey protein concentrate hydrolysate, glycomacropeptide, and lactoferrin, and evaluate their mechanical properties. The results demonstrated a high potential of this packaging to carrier these functional compounds and to enhance their bioactivity in foods.

Thus, it was observed that several functional compounds can be incorporated into edible packaging to achieve the desired effect and increase the functionality of foods.

7.6 CONCLUSIONS

As an alternative to conventional packages, edible packaging has been recognized by several authors as having great potential for protecting and adds functionality to the food.

TABLE 7.4
Functionality of Edible Coatings and Films with Other Specific Bioactive Compounds

Other Bioactive Compounds	Edible Packaging Material	Assigned Feature	Application	References
Calcium and vitamin E	Xanthan gum–based coating	Increase the calcium and vitamin E contents. Prevent moisture loss and surface whitening	Peeled baby carrots	Mei et al. (2002)
Calcium, zinc, or vitamin E	Chitosan coatings	Fortify fresh fruits and vegetables	Fresh and frozen strawberries and red raspberries	Han et al. (2004), Park and Zhao (2004)
Calcium gluconate	Chitosan coatings	Increased nutritional value	Strawberries	Hernández-Muñoz et al. (2006)
Peptide fraction from whey protein concentrate hydrolysate, glycomacropeptide, and lactoferrin	Chitosan packaging films	Carrier of functional compounds and enhance the bioactivity in foods	—	Bourbon et al. (2011)
Fish and vegetable oils, stearic, and oleic acid	Chitosan films	Decrease water vapor permeability. Higher reduction on chitosan with refined fish oil (1.3–1.8 g mm m^{-2} day^{-1} kPa^{-1})	—	Souza et al. (2011)
Sodium caseinate, glycerol, and lipids (oleic acid and/or α-tocopherol)	Starch films	Films containing lipid were more stretchable. Oxygen permeability highly increased in films with oleic acid. α-Tocopherol greatly increased the antioxidant capacity	—	Jiménez et al. (2013)

Note: –, not found.

In regard to the inherent attributes of edible coatings and films, it may also be important to refer their ability to incorporate functional compounds as antioxidants, antimicrobials, flavors (reinforcing the preexistent flavor or adding new flavors), probiotics, and nutraceuticals, without compromise their initial structure and functionality or even the consumer acceptability. These films incorporating bioactive compounds intended to be a healthy way to ingest functional compounds that positively influence the health of the consumer and may even be a way of acquiring substances that otherwise the consumer would have restricted access. In this perspective, it is predicted that they may play an important role in human health and disease prevention. Also, the addition of these compounds is intended to achieve an improvement of the nutritional, sensory, and shelf life of food characteristics. Furthermore, all these compounds are safe for health, and the vast majorities are natural and can also contribute to a valorization of a by-product.

Functional compounds are often perishable, very sensitive, and can lose their activity or even be degraded, so the edible packaging provides protection to these compounds, allowing their stability and in some cases their release under controlled conditions.

However, when the bioactive ingredients are added to the edible coatings and films, these may affect, for example, the mechanical, sensory, and functional properties.

Therefore, further studies with the aim of developing new application processes to improve the functionality of films and coating or even food compounds are required.

7.7 FUTURE TRENDS

Edible coatings and films are increasingly seen as a value proposition that add value to food and consumers. Thus, the search for new compounds to produce edible packaging or with improved functionality is increasing. To produce this unique combination many factors have to be taken in account, such as the stability of the films, the concentration of functional compounds incorporated on coatings and films or on product surface, as well as their bioavailability and gradual release.

A challenge to the use of edible coatings and films is its compatibility with other emerging stress factors such as high pressure, electric fields, ultrasound, microwave radiation, and gamma radiation. There are modifications that can be made in the structure of edible packaging without endangering the safety of the food product and the consumer. In this regard, crosslinking of the polymer, generation of composites with different fillers, use of hydrophilic biopolymers blended with lipids, may be valuable solutions.

The generation of a multilayer structure can contribute to minimize the degradation, control diffusion, gradual release, and/or adequate bioavailability of the compounds added.

Nowadays, there is already a great research on this topic, but there are different objectives that lead to different types of functional edible packaging. In the future, it will be necessary to further deepen this issue in a systematic way and to determine the magnitude of the benefit–cost ratio for each edible coating or film developed.

REFERENCES

Agati, G., Matteini, P., Goti, A., and Tattini, M. 2007. Chloroplast-located flavonoids can scavenge singlet oxygen. *New Phytologist*, *174*(1), 77–89.

Aider, M. 2010. Chitosan application for active bio-based films production and potential in the food industry: Review. *LWT—Food Science and Technology*, *43*(6), 837–842.

Ajila, C., Aalami, M., Leelavathi, K., and Rao, U. 2010. Mango peel powder: A potential source of antioxidant and dietary fiber in macaroni preparations. *Innovative Food Science & Emerging Technologies*, *11*(1), 219–224.

Al-Hassan, A., and Norziah, M. 2012. Starch-gelatin edible films: Water vapor permeability and mechanical properties as affected by plasticizers. *Food Hydrocolloids*, *26*(1), 108–117.

Arcan, I., and Yemenicioğlu, A. 2011. Incorporating phenolic compounds opens a new perspective to use zein films as flexible bioactive packaging materials. *Food Research International*, *44*(2), 550–556.

Arnon, H., Granit, R., Porat, R., and Poverenov, E. 2015. Development of polysaccharides-based edible coatings for citrus fruits: A layer-by-layer approach. *Food Chemistry*, *166*, 465–472.

Arnon, H., Zaitsev, Y., Porat, R., and Poverenov, E. 2014. Effects of carboxymethyl cellulose and chitosan bilayer edible coating on postharvest quality of citrus fruit. *Postharvest Biology and Technology*, *87*, 21–26.

Ayranci, E., and Tunc, S. 2004. The effect of edible coatings on water and vitamin C loss of apricots (*Armeniaca vulgaris* Lam.) and green peppers (*Capsicum annuum* L.). *Food Chemistry*, *87*(3), 339–342.

Azevedo, A. N., Buarque, P. R., Cruz, E. M. O., Blank, A. F., Alves, P. B., Nunes, M. L., and Santana, L. C. L. D. A. 2014. Response surface methodology for optimisation of edible chitosan coating formulations incorporating essential oil against several foodborne pathogenic bacteria. *Food Control*, *43*, 1–9.

Baldwin, E., Nisperos-Carriedo, M., and Baker, R. 1995. Edible coatings for lightly processed fruits and vegetables. *HortScience*, *30*(1), 35–38.

Baldwin, E. A., Hagenmaier, R., and Bai, J. 2012. *Edible Coatings and Films to Improve Food Quality* (2th ed.). Boca Raton, FL: CRC Press.

Beaulieu, J. C., and Lancaster, V. A. 2007. Correlating volatile compounds, sensory attributes, and quality parameters in stored fresh-cut cantaloupe. *Journal of Agricultural and Food Chemistry*, *55*(23), 9503–9513.

Beaulieu, J. C., and Lea, J. M. 2003. Volatile and quality changes in fresh-cut mangos prepared from firm-ripe and soft-ripe fruit, stored in clamshell containers and passive MAP. *Postharvest Biology and Technology*, *30*(1), 15–28.

Bigliardi, B., and Galati, F. 2013. Innovation trends in the food industry: The case of functional foods. *Trends in Food Science & Technology*, *31*(2), 118–129.

Bourbon, A. I., Pinheiro, A. C., Cerqueira, M. A., Rocha, C. M., Avides, M. C., Quintas, M. A., and Vicente, A. A. 2011. Physico-chemical characterization of chitosan-based edible films incorporating bioactive compounds of different molecular weight. *Journal of Food Engineering*, *106*(2), 111–118.

Bourtoom, T. 2008. Edible films and coatings: Characteristics and properties. *International Food Research Journal*, *15*(3), 237–248

Bourtoom, T. 2009. Edible protein films: Properties enhancement. *International Food Research Journal*, *16*(1), 1–9.

Braca, A., Siciliano, T., D'Arrigo, M., and Germanò, M. P. 2008. Chemical composition and antimicrobial activity of *Momordica charantia* seed essential oil. *Fitoterapia*, *79*(2), 123–125.

Brasil, I. M., Gomes, C., Puerta-Gomez, A., Castell-Perez, M. E., and Moreira, R. G. 2012. Polysaccharide-based multilayered antimicrobial edible coating enhances quality of fresh-cut papaya. *LWT—Food Science and Technology*, *47*(1), 39–45.

Bravin, B., Peressini, D., and Sensidoni, A. 2006. Development and application of polysaccharide-lipid edible coating to extend shelf-life of dry bakery products. *Journal of Food Engineering*, 76(3), 280–290.

Brown, J. E., and Kelly, M. F. 2007. Inhibition of lipid peroxidation by anthocyanins, anthocyanidins and their phenolic degradation products. *European Journal of Lipid Science and Technology*, 109(1), 66–71.

Campos, C. A., Gerschenson, L. N., and Flores, S. K. 2010. Development of edible films and coatings with antimicrobial activity. *Food and Bioprocess Technology*, 4(6), 849–875.

Carneiro-da-Cunha, M. G., Cerqueira, M. A., Souza, B. W. S., Souza, M. P., Teixeira, J. A., and Vicente, A. A. 2009. Physical properties of edible coatings and films made with a polysaccharide from *Anacardium occidentale* L. *Journal of Food Engineering*, 95(3), 379–385.

Chen, C.-C., and Walker, W. A. 2005. Probiotics and prebiotics: Role in clinical disease states. *Advances in Pediatrics*, 52, 77–113.

Chen, Z. 2008. Research of antioxidative capacity in essential oils of plants. *China Condiment*, 11, 40–43.

Colla, E., do Amaral Sobral, P. J., and Menegalli, F. C. 2006. Amaranthus cruentus flour edible films: Influence of stearic acid addition, plasticizer concentration, and emulsion stirring speed on water vapor permeability and mechanical properties. *Journal of Agricultural and Food Chemistry*, 54(18), 6645–6653.

Cuq, B., Gontard, N., and Guilbert, S. 1995. Edible films and coatings as active layers. In M. L. Rooney (Ed.), *Active Food Packaging* (pp. 111–142). New York, NY: Springer.

Daoud, H., and Hani, B. 2013. Lactic acid bacteria as probiotics: Characteristics, selection criteria and role in immunomodulation of human GI muccosal barrier. In J. M. Kongo (Ed.), *Biochemistry, Genetics and Molecular Biology*. New York, NY: InTech. Available at http://www.intechopen.com/books/lactic-acid-bacteria-r-d-for-food-health-and-livestock-purposes/lactic-acid-bacteria-as-probiotics-characteristics-selection-criteria-and-role-in-immunomodulation-o. Accessed December 2, 2014.

de Vos, P., Faas, M. M., Spasojevic, M., and Sikkema, J. 2010. Encapsulation for preservation of functionality and targeted delivery of bioactive food components. *International Dairy Journal*, 20(4), 292–302.

Debeaufort, F., Martin-Polo, M., and Voilley, A. 1993. Polarity homogeneity and structure affect water vapor permeability of model edible films. *Journal of Food Science*, 58(2), 426–429.

Debeaufort, F., Quezada-Gallo, J. A., and Voilley, A. 1998. Edible films and coatings: Tomorrow's packagings: A review. *Critical Reviews in Food Science and Nutrition*, 38(4), 299–313.

Del Toro-Sánchez, C. L., Ayala-Zavala, J. F., Machi, L., Santacruz, H., Villegas-Ochoa, M. A., Alvarez-Parrilla, E., and González-Aguilar, G. A. 2010. Controlled release of antifungal volatiles of thyme essential oil from β-cyclodextrin capsules. *Journal of Inclusion Phenomena and Macrocyclic Chemistry*, 67(3–4), 431–441.

Devlieghere, F., Vermeulen, A., and Debevere, J. 2004. Chitosan: Antimicrobial activity, interactions with food components and applicability as a coating on fruit and vegetables. *Food Microbiology*, 21(6), 703–714.

Dhall, R. K. 2013. Advances in edible coatings for fresh fruits and vegetables: A review. *Critical Reviews in Food Science and Nutrition*, 53(5), 435–450.

Donhowe, I. G., and Fennema, O. 1994. Edible films and coatings: Characteristics, formation, definitions, and testing methods. In J. M. Krochta, E. A. Baldwin, and M. A. Nisperos-Carriedo (Eds.), *Edible Coatings and Films to Improve Food Quality* (pp. 6–8). Lancaster, PA: Technomic Publishing Co., Inc.

Durand, L., Habran, N., Henschel, V., and Amighi, K. 2010. Encapsulation of ethylhexyl methoxycinnamate, a light-sensitive UV filter, in lipid nanoparticles. *Journal of Microencapsulation*, 27(8), 714–725.

Eswaranandam, S., Hettiarachchy, N. S., and Meullenet, J.-F. 2006. Effect of malic and lactic acid incorporated soy protein coatings on the sensory attributes of whole apple and fresh-cut cantaloupe. *Journal of Food Science*, 71(3), S307–S313.

Fabra, M. J., Chambin, O., Voilley, A., Gay, J.-P., and Debeaufort, F. 2012. Influence of temperature and NaCl on the release in aqueous liquid media of aroma compounds encapsulated in edible films. *Journal of Food Engineering*, 108(1), 30–36.

Falguera, V., Quintero, J. P., Jiménez, A., Muñoz, J. A., and Ibarz, A. 2011. Edible films and coatings: Structures, active functions and trends in their use. *Trends in Food Science & Technology*, 22(6), 292–303.

Fang, Z., and Bhandari, B. 2010. Encapsulation of polyphenols—A review. *Trends in Food Science & Technology*, 21(10), 510–523.

FAO/WHO. 2002. *Joint FAO/WHO Working Group Report on Drafting Guidelines for the Evaluation of Probiotics in Food*. London, Ontario, Canada, 30.

Fisher, K., and Phillips, C. 2008. Potential antimicrobial uses of essential oils in food: Is citrus the answer? *Trends in Food Science & Technology*, 19(3), 156–164.

Flora, S. J. 2009. Structural, chemical and biological aspects of antioxidants for strategies against metal and metalloid exposure. *Oxidative Medicine and Cellular Longevity*, 2(4), 191–206.

Fogliano, V., and Vitaglione, P. 2005. Functional foods: Planning and development. *Molecular Nutrition & Food Research*, 49(3), 256–262.

Franssen, L., Krochta, J., and Roller, S. 2003. Edible coatings containing natural antimicrobials for processed foods. In S. Roller (Ed.), *Natural Antimicrobials for the Minimal Processing of Foods* (pp. 250–262). Boca Raton, FL: CRC Press.

Franssen, L., Rumsey, T., and Krochta, J. 2004. Whey protein film composition effects on potassium sorbate and natamycin diffusion. *Journal of Food Science*, 69(5), C347–C350.

Garcia, M., Martino, M., and Zaritzky, N. 2000. Lipid addition to improve barrier properties of edible starch-based films and coatings. *Journal of Food Science*, 65(6), 941–944.

Gennadios, A., and Weller, C. L. 1990. Edible films and coatings from wheat and corn proteins. *Food Technology*, 44(10), 63–69.

Gol, N. B., Patel, P. R., and Rao, T. V. R. 2013. Improvement of quality and shelf-life of strawberries with edible coatings enriched with chitosan. *Postharvest Biology and Technology*, 85, 185–195.

Gomes, A. M., and Malcata, F. X. 1999. *Bifidobacterium* spp. and *Lactobacillus acidophilus*: Biological, biochemical, technological and therapeutic properties relevant for use as probiotics. *Trends in Food Science & Technology*, 10(4), 139–157.

Gonzalez-Aguilar, G., Robles-Sanchez, R., Martinez-Tellez, M., Olivas, G., Alvarez-Parrilla, E., and De La Rosa, L. 2008. Bioactive compounds in fruits: Health benefits and effect of storage conditions. *Stewart Postharvest Review*, 4(3), 1–10.

Gorinstein, S., Poovarodom, S., Leontowicz, H., Leontowicz, M., Namiesnik, J., Vearasilp, S., Haruenkit, R., Ruamsuke, P., Katrich, E., and Tashma, Z. 2011. Antioxidant properties and bioactive constituents of some rare exotic Thai fruits and comparison with conventional fruits: *In vitro* and *in vivo* studies. *Food Research International*, 44(7), 2222–2232.

Güçbilmez, Ç. M., Yemenicioğlu, A., and Arslanoğlu, A. 2007. Antimicrobial and antioxidant activity of edible zein films incorporated with lysozyme, albumin proteins and disodium EDTA. *Food Research International*, 40(1), 80–91.

Guilbert, S., and Biquet, B. 1996. Edible films and coatings. *Food Packaging Technology*, 1, 315–353.

Guilbert, S., Gontard, N., and Gorris, L. G. 1996. Prolongation of the shelf-life of perishable food products using biodegradable films and coatings. *LWT—Food Science and Technology*, 29(1), 10–17.

Gupta, S., and Abu-Ghannam, N. 2012. Probiotic fermentation of plant based products: Possibilities and opportunities. *Critical Reviews in Food Science and Nutrition*, 52(2), 183–199.

Hambleton, A., Debeaufort, F., Bonnotte, A., and Voilley, A. 2009. Influence of alginate emulsion-based films structure on its barrier properties and on the protection of microencapsulated aroma compound. *Food Hydrocolloids*, 23(8), 2116–2124.

Han, C., Lederer, C., McDaniel, M., and Zhao, Y. 2005. Sensory evaluation of fresh strawberries (*Fragaria ananassa*) coated with chitosan-based edible coatings. *Journal of Food Science*, 70(3), S172–S178.

Han, C., Zhao, Y., Leonard, S. W., and Traber, M. G. 2004. Edible coatings to improve storability and enhance nutritional value of fresh and frozen strawberries (*Fragaria ananassa*) and raspberries (*Rubus ideaus*). *Postharvest Biology and Technology*, 33(1), 67–78.

Han, J. H. 2000. Antimicrobial food packaging. *Food Technology*, 54, 56–65.

Han, J. H. 2005. New technologies in food packaging: Overview. In J. H. Han (Ed.), *Innovations in Food Packaging* (pp. 3–11). Oxford, UK: Elsevier Academic Press.

Hasler, C. M., and Brown, A. C. 2009. Position of the American dietetic association: Functional foods. *Journal of the American Dietetic Association*, 109(4), 735–746.

Hernandez-Izquierdo, V., and Krochta, J. 2008. Thermoplastic processing of proteins for film formation—A review. *Journal of Food Science*, 73(2), R30–R39.

Hernandez-Izquierdo, V., and Krochta, J. 2009. Thermal transitions and heat-sealing of glycerol-plasticized whey protein films. *Packaging Technology and Science*, 22(5), 255–260.

Hernández-Muñoz, P., Almenar, E., Ocio, M. J., and Gavara, R. 2006. Effect of calcium dips and chitosan coatings on postharvest life of strawberries (*Fragaria ananassa*). *Postharvest Biology and Technology*, 39(3), 247–253.

Holley, R. A., and Patel, D. 2005. Improvement in shelf-life and safety of perishable foods by plant essential oils and smoke antimicrobials. *Food Microbiology*, 22(4), 273–292.

Holzapfel, W. H., Haberer, P., Snel, J., Schillinger, U., and Huis in't Veld, J. H. 1998. Overview of gut flora and probiotics. *International Journal of Food Microbiology*, 41(2), 85–101.

Homayouni, A., Alizadeh, M., Alikhah, H., and Zijah, V. 2012. Functional dairy probiotic food development: Trends, concepts, and products. In E. C. Rigobelo (Ed.), *Probiotics* (pp. 197–212). New York, NY: InTech. Available at http://www.intechopen.com/books/probiotics/functional-dairy-probiotic-food-development-trends-concepts-and-products. Accessed December 2, 2014.

Iacopini, P., Baldi, M., Storchi, P., and Sebastiani, L. 2008. Catechin, epicatechin, quercetin, rutin and resveratrol in red grape: Content, *in vitro* antioxidant activity and interactions. *Journal of Food Composition and Analysis*, 21(8), 589–598.

Iwata, K. I., Ishizaki, S. H., Handa, A. K., and Tanaka, M. U. 2000. Preparation and characterization of edible films from fish water-soluble proteins. *Fisheries Science*, 66(2), 372–378.

Jiménez, A., Fabra, M. J., Talens, P., and Chiralt, A. 2013. Physical properties and antioxidant capacity of starch–sodium caseinate films containing lipids. *Journal of Food Engineering*, 116(3), 695–702.

Jridi, M., Hajji, S., Ayed, H. B., Lassoued, I., Mbarek, A., Kammoun, M., Souissi, N., and Nasri, M. 2014. Physical, structural, antioxidant and antimicrobial properties of gelatin-chitosan composite edible films. *International Journal of Biological Macromolecules*, 67, 373–379.

Kanmani, P., and Lim, S. T. 2013. Development and characterization of novel probiotic-residing pullulan/starch edible films. *Food Chemistry*, *141*(2), 1041–1049.
Kaushik, V., and Roos, Y. H. 2007. Limonene encapsulation in freeze-drying of gum Arabic-sucrose-gelatin systems. *LWT—Food Science and Technology*, *40*(8), 1381–1391.
Kester, J. J., and Fennema, O. R. 1986. Edible films and coatings: A review. *Food Technology*, *40*(12), 47–59.
Khanduja, K., and Bhardwaj, A. 2003. Stable free radical scavenging and antiperoxidative properties of resveratrol compared *in vitro* with some other bioflavonoids. *Indian Journal of Biochemistry and Biophysics*, *40*(6), 416–422.
Khwaldia, K., Perez, C., Banon, S., Desobry, S., and Hardy, J. 2004. Milk proteins for edible films and coatings. *Critical Reviews in Food Science and Nutrition*, *44*(4), 239–251.
Kołożyn-Krajewska, D., and Dolatowski, Z. J. 2012. Probiotic meat products and human nutrition. *Process Biochemistry*, *47*(12), 1761–1772.
Korhonen, H. 2002. Technology options for new nutritional concepts. *International Journal of Dairy Technology*, *55*(2), 79–88.
Kris-Etherton, P. M., Hecker, K. D., Bonanome, A., Coval, S. M., Binkoski, A. E., Hilpert, K. F., Griel, A. E., and Etherton, T. D. 2002. Bioactive compounds in foods: Their role in the prevention of cardiovascular disease and cancer. *The American Journal of Medicine*, *113*(9), 71–88.
Krochta, J. M., and De Mulder-Johnston, C. 1997. Edible and biodegradable polymer films: Challenges and opportunities. *Food Technology 51*(2), 61–74.
Lagaron, J. M. (14e15 July 2005). Bioactive packaging: A novel route to generate healthier foods. Paper presented at the Second Conference in Food Packaging Interactions. CAMPDEM (CCFRA), Chipping Campden, UK.
Lanciotti, R., Gianotti, A., Patrignani, F., Belletti, N., Guerzoni, M., and Gardini, F. 2004. Use of natural aroma compounds to improve shelf-life and safety of minimally processed fruits. *Trends in Food Science and Technology*, *15*(3), 201–208.
Laohakunjit, N., and Kerdchoechuen, O. 2007. Aroma enrichment and the change during storage of non-aromatic milled rice coated with extracted natural flavor. *Food Chemistry*, *101*(1), 339–344.
Lee, J., Park, H., Lee, C., and Choi, W. 2003. Extending shelf-life of minimally processed apples with edible coatings and antibrowning agents. *LWT—Food Science and Technology*, *36*(3), 323–329.
Lopez-Rubio, A., Gavara, R., and Lagaron, J. M. 2006. Bioactive packaging: Turning foods into healthier foods through biomaterials. *Trends in Food Science & Technology*, *17*(10), 567–575.
López de Lacey, A. M., López-Caballero, M. E., Gómez-Estaca, J., Gómez-Guillén, M. C., and Montero, P. 2012. Functionality of *Lactobacillus acidophilus* and *Bifidobacterium bifidum* incorporated to edible coatings and films. *Innovative Food Science & Emerging Technologies*, *16*, 277–282.
Lynch, S. R. 1997. Interaction of iron with other nutrients. *Nutrition Reviews*, *55*(4), 102–110.
Madureira, A. R., Brandão, T., Gomes, A. M., Pintado, M. E., and Malcata, F. X. 2011. Technological optimization of manufacture of probiotic whey cheese matrices. *Journal of Food Science*, *76*(2), E203–E211.
Marcuzzo, E., Sensidoni, A., Debeaufort, F., and Voilley, A. 2010. Encapsulation of aroma compounds in biopolymeric emulsion based edible films to control flavour release. *Carbohydrate Polymers*, *80*(3), 984–988.
Martin-Belloso, O., Soliva-Fortuny, R., and Baldwin, A. 2005. Conservación mediante recubrimientos comestibles. In G. A. González-Aguilar, A. A. Gardea, and F. Cuamea-Navarro (Eds.), *Nuevas Tecnologías de Conservación de Productos Vegetales Frescos Cortados* (p. 558). Mexico: Centro de Investigación en Alimentación y Desarrollo.

Martínez-Camacho, A., Cortez-Rocha, M., Ezquerra-Brauer, J., Graciano-Verdugo, A., Rodriguez-Félix, F., Castillo-Ortega, M., and Plascencia-Jatomea, M. 2010. Chitosan composite films: Thermal, structural, mechanical and antifungal properties. *Carbohydrate Polymers*, 82(2), 305–315.

Mattila-Sandholm, T., Myllärinen, P., Crittenden, R., Mogensen, G., Fonden, R., and Saarela, M. 2002. Technological challenges for future probiotic foods. *International Dairy Journal*, 12(2), 173–182.

McClements, D. J., Decker, E. A., Park, Y., and Weiss, J. 2009. Structural design principles for delivery of bioactive components in nutraceuticals and functional foods. *Critical Reviews in Food Science and Nutrition*, 49(6), 577–606.

Mei, Y., and Zhao, Y. 2003. Barrier and mechanical properties of milk protein-based edible films containing nutraceuticals. *Journal of Agricultural and Food Chemistry*, 51(7), 1914–1918.

Mei, Y., Zhao, Y., Yang, J., and Furr, H. 2002. Using edible coating to enhance nutritional and sensory qualities of baby carrots. *Journal of Food Science*, 67(5), 1964–1968.

Min, S., Harris, L. J., and Krochta, J. M. 2005. Antimicrobial effects of lactoferrin, lysozyme, and the lactoperoxidase system and edible whey protein films incorporating the lactoperoxidase system against *Salmonella enterica* and *Escherichia coli* O157:H7. *Journal of Food Science*, 70(7), m332–m338.

Min, S., and Krochta, J. M. 2005. Inhibition of *Penicillium commune* by edible whey protein films incorporating lactoferrin, lacto-ferrin hydrolysate, and lactoperoxidase systems. *Journal of Food Science*, 70(2), M87–M94.

Monedero, F. M., Hambleton, A., Talens, P., Debeaufort, F., Chiralt, A., and Voilley, A. 2010. Study of the retention and release of n-hexanal incorporated into soy protein isolate-lipid composite films. *Journal of Food Engineering*, 100(1), 133–138.

Moore, M., Han, I., Acton, J., Ogale, A., Barmore, C., and Dawson, P. 2003. Effects of antioxidants in polyethylene film on fresh beef color. *Journal of Food Science*, 68(1), 99–104.

Morillon, V., Debeaufort, F., Blond, G., Capelle, M., and Voilley, A. 2002. Factors affecting the moisture permeability of lipid-based edible films: A review. *Critical Reviews in Food Science and Nutrition*, 42(1), 67–89.

Nakatani, N. 2003. Biologically functional constituents of spices and herbs (2002's JSNFS award for excellence in research). *Journal of Japanese Society of Nutrition and Food Science*, 56(6), 389–395.

Naushad, M. E., and Stading, M. 2007. *In situ* tensile deformation of zein films with plasticizers and filler materials. *Food Hydrocolloids*, 21(8), 1245–1255.

No, H. K., Young Park, N., Ho Lee, S., and Meyers, S. P. 2002. Antibacterial activity of chitosans and chitosan oligomers with different molecular weights. *International Journal of Food Microbiology*, 74(1–2), 65–72.

Nyström, L., Achrenius, T., Lampi, A.-M., Moreau, R. A., and Piironen, V. 2007. A comparison of the antioxidant properties of steryl ferulates with tocopherol at high temperatures. *Food Chemistry*, 101(3), 947–954.

Olivas, G. I., Rodriguez, J. J., and Barbosa-Cánovas, G. V. 2003. Edible coatings composed of methylcellulose, stearic acid, and additives to preserve quality of pear wedges. *Journal of Food Processing and Preservation*, 27(4), 299–320.

Oliveira, D. M., Kwiatkowski, A., Rosa, C. I. L. F., and Clemente, E. 2012. Refrigeration and edible coatings in blackberry (*Rubus* spp.) conservation. *Journal of Food Science and Technology*, 51(9), 2120–2126.

Oms-Oliu, G., Soliva-Fortuny, R., and Martín-Belloso, O. 2008. Edible coatings with antibrowning agents to maintain sensory quality and antioxidant properties of fresh-cut pears. *Postharvest Biology and Technology*, 50(1), 87–94.

Oussalah, M., Caillet, S., Salmiéri, S., Saucier, L., and Lacroix, M. 2004. Antimicrobial and antioxidant effects of milk protein-based film containing essential oils for the preservation of whole beef muscle. *Journal of Agricultural and Food Chemistry*, 52(18), 5598–5605.

Ozimek, L., Pospiech, E., and Narine, S. 2010. Nanotechnologies in food and meat processing. *ACTA Scientiarum Polonorum Technologia Alimentaria*, 9(4), 401–412.

Ozsoy, N., Candoken, E., and Akev, N. 2009. Implications for degenerative disorders: Antioxidative activity, total phenols, flavonoids, ascorbic acid, β-carotene and β-tocopherol in *Aloe vera*. *Oxidative Medicine and Cellular Longevity*, 2(2), 99–106.

Park, S.-I., and Zhao, Y. 2004. Incorporation of a high concentration of mineral or vitamin into chitosan-based films. *Journal of Agricultural and Food Chemistry*, 52(7), 1933–1939.

Pothakamury, U. R., and Barbosa-Cánovas, G. V. 1995. Fundamental aspects of controlled release in foods. *Trends in Food Science & Technology*, 6(12), 397–406.

Pranoto, Y., Salokhe, V. M., and Rakshit, S. K. 2005. Physical and antibacterial properties of alginate-based edible film incorporated with garlic oil. *Food Research International*, 38(3), 267–272.

Rahman, K. 2007. Studies on free radicals, antioxidants, and co-factors. *Clinical Interventions in Aging*, 2(2), 219–236.

Ramos, Ó. L., Pereira, J. O., Silva, S. I., Fernandes, J. C., Franco, M. I., Lopes-da-Silva, J. A., Pintado, M. E., and Malcata, F. X. 2012. Evaluation of antimicrobial edible coatings from a whey protein isolate base to improve the shelf life of cheese. *Journal of Dairy Science*, 95(11), 6282–6292.

Rauha, J.-P., Remes, S., Heinonen, M., Hopia, A., Kähkönen, M., Kujala, T., Pihlaja, K., Vuorela, H., and Vuorela, P. 2000. Antimicrobial effects of Finnish plant extracts containing flavonoids and other phenolic compounds. *International Journal of Food Microbiology*, 56(1), 3–12.

Reid, G. 2008. Probiotics and prebiotics—Progress and challenges. *International Dairy Journal*, 18(10), 969–975.

Reineccius, G. A. 2009. *Edible Films and Coatings for Flavor Encapsulation Edible Films and Coatings for Food applications* (pp. 269–294). New York, NY: Springer.

Rhim, J., and Shellhammer, T. 2005. Lipid-based edible films and coatings. *Innovations in food packaging*, 362–383. In J. H. Han (Ed.), *Innovations in Food Packaging* (Ch. 21, pp. 362–383). Amsterdam: Elsevier Academic Press.

Roberfroid, M. 2002. Global view on functional foods: European perspectives. *British Journal of Nutrition*, 88(S2), S133–S138.

Robles-Sánchez, R. M., Rojas-Graü, M. A., Odriozola-Serrano, I., González-Aguilar, G., and Martin-Belloso, O. 2013. Influence of alginate-based edible coating as carrier of antibrowning agents on bioactive compounds and antioxidant activity in fresh-cut Kent mangoes. *LWT—Food Science and Technology*, 50(1), 240–246.

Rodgers, S. 2007. Incorporation of probiotic cultures in foodservice products: An exploratory study. *Journal of Foodservice*, 18(3), 108–118.

Rojas-Argudo, C., Del Río, M., and Pérez-Gago, M. 2009. Development and optimization of locust bean gum (LBG)-based edible coatings for postharvest storage of "Fortune" mandarins. *Postharvest Biology and Technology*, 52(2), 227–234.

Rojas-Graü, M., Tapia, M., Rodríguez, F., Carmona, A., and Martin-Belloso, O. 2007. Alginate and gellan-based edible coatings as carriers of antibrowning agents applied on fresh-cut Fuji apples. *Food Hydrocolloids*, 21(1), 118–127.

Rojas-Graü, M. A., Soliva-Fortuny, R., and Martín-Belloso, O. 2009. Edible coatings to incorporate active ingredients to fresh-cut fruits: A review. *Trends in Food Science & Technology*, 20(10), 438–447.

Rojas-Graü, M. A., Tapia, M. S., and Martín-Belloso, O. 2008. Using polysaccharide-based edible coatings to maintain quality of fresh-cut Fuji apples. *LWT—Food Science and Technology*, *41*(1), 139–147.

Romanazzi, G., Nigro, F., Ippolito, A., Divenere, D., and Salerno, M. 2002. Effects of pre- and postharvest chitosan treatments to control storage grey mold of table grapes. *Journal of Food Science*, *67*(5), 1862–1867.

Rößle, C., Auty, M. A., Brunton, N., Gormley, R. T., and Butler, F. 2010. Evaluation of fresh-cut apple slices enriched with probiotic bacteria. *Innovative Food Science & Emerging Technologies*, *11*(1), 203–209.

Rouhi, M., Sohrabvandi, S., and Mortazavian, A. 2013. Probiotic fermented sausage: Viability of probiotic microorganisms and sensory characteristics. *Critical Reviews in Food Science and Nutrition*, *53*(4), 331–348.

Salmieri, S., and Lacroix, M. 2006. Physicochemical properties of alginate/polycaprolactone-based films containing essential oils. *Journal of Agricultural and Food Chemistry*, *54*(26), 10205–10214.

Sandström, B. 2001. Micronutrient interactions: Effects on absorption and bioavailability. *British Journal of Nutrition*, *85*(Suppl. S2), S181–S185.

Shah, N., and Ravla, R. 2004. Selling the cells in desserts. *Dairy Industries International*, *69*(1), 31–32.

Shahidi, F. 2000. Antioxidants in food and food antioxidants. *Nahrung*, *44*(3), 158–163.

Shida, K., and Nanno, M. 2008. Probiotics and immunology: Separating the wheat from the chaff. *Trends in Immunology*, *29*(11), 565–573.

Silva-Weiss, A., Ihl, M., Sobral, P. J. A., Gómez-Guillén, M. C., and Bifani, V. 2013. Natural additives in bioactive edible films and coatings: Functionality and applications in foods. *Food Engineering Reviews*, *5*(4), 200–216.

Siragusa, G. R., and Dickson, J. S. 1992. Inhibition of *Listeria monocytogenes* on beef tissue by application of organic acids immobilized in a calcium alginate gel. *Journal of Food Science*, *57*(2), 293–296.

Siró, I., Kápolna, E., Kápolna, B., and Lugasi, A. 2008. Functional food. Product development, marketing and consumer acceptance—A review. *Appetite*, *51*(3), 456–467.

Song, Y., Liu, L., Shen, H., You, J., and Luo, Y. 2011. Effect of sodium alginate-based edible coating containing different anti-oxidants on quality and shelf life of refrigerated bream (*Megalobrama amblycephala*). *Food Control*, *22*(3–4), 608–615.

Soong, Y.-Y., and Barlow, P. J. 2004. Antioxidant activity and phenolic content of selected fruit seeds. *Food Chemistry*, *88*(3), 411–417.

Soukoulis, C., Yonekura, L., Gan, H.-H., Behboudi-Jobbehdar, S., Parmenter, C., and Fisk, I. 2014. Probiotic edible films as a new strategy for developing functional bakery products: The case of pan bread. *Food Hydrocolloids*, *39*, 231–242.

Souza, V. C., Monte, M. L., and Pinto, L. A. A. 2011. Preparation of biopolymer film from chitosan modified with lipid fraction. *International Journal of Food Science & Technology*, *46*(9), 1856–1862.

Sun Pan, B., Kuo, J.-M., and Wu, C.-M. 2006. Flavor compounds in foods. In Z. E. Sikorski (Ed.), *Chemical and Functional Properties of Food Components* (pp. 295–318). Boca Raton, FL: CRC Press.

Suseno, N., Savitri, E., Sapei, L., and Padmawijaya, K. S. 2014. Improving shelf-life of Cavendish banana using chitosan edible coating. *Procedia Chemistry*, *9*, 113–120.

Tajkarimi, M. M., Ibrahim, S. A., and Cliver, D. O. 2010. Antimicrobial herb and spice compounds in food. *Food Control*, *21*(9), 1199–1218.

Talens, P., and Krochta, J. M. 2005. Plasticizing effects of beeswax and carnauba wax on tensile and water vapor permeability properties of whey protein films. *Journal of Food Science*, *70*(3), E239–E243.

Tanada-Palmu, P. S., and Grosso, C. R. 2005. Effect of edible wheat gluten-based films and coatings on refrigerated strawberry (*Fragaria ananassa*) quality. *Postharvest Biology and Technology*, *36*(2), 199–208.

Tapia, M. S., Rojas-Graü, M. A., Rodríguez, F. J., Ramírez, J., Carmona, A., and Martin-Belloso, O. 2007. Alginate- and gellan-based edible films for probiotic coatings on fresh-cut fruits. *Journal of Food Science*, *72*(4), E190–E196.

Thankachan, P., Walczyk, T., Muthayya, S., Kurpad, A. V., and Hurrell, R. F. 2008. Iron absorption in young Indian women: The interaction of iron status with the influence of tea and ascorbic acid. *The American Journal of Clinical Nutrition*, *87*(4), 881–886.

Torres, J. A., Motoki, M., and Karel, M. 1985. Microbial stabilization of intermediate moisture food surfaces I. Control of surface preservative concentration. *Journal of Food Processing and Preservation*, *9*(2), 75–92.

Traka, M. H., and Mithen, R. F. 2011. Plant science and human nutrition: Challenges in assessing health-promoting properties of phytochemicals. *The Plant Cell Online*, *23*(7), 2483–2497.

Vargas, M., Pastor, C., Chiralt, A., McClements, D. J., and Gonzalez-Martinez, C. 2008. Recent advances in edible coatings for fresh and minimally processed fruits. *Critical Reviews in Food Science and Nutrition*, *48*(6), 496–511.

Vásconez, M. B., Flores, S. K., Campos, C. A., Alvarado, J., and Gerschenson, L. N. 2009. Antimicrobial activity and physical properties of chitosan-tapioca starch based edible films and coatings. *Food Research International*, *42*(7), 762–769.

Vasiljevic, T., and Shah, N. P. 2008. Probiotics—From metchnikoff to bioactives. *International Dairy Journal*, *18*(7), 714–728.

Viskupicova, J., Danihelova, M., Ondrejovic, M., Liptaj, T., and Sturdik, E. 2010. Lipophilic rutin derivatives for antioxidant protection of oil-based foods. *Food Chemistry*, *123*(1), 45–50.

Wettasinghe, M., and Shahidi, F. 1999. Antioxidant and free radical-scavenging properties of ethanolic extracts of defatted borage (*Borago officinalis* L.) seeds. *Food Chemistry*, *67*(4), 399–414.

Wittaya, T. 2012. Protein-based edible films: Characteristics and improvement of properties. In A. A. Eissa (Ed.), *Structure and Function of Food Engineering* (pp. 43–70). New York, NY: InTech. Available at http://www.intechopen.com/books/structure-and-function-of-food-engineering/protein-based-edible-films-characteristics-and-improvement-of-properties. Accessed December 2, 2014.

Wu, Y., Weller, C., Hamouz, F., Cuppett, S., and Schnepf, M. 2001. Moisture loss and lipid oxidation for precooked ground-beef patties packaged in edible starch-alginate-based composite films. *Journal of Food Science*, *66*(3), 486–493.

Zubillaga, M., Weill, R., Postaire, E., Goldman, C., Caro, R., and Boccio, J. 2001. Effect of probiotics and functional foods and their use in different diseases. *Nutrition Research*, *21*(3), 569–579.

8 Antimicrobial Edible Packaging

Lorenzo M. Pastrana, Maria L. Rúa, Paula Fajardo, Pablo Fuciños, Isabel R. Amado, and Clara Fuciños

CONTENTS

Abstract 244
8.1 Physical Properties of Materials Forming Antimicrobial Edible
 Films and Coatings for Food Packaging..244
    8.1.1 Polysaccharides..245
        8.1.1.1 Starch 245
        8.1.1.2 Cellulose and Derivatives 245
        8.1.1.3 Chitin/Chitosan...245
        8.1.1.4 Carrageenan 246
        8.1.1.5 Alginate...246
        8.1.1.6 Pectin 246
        8.1.1.7 Galactomannans..247
    8.1.2 Proteins...247
        8.1.2.1 Gelatin...247
        8.1.2.2 Zein Protein 248
        8.1.2.3 Wheat Gluten 248
        8.1.2.4 Soy Protein..248
        8.1.2.5 Milk Proteins 248
    8.1.3 Lipids and Waxes...249
8.2 Antimicrobial Compounds Used in Edible Films and Coatings
 for Food Packaging..249
8.3 Films and Coatings with Antimicrobial Properties Using Chitosan
 and Other Antimicrobial Compounds 253
    8.3.1 Factors Affecting the Antimicrobial Properties of Chitosan 253
    8.3.2 Methods to Prepare Antimicrobial Edible Films and Coatings
        for Food Active Packaging Purposes...254
    8.3.3 Improving the Performance of Chitosan Edible Films
        and Coatings in Food Active Packaging..255
    8.3.4 Improving the Physical Properties of Chitosan Films
        by Blending 255
    8.3.5 Use of Chitosan Films to Deliver Antimicrobial Compounds 257

8.4 Food Applications of Antimicrobial Edible Films and Coatings 258
    8.4.1 Antimicrobial Edible Films and Coatings for Meat, Poultry, and Fish 258
        8.4.1.1 Films with Chitosan 260
        8.4.1.2 Films with Pectin 260
        8.4.1.3 Films with Proteins 261
    8.4.2 Antimicrobial Edible Films and Coatings for Fruits and Vegetables 262
    8.4.3 Antimicrobial Edible Films and Coatings for Cheese and Other Foods 267
8.5 Conclusions and Future Trends 273
References 273

ABSTRACT

Antimicrobial edible packaging allows preserving food quality by avoiding or reducing the growth of food spoilage pathogen microorganism. Biopolymers used for this purpose include polysaccharides, proteins, and lipids. The mechanical and antimicrobial properties of biopolymers can be improved, respectively, by blending with other polymers or adding antimicrobial compounds (such as plant extracts and their essential oils) that are delivered from the matrix. Chitosan has by itself antimicrobial activity depending on pH, deacetylation degree, and molecular weight. The main applications to extend the shelf life are meat, poultry, fish and cheese products, and fruits and vegetables. In all cases, the protective effect of antimicrobial edible coating is complemented by the reduction of water loss and prevention of maturation and senescence or protection against oxidation.

8.1 PHYSICAL PROPERTIES OF MATERIALS FORMING ANTIMICROBIAL EDIBLE FILMS AND COATINGS FOR FOOD PACKAGING

An edible film is a thin continuous sheet formed from a biopolymer matrix that is cohesive enough and has the physical integrity to stand-alone. The thickness of an edible film is typically 2–10 mils (0.050–0.250 mm) (Jooyandeh, 2011). Edible coatings are edible films formed directly on the surface of a food or material. They are typically thinner than stand-alone edible films. The main purpose for edible films from biopolymers is to control mass transfer of multiple compounds including gas, aroma, oil, and water vapor into or out of a food, preserving food quality. Edible films must also be both strong and flexible to withstand forces experienced during handling and processing (Jooyandeh, 2011).

Biopolymer films, including both edible films and coatings, are made from biological materials such as polysaccharides (starch, starch derivatives, cellulose, pectin, and alginate), proteins (gelatin, casein, wheat gluten, zein, and soy protein), and lipids (beeswax, acetylated monoglycerides, fatty alcohols, and fatty acids) or the combination of these components. Minor components are also included in the formulation of both films and coatings, which include polyols acting as plasticizers (such as glycerol or polyethylene glycol) or acid/base compounds used to regulate pH, such as acetic or lactic acid (Bravin et al., 2006; Falguera et al., 2011).

8.1.1 Polysaccharides

Polysaccharides that have been used to form films/coatings include starch and starch derivatives, cellulose derivatives, alginates, carrageenan, various plant and microbial gums, chitosan, and pectinates (Lin and Zhao, 2007; Rinaudo, 2008). Their hydrophilic properties provide a good barrier to carbon dioxide and oxygen under certain conditions but a poor barrier to water vapor and deficient mechanical properties (Guilbert et al., 1996). Although most polysaccharide coatings present poor water barrier properties, the addition of fat to get emulsified films is assumed to significantly reduce water transfer through the film.

8.1.1.1 Starch

Starch is the main component. Starch is a natural polymer that can readily be cast into films. It is composed of amylose (poly-α-1,4-D-glucopyranoside) and amylopectine (poly-α-1,4-D-glucopyranoside and α-1,6-D-glucopyranoside). Amylopectin is a highly branched, very high-molecular-weight (MW) polymer (5000–30,000 kDa), whereas amylose is a linear chain polysaccharide and has a lower molecular size (20–800 kDa) (Peressini et al., 2004). Amylose accounts for about 20%–25% of most granular starch. Despite their ease of preparation, starch films have poor physical properties that have been improved by blending with synthetic and natural polymers such as cellulose (Peressini et al., 2004) and proteins (Chinma et al., 2013; Jagannath et al., 2003).

8.1.1.2 Cellulose and Derivatives

Cellulose together with starch is the most important raw material for the preparation of films. Cellulose is a high-MW polymer of β-(1,4)-linked glucose residues that appear as a cold water–insoluble crystalline polymer (Gennadios et al., 1997). Water solubility can be increased by treating cellulose with alkali to swell the structure, followed by reaction with several reactants to obtain a number of derivatives (ethers and esters), such as carboxymethyl cellulose, methyl cellulose (MC), hydroxypropyl methyl, or hydroxypropyl cellulose. These cellulose derivatives possess good film-forming characteristics, such as generally odorless and tasteless, flexible and of moderate strength, transparent, resistance to oil and fats, water soluble, moderate to moisture, and oxygen transmission (Bourtoom, 2008; Krochta and Mulder-Johnson, 1997). MC is most resistant to water, and it is the lowest hydrophilic cellulose derivative (Kester and Fennema, 1986).

8.1.1.3 Chitin/Chitosan

Chitosan is a natural cationic polysaccharide formed by the N-deacetylation (DA) of chitin, the structural component of the exoskeletons of arthropods (including crustaceans and insects), in marine diatoms and algae, as well as in some fungal cell walls (Raafat and Sahl, 2009). Chitosan is a linear binary copolymer that consists of β(1,4)-linked 2-acetoamido-2-deoxy-β-D-glucopyranose (Glc-NAc; A-unit) and 2-amino-2-deoxy-β-D-glucopyranose (GlcN; D-unit) (Carneiro-da-Cunha et al., 2010). The relative amount of the two monosaccharides in chitosan may vary, giving samples of different degrees of DA (75%–95%), MWs (50–2000 kDa), viscosities, and pK_a values (Singla and Chawla, 2001; Tharanathan and Kittur, 2003). Chitosan is especially of interest for its antifungal and antimicrobial activities which are believed to emerge from its polycationic nature that cause a membrane-disrupting effect (Kim et al., 2003; Elsabee and Abdou, 2013).

Recently, there has been a great interest in edible films and coatings from chitosan (see below), due to their good transport and mechanical properties, although chitosan films, as from other polysaccharides, are highly permeable to water vapor which limits their use (Casariego et al., 2009; Souza et al., 2009; Vargas et al., 2009).

8.1.1.4 Carrageenan

Carrageenans are water-soluble polymers extracted from the cell walls of various red seaweeds (Rhodophyceae), which present high potentiality as film-forming material. They are anionic linear sulfated polysaccharides composed of D-galactopyranose residues bonded by regularly alternating α-(1,3) and β-(1,4) bonds (Karbowiak et al., 2007). The number and position of sulfate groups on the disaccharide repeating unit together with an anhydrogalactose bridge determine classification in three major types: κ, ι, and λ. The κ-, ι-, and λ-carrageenans exhibit one, two, and three sulfate ester groups per dimeric unit. κ- and ι-carrageenans contain the 3,6-anhydro units and are used as gelling agents and form thermoreversible gels on cooling below the critical temperature. Carrageenan film formation includes this gelation mechanism during moderate drying, leading to a three-dimensional network formed by polysaccharide double helices and to a solid film after solvent evaporation (Karbowiak et al., 2006, 2007).

8.1.1.5 Alginate

Alginates are hydrophilic colloidal carbohydrates extracted with dilute alkali from various species of brown seaweeds (Phaeophyceae). Alginates are linear water-soluble polysaccharides formed by sequences of α-(1,4)-linked units of β-D-mannuronate (M) and α-L-guluronate (G) at different proportions and different distributions in the chain. The chemical composition and sequence of the M and G residues depend on the biological source and the state of maturation of the plant (da Silva et al., 2009). The ability of alginates to react with di- and trivalent cations is utilized in alginate film formation (Hambleton et al., 2009). Calcium-induced gelation has been demonstrated to result from specific and strong interactions between calcium ions and guluronate and galacturonate blocks in alginate. The mechanism that better describes gel formation in alginate, and LM pectin in the presence of calcium ions is the so-called "egg-box" model (Bryce et al., 1974; da Silva et al., 2009). Edible films prepared from alginate form strong films and exhibit poor water resistance because of their hydrophilic nature.

8.1.1.6 Pectin

Pectin is a heterogeneous grouping of acidic structural polysaccharides, found in fruit and vegetables and mainly prepared from citrus peel and apple pomace. This complex anionic polysaccharide is composed of β-1,4-linked D-galacturonic acid residues, wherein the uronic acid carboxyls are either fully (high methoxy pectin) or partially (low methoxy pectin) methyl esterified. As alginate, LM pectin undergoes chain-to-chain association and forming hydrogels upon addition of divalent cations (e.g., Ca^{2+}) (Fang et al., 2008). Pectin and alginate tend to form strong films but with poor resistance to water due to their hydrophilic nature (da Silva et al., 2009).

8.1.1.7 Galactomannans

Galactomannans are neutral polysaccharides isolated from seeds (carob, guar, locust bean, and tara). The main chain is made of [4]-β-D-Man-(1] (Man = mannose M) with different degrees of substitution on O-6 with α-D-galactopyranosyl units (G) (Rinaudo, 2008). The solubility of the galactomannan depends on the M/G ratio and on the distribution of galactose units along the mannan backbone chain: the larger the galactose content, the higher the solubility in water. Some works have already used galactomannans from commercial and nontraditional origins as a source for films and coatings production (Cerqueira et al., 2009, 2010; Chen and Nussinovitch, 2001).

8.1.2 Proteins

Both fibrous and globular proteins can form films. Of the fibrous proteins, collagen has received the most attention in the production of edible films. Several globular proteins, including caseins and whey proteins (from animal sources) and corn zein, wheat gluten, or soy protein (from plant sources), have been investigated for their film properties.

Fibrous proteins are fully extended and associated closely with each other in parallel structures, generally through hydrogen bonding, to form fibers. Generally, globular proteins must be denatured by heat, acid, base, and/or solvents to form more extended structures that are required to form films. Once extended, protein chains can associate through hydrogen, ionic, hydrophobic, and covalent bonding. The chain-to-chain interaction that produces a cohesive film is affected by the degree of chain extension (protein structure) and the nature and sequence of amino acid residues. Uniform distribution of polar, hydrophobic, and/or thiols along the polymer chain increases the likelihood of the respective interaction. Polymers with abundance of groups that can associate through hydrogen or ionic bonding, form films that are an excellent oxygen barrier but that are susceptible to moisture. Contrarily, those polymers, where the hydrophobic groups are dominant, are poor oxygen barriers but excellent moisture barriers (Kester and Fennema, 1986). In general, when a protein-based film with high water vapor permeability (WVP) is required, addition of lipid components is needed. Often low MW plasticizers must be added to protein films to increase flexibility (decrease brittleness) by reducing chain-to-chain interactions, although this generally increases film permeability as well. Plasticizers that have been used for edible films include mono-, di-, and oligosaccharides, polyols, and lipids (Kester and Fennema, 1986).

8.1.2.1 Gelatin

Gelatin is prepared by the thermal denaturation of collagen, isolated from animals' kin, bones, and fish skins. The properties and film-forming ability of gelatins are directly related to the MW, that is, the higher the average MW, the better the quality of the film. The MW distribution depends mainly on the degree of collagen cross-linking and the extraction procedure (Gómez-Guillén et al., 2007).

Traditionally, gelatins from mammalian sources have been mostly used. However, in the year 2000, the gelatin films forming principally with fish gelatin have returned to the attention of researchers (Gómez-Estaca et al., 2009; Simon-Lukasik and Ludescher, 2004). Films from tuna skin gelatin plasticized with glycerol presented lower WVP compared to values reported for pigskin gelatin (Gómez-Guillén et al., 2007).

8.1.2.2 Zein Protein

Zein is the prolamin fraction (soluble in 70% ethanol) of corn gluten, making up to 70% of the corn gluten and it is mainly isolated from corn gluten meal, the by-product of corn wet milling. The average MW of zein ranges from 9 to 26 kDa and is rich in nonpolar aminoacids, for example, leucine, alanine, and proline (Anderson and Lamsal, 2011; Lawton, 2002). Four distinct types of zein (α-, β-, γ-, and δ-zein) have been described based on their MWs and solubility in aqueous alcohol and reducing agents. α-Zein, the most abundant type, is the predominant fraction present in commercial zein (Lawton, 2002; Wilson, 1988).

8.1.2.3 Wheat Gluten

Gluten proteins are storage proteins of the wheat endosperm with unique viscoelastic properties. They consist of monomeric gliadins with a MW between 30 and 60 kDa and a mixture of glutenin polymers with a MW ranging from about 80 kDa to several million (Veraverbeke and Delcour, 2002). Gliadins represent a heterogeneous mixture of single-chained or monomeric gluten proteins (existing α-, γ-, and ω-types), while glutenin is a highly heterogeneous mixture of polymers consisting of a number of different high- and low-MW glutenin subunits linked by disulfide bonds. Glutenin large size makes them partly insoluble in most common solvents. Gliadins instead are readily extractable in aqueous alcohols (Bourtoom, 2008; Lagrain et al., 2010).

8.1.2.4 Soy Protein

Soy protein–based edible film is most commonly prepared from soy protein isolates (SPIs). SPIs are produced from defatted soy meal by alkali extraction followed by acid precipitation (pH 4.5) and have a protein content >90% (Cho, 2007). The most common way to classify soy proteins is based on their sedimentation rate in fractional ultracentrifugation. Larger Svedberg (S) numbers indicate a larger protein. Soy protein fractions include 2S, 7S, 11S, and 15S (Hernandez-Izquierdo and Krochta, 2008), but the main components are glycinin (11S protein) and β-conglycinin (7S protein). The latter has a MW of 180 kDa and is rich in asparagine, glutamine, leucine, and arginine residues. Unlike glycinin, which contains 20 intramolecular disulfide bonds, disulfide cross-linking in conglycinin is limited (Krochta, 1997; Khorshid et al., 2007).

8.1.2.5 Milk Proteins

Milk proteins can be divided into casein and whey protein. Casein represents 80% of the total milk protein and consists of α-, β-, and κ-casein with MWs between 19 and

25 kDa, and most of them exist in a colloidal particle known as the casein micelle. The low cysteine levels and the high proline content in casein result in little disulfide cross-linking and an open, random-coil structure (Khwaldia et al., 2004). Whey protein comprises 20% of the milk protein and is the protein that remains soluble after casein has been precipitated at pH 4.6. Whey includes heat-sensitive, globular, water-soluble proteins, and enzymes (Goff and Hill, 1993; Raikos, 2010). Whey protein is composed mainly by β-lactoglobulin (β-Lg) (~50%), in milk of ruminants and many other mammals, and α-lactalbumin (~20%), with minor contents of immunoglobulins, lactoferrin, lactoperoxidase, albumin, lysozyme, and several peptides (Zadow, 1994).

β-Lactoglobulin is a globular protein from the lipocalin structural family. It has a MW, which differs slightly around 18.3 kDa for the various genetic variants (A to G) (Phillips et al., 1994). The protein has complex equilibrium aggregation behavior depending on environment (Sawyer and Kontopidis, 2000) and is normally found as a dimer under physiological conditions. α-Lactalbumin is a smaller globular metalloprotein (14 kDa), structurally homologous to lysozyme, and requires calcium to assume its functional fold. Whey protein isolate has been recognized to produce flexible, transparent films with excellent oxygen-, aroma-, and oil-barrier properties at low relative humidity (Janjarasskul et al., 2011).

8.1.3 Lipids and Waxes

Lipid compounds include neutral lipids of glycerides (esters of glycerol and fatty acids) and waxes (esters of long-chain monohydric alcohols and fatty acids). The primary function of a lipid coating is to block transport of moisture. Waxes are the most efficient substances to reduce moisture permeability. Their high hydrophobicity is a consequence of a high content in esters of long-chain fatty alcohols and acids, as well as long-chain alkanes (Kester and Fennema, 1986). In contrast, the hydrophobic characteristic of lipid forms thicker and more brittle films. Generally, WVP decreases when the concentration of hydrophobicity phase increases. Lipid-based films are often supported on a polymer structure matrix, usually a polysaccharide, to provide mechanical strength (Bourtoom, 2008; Debeaufort et al., 1993).

Among waxes, paraffin (derived from distillate fraction of crude petroleum), carnauba (exudate from palm tree leaves *Copoernica cerifera*), beewax, and candelilla obtained from candelilla plant are the most efficient edible compounds providing a humidity barrier (Bourtoom, 2008).

8.2 ANTIMICROBIAL COMPOUNDS USED IN EDIBLE FILMS AND COATINGS FOR FOOD PACKAGING

The incorporation of antimicrobial agents into polymeric films allows industry to combine the preservative functions of antimicrobials with the protective functions of the classical packaging concept (Persico et al., 2009), thus offering several advantages compared with the direct addition of preservatives into food products; the incorporation of antimicrobial agents directly onto food surface without any matrix may result in the partial inactivation of the active substance (Reps et al., 2002) and a rapid diffusion within the bulk of food (Ouattar et al., 2000).

First, for the selection of an antimicrobial, it must be considered the effectiveness against the target microorganism and the possible interactions among the antimicrobial, the film-forming biopolymer, and other food components that can modify the antimicrobial activity (Campos et al., 2011).

Antimicrobial activity of both natural agents and edible films and coatings is commonly evaluated against spoilage and typical food-borne bacteria. *Staphylococcus aureus* (Gram-positive) and *Escherichia coli* (Gram-negative) bacteria were sensitive to thymol, carvacrol (Ramos et al., 2012), and silver nanoparticles (López-Carballo et al., 2013). Rosemary improved antimicrobial activity of chitosan films against *Pseudomonas putida*, *Streptococcus agalactiae*, and *Lactococcus lactis* (Abdollahi et al., 2012). Carvacrol had strong anti-*Listeria monocytogenes* activity (Veldhuizen et al., 2007); however, some authors use the nonpathogen bacteria *Listeria innocua* as antilisterial target, sensitive to thymol, carvacrol (Guarda et al., 2011), and lysozyme (Fajardo et al., 2014b).

The inhibition of yeast and fungi is also a target for the development of antimicrobial films and coatings. Plastic flexible films with a coating of microcapsules containing carvacrol and thymol presented antimicrobial properties against *Saccharomyces cerevisiae* (Guarda et al., 2011). Gliadin films incorporating cinnamaldehyde were tested against food-spoilage fungi as *Penicillium expansum* and *Aspergillus niger* (Balaguer et al., 2013), and *Alternaria solani* and *Colletotrichum acutatum* (Balaguer et al., 2014). Chitosan-based coatings/films containing natamycin were used to inhibit growth of *A. niger, Penicillium crustosum, Penicillium commune*, and *Penicillium roqueforti* in cheese (Fajardo et al., 2010).

Sometimes food materials also act as viral pathogen vectors for virus such as hepatitis A and norovirus. Norovirus is currently recognized as the predominant agent of nonbacterial gastroenteritis in humans, which causes food-borne gastroenteritis outbreaks. For this reason, recently researchers are interested in the virucidal properties of natural compounds. Two human norovirus surrogates, murine norovirus and feline calicivirus, were used to investigate the antiviral activity of bacteriocins (Fajardo et al., 2014a) and silver ions incorporated into polylactide films (Martínez-Abad et al., 2013). However, very scarce information is available about packaging materials with both antibacterial and virucidal properties (Martínez-Abad et al., 2013).

Incorporation of natural additives to active packaging systems or biopolymer-based edible films can modify the film structure, functionality, and application to foods (Silva-Weiss et al., 2013). Physical, thermal, and microstructural properties of films and coatings could be altered. Then, once the concentration of natural compound that inhibits the microorganism studied is found, modifications of characteristics of the film due to the incorporation of antimicrobial compounds have to be studied. Silva-Weiss reviewed the effect of the incorporation of natural additives from plant extracts or their isolated active compounds on the functionality of edible films and their application to foods.

The natural compound has to be miscible with the film-forming solutions and provides a homogeneous dispersion. The color of the film could change with the incorporation of the natural compounds, and this could affect the final application of the film. For instance, increasing the opacity of the film could protect the food from the UV deterioration, but sometimes film transparency is a priority to maintain the food appearance.

Mechanical properties such as maximum tensile strength (σ_m), the percentage of elongation at break (ε_b), and Young's modulus (E) can be modified by the addition of the natural compound and these changes can be different for each compound and polymer combination. When lysozyme was incorporated into gliadin films, a denser and tighter polymer network was achieved (Fajardo et al., 2014b). However, when lysozyme was added to chitosan films, the σ_m decreased (Park et al., 2004). In the same way, a decrease in σ_m and ε_b of Na-alginate- and κ-carrageenan-based films was reported with lysozyme incorporation in film matrix (Cha et al., 2002). For caseinate films containing lysozyme, a negligible decrease in σ_m and ε_b was reported (de Souza et al., 2010).

The effect of adding antimicrobial compounds to edible films on barrier properties (water vapor, O_2, and CO_2 permeabilities) is also commonly studied. Films with a proper oxygen barrier can improve food quality and extend food shelf life but different behavior on this property was reported for different antimicrobial compounds and films. The value of O_2P for chitosan films increased due to the incorporation of natamycin (Fajardo et al., 2010). Natamycin may have created additional sites for the dissolution of oxygen, increasing mobility of O_2 molecules within the polymer and consequently O_2P values. Similarly, the incorporation of natamycin reduced the barrier properties of gliadin films cross-linked with 5% of cinnamaldehyde (Balaguer et al., 2014). This result was also associated with a looser packing of the protein chains as a consequence of the presence of natamycin and lactose, which increased the free volume of the polymeric structure and thus enhanced permeability. However, O_2P values for carboxymethylcellulose films with extract of murta leaves were lower than those without the extract (Bifani et al., 2007). The extract of murta leaves showed a high concentration of some polyphenolic species, like myricetin and quercetin glycosides. Some of the B ring of the flavonols myricetin and quercetin could be hydrolyzed in the heat treatment used for the extraction of components, and these smaller molecules could act filling part of the pores of the carboxymethylcellulose-based films, reducing their sizes, avoiding some of the free passage of gases. On the contrary, the addition of lysozyme in gliadin films did not have a significant effect on the barrier properties (Fajardo et al., 2014b).

The more commonly used antimicrobials are mainly plant extracts and their essential oils (Efrati et al., 2014), organic acids and its salts (Pérez et al., 2014), nisin (Guo et al., 2014; Martins et al., 2010), and the lactoperoxidase system (Lee and Min, 2014).

The essential oils are extracted from different parts of plants and composed of low MW molecules which may have different chemical structures (Del Nobile et al., 2008). The major limitation of essential oils as antimicrobial agents incorporated into a polymeric matrix is their low thermal stability and high volatility (Efrati et al., 2014). Currently, there are a small number of plant-derived antimicrobial agents available commercially because the commercial uptake of these agents is hampered by the fact that they are relatively expensive in comparison with synthetic agents (Kenny et al., 2014).

Several studies evaluated the incorporation of various essential oils for the development of edible films with antibacterial properties, including thymol (Del Nobile et al., 2008), carvacrol (Arrieta et al., 2014; Peretto et al., 2014a; Zhu et al., 2014), eugenol (Zhang and Jiang, 2012), cinnamaldehyde (Balaguer et al., 2013, 2014), rosemary (Abdollahi et al., 2012), and citral (Rojas-Graü et al., 2007a).

Apple-based edible films containing carvacrol and cinnamaldehyde showed potential to reduce *Campylobacter jejuni* on chicken and therefore, the risk of campylobacteriosis (Mild et al., 2011). Wine grape pomace extract–based films showed antibacterial activity against both *E. coli* and *L. innocua* (Deng and Zhao, 2011). Edible rapeseed protein–gelatin film containing antimicrobial grapefruit seed extract (Jang et al., 2011) and sweet potato starch–based film containing origanum oil (Ehivet et al., 2011) inhibited the growth of pathogenic bacteria such as *E. coli* O157:H7 and *L. monocytogenes*. Chitosan edible films incorporating garlic oil were found to have antimicrobial activity against *S. aureus, L. monocytogenes*, and *Bacillus cereus* (Pranoto et al., 2005).

Nisin and natamycin, two natural food preservatives produced by microorganisms and classified as generally recognized as safe (GRAS) by the U.S. Food and Drug Administration, have been widely used as antimicrobial compounds in food packaging. Nisin was incorporated in different polymers as antimicrobial agent for Gram-positive bacteria, including gelatin (Min et al., 2010), chitosan (Cai et al., 2010), and starch (Resa et al., 2014). In addition, galactomannan-based edible coatings combined with nisin reduced *L. monocytogenes* postcontamination on cheese products during storage (Martins et al., 2010).

Natamycin is allowed as food additive in cheese and cured and dried nonheat-treated processed meat (Codex alimentarius, 2014). The effectiveness of edible films incorporating natamycin for preservation of cheese has been demonstrated. Natamycin incorporated in tapioca starch films controlled growth of *S. cerevisiae* on the surface of Port Salut cheese during storage (Ollé Resa et al., 2014). In addition, this preservative included in wheat gluten films eliminated *A. niger* on the surface of fresh kashar cheese (Ture et al., 2011). Chitosan-based coatings/films were used as release system containing natamycin to create an additional hurdle for molds/yeasts in cheese, thus contributing to extend its shelf life (Fajardo et al., 2010). Moreover, natamycin was successfully loaded in a smart delivery device to control its release and added to edible films without changing their main packaging properties (Cerqueira et al., 2014a).

Some works reported the use of proteins with antimicrobial activity in edible films. Lactoferrin was incorporated in chitosan films (Bourbon et al., 2011). The positive amino acids in lactoferrin can interact with anionic molecules on some bacterial, viral, fungal, and parasite surfaces, causing cell lysis (González-Chávez et al., 2009). Gliadin films cross-linked with cinnamaldehyde were used as systems for the release of lysozyme (Fajardo et al., 2014b). These films containing 10% (w/w) of lysozyme significantly ($p \leq 0.05$) reduced the growth of *L. innocua*.

Antimicrobial activity can be achieved using polymeric materials with antimicrobial properties, as chitosan, or adding antimicrobial compounds during their manufacturing. Incorporation of antimicrobial agents can be done adding the compounds directly into polymers, coating antimicrobials onto polymer surfaces, or immobilizing antimicrobials by chemical grafting (Persico et al., 2009). Immobilization systems suppress the growth of microorganisms mainly at the contact surface and release systems allow migration of the antimicrobial agent into the food (Fajardo et al., 2014b), or the headspace inside the package (Balaguer et al., 2014) inhibiting the growth of microorganisms.

The amount of diffusion of active substance from the film package to the product is crucial and must be determined. While high concentrations of a released compound in food could cause sensorial or toxicological problems, the low concentrations would not be effective against microbial contamination (Conte et al., 2006). In this connection, a considerable effort is being conducted on the development of sustained release systems of active compounds for food packaging (Chen et al., 2006; Fajardo et al., 2014b; Mastromatteo et al., 2010).

Traditionally, active components were incorporated into a single material film or a laminate. However, recent innovations in antimicrobial edible films and coatings include microencapsulation of antimicrobial compounds (Guarda et al., 2011; Otoni et al., 2014), the reinforcement of edible polymers with other materials (Ruiz et al., 2013), and use of nanotechnology (Morsy et al., 2014) to increase effectiveness of antimicrobial food packaging. Nanolayers and nanostructured layers made of biopolymers are able to retain transparency and enhance barrier properties of biodegradable and renewable materials without significant changes in mechanical performance (Fabra et al., 2013). Durán and Marcato (2013) have reviewed the application and the benefits of nanotechnology in different areas of food industry, including bioactive nanoencapsulation, edible thin films, and packages. In addition, Cerqueira et al. (2014b) highlight the potentialities of nanotechnology for food industry, the principles of design and production of bio-nanosystems for oral delivery and their behavior within the human gastrointestinal tract.

8.3 FILMS AND COATINGS WITH ANTIMICROBIAL PROPERTIES USING CHITOSAN AND OTHER ANTIMICROBIAL COMPOUNDS

Chitosan has been referred to have interesting technological and functional properties as well as biological activities. This polysaccharide is nontoxic, edible, biodegradable, biocompatible, and able to form films. In addition, chitosan presents antimicrobial, antitumoral, and hypocholesterolemic activities (No et al., 2002).

8.3.1 Factors Affecting the Antimicrobial Properties of Chitosan

The antimicrobial activity of chitosan and chitosan oligomers against different groups of microorganisms includes mainly Gram-negative bacteria and also Gram positive and fungi (Limam et al., 2011). It is accepted that antifungal activity of chitosan is less effective when compared with its activity against bacteria (Bautista Baños et al., 2006). Chitosan is considered as bacteriostatic rather than bactericidal (Coma et al., 2002), but the exact antimicrobial mechanism of action is, until now, not fully understood. One proposed acceptable model is based on the electrostatic interaction between positively charged chitosan molecules and negatively charged microbial cell membranes that cause internal osmotic imbalances and/or the leakage of intracellular electrolytes due to changes in membrane and wall permeability (Goy et al., 2009). Zheng and Zhu (2003) propose that low MW chitosan enters into the microbial interfering with the metabolism of *E. coli*. Other less probable mechanisms are based on the chelation of metals (Cuero et al., 1991), as well as on the inhibition of the mRNA and protein synthesis because the chitosan oligomers are able to pass through the bacterial-cell-wall-binding microbial DNA (Liu et al., 2004).

It has been reported that antimicrobial activity is dependent on pH, DA degree, and MW (Devlieghere et al., 2004; Liu et al., 2004). In general, higher activity was found at lower pH value (No et al., 2002). High DA leads to high antimicrobial activity (Benhabiles et al., 2012). Relation between antimicrobial activity and MW is not always lineal. In general, it is accepted that high MW decreases their antimicrobial activity but some exceptions are reported. Thus, chitosan has more antibacterial activity than chitosan oligomers although inhibitory effects differ with MWs of chitosan and of bacteria (No et al., 2002). In the same way for *S. aureus*, a Gram-positive bacteria, as the MW of chitosan increased, the antimicrobial effect was enhanced while the effect was inverse for *E. coli* a Gram-negative bacteria (Zheng and Zhu, 2003).

Solubility also affects to antimicrobial activity of chitosan. Benhabiles et al. (2012) observed high activity for water-soluble chitosan oligomers. However, Qin et al. (2006) reported that the water-soluble half-acetylated chitosan and chitosan oligomers had no significant antimicrobial activity and, even, promoted the growth of *Candida albicans*. On the contrary, the same authors observed inhibitory effect against *S. aureus, E. coli*, and *C. albicans* for water-insoluble chitosan in acidic medium.

8.3.2 Methods to Prepare Antimicrobial Edible Films and Coatings for Food Active Packaging Purposes

Owing to their film-forming properties, chitosan can be used to prepare antimicrobial edible films and coatings. Solution casting is one of the most common methods to prepare edible films. Coatings made with low- and high-MW chitosan dissolved in lactic or acetic acids were effective against *L. monocytogenes* on the surface of ready-to-eat roast beef (Beverlya et al., 2008). In the same way, both low- and medium-MW chitosan exhibited highest antimicrobial activity than commercial mixtures of organic acids used as coatings for meat products (Cruz Romero et al., 2013). Contrary, when chitosan with high DA and low MW was added to the culture media, *L. monocytogenes* and different lactic acid bacteria were less susceptible than *Brochothrix thermosphacta* and *Ba. cereus* (Devlieghere et al., 2004).

The coatings with chitosan homopolymer inhibited the growth of microorganisms as well as retarded the drop in sensory quality, thus extending the shelf life of sliced mango (Chien et al., 2007), strawberry (Hernández Muñoz et al., 2008), and blueberry (Duan et al., 2011) fruits. A chitosan coating applied to minimally processed broccoli was able to inhibit growth of total coliforms, psychrotrophic and mesophilic aerobes during storage, as well as avoid the risk of an accidental contamination with *E. coli* (even the O157:H7 strain). Broccoli samples coated with chitosan-based coating kept the same appearance than uncoated control samples (Moreira et al., 2011). In nonvegetal products, it was reported the use of higher DA chitosan coatings to inhibit growth of spoilage bacteria of filleted Indian oil sardine (*Sardinella longiceps*) during chilled storage. Additionally, chitosan coating improved the quality and extended the shelf life of sardine due to its effectiveness to reduce the formation of volatile bases and oxidation products, as well as improve the water-holding capacity, drip loss, and textural properties (Mohan et al., 2012). In Emmental cheese, the use of chitosan films showed a strong (10 times) bactericidal effect against *L. innocua* comparing with nontreated cheese, and no colonies were observed after 5 days of storage at 37°C (Coma et al., 2002).

8.3.3 Improving the Performance of Chitosan Edible Films and Coatings in Food Active Packaging

Two strategies can be used to improve the performance of chitosan films and coatings to be applied to preserve foods. The first is to improve the physical properties of the coatings, particularly those relating with the WVP, flexibility, and stretchability. Considering the stress during the shipping and handling of the food, good mechanical properties (i.e., high tensile strength) are generally required for edible active packaging. The second strategy is to enhance the antimicrobial activity of chitosan-based films by adding other substances to the film, with antibacterial and/or antifungal activities.

Blends of chitosan with other molecules such as hydrocolloids (proteins: milk proteins, soy protein, collagen, and gelatin and polysaccharides: starch, alginate, and cellulose) or lipids (fatty acids, acylglycerols, or waxes) are used to improve the mechanical and physical properties of these bio-based films (Elsabee and Abdou, 2013).

8.3.4 Improving the Physical Properties of Chitosan Films by Blending

Properties of chitosan–starch films are different as function of type and origin of starch. It was described blends using Hylon VII solutions plasticized with glycerol or erythritol (Cervera et al., 2004), cassava starch dispersion containing glycerol (Bangyekan et al., 2006), and tapioca starch (Chillo et al., 2008; Vásconez et al., 2009). In general, the addition of chitosan improved the physical properties of starch films, reduce WVP and solubility of starch films, increase the gloss values and transparency, and decrease the wettability of the coated blends (Elsabee and Abdou, 2013).

Mechanical properties are also improved when starch is blended with chitosan. Therefore, chitosan improves the tensile stress at maximum load and tensile modulus of coated starch films (Bangyekan et al., 2006) and chitosan tapioca starch films (Chillo et al., 2008). Film flexibility and stretchability must be estimated by measure of elongation at the break value in each case. Not always an increase in chitosan concentrations in the blend improves these parameters. Thus, elongation at the break value decreases in blends rice starch/chitosan when the ratio decreases (Bourtoom and Chinnan, 2008). On the contrary, the addition of chitosan significantly improved both tensile strength and elongation at the break of sweet potato starch film (Shen et al., 2010). In addition, using water chestnut starch–chitosan blends, good mechanical properties were obtained and when the films incorporated a fruit extract and nisin were able to reduce the number of *E. coli* O157:H7, *St. aureus,* and *L. monocytogenes* (Mei et al., 2013). Antibacterial activity of starch chitosan films can be modified by changing their structure through physical procedures. Thus, it was reported that irradiation enhances the antimicrobial properties of starch/chitosan blends due to chitosan degradation (Zhai et al., 2004).

Physical and mechanical properties of starch blend films improved with the incorporation of chitosan. However, the antimicrobial performance of coatings chitosan–starch films not always provides better results than chitosan alone. For example, Vásconez et al. (2009) reported that in salmon slides, chitosan–tapioca starch-based coatings were less effective to reduce aerobic mesophilic and psychrophilic cell count than the chitosan-based coating alone. In spite of this, starch–chitosan blends showed antimicrobial activity in the film form (Tripathi et al., 2008) and have proved to be effective for control of spoilage microbiota in other minimally processed vegetables such as carrot (Durango et al., 2006).

The ratio of starch/chitosan in the blend affects the mechanical and antimicrobial properties of the films. Bonilla et al. (2013) showed that the mechanical properties improved as the chitosan ratio increased in the films, as well as chitosan seems to inhibit starch retrogradation. Nevertheless, when the blends were used to coat minced pork samples, a degree of substitution with starch below 50% was necessary to obtain antimicrobial activity equivalent to chitosan film alone.

Blends of chitosan with other polysaccharides such as pectins, alginates, and carrageenans have been scarcely explored. The coatings with the two last polymers enable in reducing moisture loss in fruits and vegetables, and in combination with food grade antimicrobials such as chitosan, it also prevents the microbial spoilage during storage (Elsabe and Abdou, 2013).

Chitosan–protein blends have also been studied, being the most used in blends: casein, soy protein, zein, gluten, keratin, and albumin (de Souza et al., 2010). Addition of fish gelatin to chitosan enhanced mechanical properties of the films when compared with the properties of the film of both polymers separately. Thus, composites gelatin–chitosan allowed to obtain very transparent and stronger films with increased tensile strength and elastic modulus, as well as improved barrier properties against UV light and reduced WVP (Hosseini et al., 2013).

Comparing with other filmogenic compounds, zein has some advantages: good oxygen and water vapor barrier, as well as low water solubility. These properties were used to improve the physical properties of chitosan, yielding blends rougher, more elastic, and less hard film structures (Escamilla García et al., 2013). Other proteins such as quinoa allow making edible films with chitosan with good gas barrier properties without using a plasticizer (Abugoch et al., 2011). It was recently reported that the water vapor barrier of this blend can be performed adding sunflower oil (Valenzuela et al., 2013).

In fact, addition of lipids to chitosan-based films or their blends is a way to improve the water barrier properties of the films. Edible coatings based on high MW chitosan combined with sunflower oil were applied to the surface of pork meat hamburgers, reducing the microbial counts during storage (Vargas et al., 2011). In addition, coatings with high-MW chitosan and oleic acid were used to preserve quality of cold-stored strawberries (Vargas et al., 2006). This fruit was also treated with composites beeswax–chitosan-based coatings, showing to be effective to avoid fungal infection and to extend the shelf life improving the main quality parameters (Velickova et al., 2013).

More complex mixtures such as formulations containing chitosan, beeswax, and lime essential oil were successfully tested by Ramos García et al. (2012) to control *Rhizopus stolonifer* and *E. coli* DH5a during the storage of tomatoes. No growth of both microorganisms was observed and a synergistic effect greater than the sum of the effects of applying individual alternatives was suggested by the authors.

8.3.5 Use of Chitosan Films to Deliver Antimicrobial Compounds

Finally, it must be pointed that, due to their properties as carrier, chitosan coatings and films can be used for edible antimicrobial active packaging purposes. Thus, adding antimicrobial compounds that are released from the matrix to the food, the antimicrobial properties of chitosan films can be enhanced. In addition, this strategy has other advantages such as the use of small antimicrobial concentrations and low diffusion rates.

Essential oils have been widely used as antimicrobials in different formulations of chitosan-based edible coatings. Avila Sosa et al. (2012) reported the incorporation of Mexican oregano or cinnamon essential oils to chitosan edible films. These films were able to inhibit *A. niger* and *Penicillium digitatum* growth by vapor contact. In these experiments, since chitosan edible films improved the release of the volatile compounds of essential oils, it was possible to reduce essential oil concentrations required to obtain high antimicrobial activity. In the same way, antimicrobial as well as antioxidant properties of starch–chitosan films were improved by addition of *Thymus kotschyanus* essential oil (Mehdizadeh et al., 2012).

The botanical origin of essential oil used in the formulations has an important effect on the antimicrobial properties of the edible film, even when the same genus employed. For example, incorporation of *Thymus piperella* and *Thymus moroderi* essential oils to chitosan edible films yielded different activity against some important pathogens. Films containing *T. piperella* essential oil were more effective against *Serratia marcescens* and *L. innocua* than those containing *T. moroderi* essential oil (Ruiz-Navajas et al., 2013).

Lime and thyme essential oils were added to a blend containing chitosan, beeswax, and oleic acid. These coatings have rendered a synergistic effect on the control of pathogens during the storage of tomatoes. Thus, the antimicrobial effect of the mixtures of lime oil with chitosan films was greater than the sum of the effects of applying individual alternatives (Ramos-García et al., 2012).

A limitation of the use of essential oils in films made with chitosan is their strong taste that could give a particular flavor and aroma to the food. For this reason, complementary studies relating the customers acceptability of the coated food with chitosan essential oil blends should be addressed. Thus, it has been reported that some solutions extending the shelf life could be interesting in terms of taste such as sweet pepper with cinnamon (Xing et al., 2011), grapes with oregano or fresh-cut pears with rosemary (Xiao et al., 2010), dried fish (*Decapterus maruadsi*) with cinnamon, clove, anise, turmeric, guava leaf, nutmeg, and lime oils (Matan, 2012).

No common essential oils have been also assayed to enhance antimicrobial properties of chitosan-based coatings. *Lippia gracilis* Schauer essential oil (a Brazilian plant with essential oil containing thymol and carvacrol as antimicrobials) was added to a cassava starch—chitosan coating and showed to be able to keep below the maximum limit recommended for total psychrophilic aerobic bacteria, yeast, and mold counts in strawberries during a week of storage under refrigeration (Azevedo et al., 2014).

In summary, chitosan is a suitable edible active packaging material because it is affordable, reliable, natural, nontoxic, and biodegradable. Their mechanical and antimicrobial properties can be improved by blending with other polymers or adding antimicrobial compounds that are delivered from the matrix. For these reasons, chitosan-based coatings are employed in a variety of applications in the food industry.

8.4 FOOD APPLICATIONS OF ANTIMICROBIAL EDIBLE FILMS AND COATINGS

8.4.1 Antimicrobial Edible Films and Coatings for Meat, Poultry, and Fish

Typically animals act as healthy carriers of pathogens and then these microorganisms are transferred to humans through production, handling, and consumption (Nørrung and Buncic, 2008). Major causes of concern associated with fresh meat products are *Campylobacter*, *E. coli* O157:H7, and related enteric pathogens such as *Salmonella*, while *L. monocytogenes* is more associated to ready-to-eat meat and poultry products (Sofos, 2008). On the contrary, pathogens responsible for fish and seafood infections include in addition to *Salmonella*, *Clostridium botulinum*, *S. aureus*, *Clostridium perfringens*, *B. cereus*, and species of the genus *Vibrio* and *Shigella*.

All the above-mentioned bacteria are found in the group of pathogens traditionally associated to meat products also including *Yersinia enterocolitica* and *Cl. botulinum* (Mor-Mur and Yuste, 2010). However, current research focusses on emerging (including new, emerging, and evolving) bacterial pathogens in meat, poultry, and derived products (Mor-Mur and Yuste, 2010; Sofos, 2008). These emerging microorganisms include *C. jejuni*, *Salmonella typhimurium* DT104, *E. coli* O157:H7, and other enterohemorrhagic *E. coli* (EHEC), *L. monocytogenes*, *Arcobacter butzleri*, *Mycobacterium avium* ssp. *paratuberculosis*, and *Aeromonas hydrophila* (Mor-Mur and Yuste, 2010). In addition, seafood has been identified as an important source of emerging food-borne diseases including bacteriological etiological agents and parasitic infections such as toxoplasmosis, giardiosis, and cysticercosis (Broglia and Kapel, 2011; Nawa et al., 2005).

The overall trend of food consumption habits goes now toward more fresh, natural (with less or none additives), locally produced, and/or ready-to-eat quality foods but, at the same time, with an extended shelf life. The problem of fresh-consumed meat or fish is that it cannot be subjected to the usual antimicrobial treatments of the food industry such as thermal processing, freezing, and irradiation. So, to meet all consumers' demands without compromising safety, in the last decades alternative nonthermal preservation technologies such as high hydrostatic pressure, irradiation, light pulses, natural biopreservatives, and active packaging have been investigated (Aymerich et al., 2008). Treatments commonly used to preserve ready-to-eat meats include the application of vacuum packaging or modified atmospheres. However, these conservation methods are not entirely effective because of the risk of development of microaerophilic and/or psychrophilic microorganisms. Although few, there are pathogens belonging to this last group such as *L. monocytogenes* and *Y. enterocolitica*. In addition, *C. jejuni* is a microaerophilic microorganism able to grow in environments of reduced oxygen content (2%–10%) and so capable of tolerating modified atmospheres better than other microorganisms such as *Pseudomonas* (Coma, 2008).

As a result of the above, research on the use of active packaging systems for meat and meat products has increased in recent years. Active packaging consists on the incorporation of additives into packaging systems with the aim of maintaining or extending meat product quality and shelf life. In regard to the latter, these materials must be able to delay or inhibit microbial growth. To confer these properties, antimicrobial agents might be coated, incorporated, immobilized, or surface modified onto package materials (Suppakul et al., 2003). In addition to the direct incorporation, the antimicrobial can be released from a carrier matrix covering the package or from a pouch that is attached to the packaging and from which the active compounds are released (Coma, 2008). Recently, the utilization of inherently antimicrobial polymers exhibiting film-forming properties or polymers which are chemically modified to produce bioactive properties (Coma, 2008) has also been addressed as an alternative or complementary strategy to the utilization of exogenous antimicrobial compounds. Finally, due to environmental concerns about the biodegradability of some synthetic polymers, problems of storage coupled with the development of new market opportunities for underutilized agricultural products with film-forming properties, there is a growing interest in edible coatings with antimicrobial activity (Quintavalla and Vicini, 2002).

According to Gennadios et al. (1997), there are several potential benefits on the use of edible coatings in the meat, poultry, and fisheries industries. These advantages include the prevention of moisture loss, reducing lipid oxidation, and discoloration and enhancement of product appearance. But edible films can also act as carriers for different food additives, such antioxidants, antimicrobials, antifungal agents, colorants, and other nutrients (Han, 2000). However, many factors should be considered when designing antimicrobial packaging systems. Among them, the selection of active compounds should be carefully addressed, since these additives can modify coating functionality depending on its concentration, stability, chemical structure, degree of dispersion, and degree of interaction with the polymer (Suppakul et al., 2003). Besides, the choice of active agents must be limited to food-grade substances, since they have to be consumed along with the edible films or coatings (Cerqueira et al., 2011). Enzymes, natural bacteriocins or essential oils, and synthetic antimicrobial agents (e.g., quaternary ammonium salts, EDTA, propionic, benzoic, and sorbic acids) are among the antimicrobials commonly used in active packaging (Suppakul et al., 2003). As follows, a number of applications of antimicrobial packaging systems will be reviewed according to the main material utilized for their preparation.

8.4.1.1 Films with Chitosan

Pranoto et al. (2005) reported that common contaminant bacteria of meat products, such as *E. coli*, *S. aureus*, *S. typhimurium*, *L. monocytogenes,* and *B. cereus*, were inhibited by chitosan edible films incorporating garlic oil, potassium sorbate, or nisin. However, Gram positive are more sensitive than Gram-negative bacteria due to the protective effect of cell wall (i.e., lipopolysaccharide in the latter group of bacteria) (Russel, 1991). This enhanced antimicrobial activity against Gram-positive bacteria, such as *S. aureus* and *B. cereus*, was observed by Kanatt et al. (2008) in minced lamb meat treated with mint chitosan films. These authors reported a less pronounced effect of these films against *E. coli*, *Pseudomonas fluorescens*, and *S. typhimurium* growth. *L. monocytogenes* is among the most studied bacterial pathogens in meat products, which inhibition in meat treated with chitosan films incorporating different antimicrobials such as organic acids (Beverlya et al., 2008), essential oils (Moradi et al., 2011), and other bioactive compounds (Vodnar, 2012) has been described in the literature (Table 8.1).

Chitosan has also been studied in combination with other biopolymers. Moreira et al. (2011) studied the effectiveness of bioactive packaging materials composed of chitosan and sodium caseinate in salami and other foods (carrots and cheese). This combination was intended to take advantage of the intrinsic antimicrobial properties of chitosan and the good thermoplastic and film-forming properties of sodium caseinate. The results of this study showed a significant bactericidal effect on mesophilic, psychrotrophic, yeasts, and mold counts in salami samples of both chitosan and chitosan/sodium caseinate, although an improvement of this activity was observed in blended films. In addition, chitosan–gelatin (1:1 ratio) films have shown improved bacteriostatic effect against *B. cereus*, *S. aureus*, and *E. coli* in different meat products (veal and rabbit, boiled sausages, smoked sausages, and smoked–boiled pork brisket) in contrast to coatings developed with only one of these structural components (Baranenko et al., 2013).

8.4.1.2 Films with Pectin

Other very interesting polysaccharide with good film-forming properties is pectin, a naturally occurring, hydrophilic polymer found in plant cell walls. Recently, the antimicrobial and physical–mechanical properties of different edible composite films containing pectin alone or in combination with fruit puree and essential oils have been studied (Otoni et al., 2014). In this paper, papaya puree and cinnamaldehyde were used to adequately plasticize pectin films, reducing their rigidity while increasing their extensibility, and also to optimize the water barrier properties of pectin films. Pectin (apple)-based edible films containing plant antimicrobials (cinnamaldehyde or carvacrol) were effective against *Salmonella enterica* or *E. coli* O157:H7 on chicken breasts and *L. monocytogenes* on ham after storage under refrigerated or ambient temperature conditions (Ravishankar et al., 2009). Other polysaccharide-based films that have been applied in meat and fish products, as microbial growth inhibitors include agarose from the algae *Gelidium corneum*

TABLE 8.1
Polysaccharide Edible Films/Coating for Meat, Poultry, and Fish Products

Film/Coating	Antimicrobial Compounds	Meat Product	Antimicrobial Activity	References
Chitosan	Mint oil	Minced lamb meat	*S. aureus*/*B. cereus*	Kanatt et al. (2008)
	Zataria multiflora Boiss essential oil and grape seed extract	Mortadella sausages	*L. monocytogenes*	Moradi et al. (2011)
	Green and clack tea bioactive compounds	Vacuum-packed ham steak	*L. monocytogenes*	Vodnar (2012)
	Acetic/lactic acid	Ready-to-eat roasted beef	*L. monocytogenes*	Beverlya et al. (2008)
Chitosan/chitosan–sodium caseinate	None	Salami	Native microflora	Moreira et al. (2011)
Chitosan–gelatin	Organic acids	Veal and rabbit, boiled sausages, smoked sausages, and smoked–boiled pork brisket	*B. cereus, S. aureus,* and *E. coli*	Baranenko et al. (2013)
Pectin	Cinnamaldehyde or carvacrol	Ham and chicken breasts	*S. enterica*/*E. coli* O157:H7/ *L. monocytogenes*	Ravishankar et al. (2009)
Agarose	Carvacrol	Ham	*E. coli* O157:H7 and *L. monocytogenes*	Lim et al. (2010)
	Catechin	Sausages	*E. coli* O157:H7 and *L. monocytogenes*	Ku et al. (2008)
Mucilage	Oregano/thyme essential oil	Rainbow trout fillets	*Pseudomonas* spp., Enterobacteriaceae, and lactic acid bacteria	Jouki et al. (2014)
κ-Carrageenan	Ovotransferrin and EDTA	Chicken breast	*E. coli*	Seol et al. (2009)

(Ku et al., 2008; Lim et al., 2010) mucilage from quince seeds (Jouki et al., 2014) and κ-carrageenan (Seol et al., 2009).

8.4.1.3 Films with Proteins

Films prepared with animal and plant proteins, such as collagen, gelatin, milk proteins, wheat gluten, soy protein, and corn zein, have been recently reviewed (Jooyandeh, 2011; Kushwaha and Kawtikwar, 2013; Phadke et al., 2011; Ramos et al., 2012).

In general, protein films provide mechanical stability when compared with polysaccharide or lipid components. Preparation of some protein films occurs by heating which causes that intramolecular and intermolecular disulfide bonds are cleaved and reduced to sulfhydryl groups during denaturation (Okamoto, 1978). When casting the film-forming solution, disulfide bonds are newly formed joining the polypeptide chains together to produce the film structure and with the aid of hydrogen bonding and hydrophobic interactions (Cagri et al., 2004). Improved films can be obtained by adjusting coating solution pH in relation to the protein isoelectric point, where proteins become least soluble.

As summarized in Table 8.2, protein films incorporating different antimicrobial compounds have been applied to different meat and fish products. Nevertheless, a problem associated to protein-based coatings applied to raw meat, poultry, and seafood is the susceptibility of these proteins to proteolytic enzymes present in these foods. In addition, the potential allergenic effects of some of these proteins, including milk, egg white, peanuts, soybeans, rice, and wheat gluten must be considered in terms of consumer acceptance and food labeling requirements (Gennadios et al., 1997).

8.4.2 Antimicrobial Edible Films and Coatings for Fruits and Vegetables

Fresh and minimally processed fruits and vegetables are highly perishable food products. From the harvest until the moment they are consumed, fruits and vegetables undergo intense changes, mainly due to their intense physiological postharvest activity, their high water content (80%–90% w/w) and the frequent contamination with spoilage microorganisms that may grow during the storage (Campos et al., 2013).

TABLE 8.2
Protein Edible Films/Coating for Meat, Poultry, and Fish Products

Film/ Coating	Antimicrobial Compounds	Meat Product	Antimicrobial Activity	References
Cellulose	Nisin	Vacuum-packaged frankfurters	*L. monocytogenes*	Nguyen et al. (2008)
Sodium caseinate	*Lactobacillus sakei*	Beef	*L. monocytogenes*	Gialamas et al. (2010)
Whey protein isolate	Oregano oil	Beef	Total flora and *Pseudomonas*	Zinoviadou et al. (2009)
	Nisin		*B. thermosphacta*	Rossi-Márquez et al. (2009)
Soy protein	Thyme and oregano oil	Ground beef patties	*E. coli* O157:H7 and *S. aureus*	Emiroglu et al. (2010)
Gelatin–chitosan	Clove essential oil	Dolphinfish homogenate	Enterobacteria	Gómez-Estaca et al. (2010)
Sunflower protein	Clove essential oil	Sardine patties	Total mesophiles	Salgado et al. (2013)

During the harvesting, and the subsequent preparation for marketing and consumption, significant losses in quality occur as a consequence of the high respiration rate and loss of water that leads fruits and vegetables to an accelerated senescence. During this period, even minimal processing operations such as washing, sorting, or trimming may negatively impact the product quality. Especially in the case of fresh-cut fruits and vegetables, operations such as peeling, slicing, or coring promote an intense water evaporation, cell tissue disruption and membrane collapse, resulting in a reduced product shelf life (Campos et al., 2011; Dhall, 2013; Peretto et al., 2014a; Rojas-Graü et al., 2009). The epidermis represents a physical and chemical barrier. Therefore, when the cuticle is removed, during handling and processing operations, fruits and vegetables lose their natural protection. As a consequence, cross contamination from the product surface may favor the growth of spoilage and pathogen microorganisms. Thus, causing adverse effects on product quality such as browning or discoloration, off-flavors development, and texture changes (Martín-Belloso and Fortuny, 2010; Rojas-Graü et al., 2009). In addition, it may also increase the risk of outbreaks caused by food-borne pathogens such as *Salmonella* spp., different strains of *E. coli* (mainly serotype O157:H7, but also O104:H4), and *L. monocytogenes* (Lynch et al., 2009; Sant'Ana et al., 2014; Soon et al., 2013).

Therefore, due to their intrinsic perishability, fresh, and minimally processed fruits and vegetables require the application of adequate techniques to extend their shelf life (Ponce et al., 2008). For this purpose, three are critical parameters (1) to reduce the water loss, (2) to delay maturation and senescence, and (3) to reduce the microbial growth (Erbil and Muftugil, 1986).

Edible films and coatings are a useful approach to enhance the food quality, safety, and stability of fresh fruits and vegetables. They can act as a barrier for moisture, gas, and solutes reducing the weight loss, the respiration rates, and oxidative reactions, and also the occurrence of physiological disorders, leading to a delay in the senescence (Peretto et al., 2014a; Rojas-Graü et al., 2009; Valencia-Chamorro et al., 2011).

Different types of edible films and coatings have been used for fresh fruits and vegetables such as apple, grapes, melon, papaya, strawberry, bamboo shoots, broccoli, garlic, lettuce, tomato, or asparagus (Table 8.3). The most commonly used materials are hydrocolloids (including polysaccharides and proteins), lipids (including waxes, acylglycerols, and fatty acids), and composites made with mixtures of both hydrocolloids and lipids. Nevertheless, fruit-based coatings, prepared from materials such as tomato or apple puree, have also been used (Valencia-Chamorro et al., 2011). In some cases, plasticizers (e.g., sucrose, glycerol, sorbitol, propylene glycol, polyethylene glycol, fatty acids, and monoglycerides) and emulsifiers (e.g., fatty acids, ethylene glycol monostearate, glycerol monostearate, esters of fatty acids, lecithin, sucrose ester, and sorbitan monostearate or polysorbates) are also added to the mixture to improve the flexibility and reduce brittleness (Badwaik et al., 2014). Film solutions made with these materials can be applied to fruits and vegetables by several methods such as dipping, spraying, brushing, and panning to form the protective barrier (Dhall, 2013). In addition, edible coatings may also improve the appearance of fresh fruits and vegetables and, since they can partially replace plastic packaging

TABLE 8.3
Antimicrobial Edible Films and Coatings Used in Fresh and Minimally Processed Foods and Vegetables

Coating Material	Product (Fresh Fruit or Vegetable)	Antimicrobial Compound	Target Microorganism	References
Polysaccharides				
Alginate	Apple	Malic acid, cinnamon, clove, lemongrass essential oils and their active compounds, cinnamaldehyde, eugenol, and citral	*E. coli* O157:H7	Raybaudi-Massilia et al. (2008)
	Strawberries	Carvacrol and methyl cinnamate	*E. coli* O157:H7 and *Botrytis cinerea*	Peretto et al. (2014a)
Carboxymethyl cellulose	Fresh pistachios	Potassium sorbate	Mycotoxigenic *Aspergillus* species	Sayanjali et al. (2011)
Chitosan	Broccoli	Chitosan, natural extracts (tea tree, pollen and propolis, pomegranate, and resveratrol)	*E. coli*, *L. monocytogenes*, total mesophilic, and psychrotrophic counts	Alvarez et al. (2013)
	Broccoli	Chitosan	*E. coli* O157:H7 and native microflora (mesophilic, psychrotrophic, yeast, molds, lactic acid bacteria, and coliforms)	Moreira et al. (2011)
	Papaya	Chitosan, peppermint essential oil	Fungi	Picard et al. (2013)
	Papaya	Chitosan	Fungi	González-Aguilar et al. (2009)
	Strawberries	Chitosan	Fungi	Han et al. (2006)
	Strawberries	Chitosan	Total psychrotrophic and mesophilic bacteria	Campaniello et al. (2008)
	Strawberries, cucumber, and bell pepper	Chitosan	Fungi	Ghaouth et al. (1991a, 1991b)

(Continued)

TABLE 8.3 (Continued)
Antimicrobial Edible Films and Coatings Used in Fresh and Minimally Processed Foods and Vegetables

Coating Material	Product (Fresh Fruit or Vegetable)	Antimicrobial Compound	Target Microorganism	References
Pullulan	Apple	Meadowsweet flower extracts	*Rhizopus arrhizus*	Gniewosz et al. (2014)
	Baby carrot	Caraway essential oil	*Salmonella enteritidis*, *S. aureus*, *S. cerevisiae*, and *A. niger*	Gniewosz et al. (2013)
	Pepper and apple	*Satureja hortensis* extract	*S. aureus*, *Bacillus subtilis*, *S. enteritidis*, *E. coli*, *A. niger*, *P. expansum*, and *R. arrhizus*	Kraśniewska et al. (2014)
Starch	Strawberries	Potassium sorbate	Total counts of mesophilic aerobes, molds, and yeasts	García et al. (2001)
	Tomatoes	Coconut oil and tea leaf extract	Total microbial counts	Das et al. (2013)
Proteins				
γ-Irradiated milk proteins	Strawberries	*Quillaja saponaria* extract	Fungi	Zúñiga et al. (2012)
Fruit or Vegetable-Based Films				
Aloe vera gel	Grapes	*Aloe vera*	Total counts of mesophilic aerobic bacteria, yeasts, and molds	Valverde et al. (2005)
	Papaya	*Aloe vera*, papaya leaf	Fungi	Marpudi et al. (2011)
	Leafy greens (organic romaine and iceberg lettuce, and mature and baby spinach)	Carvacrol and cinnamaldehyde	*S. enterica* serovar Newport	Zhu et al. (2014)
Apple, carrot, and hibiscus puree				
Strawberry puree	Strawberries	Carvacrol and methyl cinnamate	*Trichoderma* sp., *Cladosporium silenes*, and *B. cinerea*	Peretto et al. (2014b)

(*Continued*)

TABLE 8.3 (Continued)
Antimicrobial Edible Films and Coatings Used in Fresh and Minimally Processed Foods and Vegetables

Coating Material	Product (Fresh Fruit or Vegetable)	Antimicrobial Compound	Target Microorganism	References
		Composite Films		
Agar-agar/chitosan	Garlic	Chitosan, acetic acid	Filamentous fungi and aerobic mesophilic bacteria	Geraldine et al. (2008)
Alginate/apple puree	Apples	Vanillin, lemongrass oil, and oregano oil	*L. innocua*, total psychrophilic aerobes, molds, and yeasts	Rojas-Graü et al. (2007b)
Alginate/starch/ carboxymethyl cellulose	Bamboo shoots	Extracts from a lactobacillus fermentation	*E. coli*, *S. aureus*, and *B. cereus*	Badwaik et al. (2014)
Banana flour/ chitosan	Asparagus, baby corn, and Chinese cabbage	Chitosan	*S. aureus*	Pitak and Rakshit (2011)
Carnauba wax/ dextrin	Mango	*Lippia sidoides* and *Piper aduncum* oils	*Lasiodiplodia theobromae*, *Botryosphaeria dothidea*	de Menezes Cruz et al. (2012)
Chitosan/pectin	Melon	*trans*-cinnamaldehyde	Total aerobic counts, psychrotrophic microorganisms, yeasts, and molds	Fagundes et al. (2013)
Hydroxypropyl methylcellulose/ lipid	Cherry tomatoes	Potassium carbonate, ammonium phosphate, potassium bicarbonate, ammonium carbonate, sodium methylparaben, sodium ethylparaben, and sodium propylparaben	*B. cinerea* and *Alternaria alternata*	Fagundes et al. (2013)
Rapeseed protein/ gelatin	Strawberries	Grapefruit seed extract	*E. coli* O157:H7, *L. monocytogenes*	Jang et al. (2011)
Tapioca starch/ decolorized hsian-tsao leaf gum	Lettuce hearts, fruit-based salads	Green tea extracts	*S. aureus*, *B. cereus*, *L. monocytogenes*, total yeasts, and molds	Chiu and Lai (2010)
Xanthan gum/ chitosan/guar gum	Papaya	Chitosan	Total psychrophilic, *Salmonella* spp., total coliforms, and thermotolerants	Cortez-Vega et al. (2013)

and can be consumed together with the food, also contribute to reduce the waste disposal (Dhall, 2013).

Regarding the use of antimicrobials, they were traditionally applied as dipping aqueous solutions. However, their direct application on the food surface may reduce the antimicrobial activity, as the active substances can be quickly neutralized, washed, or diffused into the food, thus reducing their concentration and ability to protect against spoilage and pathogenic bacteria (Dhall, 2013; Kraśniewska and Gniewosz, 2012). The incorporation of antimicrobials into edible films and coatings provides a slow release rate that maintains an effective concentration of the antimicrobial compounds on the food surfaces during extended storage periods (Kraśniewska and Gniewosz, 2012).

Antimicrobials that have been successfully used in edible films for fresh fruits and vegetables include salts (e.g., sodium bicarbonate, nitrates, and nitrites), organic acids (e.g., acetic, benzoic, lactic, propionic, and sorbic), fatty acid esters (e.g., glyceryl monolaurate), enzymes (e.g., lysozyme, peroxidase, and lactoferrin), and bioactive peptides (e.g., nisin, pediocin, and sakacin A) (Dhall, 2013; Kraśniewska and Gniewosz, 2012). Nevertheless, due to the consumers' demand for natural products, in the last few years most researches were focused on the use of plant essential oils such as garlic, oregano, cinnamon, and lemongrass, among others (Burt, 2004; Rojas-Graü et al., 2009) (Table 8.3).

On the contrary, some film-forming materials such as chitosan and *Aloe vera* gels have been reported to have antimicrobial activity themselves, with effectiveness against different bacteria and fungi. Therefore, they have been used to preserve the quality and extend the shelf life of a variety of fresh fruits (melon, grapes, and papaya) and vegetables (broccoli, asparagus, corn, cabbage, and garlic) (Geraldine et al., 2008; Marpudi et al., 2011; Martiñon et al., 2014; Moreira et al., 2011; Valverde et al., 2005).

Despite the promising results obtained so far, some drawbacks have also arisen, and multiple factors must be considered in the use of antimicrobial edible coatings for fruits and vegetables. For instance, in some cases, the addition of antimicrobial compounds may have an adverse effect on the barrier, mechanical, and optical properties of the coatings. Particularly, essential oils have a strong flavor that may change the original flavor of fruits and vegetables, which affects the product acceptability (Dhall, 2013; Peretto et al., 2014a; Valencia-Chamorro et al., 2011). For this reason, to obtain commercial applications, further studies must be conducted. Each particular combination of fruit or vegetable, coating material and antimicrobial compound should be tested to assess their inhibitory activity against the target pathogens or spoilage microorganisms and also to evaluate the effects that the antimicrobials have on the physical properties of the coating and the organoleptic characteristics of the product.

8.4.3 Antimicrobial Edible Films and Coatings for Cheese and Other Foods

Cheese is one of the ready-to-eat products considered as "potentially dangerous" (Cao-Hoang et al., 2010). The European Food Safety Authority (EFSA, 2008) noted that in 2006, the consumption of contaminated cheese accounted for 0.4% of the total food-borne outbreaks in Europe (Kousta et al., 2010).

In the main food-borne outbreaks caused by the consumption of cheeses, were contaminated by *L. monocytogenes*, *Salmonella* spp., *E. coli* O157:H7, and *S. aureus*, with more than 4600 cases and over 110 deaths, around the world (Kousta et al., 2010). *L. monocytogenes* represents the highest risk due to both its higher severity and the probability of the *Listeria* development is relatively high compared to other food-borne diseases associated with cheese (Cao-Hoang et al., 2010).

Contamination of cheese during its production can occur in several stages from raw milk, surfaces and equipment, and also from workers at the processing plant (Kousta et al., 2010). The risk of pathogenic bacteria contamination during cheese processing can be removed by pasteurization of raw milk, the storage temperature control, and the length of maturation, together with the intrinsic cheese properties such as pH, water activity, and presence of antimicrobial compounds produced by starter culture (Kousta et al., 2009).

However, the control once the product leaves the factory is much more difficult. At this point, the type of cheese packaging or coating becomes important. Many authors suggest the use of modified atmospheres with N_2 and/or CO_2, by reducing the O_2 within the package. However, pollution caused by fungi and bacteria can occur even at very low levels of O_2 and high levels of CO_2. Thus, depending on the type of cheese it is necessary to design the type of packaging whose permeability allows an adequate balance of O_2/CO_2, as during cheese ripening water activity decreases until the surface is in equilibrium with the surrounding atmosphere, which will affect the chemical and microbiological evolution of the cheese (Cerqueira et al., 2010; Saurel et al., 2004). Additionally, environmental factors (e.g., light, relative humidity, and temperature) should also be taken into account, keeping the package in good condition without affecting the organoleptic characteristics of the cheese to extend their storage time (Cerqueira et al., 2009, 2010). Cerqueira et al. (2009) listed the criteria to follow for choosing a suitable coating for cheese:

- *Wettability*: This is one of the most important properties when the ability of a solution to be used as coating is evaluated. The closer to zero the values of wettability are, the better a surface will be coated.
- *Gas transport properties*:
  - WVP is the most studied property of edible films, mainly due to the importance of water in spoilage reactions. Thus, low WVP values are optimal to reduce water loss during storage of the cheese.
  - O_2 permeability (O_2P) is the key factor in cheese preservation. Edible films that provide an adequate O_2 barrier can help improve food quality and extend its shelf life. Thus, low O_2P values prevent fat oxidation and reduce cheese spoilage with undesirable microorganisms.
  - CO_2 permeability (CO_2P) should be high to increase the shelf life of cheese by increasing the lag phase for the growth of coliform and other contaminating Gram-negative bacteria, yeast, and mold.
- *Opacity*: Reducing the incidence of light on the cheese will yield the reduction in their oxidation.

The increasing consumer's demand for fresh food with high quality and free of artificial preservatives (Kuorwel et al., 2011), together with the increased consumer's concern for environmental protection made of biodegradable coatings an excellent alternative to the packaging materials used in cheese so far (Fajardo et al., 2010; Henriques et al., 2013), such as cellophane, cellophane–polyethylene, saran, parakote, pliofilm, cryovac, aluminum foil, or aqueous dispersions of butyl rubber and a copolymer of vinyl and vinylidene chloride (Cerqueira et al., 2010; Kampf and Nussinovitch, 2000). This is why in the recent years it has significantly increased the number of works that address the use of edible coatings for cheese packaging (Table 8.4). The most common materials are those of polysaccharide origin, such as chitosan, galactomannan, or agar, followed by those obtained from a protein source, such as whey protein concentrate, sodium caseinate, or corn zein (Table 8.4).

The protective effect of edible coating is determined by itself because on one hand allows the creation of a modified atmosphere similar to that obtained by controlled or modified atmosphere with gases, protecting cheeses from the moment it is applied, during their transport, in their final retail destination, and even in the home of the consumer (Cerqueira et al., 2009). The antimicrobial effect of certain materials such as chitosan against fungi, bacteria, and virus has already been reported (Rabea et al., 2003).

On the other hand, the above properties can be added to the combined effect of an antimicrobial agent, usually naturally occurring, such as nisin or natamycin (Table 8.4). These compounds can be added directly to food to inhibit the growth of undesirable microorganisms during storage. However, in these cases, it may result in a rapid reduction in the microbial populations due to its rapid diffusion into the food bulk and also in a partial inactivation of the active substance making these compounds less effective in long-term storage. To solve this, the manufacturers usually increase the amount of preservatives initially added to food bulk (Cao-Hoang et al., 2010; Fajardo et al., 2010; Fuciños et al., 2012). Therefore, the incorporation of these compounds into edible coatings will allow maintaining their activity, protecting them from degrading environment in food and a slower release of the antimicrobial agents. This system permits slowing down the growth of undesirable microorganisms for long period after packaging, maintaining the antimicrobial efficiency during storage phase and food (Hoffman et al., 2001).

The methods of application of antimicrobial edible coatings may involve dipping, fluidized bed, spraying, electrostatic spraying, panning, and enrobing. Bellow follows a brief description of these methods found in the literature, including their advantages and main drawbacks (Andrade et al., 2012; Dewettinck and Huyghebaert, 1999; Henriques et al., 2013; Zhong et al., 2014):

- *Dipping*: It is the most used method at lab scale. It involves immersing the food into a vat containing the coating solution. Main advantages are its facility to be applied; it has a low cost and provides good coverage even on uneven surfaces. The main disadvantages are the dilution of the coating solution, the accumulation of a large amount of residual coating solution, and the possible growth of contaminating microorganisms in the dipping tank.

TABLE 8.4
Edible Coatings for Cheese Products

Coating Material	Type of Cheese/Food	Treatments or Additives to Improve Film	Effect on Cheese Quality	References
Protein-Based Films				
Sodium caseinate	Semi-soft cheese: mini red Babybel® cheese	Nisin as antimicrobial agent Sorbitol as plasticizer	*L. innocua* inhibition	Cao-Hoang et al. (2010)
Whey protein concentrate	Desert: Mustafakemalpasa cheese sweets	Glycerol as plasticizer	Mold and yeast inhibition increasing the shelf life keeping their sensory properties	Guldas et al. (2010)
	Semi-hard bovine cheese	Lactic acid and natamycin as antimicrobial agents Glycerol as plasticizer Tween 20 as surfactant Sunflower oil to decrease water affinity Guar gum as thickener and emulsifier	Similar values in terms of physicochemical, microbiological, and sensorial properties when compared with commercial coatings	Henriques et al. (2013)
Corn zein	Desert: Mustafakemalpasa cheese sweets		Mold and yeast inhibition increasing the shelf life keeping their sensory properties	Guldas et al. (2010)
Polysaccharide-Based Films				
Chitosan	Semi-hard cheese: *Regional Saloio* cheese	Glycerol/sorbitol as plasticizers Tween 80 as surfactant Corn oil to decrease water affinity	Reduce gas transfer rates Decrease the relative weight loss Delay mold growth	Cerqueira et al. (2009)
		Glycerol/sorbitol as plasticizers	Reduce water loss Decrease the color change Decrease microbial counts	Cerqueira et al. (2010)
		Natamycin as antimicrobial agent Glycerol as plasticizer Tween 80 as surfactant	Mold and yeast inhibition Increase in O_2 and CO_2 permeability	Fajardo et al. (2010)
	Fresh cheese: *Fior di latte* cheese	Lysozyme and Na_2–EDTA as antimicrobial agents Modified atmosphere	Increase the shelf life	Del Nobile et al. (2009)

(Continued)

TABLE 8.4 (Continued)
Edible Coatings for Cheese Products

Coating Material	Type of Cheese/Food	Treatments or Additives to Improve Film	Effect on Cheese Quality	References
Galactomannan of *Gleditsia triacanthos*	Semi-soft cheese: Mozzarella cheese	Lysozyme as antimicrobial agent Glycerol as plasticizer	Mold and pathogenic bacteria (*P. fluorescens* and *L. monocytogenes*) inhibition	Duan et al. (2007)
	Semi-hard cheese: *Regional Saloio* cheese	Glycerol/sorbitol as plasticizers Corn oil to decrease water affinity	Lower gas transfer rates Decrease the relative weight loss Delay mold growth	Cerqueira et al. (2009)
	Semi-hard cheese: *Regional Saloio* cheese	Glycerol as plasticizer Corn oil to decrease water affinity	Reduce water loss Decrease the color change Decrease microbial counts	Cerqueira et al. (2010)
Agar of *Gleditsia birdiae*	Soft cheese: *Ricotta* cheese	Nisin as antimicrobial agent	Reduce *L. monocytogenes* post contamination during storage	Martins et al. (2010)
	Semi-hard cheese: *Regional Saloio* cheese	Glycerol/sorbitol as plasticizers Corn oil to decrease water affinity	Lower gas transfer rates Decrease the relative weight loss Delay mold growth	Cerqueira et al. (2009)
Cellulose	Semi-soft cheese: sliced mozzarella cheese	Nisin and natamycin as antimicrobial agents	Molds and yeasts, and psychrotrophic bacteria inhibition	dos Santos Pires et al. (2008)
Starch	Hard cheese: *Cheddar* cheese	Linalool, carvacrol, and thymol as antimicrobial agents Polyethylene glycol (PEG) as a plasticizer Methylcellulose and hydroxypropyl methylcellulose	*S. aureus* inhibition	Kuorwel et al. (2011)
Composite Films				
Chitosan/whey protein	Soft cheese: *Ricotta* cheese	Modified atmosphere	Increase water vapor permeability Reduce the growth of microbial contaminants Delay the development of undesirable acidity Maintain better the texture without modifying sensory characteristics	Di Pierro et al. (2011)

- *Fluidized-bed processing*: This method is normally used to apply a very thin film layer onto dry particles of very low density and/or small size. During the coating application, food particles are fluidized with hot air while being sprayed with the coating solution. The main advantages are a high coating capacity and large energy savings because, as the product remains dry, a drying step is not necessary. The main disadvantages are the premature drying of the coated particles, particle agglomeration, damage of heat-sensitive products, or the development of undesired odors.
- *Spraying*: This method of application is very useful for successive coatings with various types of solutions, for example, combining a hydrophobic and a hydrophilic solution, without contaminating the coating solutions. Other advantages are good control of the thickness of the coating layer and this method allows working with large surface areas. As a disadvantage, it might be noted that the application of the coating on the bottom of the product requires the food turned after a preliminary drying step.
- *Electrostatic spraying*: There is a growing interest in adapting this method, from the paint industry, to the food industry. The main advantages of this method compared to traditional spraying method include the control of the droplets size, increasing their coverage, and deposition and thus allowing a more homogeneous distribution of the coating. This application method also allows reducing wastage.
- *Panning*: This method consists of placing the food to be coated in a large rotating pan while the coating solution is directly added or sprayed, keeping the food product in motion to achieve an uniform coating. Forced air, either at room temperature or at high temperature is applied to dry the coating. Although the heating occurs due to the friction of food during tumbling, it is necessary to apply a stream of cold air. Another disadvantage of this method is that it requires a very intense tumbling to ensure a good distribution of the coating.
- *Enrobing*: In this method, the coating solution flows vertically to the treated food. To achieve good coating, it will be necessary a good control of the viscosity of the coating solution and the surface of food should be flat.

The selection of an appropriate coating method not only affects the efficiency of the coatings in food preservation but also determines the production cost and the process efficiency (Zhong et al., 2014). Given the shape and dimensions of the different types of cheese, the most used application methods, both in laboratory and industrial scale, are spraying and dipping.

The increasing interest in developing antimicrobial edible coatings makes that the study of its application goes beyond the foods mentioned so far (fruits, vegetables, meat, fish, and cheese), although there are still a few additional applications. Altamirano-Fortoulet al. (2012) encapsulated *Lactobacillus acidophilus* into starch-based edible coatings to be applied onto bread. In this case, it was a probiotic coating but it could extend the shelf life of bread if an antifungal agent is incorporated into the coating. Guldas et al. (2010) used edible films based on corn zein to coat a traditional Turkish dessert (Mustafakemalpasa cheese sweets), getting molds and yeasts inhibition and thus increasing the shelf life of this product without altering its sensory properties (Table 8.4).

8.5 CONCLUSIONS AND FUTURE TRENDS

The recent developments in antimicrobial edible coating demonstrate the feasibility of this technology to preserve food. A broad range of applications covering the main food families allow to market safer foods and extend their shelf life. Food grade biopolymers from natural origin can be used to obtain films and coatings with suitable mechanical and technological properties. The antimicrobial activity of these coatings can be obtained by adding natural antimicrobial compounds or using blends containing antimicrobial polymers such as chitosan.

Since coating materials and antimicrobial compounds are affordable and the technology to apply this film in foods is easy to implement at real scale, it could be expected an increase in the industrial applications of antimicrobial edible films in the near future.

Before this, in the next years, it will be to face some challenges relating a better understanding of how the antimicrobial coating films work in a food and the relation between the chemical composition of the blends, their structure and their mechanical and functional properties.

REFERENCES

Abdollahi, M., Rezaei, M., and Farzi, G. 2012. A novel active bionanocomposite film incorporating rosemary essential oil and nanoclay into chitosan. *Journal of Food Engineering*, 111, 343–350.

Abugoch, L., Tapia, C., Villamán, M., Pedram, M. Y., and Díaz, M. 2011. Characterization of quinoa protein–chitosan blend edible films. *Food Hydrocolloids*, 25, 879–886.

Altamirano-Fortoul, R., Moreno-Terrazas, R., Quezada-Gallo, A., and Rosell, C. M. 2012. Viability of some probiotic coatings in bread and its effect on the crust mechanical properties. *Food Hydrocolloids*, 29, 166–174.

Alvarez, M. V., Ponce, A. G., and Moreira, M. D. R. 2013. Antimicrobial efficiency of chitosan coating enriched with bioactive compounds to improve the safety of fresh cut broccoli. *LWT—Food Science and Technology*, 50, 78–87.

Anderson, T. J., and Lamsal, B. P. 2011. Zein extraction from corn, corn products, and coproducts and modifications for various applications: A review. *Cereal Chemistry*, 88, 159–173.

Andrade, R. D., Skurtys, O., and Osorio, F. A. 2012. Atomizing spray systems for application of edible coatings. *Comprehensive Reviews in Food Science and Food Safety*, 11, 323–337.

Arrieta, M. P., Peltzer, M. A., López, J., Garrigós, M. D. C., Valente, A. J. M., and Jiménez, A. 2014. Functional properties of sodium and calcium caseinate antimicrobial active films containing carvacrol. *Journal of Food Engineering*, 121, 94–101.

Avila Sosa, R., Palou, E., Jiménez Munguía, M., Nevárez Moorillón, G., Navarro Cruz, A., and López Malo A. 2012. Antifungal activity by vapor contact of essential oils added to amaranth, chitosan, or starch edible films. *International Journal of Food Microbiology*, 153, 66–72.

Aymerich, T., Picouet, P. A., and Monfort, J. M. 2008. Decontamination technologies for meat products. *Meat Science*, 78, 114–129.

Azevedo, A., Buarque, P., Oliveira Cruz, E. M., Blank, A., Alves, P., Nunes, M., and de Aquino S. L. C. L. 2014. Response surface methodology for optimisation of edible chitosan coating formulations incorporating essential oil against several foodborne pathogenic bacteria. *Food Control*, 43, 1–9.

Badwaik, L. S., Borah, P. K., and Deka, S. C. 2014. Antimicrobial and enzymatic antibrowning film used as coating for bamboo shoot quality improvement. *Carbohydrate Polymers*, 103, 213–220.

Balaguer, M. P., Fajardo, P., Gartner, H., Gomez-Estaca, J., Gavara, R., Almenar, E., and Hernandez-Munoz, P. 2014. Functional properties and antifungal activity of films based on gliadins containing cinnamaldehyde and natamycin. *International Journal of Food Microbiology*, 173, 62–71.

Balaguer, M. P., Lopez-Carballo, G., Catala, R., Gavara, R., and Hernandez-Munoz, P. 2013. Antifungal properties of gliadin films incorporating cinnamaldehyde and application in active food packaging of bread and cheese spread foodstuffs. *International Journal of Food Microbiology*, 166, 369–377.

Bangyekan, C., Aht Ong, D., and Srikulkit, K. 2006. Preparation and properties evaluation of chitosan-coated cassava starch films. *Carbohydrate Polymers*, 63, 61–71.

Baranenko, D. A., Kolodyaznaya, V. S., and Zabelina, N. A. 2013. Effect of composition and properties of chitosan-based edible coatings on microflora of meat and meat products. *Acta Scientiarum Polonorum, Technologia Alimentaria*, 12, 149–157.

Bautista Baños, S., Hernández Lauzardo, A. N., Velázquez-del Valle, M. G., Hernández López, M., Ait Barka, E., Bosquez Molina, E., and Wilson C. L., 2006. Chitosan as a potential natural compound to control pre and postharvest diseases of horticultural commodities. *Crop Protection*, 25, 108–118.

Benhabiles, M. S., Salah, R., Lounici, H., Drouiche, N., Goosen, M. F. A., and Mameri, N. 2012. Antibacterial activity of chitin, chitosan and its oligomers prepared from shrimp shell waste. *Food Hydrocolloids*, 29, 48–56.

Beverlya, R. L., Janes, M. E., Prinyawiwatkula, W., and No, H. K. 2008. Edible chitosan films on ready-to-eat roast beef for the control of *Listeria monocytogenes*. *Food Microbiology*, 25, 534–537.

Bifani, V., Ramírez, C., Ihl, M., Rubilar, M., García, A., and Zaritzky, N. 2007. Effects of murta (*Ugni molinae* Turcz) extract on gas and water vapor permeability of carboxymethylcellulose-based edible films. *LWT—Food Science and Technology*, 40, 1473–1481.

Bonilla, J., Talón, E., Atarés, L., Vargas, M., and Chiralt, A. 2013. Effect of the incorporation of antioxidants on physicochemical and antioxidant properties of wheat starch–chitosan films. *Journal of Food Engineering*, 118, 271–278.

Bourbon, A. I., Pinheiro, A. C., Cerqueira, M. A., Rocha, C. M. R., Avides, M. C., Quintas, M. A. C., and Vicente, A. A. 2011. Physico-chemical characterization of chitosan-based edible films incorporating bioactive compounds of different molecular weight. *Journal of Food Engineering*, 106, 111–118.

Bourtoom, T. 2008. Edible films and coatings: Characteristics and properties. *International Food Research Journal*, 15, 237–248.

Bourtoom, T., and Chinnan, M. 2008. Preparation and properties of rice starch–chitosan blend biodegradable film. *Food Science & Technology*, 41, 1633–1641.

Bravin, B., Peressini, D., and Sensidoni, A. 2006. Development and application of polysaccharide–lipid edible coating to extend shelf life of dry bakery products. *Journal of Food Engineering*, 76, 280–290.

Broglia, A., and Kapel, C. 2011. Changing dietary habits in a changing world: Emerging drivers for the transmission of foodborne parasitic zoonoses. *Veterinary Parasitology*, 182, 2–13.

Bryce, T. A., McKinnon, A. A., Morris, E. R., Rees, D. A., and Thom, D. 1974. Chain conformations in the sol–gel transitions for polysaccharide systems, and their characterization by spectroscopic methods. *Faraday Discussions Chemical Society*, 57, 221–229.

Burt, S. 2004. Essential oils: Their antibacterial properties and potential applications in foods—A review. *International Journal of Food Microbiology*, 94, 223–253.

Cagri, A., Ustunol, Z., and Ryser, E. T. 2004. Antimicrobial edible films and coatings. *Journal of Food Protection*, 67, 833–848.

Cai, J., Yang, J., Wang, C., Hu, Y., Lin, J., and Fan, L. 2010. Structural characterization and antimicrobial activity of chitosan (CS-40)/nisin complexes. *Journal of Applied Polymer Science*, 116, 3702–3707.

Campaniello, D., Bevilacqua, A., Sinigaglia, M., and Corbo, M. 2008. Chitosan: Antimicrobial activity and potential applications for preserving minimally processed strawberries. *Food Microbiology*, 25, 992–1000.

Campos, C. A., Gerschenson, L. N., and Flores, S. K. 2011. Development of edible films and coatings with antimicrobial activity. *Food and Bioprocess Technology*, 4, 849–875.

Cao-Hoang, L., Grégoire, L., Chaine, A., and Waché, Y. 2010. Importance and efficiency of in-depth antimicrobial activity for the control of *Listeria* development with nisin-incorporated sodium caseinate films. *Food Control*, 21, 1227–1233.

Carneiro-da-Cunha, M. G., Cerqueira, M. A., Souza, B. W. S., Carvalho, S., Quintas, M. A. C. Teixeira, J. A., and Vicente, A. A. 2010. Physical and thermal properties of a chitosan/alginate nanolayered PET film. *Carbohydrate Polymers*, 82, 153–159.

Carpenter, C. E., Cornforth, D. P., and Whittier, D. 2001. Consumer preferences for beef color and packaging did not affect eating satisfaction. *Meat Science*, 57, 359–363.

Casariego, A., Souza, B. W. S., Cerqueira, M. A., Teixeira, J. A., Cruz, L., Díaz, R., and Vicente, A. A. 2009. Chitosan/clay films' properties as affected by biopolymer and clay micro/nanoparticles' concentrations. *Food Hydrocolloids*, 23, 1895–1902.

Cerqueira, M. A., Bourbon, A. I., Pinheiro, A. C., Martins, J. T., Souza, B. W. S., Teixeira, J. A., and Vicente, A. A. 2011. Galactomannans use in the development of edible films/coatings for food applications. *Trends in Food Science and Technology*, 22, 662–671.

Cerqueira, M. A., Costa, M. J., Fuciños, C., Pastrana, L. M., and Vicente, A. A. 2014a. Development of active and nanotechnology-based smart edible packaging systems: Physical-chemical characterization. *Food and Bioprocess Technology*, 7, 1472–1482.

Cerqueira, M. A., Lima, A. M., Souza, B. W. S., Teixeira, J. A., Moreira, R. A., and Vicente, A. A. 2009. Functional polysaccharides as edible coatings for cheese. *Journal of Agricultural and Food Chemistry*, 57, 1456–1462.

Cerqueira, M. A., Pinheiro, A. C., Silva, H. D., Ramos, P. E., Azevedo, M. A., Flores-López, M. L., Rivera, M. C., Bourbon, A. I., Ramos, Ó. L., and Vicente, A. A. 2014b. Design of bio-nanosystems for oral delivery of functional compounds. *Food Engineering Reviews*, 6, 1–19.

Cerqueira, M. A., Sousa-Gallagher, M. J., Macedo, I., Rodriguez-Aguilera, R., Souza, B. W. S., Teixeira, J. A., and Vicente, A. A. 2010. Use of galactomannan edible coating application and storage temperature for prolonging shelf-life of "Regional" cheese. *Journal of Food Engineering*, 97, 87–94.

Cervera, M., Karjalainen, M., Airaksinen, S., Rantanen, J., Krogars, K., Heinämäki, J., Colarte, A., and Yliruusi. J. 2004. Physical stability and moisture sorption of aqueous chitosan–amylose starch films plasticized with polyols. *European Journal of Pharmaceutics and Biopharmaceutics*, 58, 69–76.

Cha, D. S., Choi, J. H., Chinnan, M. S., and Park, H. J. 2002. Antimicrobial films based on Na-alginate and kappa-carrageenan. *Lebensmittel-Wissenschaft Und-Technologie-Food Science and Technology*, 35, 715–719.

Chen, L. Y., Remondetto, G. E., and Subirade, M. 2006. Food protein-based materials as nutraceutical delivery systems. *Trends in Food Science & Technology*, 17, 272–283.

Chen, S., and Nussinovitch, A. 2001. Permeability and roughness determinations of wax hydrocolloid coatings, and their limitations in determining citrus fruit overall quality. *Food Hydrocolloids*, 15, 127–137.

Chien, P. J., Sheu, F., and Yang, F. H. 2007. Effects of edible chitosan coating on quality and shelf life of sliced mango fruit. *Journal of Food Engineering*, 78, 225–229.

Chillo, S., Flores, S., Mastromatteo, M., Conte, A., Gerschenson, L., and Del Nobile, M.A. 2008. Influence of glycerol and chitosan on tapioca starch-based edible film properties. *Journal of Food Engineering*, 88, 159–168.

Chinma, C. E., Ariahu, C. C., and Alakali, J. S. 2013. Effect of temperature and relative humidity on the water vapour permeability and mechanical properties of cassava starch and soy protein concentrate based edible. *Journal of Food Science and Technology*, 1, 7.

Chiu, P. E., and Lai, L. S. 2010. Antimicrobial activities of tapioca starch/decolorized Hsiantsao leaf gum coatings containing green tea extracts in fruit-based salads, romaine hearts and pork slices. *International Journal of Food Microbiology*, 139, 23–30.

Cho, S. Y. 2007. Edible films made from membrane processed soy protein concentrates. *Food Science and Technology*, 40, 418–423.

Codex Alimentarius 2014. FAO/WHO Food Standards. Updated up to the 37th Session of the Codex Alimentarius Commission. Available at http://www.codexalimentarius.net/gsfaonline/additives/details.html?id=208

Coma, V. 2008. Bioactive packaging technologies for extended shelf life of meat-based products, *Meat Science*, 78, 90–103.

Coma, V., Martial-Gros, A., Garreau, S., Copinet, A., Salin, F., and Deschamps, A. 2002. Edible antimicrobial films based on chitosan matrix. *Journal of Food Science*, 67, 1162–1169.

Conte, A., Buonocore, G. G., Bevilacqua, A., Sinigaglia, M., and Del Nobile, M. A. 2006. Immobilization of lysozyme on polyvinylalcohol films for active packaging applications. *Journal of Food Protection*, 69, 866–870.

Cortez-Vega, W. R., Piotrowicz, I. B. B., Prentice, C., and Borges, C. D. 2013. Conservation of papaya minimally processed with the use of edible coating based on xanthan gum. *Semina: Ciencias Agrarias*, 34, 1753–1764.

Cruz Romero, M. C., Murphy, T., Morris, M., Cummins, E., and Kerry, J. P. 2013. Antimicrobial activity of chitosan, organic acids and nano-sized solubilisates for potential use in smart antimicrobially-active packaging for potential food applications. *Food Control*, 34, 393–397.

Cuero, R. G., Osuji, G., and Duffus, E. 1991. N-carboxymethylchitosan: Uptake and effect on chlorophyll production, water potential and biomass in tomato plants. *Food Biotechnology*, 5, 95–103.

da Silva, M. A., Bierhalz, A. C. K., and Kieckbusch, T. G. 2009. Alginate and pectin composite films crosslinked with Ca^{2+} ions: Effect of the plasticizer concentration. *Carbohydrate Polymers*, 77, 736–742

Das, D. K., Dutta, H., and Mahanta, C. L. 2013. Development of a rice starch-based coating with antioxidant and microbe-barrier properties and study of its effect on tomatoes stored at room temperature. *LWT—Food Science and Technology*, 50, 272–278.

de Menezes Cruz, M., de Oliveira Lins, S. R., de Oliveira, S. M. A., and Barbosa, M. A. G. 2012. Effects of essencial oils and edible coatings on postharvest rot of mango, cv. Kent. *Revista Caatinga*, 25, 1–6.

de Souza, P. M., Fernandez, A., Lopez-Carballo, G., Gavara, R., and Hernandez-Munoz, P. 2010. Modified sodium caseinate films as releasing carriers of lysozyme. *Food Hydrocolloids*, 24, 300–306.

Debeaufort, F., Martin-Polo, M., and Voilley, A. 1993. Polarity homogeneity and structure affect water vapor permeability of model edible films. *Journal of Food Science*, 58, 426–434.

Del Nobile, M. A., Conte, A., Incoronato, A. L., and Panza, O. 2008. Antimicrobial efficacy and release kinetics of thymol from zein films. *Journal of Food Engineering*, 89, 57–63.

Del Nobile, M. A., Gammariello, D., Conte, A., and Attanasio, M. 2009. A combination of chitosan, coating and modified atmosphere packaging for prolonging Fior di latte cheese shelf life. *Carbohydrate Polymers*, 78, 151–156.

Deng, Q., and Zhao, Y. 2011. Physicochemical, nutritional, and antimicrobial properties of wine grape (cv. Merlot) pomace extract-based films. *Journal of Food Science*, 76, 309–317.

Devlieghere, F., Vermeulen, A., and Debevere, J. 2004. Chitosan: Antimicrobial activity, interactions with food components and applicability as a coating on fruit and vegetables. *Food Microbiology*, 21, 703–714.

Dewettinck, K., and Huyghebaert, A. 1999. Fluidized bed coating in food technology. *Trends in Food Science and Technology*, 10, 163–168.

Dhall, R. K. 2013. Advances in edible coatings for fresh fruits and vegetables: A review. *Critical Reviews in Food Science and Nutrition*, 53, 435–450.

Di Pierro, P., Sorrentino, A., Mariniello, L., Giosafatto, C. V. L., and Porta, R. 2011. Chitosan/whey protein film as active coating to extend ricotta cheese shelf-life. *LWT—Food Science and Technology*, 44, 2324–2327.

dos Santos Pires, A. C., De Fátima Ferreira Soares, N., De Andrade, N. J., Da Silva, L. H. M., Camilloto, G. P., and Bernardes, P. C. 2008. Development and evaluation of active packaging for sliced mozzarella preservation. *Packaging Technology and Science*, 21, 375–383.

Duan, J., Park, S., Daeschel, M. A., and Zhao, Y. 2007. Antimicrobial chitosan-lysozyme (CL) films and coatings for enhancing microbial safety of mozzarella cheese. *Journal of Food Science*, 72, M355–M362.

Duan, J., Wu, R., Strik, B., and Zhao, Y. 2011. Effect of edible coatings on the quality of fresh blueberries (Duke and Elliott) under commercial storage conditions. *Postharvest Biology and Technology*, 59, 71–79.

Durán, N., and Marcato, P. D. 2013. Nanobiotechnology perspectives. Role of nanotechnology in the food industry: A review. *International Journal of Food Science and Technology*, 48, 1127–1134.

Durango, A. M., Soares, N. F. F., and Andrade, N. J. 2006. Microbiological evaluation of an edible antimicrobial coating on minimally processed carrots. *Food Control*, 17, 336–341.

Efrati, R., Natan, M., Pelah, A., Haberer, A., Banin, E., Dotan, A., and Ophir, A. 2014. The combined effect of additives and processing on the thermal stability and controlled release of essential oils in antimicrobial films. *Journal of Applied Polymer Science*, 131, 40564.

Ehivet, F. E., Min, B., Park, M. K., and Oh, J. H. 2011. Characterization and antimicrobial activity of sweetpotato starch-based edible film containing origanum (*Thymus capitatus*) oil. *Journal of Food Science*, 76, C178–C184.

Elsabee, M. Z., and Abdou, E. S. 2013. Chitosan based edible films and coatings: A review. *Materials Science and Engineering C*, 33, 1819–1841.

Emiroglu, Z. K., Yemis, G. P., Coskun, B. K., and Candogan, K. 2010. Antimicrobial activity of soy edible films incorporated with thyme and oregano essential oils on fresh ground beef patties. *Meat Science*, 86, 283–288.

Erbil, H. Y., and Muftugil, N. 1986. Lengthening the postharvest life of peaches by coating with hydrophobic emulsions. *Journal of Food Processing and Preservation*, 10, 269–279.

Escamilla García, M., Calderón Domínguez, G., Chanona Pérez, J. J., Farrera Rebollo, R. R., Andraca Adame, J. A., Arzate Vázquez, I., Mendez Mendez, J. V., and Moreno Ruiz, L. A. 2013. Physical and structural characterisation of zein and chitosan edible films using nanotechnology tools. *International Journal of Biological Macromolecules*, 61, 196–203.

European Comission 2013. Prospects for agricultural markets and income in the EU 2013–2023. Available at http://ec.europa.eu/agriculture/markets-and-prices/medium-term-outlook/2013/fullrep_en.pdf

Fabra, M. J., Busolo, M. A., Lopez-Rubio, A., and Lagaron, J. M. 2013. Nanostructured biolayers in food packaging. *Trends in Food Science and Technology*, 31, 79–87.

Fagundes, C., Pérez-Gago, M. B., Monteiro, A. R., and Palou, L. 2013. Antifungal activity of food additives *in vitro* and as ingredients of hydroxypropyl methylcellulose-lipid edible coatings against *Botrytis cinerea* and *Alternaria alternata* on cherry tomato fruit. *International Journal of Food Microbiology*, 166, 391–398.

Fajardo, P., Atanassova, M., Garrido-Maestu, A., Wortner-Smith, T., Cotterill, J., and Cabado, A. G. 2014a. Bacteria isolated from shellfish digestive gland with antipathogenic activity as candidates to increase the efficiency of shellfish depuration process. *Food Control*, 46, 272–281.

Fajardo, P., Balaguer, M.P., Gomez-Estaca, J., Gavara, R., and Hernandez-Munoz, P., 2014b. Chemically modified gliadins as sustained release systems for lysozyme. *Food Hydrocolloids*, 41, 53–59.

Fajardo, P., Martins, J. T., Fuciños, C., Pastrana, L., Teixeira, J. A., and Vicente, A. A. 2010. Evaluation of a chitosan-based edible film as carrier of natamycin to improve the storability of Saloio cheese. *Journal of Food Engineering*, 101, 349–356.

Falguera, V. Quintero, J. P. Jiménez, A., Muñoz, J. A., and Ibarz, A. 2011. Edible films and coatings: Structures, active functions and trends in their use. *Trends in Food Science and Technology*, 22, 292–303.

Fang, Y., Al-Assaf, S., Phillips, G. O., Nishinari, K., Funami, T., and Williams, P.A. 2008. Binding behavior of calcium to polyuronates: Comparison of pectin with alginate. *Carbohydrate Polymers*, 72, 334–341.

Fuciños, C., Guerra, N. P., Teijón, J. M., Pastrana, L. M., Rúa, M. L., and Katime, I. 2012. Use of poly(N-isopropylacrylamide) nanohydrogels for the controlled release of pimaricin in active packaging. *Journal of Food Science*, 77, N21–N28.

García, M. A., Martino, M. N., and Zaritzky, N. E. 2001. Composite starch-based coatings applied to strawberries (*Fragaria ananassa*). *Die Nahrung*, 45, 267–272.

Gennadios, A., Hanna, M. A., and Kurth, L. B. 1997. Application of edible coatings on meat, poultry and seafoods: A review. *Lebensmittel-Wissenschaft und Technologie*, 30, 337–350.

Geraldine, R. M., Soares, N. D. F. F., Botrel, D. A., and de Almeida Gonçalves, L. 2008. Characterization and effect of edible coatings on minimally processed garlic quality. *Carbohydrate Polymers*, 72, 403–409.

Ghaouth, A., Arul, J., Ponnampalam, R., and Boulet, M. 1991a. Chitosan coating effect on storability and quality of fresh strawberries. *Journal of Food Science*, 56, 1618–1620.

Ghaouth, A., Arul, J., Ponnampalam, R., and Boulet, M. 1991b. Use of chitosan coating to reduce water loss and maintain quality of cucumber and bell pepper fruits. *Journal of Food Processing and Preservation*, 15, 359–368.

Gialamas, H., Zinoviadou, K. G., Biliaderis, C. G., and Koutsoumanis, K. P. 2010. Development of a novel bioactive packaging based on the incorporation of *Lactobacillus sakei* into sodium–caseinate films for controlling *Listeria monocytogenes* in foods. *Food Research International*, 43, 2402–2408.

Gniewosz, M., Kraśniewska, K., Woreta, M., and Kosakowska, O. 2013. Antimicrobial activity of a pullulan-caraway essential oil coating on reduction of food microorganisms and quality in fresh baby carrot. *Journal of Food Science*, 78, M1242–M1248.

Gniewosz, M., Synowiec, A., Kraśniewska, K., Przybył, J. L., Bączek, K., and Węglarz, Z. 2014. The antimicrobial activity of pullulan film incorporated with meadowsweet flower extracts (*Filipendulae ulmariae flos*) on postharvest quality of apples. *Food Control*, 37, 351–361.

Goff, H. D., and Hill, A. R. 1993. Chemistry and physics. In Y. H. Hui (Ed.), *Dairy Science and Technology Handbook. Principles and Properties* (pp. 1–82). New York: VCH Publishers.

Gómez-Estaca, J., López, de Lace, A., López-Caballero, M. E., Gómez-Guillén, M. C., and Montero, P. 2010. Biodegradable gelatin–chitosan films incorporated with essential oils as antimicrobial agents for fish preservation. *Food Microbiology*, 27, 889–896.

Gómez-Estaca, J., Montero, P., Fernández-Martín, F., and Gómez-Guillén, M. C. 2009. Physico-chemical and film forming properties of bovine-hide and tuna-skin gelatin: A comparative study. *Journal of Food Engineering*, 90, 480–486.

Gómez-Guillén, M. C., Ihl, M., Bifani, V., Silva, A., and Montero, P. 2007. Edible films made from tuna-fish gelatin with antioxidant extracts of two different murta ecotypes leaves (*Ugni molinae* Turcz). *Food Hydrocolloid*, 21, 1133–1143.

González-Aguilar, G. A., Valenzuela-Soto, E., Lizardi-Mendoza, J., Goycoolea, F., Martínez-Téllez, M. A., Villegas-Ochoa, M. A., Monroy-García, I. N., and Ayala-Zavala, J. F. 2009. Effect of chitosan coating in preventing deterioration and preserving the quality of fresh-cut papaya "Maradol." *Journal of the Science of Food and Agriculture*, 89, 15–23.

González-Chávez, S. A., Arévalo-Gallegos, S., and Rascón-Cruz, Q. 2009. Lactoferrin: Structure, function and applications. *International Journal of Antimicrobial Agents*, 33, 301.e1–301.e8

Goy, R. C., de Britto, D., and Assis, O. B. G., 2009. A review of the antimicrobial activity of chitosan. *Polímeros*, 19, 241–247.

Guarda, A., Rubilar, J. F., Miltz, J., and Galotto, M. J. 2011. The antimicrobial activity of microencapsulated thymol and carvacrol. *International Journal of Food Microbiology*, 146, 144–150.

Guilbert, S., Gontard, N., and Gorris, L. G. M. 1996. Prolongation of the shelf-life of perishable food products using biodegradable films and coatings. *LWT-Food Science and Technology*, 29, 10–17.

Guldas, M., Akpinar-Bayizit, A., Ozcan, T., and Yilmaz-Ersan, L. 2010. Effects of edible film coatings on shelf-life of mustafakemalpasa sweet, a cheese based dessert. *Journal of Food Science and Technology*, 47, 476–481.

Guo, M., Jin, T. Z., Wang, L., Scullen, O. J., and Sommers, C. H. 2014. Antimicrobial films and coatings for inactivation of *Listeria innocua* on ready-to-eat deli turkey meat. *Food Control*, 40, 64–70.

Hambleton, A., Debeaufort, F., Bonnotte, A., and Voilley, A. 2009. Influence of alginate emulsion-based films structure on its barrier properties and on the protection of microencapsulated aroma compound. *Food Hydrocolloids*, 23, 2116–2124.

Han, J. H. 2000. Antimicrobial food packaging. *Food Technology*, 54, 56–65.

Han, C., Lederer, C., McDaniel, M., and Zhao, Y. 2006. Sensory evaluation of fresh strawberries (*Fragaria ananassa*) coated with chitosan-based edible coatings. *Journal of Food Science*, 70, S172–S178.

Henriques, M., Santos, G., Rodrigues, A., Gomes D., Pereira, C., and Gil, M. 2013. Replacement of conventional cheese coatings by natural whey protein edible coatings with antimicrobial activity. *Journal of Hygienic Engineering and Design*, 3, 34–47.

Hernández Muñoz, P., Almenar, E., Valle, V., Vélez, D., and Gavara, R. 2008. Effect of chitosan coating combined with postharvest calcium treatment on strawberry (*Fragaria ananassa*) quality during refrigerated storage. *Food Chemistry*, 110, 428–435.

Hernandez-Izquierdo, V. M. and Krochta, J. M. 2008. Thermoplastic processing of proteins for film formation. A review. *Journal of Food Science*, 73, 30–39.

Hoffman, K. L., Han, I. Y., and Dawson, P. L. 2001. Antimicrobial effects of corn zein films impregnated with nisin, lauric acid, and EDTA. *Journal of Food Protection*, 64, 885–889.

Hosseini, F., Rezaei, M., Zandi, M., and Ghavi, F. F. 2013. Preparation and functional properties of fish gelatin–chitosan blend edible films. *Food Chemistry*, 136, 1490–1495.

Jagannath, J. H., Nanjappa, C., Das Gupta, D. K., and Bawa, A. S. 2003. Mechanical and barrier properties of edible starch–protein-based films. *Journal of Applied Polymer Science*, 88, 64–71.

Jang, S. A., Shin, Y. J., and Song, K. B. 2011. Effect of rapeseed protein-gelatin film containing grapefruit seed extract on "Maehyang" strawberry quality. *International Journal of Food Science and Technology*, 46, 620–625.

Janjarasskul, T., Min, S. C., and Krochta, J. M. 2011. Storage stability of ascorbic acid incorporated in edible whey protein films. *Journal of Agricultural and Food Chemistry*, 59, 12428–12432.

Jooyandeh, H. 2011. Whey protein films and coatings: A review. *Pakistan Journal of Nutrition*, 10, 293–301.

Jouki, M., Yazdia, F. T., Mortazavia, S. A., Koocheki, A., and Khazaei, N. 2014. Effect of quince seed mucilage edible films incorporated with oregano or thyme essential oil on shelf life extension of refrigerated rainbow trout fillets. *International Journal of Food Microbiology*, 174, 88–97.

Kampf, N., and Nussinovitch, A. 2000. Hydrocolloid coating of cheeses. *Food Hydrocolloids*, 14, 531–537.

Kanatt, S. R., Chander, R., and Sharma, A. 2008. Chitosan and mint mixture: A new preservative for meat and meat products. *Food Chemistry*, 107, 845–852.

Karbowiak, T., Debeaufort, F., Champion, D., and Voilley, A. 2006. Wetting properties at the surface of iota-carrageenan-based edible films. *Journal of Colloid and Interface Science*, 294, 400–410.

Karbowiak, T., Debeaufort, F., and Voilley, A. 2007. Influence of thermal process on structure and functional properties of emulsion-based edible films. *Food Hydrocolloid*, 21, 879–888.

Kenny, O., Smyth, T. J., Walsh, D., Kelleher, C. T., Hewage, C. M., and Brunton, N. P. 2014. Investigating the potential of under-utilised plants from the Asteraceae family as a source of natural antimicrobial and antioxidant extracts. *Food Chemistry*, 161, 79–86.

Kester, J. J., and Fennema, O. R. 1986. Edible films and coatings: A review. *Food Technology*, 40, 47–59.

Khorshid, N., Hossain, M., and Farid, M. M. 2007. Precipitation of food protein using high-pressure carbon dioxide. *Journal of Food Engineering*, 79, 1214–1220.

Khwaldia, K., Perez, C., Banon, S., Desobry, S., and Hardy, J. 2004. Milk proteins for edible films and coatings. *Critical Reviews in Food Science and Nutrition*, 44, 239–251.

Kim, K. W., Thomas, R. L., Lee, C., and Park, H. J. 2003. Antimicrobial activity of native chitosan, degraded chitosan, and o-carboxymethylated chitosan. *Journal of Food Protection*, 66, 1495–1498.

Kousta, M., Mataragas, M., Skandamis, P., and Drosinos, E. H. 2010. Prevalence and sources of cheese contamination with pathogens at farm and processing levels. *Food Control*, 21, 805–815.

Kraśniewska, K., and Gniewosz, M. 2012. Substances with antibacterial activity in edible films—A review. *Polish Journal of Food and Nutrition Sciences*, 62, 199–206.

Kraśniewska, K., Gniewosz, M., Synowiec, A., Przybył, J. L., Baczek, K., and Weglarz, Z. 2014. The use of pullulan coating enriched with plant extracts from *Satureja hortensis* L. to maintain pepper and apple quality and safety. *Postharvest Biology and Technology*, 90, 63–72.

Krochta, J. M. 1997. Edible protein films and coatings. In S. Damodaran and A. Paraf (Eds.), *Food Proteins and Their Applications* (pp. 529–550). New York: Marcel Dekker Inc.

Krochta, J. M., and Mulder-Johnston, C. D. 1997. Edible and biodegradable polymer films: Challenges and opportunities. *Food Technology*, 51, 61–74.

Ku, K.-J., Hong, Y.-H., and Song, K. B. 2008. Mechanical properties of a *Gelidium corneum* edible film containing catechin and its application in sausages. *Journal of Food Science*, 73, C217–C221.

Kuorwel, K. K., Cran, M. J., Sonneveld, K., Miltz, J., and Bigger, S. W. 2011. Antimicrobial activity of natural agents coated on starch-based films against *Staphylococcus aureus*. *Journal of Food Science*, 76, M531–M537.

Kushwaha, A. K., and Kawtikwar, P. S. 2013. Zein as a natural film forming agent: A review. *International Journal of Pharmacy and Technology*, 5, 2578–2593.

Lagrain, B., Goderis B., Brijs, K., and Delcour, J. A. 2010. Molecular basis of processing wheat gluten toward biobased materials. *Biomacromolecules*, 11, 533–541.

Lawton, J. W. 2002. Zein: A history of processing and use. *Cereal Chemistry*, 79, 1–18.

Lee, H., and Min, S. C. 2014. Development of antimicrobial defatted soybean meal-based edible films incorporating the lactoperoxidase system by heat pressing. *Journal of Food Engineering*, 120, 183–190.

Lim, G. O., Hong, Y. H., and Song, K. B. 2010. Application of *Gelidium corneum* edible films containing carvacrol for ham packages. *Journal of Food Science*, 75, C90–C93.

Limam, Z., Selmi, S., Sadock, S., and El-abed, A. 2011. Extraction and characterization of chitin and chitosan from crustacean by-products: Biological and physicochemical properties. *African Journal of Biotechnology*, 10, 640–647.

Lin, D., and Zhao, Y. 2007. Innovations in the development and application of edible coatings for fresh and minimally processed fruits and vegetables. *Comprehensive Food Science and Food Safety*, 6, 60–75.

Liu, H., Du, Y., Wang, X., and Sun, L. 2004. Chitosan kills bacteria through cell membrane damage. *International Journal of Food Microbiology*, 95, 147–155.

López-Carballo, G., Higueras, L., Gavara, R., and Hernández-Muñoz, P. 2013. Silver ions release from antibacterial chitosan films containing *in situ* generated silver nanoparticles. *Journal of Agricultural and Food Chemistry*, 61, 260–267.

Lynch, M. F., Tauxe, R. V., and Hedberg, C. W. 2009. The growing burden of foodborne outbreaks due to contaminated fresh produce: Risks and opportunities. *Epidemiology and Infection*, 137, 307–315.

Marpudi, S. L., Abirami, L. S. S., Pushkala, R., and Srividya, N. 2011. Enhancement of storage life and quality maintenance of papaya fruits using *Aloe vera* based antimicrobial coating. *Indian Journal of Biotechnology*, 10, 83–89.

Martín-Belloso, O., and Fortuny, R. S. 2010. *Advances in Fresh-Cut Fruits and Vegetables Processing*. Boca Raton, FL: CRC Press.

Martínez-Abad, A., Ocio, M. J., Lagarón, J. M., and Sánchez, G. 2013. Evaluation of silver-infused polylactide films for inactivation of *Salmonella* and feline calicivirus *in vitro* and on fresh-cut vegetables. *International Journal of Food Microbiology*, 162, 89–94.

Martins, J. T., Cerqueira, M. A., Souza, B. W., Carmo Avides, M., and Vicente, A. A. 2010. Shelf life extension of ricotta cheese using coatings of galactomannans from nonconventional sources incorporating nisin against *Listeria monocytogenes*. *Journal of Agricultural and Food Chemistry*, 58, 1884–1891.

Martiñon, M. E., Moreira, R. G., Castell-Perez, M. E., and Gomes, C. 2014. Development of a multilayered antimicrobial edible coating for shelf-life extension of fresh-cut cantaloupe (*Cucumis melo* L.) stored at 4°C. *LWT—Food Science and Technology*, 56, 341–350.

Mastromatteo, M., Conte, A., and Del Nobile, M. A. 2010. Advances in controlled release devices for food packaging applications. *Trends in Food Science & Technology*, 21, 591–598.

Matan, N. 2012. Antimicrobial activity of edible film incorporated with essential oils to preserve dried fish (*Decapterus maruadsi*). *International Food Research Journal*, 19, 1733–1738.

Mehdizadeh, T., Tajik, H., Razavi Rohani, S. M., and Oromiehie, A. R. 2012. Antibacterial, antioxidant and optical properties of edible starch–chitosan composite film containing *Thymus kotschyanus* essential oil. *Veterinary Research Forum*, 3, 167–173.

Mei, J., Yuan, Y., Qizhen, G., Yan W., Li, Y., and Yu, H., 2013. Characterization and antimicrobial properties of water chestnut starch–chitosan edible films. *International Journal of Biological Macromolecules*, 61, 169–174.

Mild, R. M., Joens, L. A., Friedman, M., Olsen, C. W., McHugh, T. H., Law, B., and Ravishankar, S. 2011. Antimicrobial edible apple films inactivate antibiotic resistant and susceptible *Campylobacter jejuni* strains on chicken breast. *Journal of Food Science*, 76, M163–M168.

Min, B. J., Han, I. Y., and Dawson, P. L. 2010. Antimicrobial gelatin films reduce *Listeria monocytogenes* on turkey bologna. *Poultry Science*, 89, 1307–1314.

Mohan, C. O., Ravishankar, C. N., Lalitha, K. V., and Gopal, T. K. 2012. Effect of chitosan edible coating on the quality of double filleted Indian oil sardine (*Sardinella longiceps*) during chilled storage. *Food Hydrocolloids*, 26, 167–174.

Mor-Mur, M., and Yuste, J. 2010. Emerging bacterial pathogens in meat and poultry: An Overview. *Food and Bioprocess Technology*, 3, 24–35.

Moradi, M., Tajik, H., Razavi Rohani, S. M., and Oromiehie, A. R. 2011. Effectiveness of *Zataria multiflora* Boiss essential oil and grape seed extract impregnated chitosan film on ready-to-eat mortadella-type sausages during refrigerated storage. *Journal of the Science of Food and Agriculture*, 91, 2850–2857.

Moreira, M. D. R., Pereda, M., Marcovich, N. E., and Roura, S. I. 2011. Antimicrobial effectiveness of bioactive packaging materials from edible chitosan and casein polymers: Assessment on carrot, cheese, and salami. *Journal of Food Science*, 76, M54–M63.

Moreira, M. D. R., Roura, S. I., and Ponce, A. 2011. Effectiveness of chitosan edible coatings to improve microbiological and sensory quality of fresh cut broccoli. *LWT—Food Science and Technology*, 44(10), 2335–2341.

Morsy, M. K., Khalaf, H. H., Sharoba, A. M., El-Tanahi, H. H., and Cutter, C. N. 2014. Incorporation of essential oils and nanoparticles in pullulan films to control foodborne pathogens on meat and poultry products. *Journal of Food Science*, 79, M675–M684.

Nawa, Y., Hatz, C., and Blum, J. 2005. Sushi delights and parasites: The risk of fishborne and foodborne parasitic zoonoses in Asia. *Clinical Infectious Disases*, 41, 1297–1303.

Nguyen, V. T., Gidley, M. J., and Dykes, G. A. 2008. Potential of a nisin-containing bacterial cellulose film to inhibit *Listeria monocytogenes* on processed meats. *Food Microbiology*, 25, 471–478.

No, H. K., Park, N. Y., Lee, S. H., Hwang, H. J., and Meyers, S. P. 2002. Antibacterial activities of chitosans and chitosan oligomers with different molecular weights on spoilage bacteria isolated from tofu. *Journal of Food Science*, 67, 1511–1514.

Nørrung, B., and Buncic, S. 2008. Microbial safety of meat in the European Union. *Meat Science*, 78, 14–24.

Okamoto, S. 1978. Factors affecting protein film formation. *Cereals Food World*, 23, 256–262.

Resa, C. P. O., Gerschenson, L. N., and Jagus, R. J. 2014. Natamycin and nisin supported on starch edible films for controlling mixed culture growth on model systems and Port Salut cheese. *Food Control*, 44, 146–151.

Otoni, C. G., Moura, M. R. d., Aouada, F. A., Camilloto, G. P., Cruz, R. S., Lorevice, M. V., Soares, N. d. F. F., and Mattoso, L. H. C. 2014. Antimicrobial and physical-mechanical properties of pectin/papaya puree/cinnamaldehyde nanoemulsion edible composite films. *Food Hydrocolloids*, 41, 188–194.

Otoni, C. G., Pontes, S. F. O., Medeiros, E. A. A., and Soares, N. D. F. 2014. Edible films from methylcellulose and nanoemulsions of clove bud (*Syzygium aromaticum*) and oregano (*Origanum vulgare*) essential oils as shelf life extenders for sliced bread. *Journal of Agricultural and Food Chemistry*, 62, 5214–5219.

Ouattar, B., Simard, R. E., Piett, G., Bégin, A., and Holley, R. A. 2000. Inhibition of surface spoilage bacteria in processed meats by application of antimicrobial films prepared with chitosan. *International Journal of Food Microbiology*, 62, 139–148.

Park, S., Daeschel, M., and Zhao, Y. 2004. Functional properties of antimicrobial lysozyme–chitosan composite films. *Journal of Food Science*, 69, M215–M221.

Peressini, D., Bravin, B., and Sensidoni, A. 2004. Tensile properties, water vapor permeabilities and solubilities of starch–methylcellulose-based edible films. *Italian Journal of Food Science*, 16, 5–16.

Peretto, G., Du, W. X., Avena-Bustillos, R. J., Berrios, J. D. J., Sambo, P., and McHugh, T. H. 2014a. Optimization of antimicrobial and physical properties of alginate coatings containing carvacrol and methyl cinnamate for strawberry application. *Journal of Agricultural and Food Chemistry*, 62, 984–990.

Peretto, G., Du, W. X., Avena-Bustillos, R. J., Sarreal, S. B. L., Hua, S. S. T., Sambo, P., and McHugh, T. H. 2014b. Increasing strawberry shelf-life with carvacrol and methyl cinnamate antimicrobial vapors released from edible films. *Postharvest Biology and Technology*, 89, 11–18.

Pérez, L. M., Soazo, M. D. V., Balagué, C. E., Rubiolo, A. C., and Verdini, R. A. 2014. Effect of pH on the effectiveness of whey protein/glycerol edible filmscontaining potassium sorbate to control non-O157 shiga toxin-producing *Escherichia coli* in ready-to-eat foods. *Food Control*, 37, 298–304.

Persico, P., Ambrogi, V., Carfagna, C., Cerruti, P., Ferrocino, I., and Mauriello, G. 2009. Nanocomposite polymer films containing carvacrol for antimicrobial active packaging. *Polymer Engineering and Science*, 49, 1447–1455.

Phadke, G. G., Pagarkar, A. U., Sehgal, K., and Mohanta, K. N. 2011. Application of edible and biodegradable coatings in enhancing seafood quality and storage life: A review. *Ecology, Environment and Conservation*, 17, 619–623.

Phillips, L. G., Whitehead, D. M., and Kinsella, J. 1994. *Structure–Function Properties of Food Proteins*. San Diego, CA: Academic Press.

Picard, I., Hollingsworth, R. G., Wall, M., Nishijima, K., Salmieri, S., Vu, K. D., and Lacroix, M. 2013. Effects of chitosan-based coatings containing peppermint essential oil on the quality of post-harvest papaya fruit. *International Journal of Postharvest Technology and Innovation*, 3, 178–189.

Pitak, N., and Rakshit, S. K. 2011. Physical and antimicrobial properties of banana flour/chitosan biodegradable and self sealing films used for preserving fresh-cut vegetables. *LWT—Food Science and Technology*, 44, 2310–2315.

Ponce, A. G., Roura, S. I., del Valle, C. E., and Moreira, M. R. 2008. Antimicrobial and antioxidant activities of edible coatings enriched with natural plant extracts: In vitro and in vivo studies. *Postharvest Biology and Technology*, 49, 294–300.

Pranoto, Y., Rakshit, S. K., and Salokhe, V. M. 2005. Enhancing antimicrobial activity of chitosan films by incorporating garlic oil, potassium sorbate and nisin. *LWT—Food Science and Technology*, 38, 859–865.

Qin, C., Li, H., Xiao, Q., Liu, Y., Zhu, J., and Du, Y., 2006. Water-solubility of chitosan and its antimicrobial activity. *Carbohydrate Polymers*, 63, 367–374.

Quintavalla, S., and Vicini, L. 2002. Antimicrobial food packaging in meat industry. *Meat Science*, 62(3), 373–380.

Raafat, D., and Sahl, H. G. 2009. Chitosan and its antimicrobial potential—A critical literature survey. *Microbial Biotechnology*, 2, 186–201.

Rabea, E. I., Badawy, M. E., Stevens, C. V., Smagghe, G., and Steurbaut, W. 2003. Chitosan as antimicrobial agent: Applications and mode of action. *Biomacromolecules*, 4, 1457–1465.

Raikos, V. 2010. Effect of heat treatment on milk protein functionality at emulsion interfaces. A review. *Food Hydrocolloids*, 24, 259–265.

Ramos, Ó. L., Fernandes, J. C., Silva, S. I., Pintado, M. E., and Malcata, F. X. 2012. Edible films and coatings from whey proteins: A review on formulation, and on mechanical and bioactive properties. *Critical Reviews in Food Science and Nutrition*, 52, 533–552.

Ramos, M., Jiménez, A., Peltzer, M., and Garrigós, M. C., 2012. Characterization and antimicrobial activity studies of polypropylene films with carvacrol and thymol for active packaging. *Journal of Food Engineering*, 109, 513–519.

Ramos García, M., Bosquez Molina, E., Hernández Romano, J., Zavala Padilla, G., Terrés Rojas, E., Alia-Tejacal, L., Barrera-Necha, L., Hernandez-Lopez, M., and Bautista-Baños S. 2012. Use of chitosan-based edible coatings in combination with other natural compounds, to control *Rhizopus stolonifer* and *Escherichia coli* DH5a in fresh tomatoes. *Crop Protection*, 38, 1–6.

Ravishankar, S., Zhu, L., Olsen, C. W., McHugh, T. H., and Friedman, M. 2009. Edible apple film wraps containing plant antimicrobials inactivate foodborne pathogens on meat and poultry products. *Journal of Food Science*, 74, M440–M445.

Raybaudi-Massilia, R. M., Rojas-Graü, M. A., Mosqueda-Melgar, J., and Martín-Belloso, O. 2008. Comparative study on essential oils incorporated into an alginate-based edible coating to assure the safety and quality of fresh-cut Fuji apples. *Journal of Food Protection*, 71, 1150–1161.

Reps, A., Drychowski, L. J., Tomasik, J., and Winiewska, K. 2002. Natamycin in ripening cheeses. *Pakistan Journal of Nutrition*, 1, 243–247.

Rinaudo, M. 2008. Main properties and current applications of some polysaccharides as biomaterials. *Polymer International*, 57, 397–430.

Rojas-Graü, M. A., Avena-Bustillos, R. J., Olsen, C., Friedman, M., Henika, P. R., Martín-Belloso, O., Pan, Z., and McHugh, T. H. 2007a. Effects of plant essential oils and oil compounds on mechanical, barrier and antimicrobial properties of alginate-apple puree edible films. *Journal of Food Engineering*, 81, 634–641.

Rojas-Graü, M. A., Raybaudi-Massilia, R. M., Soliva-Fortuny, R. C., Avena-Bustillos, R. J., McHugh, T. H., and Martín-Belloso, O. 2007b. Apple puree-alginate edible coating as carrier of antimicrobial agents to prolong shelf-life of fresh-cut apples. *Postharvest Biology and Technology*, 45, 254–264.

Rojas-Graü, M. A., Soliva-Fortuny, R., and Martín-Belloso, O. 2009. Edible coatings to incorporate active ingredients to fresh-cut fruits: A review. *Trends in Food Science & Technology*, 20, 438–447.

Rossi-Márquez, G., Han, J. H., García-Almendárez, B. C.-T., E., and Regalado-González, C. 2009. Effect of temperature, pH and film thickness on nisin release from antimicrobial whey protein isolate edible films. *Journal of the Science of Food and Agriculture*, 89, 2492–2497.

Ruiz, H. A., Cerqueira, M. A., Silva, H. D., Rodríguez-Jasso, R. M., Vicente, A. A., and Teixeira, J. A. 2013. Biorefinery valorization of autohydrolysis wheat straw hemicellulose to be applied in a polymer-blend film. *Carbohydrate Polymers*, 92, 2154–2162.

Ruiz-Navajas, Y., Viuda Martos, M., Sendra, E., Perez Alvarez, J. A., and Fernández López, J. 2013. *In vitro* antibacterial and antioxidant properties of chitosan edible films incorporated with *Thymus moroderi* or *Thymus piperella* essential oils. *Food Control*, 30, 386–392.

Russel, A. D. 1991. Mechanisms of bacterial resistance to non-antibiotics: Food additives and food pharmaceutical preservatives. *Journal of Applied Bacteriology*, 71, 191–201.

Salgado, P. R., López-Caballero, M. E., Gómez-Guillén, M. C., Mauri, A. N., and Montero, M. P. 2013. Sunflower protein films incorporated with clove essential oil have potential application for the preservation of fish patties. *Food Hydrocolloids*, 33, 74–84.

Sant'Ana, A. S., Franco, B. D. G. M., and Schaffner, D. W. 2014. Risk of infection with *Salmonella* and *Listeria monocytogenes* due to consumption of ready-to-eat leafy vegetables in Brazil. *Food Control*, 42, 1–8.

Saurel, R., Pajonk, A., and Andrieu, J. 2004. Modelling of French Emmental cheese water activity during salting and ripening periods. *Journal of Food Engineering*, 63, 163–170.
Sawyer, L., and Kontopidis, G. 2000. The core lipocalin, bovine beta lactoglobulin. *Biochimica et Biophysica Acta*, 1482, 136–148.
Sayanjali, S., Ghanbarzadeh, B., and Ghiassifar, S. 2011. Evaluation of antimicrobial and physical properties of edible film based on carboxymethyl cellulose containing potassium sorbate on some mycotoxigenic *Aspergillus* species in fresh pistachios. *LWT—Food Science and Technology*, 44, 1133–1138.
Seol, K.-H., Lim, D.-G., Jang, A., Jo, C., and Lee, M. 2009. Antimicrobial effect of κ-carrageenan-based edible film containing ovotransferrin in fresh chicken breast stored at 5°C. *Meat Science*, 83, 479–483.
Shen, X., Wu, J., Chen, Y., and and Zhao, G. 2010. Antimicrobial and physical properties of sweet potato starch films incorporated with potassium sorbate or chitosan. *Food Hydrocolloids*, 24, 285–290.
Silva-Weiss, A., Ihl, M., Sobral, P. J. A., Gómez-Guillén, M. C., and Bifani, V. 2013. Natural additives in bioactive edible films and coatings: Functionality and applications in foods. *Food Engineering Reviews*, 5, 200–216.
Simon-Lukasik, K. V., and Ludescher, R. D. 2004. Erythrosin B phosphorescence as a probe of oxygen diffusion in amorphous gelatin films. *Food Hydrocolloids*, 18, 621–630.
Singla A. K., and Chawla M. 2001. Chitosan: Some pharmaceutical and biological aspects—An update. *Journal of Pharmacy and Pharmacology*, 53, 1047–1067.
Sofos, J. N. 2008. Challenges to meat safety in the 21st century. *Meat Science*, 78, 3–13.
Soon, J. M., Seaman, P., and Baines, R. N. 2013. *Escherichia coli* O104:H4 outbreak from sprouted seeds. *International Journal of Hygiene and Environmental Health*, 216, 346–354.
Souza, B. W. S., Cerqueira, M. A., Casariego, A., Lima, A. M. P., Teixeira, J. A., and Vicente, A. 2009. Effect of moderate electric fields in the permeation properties of chitosan coatings. *Food Hydrocolloids*, 23, 2110–2115.
Suppakul, P., Miltz, J., Sonneveld, K., and Bigger, S. W. 2003. Active packaging technologies with an emphasis on antimicrobial packaging and its applications. *Journal of Food Science*, 68, 408–420.
Tharanathan, R. N., and Kittur, F. S. 2003. Chitin—The undisputed biomolecule of great potential. *Critical Reviews in Food Science and Nutrition*, 43, 61–87.
Tripathi, S., Mehrotra, G. K., and Dutta, P. K. 2008. Chitosan based antimicrobial films for food packaging applications. *e-Polymers*, 8, 1082–1088.
Ture, H., Eroglu, E., Ozen, B., and Soyer, F. 2011 Effect of biopolymers containing natamycin against *Aspergillus niger* and *Penicillium roqueforti* on fresh kashar cheese. *International Journal of Food Science and Technology*, 46, 154–160.
Valencia-Chamorro, S. A., Palou, L., del Río, M. A., and Pérez-Gago, M. B. 2011. Antimicrobial edible films and coatings for fresh and minimally processed fruits and vegetables: A review. *Critical Reviews in Food Science and Nutrition*, 51, 872–900.
Valenzuela, C., Abugoch, L., and Tapia, C. 2013. Quinoa protein–chitosan–sunflower oil edible film: Mechanical, barrier and structural properties. *Food Science and Technology*, 50, 531–537.
Valverde, J. M., Valero, D., Martínez-Romero, D., Guillén, F., Castillo, S., and Serrano, M. 2005. Novel edible coating based on *Aloe vera* gel to maintain table grape quality and safety. *Journal of Agricultural and Food Chemistry*, 53, 7807–7813.
Vargas, M., Albors, A., and Chiralt, A. 2011. Application of chitosan–sunflower oil edible films to pork meat hamburgers. *Procedia Food Science*, 1, 39–43.
Vargas, M., Albors, A., Chiralt, A., and González Martínez, C. 2006. Quality of cold-stored strawberries as affected by chitosan–oleic acid edible coatings. *Postharvest Biology and Technology*, 41, 164–171.

Vargas, M., Albors, A., Chiralt, A., and González-Martínez, C. 2009. Characterization of chitosan–oleic acid composite films. *Food Hydrocolloids*, 23, 536–547.

Vásconez, M., Flores, S., Campos, C., Alvarado, J., and Gerschenson, L. 2009. Antimicrobial activity and physical properties of chitosan–tapioca starch based edible films and coatings. *Food Research International*, 42, 762–769.

Veldhuizen, E. J. A., Creutzberg, T. O., Burt, S. A., and Haagsman, H. P. 2007. Low temperature and binding to food components inhibit the antibacterial activity of carvacrol against *Listeria monocytogenes* in steak tartare. *Journal of Food Protection*, 70, 2127–2132.

Velickova, E., Winkelhausen, E., Kuzmanova, S., Alves, V., and Moldão Martins, M. 2013. Impact of chitosan–beeswax edible coatings on the quality of fresh strawberries (*Fragaria ananassa* cv. *camarosa*) under commercial storage conditions. *Food Science & Technology*, 52, 80–92.

Veraverbeke, W. S., and Delcour, J. A. 2002. Wheat protein composition and properties of wheat glutenin in relation to breadmaking functionality. *Critical Reviews in Food Science and Nutrition*, 42, 179–208.

Vodnar, D. C. 2012. Inhibition of *Listeria monocytogenes* ATCC 19115 on ham steak by tea bioactive compounds incorporated into chitosan-coated plastic films. *Chemistry Central Journal*, 6, 74.

Wilson, C. M. 1988. Electrophoretic analyses of various commercial and laboratory-prepared zeins. *Cereal Chemistry*, 65, 72–73.

Xiao, C., Zhu, L., Luo, W., Song, X., and Deng, Y. 2010. Combined action of pure oxygen pretreatment and chitosan coating incorporated with rosemary extracts on the quality of fresh-cut pears. *Food Chemistry*, 121, 1003–1009.

Xing, Y., Li, X., Xu, Q., Yun, J., Lu, Y., and Tang, Y. 2011. Effects of chitosan coating enriched with cinnamon oil on qualitative properties of sweet pepper (*Capsicum annuum* L.). *Food Chemistry*, 124, 1443–1450.

Zadow, J. G. 1994. Utilization of milk components: Whey. In R. K. Robinson (Ed.), *Modern Dairy Technology, Advances in Milk Processing*, vol. 1 (2nd ed.), London, UK: Chapman & Hall.

Zhai, M., Zhao, L., Yoshii, F., and Kume, T. 2004. Study on antibacterial starch/chitosan blend film formed under the action of irradiation. *Carbohydrate Polymers*, 57, 83–88.

Zhang, Y., and Jiang, J. 2012. Physical properties and antimicrobial activities of soy protein isolate edible films incorporated with essential oil monomers. *Advanced Materials Research*, 560–561, 361–367.

Zheng, L. Y., and Zhu, J. F. 2003. Study on antimicrobial activity of chitosan with different molecular weights. *Carbohydrate Polymers*, 54, 527–530.

Zhong, Y., Cavender, G., and Zhao, Y. 2014. Investigation of different coating application methods on the performance of edible coatings on Mozzarella cheese. *LWT—Food Science and Technology*, 56, 1–8.

Zhu, L., Olsen, C., Mchugh, T., Friedman, M., Jaroni, D., and Ravishankar, S. 2014. Apple, carrot, and hibiscus edible films containing the plant antimicrobials carvacrol and cinnamaldehyde inactivate *Salmonella* Newport on organic leafy greens in sealed plastic bags. *Journal of Food Science*, 79, M61–M66.

Zinoviadou, K. G., Koutsoumanis, K. P., and Biliaderis, C. G. 2009. Physico-chemical properties of whey protein isolate films containing oregano oil and their antimicrobial action against spoilage flora of fresh beef. *Meat Science*, 82, 338–345.

Zúñiga, G. E., Junqueira-Gonçalves, M. P., Pizarro, M., Contreras, R., Tapia, A., and Silva, S. 2012. Effect of ionizing energy on extracts of *Quillaja saponaria* to be used as an antimicrobial agent on irradiated edible coating for fresh strawberries. *Radiation Physics and Chemistry*, 81, 64–69.

9 Nanotechnology in Edible Packaging

*Marthyna Pessoa de Souza
and Maria G. Carneiro-da-Cunha*

CONTENTS

Abstract	288
9.1 Introduction	288
9.2 Nanotechnology	289
9.2.1 A Research Area with Multidisciplinary Approaches and Materials	289
9.2.2 New Knowledge Level with Scientific and Economic Impacts	290
9.2.3 Expectations of Forthcoming Developments	291
9.2.4 Nanomaterials	291
9.3 Nanotechnology in Food Packaging	292
9.3.1 Active Packaging	292
9.3.2 Applications of Nanomaterials and Nanotechnologies	293
9.3.3 Naturally Occurring Nanomaterials	293
9.3.3.1 Polysaccharides	293
9.3.3.2 Proteins	293
9.3.3.3 Lipids	294
9.3.3.4 General Properties of These Nanomaterials	294
9.3.4 Nanolaminates	294
9.3.5 Electrostatic LbL Self-Assembly Technique	295
9.3.5.1 Dipping LbL Method	295
9.3.5.2 Spray LbL Method	296
9.3.5.3 Advantages and Disadvantages of LbL Technique	296
9.3.6 Control of Multilayer Characteristics	297
9.3.7 Main Factors Affecting the Assembly of Adsorbed Multilayer Films	297
9.3.7.1 pH	297
9.3.7.2 Ionic Strength	298
9.3.7.3 Concentration	298
9.3.7.4 Other Factors	298
9.3.8 Applications of Multilayers	299
9.4 Encapsulation of Bioactive Molecules	301
9.4.1 Nanoparticles	301
9.4.2 Nanoemulsions	302
9.4.3 Bio-Nanocomposites	303

9.5	Tools for Characterization of Nanosystems	305
	9.5.1 Light Scattering	306
	9.5.2 Zeta Potential	307
	9.5.3 Ultraviolet-Visible Spectroscopy	307
	9.5.4 Contact Angle	307
	9.5.5 Nanoindentation	308
	9.5.6 Electron Microscopy	308
	9.5.6.1 Scanning Electron Microscopy	309
	9.5.6.2 Transmission Electron Microscopy	309
	9.5.7 Atomic Force Microscopy	309
	9.5.8 Confocal Laser Scanning Microscopy	310
9.6	Nanotechnology: Risk, Regulation, and Future Trends	310
Acknowledgment		311
References		311

ABSTRACT

This chapter provides an overview of the latest food packaging evolutions and expectations of forthcoming developments involving the use of nanotechnology, with a specific focus on edible packaging. Food protection with environmental and health concerns, the concept of nanotechnology, the interest in the development of biodegradable and renewable materials for food packaging, applications of nanomaterials and nanotechnologies, tools for coating systems characterization, some of the latest examples, and other topics are pointed out.

9.1 INTRODUCTION

Food preservation is a very ancient practice and has always been a challenge for human being. Since ancient times, not only through dehydration by the sun, smoking, and salting, but also by preservation in containers with honey, olive oil, wine, vinegar, fat, clay, and wax up to current times with other technologies, food packaging turned out to become important in this continuous route. At present, food packaging is not always used only as a container but also to maintain the quality and ensure the safety of food products during transport and storage and to extend the preservation time of perishable food products.

Over the years, in this route, different physical and chemical techniques have been developed to extend the shelf life of minimally or not processed food produce. Processes such as refrigeration, disinfection, ethylene absorption, gamma irradiation, edible coating, chemical dipping, and controlled/modified atmosphere were involved. The fact is that, for instance, because different fruits react differently to different treatments, it was imperative to search for the right combination of treatments to extend their shelf life (Bico et al., 2009). Great losses (from 20% to 80%) in the quality of fresh fruits occurring from harvesting to final consumption allied to fruits' short shelf life were of great concern to distribution chains. On the contrary, consumers around the world began to seek high-quality foods, without chemical preservatives and with extended shelf life. As a consequence, increased efforts have been made to find natural and antimicrobial preservatives and new packaging materials derived from renewable resources were developed, imposing a modified atmosphere surrounding the commodity (Cerqueira et al., 2009).

Recently, a new class of materials represented by bio-nanocomposites has been considered as a promising option in improving the properties of biopolymer-based packaging materials. Bio-nanocomposites opened an opportunity for the use of new, high performance, light weight, green nanocomposite materials to replace conventional nonbiodegradable petroleum-based plastic packaging materials. Such biodegradable packaging materials should be obtained from renewable biological resources (Rhim et al., 2013).

Meanwhile, coating operations have been widely used in various industries. There are many advantages for the use of coatings also in food systems, such as to extend the shelf life of material by retarding the transfer of moisture and volatiles and to improve the consumer acceptability by altering the appearance and color of foods, including creating a glossy surface, and masking undesirable tastes. Coatings can also be used to develop new types of foods (e.g., multilayered foods with different tastes) and to incorporate healthy additives. Edible coatings have been applied to a variety of foodstuffs, using diverse techniques. All these techniques exhibit several advantages and disadvantages and the selection of an appropriate method depends on the characteristics of food, coating materials, intended effect of the coating and cost (Andrade et al., 2013).

Increased consumer demands for minimally processed and ready-to-eat fresh foods have motivated researchers to develop alternative new technologies for securing food safety and providing healthy food (Kanmani and Rhim, 2014). Active packaging has been one of the most innovative developments in food packaging during the last decades. Active packaging provides extension of food shelf life and the maintenance or even improvement of their quality due to its interaction with food and/or environment (Noronha et al., 2014).

This chapter will help the reader to understand the use of edible coatings in food protection with environmental and health concerns for humans, the contribution and the importance of nanotechnology in materials and technologies used in the construction and characterization of those coatings, the models and forms used, examples of application and expectations regarding future developments.

9.2 NANOTECHNOLOGY

9.2.1 A Research Area with Multidisciplinary Approaches and Materials

In nanotechnology, the matter is manipulated at the atomic and molecular scale, creating new materials and processes with different functional characteristics from common materials. A nanometer is one billionth of a meter or one millionth of a millimeter. It is not just the study of the very small, but it is the practical application of this knowledge.

At the atomic level, there are new forces, new kinds of possibilities, new effects occurring, different laws, being so expected different situations, and behaviors (Feynman, 2003). Extreme surface-to-volume ratios of the particles are characteristic of nanoscaled materials. Compared with macroscaled materials, this results in entirely different physical and chemical properties (Weiss and Gibis, 2013).

Nanotechnology is a very broad and multidisciplinary area of research and development that incorporates many scientific and engineering disciplines including physics, chemistry, biology, biotechnology, material science, molecular biology, and medicine (Schummer, 2004). Being a highly multidisciplinary field, it involves a number of fields such as applied physics, materials science, device physics, chemistry, chemical and textile engineering, polymer engineering, electronic engineering, biological engineering, among others.

Nanotechnology, as an advanced technology, is seen as a new scientific revolution that promises to solve many problems of social life, such as those relating to the environment, health, food, genetic engineering, modernization of the artifacts of war, and so on (Silva et al., 2012).

Nanostructured materials are already used in luxury goods such as golf and bowling balls, in the fabrication of high-performance tires and stain- and wrinkle-resistant fabrics, in cosmetics and new therapeutic treatments, in filters and membranes for water purification and other environmental "solutions," in the improvement of production processes through the introduction of more resistant or efficient materials, or in the design of new materials for uses that range from electronics and practically the entire transportation industry to inputs used in chemical–biological weapon detectors or for the fabrication of more sophisticated weapons (Arnaldi et al., 2011).

Nanotechnology is also being used already in some countries in the production of agricultural products, processed foods and drinks, and in food packaging (Coles and Frewer, 2013). The concepts of potential use of nanomaterials in food and the implied benefits for stakeholders including consumers have not changed significantly since 2009. New products are being developed and claimed to enter in the market, but the available data from published sources and databases do not allow verifying whether product ideas are just concepts, unsubstantiated claims, or already resulting in exposure of consumers to food being produced with nanotechnology/nanomaterials at any significant rate (Takeuchi et al., 2014).

9.2.2 New Knowledge Level with Scientific and Economic Impacts

Nanotechnology represents a new level of knowledge, with immense and yet not properly measured scientific and economic impacts, and has mainly led the leading countries such as the United States, Japan, and the European Community, for national or regional initiatives to encourage and privilege funding to the area, seeking new levels of competitiveness for their companies. This type of behavior also has already been reproduced by the emerging economies. Thus, nanotechnology plays a role with transformative potential, inducing innovation, competitive restructuring of the economy, and conditioning of changes in the way individuals interact with the technological world. This technological platform brings a quickening breath, broad spectrum, for the industrial structure and technological services, through the incorporation of processes, productive segments, and new products based on nanoscale operations.

According to Roco (2011), the global nanotechnology R&D annual investment from private and public sources showed an average annual growth rate worldwide of approximately 35% between 2000 and 2008 and reached about $1.4 billion in 2008. However, venture capital funds decreased about 40% during the 2009 financial crisis. The value of products incorporating nanotechnology as the key component reached about $200 billion in value worldwide in 2008. One may estimate that by 2020, there will be about $3 trillion in products that incorporate nanotechnology, and the nanotechnology markets and related jobs are expected to double every 3 years.

9.2.3 Expectations of Forthcoming Developments

From the report "Food Nano 2040" consultants "Helmut Kaiser Consulting" on "Nanotechnology in food, food processing, packaging and consumption, state of science, technology, market, applications and developments for 2015 and 2040," extracts up succinctly that nanotechnology will have an impact on the food industry, growth and production of foods, as these will be packed, transported, and consumed. Companies are developing nanomaterials that will make a difference not only in the taste of the food but also in the safety of it and distributed food with health benefits. Nanotechnology will transform completely the food industry over the next 20 years. In agriculture, it promises various applications aimed at the reduction in uses of pesticides and water, improving the reproduction of plants and animals, and creating nano-bio-industrial products (Kaiser, 2014). The same report also emphasizes that in "2040 will be common the use of nanoproduced food, which has the correct nutritional composition and the same taste and texture of organically produced food, meaning that the availability of food is no longer affected by limited resources, bad crop weather, water problems or others."

9.2.4 Nanomaterials

Nanomaterials consist of structures with one or more external dimensions in the size range of 1–100 nm (Verleysen et al., 2014). Often only one or two dimensions are in the nanoscale, as in quantum wells and nanowires, but sometimes all three dimensions are nanoscale, as in quantum dots and nanocrystals (Adams and Barbante, 2013). If only one of the dimensions is restricted, we have a layered shape or 2-D material; if two dimensions are limited in size, we will have a wired or 1-D material; if all dimensions are in the range of a few nanometers, we have 0-D materials (Adams and Barbante, 2013; Huang et al., 2013). However, these nanomaterials do not necessarily have nanosize but have in their composition nanoscale structures that generate new properties and applications. According to the European Commission Recommendation (2011/696/EU), a nanomaterial is defined as follows:

1. A natural, incidental, or manufactured material containing particles, in an unbound state or as an aggregate or an agglomerate, and whose number–size distribution, 50% or more of the particles have one or more external dimensions in the size range between 1 and 100 nm.
2. In specific cases and always, it is justified due to environmental and related health, safety, and competitiveness concerns, the threshold of the number–size distribution of 50% may be replaced by a threshold between 1% and 50%.

3. It is considered that a material falls within the definition of nanomaterials (point 1) if its specific surface area per volume exceeds 60 m^2 cm^{-3}. It should, however, be regarded as a material that, according to their number–size distribution, constitutes a nanomaterial corresponds to the definition in point 1, although its specific surface area be less than 60 m^2 cm^{-3}.

9.3 NANOTECHNOLOGY IN FOOD PACKAGING

9.3.1 Active Packaging

Active packaging is an extension of the protection function of traditional food packaging enriched with components that enable the release or absorption of substances into or from the packaged food or the environment surrounding the food. Although the concepts of active and intelligent packaging are often used interchangeably in literature, they are not the same. Intelligent packaging is a system in which the product, the package, and the environment interact in a positive way to extend shelf life, improve the packaged food condition, or achieve some characteristics that cannot be obtained otherwise. Intelligent packaging and active packaging are not mutually exclusive. Both packaging systems can work synergistically to realize so-called smart packaging. Smart packaging provides a total packaging solution that on the one hand monitors changes in the product or the environment (intelligent) and on the other hand acts upon these changes (active) (Vanderroost et al., 2014).

Nanotechnology has offered different possibilities to provide foods with longer packaging life, safer packaging, better traceability, and healthier characteristics (Brody, 2003; Chaudhry et al., 2008; Neethirajan and Jayas, 2011; Weiss et al., 2006). Its applications in the food industry range from intelligent packaging to the creation of on-demand interactive foods that allow consumers to modify them, depending on the nutritional needs and tastes. The key to future advances in flexibility of active and intelligent packaging is in polymer nanocomposite technology (Neethirajan and Jayas, 2011).

Intelligent food packaging can sense when its contents are spoiling, detecting the type of spoilage source and the type of deterioration and warning the consumer about it. This is the case of packaging containing nanosensors extremely sensitive to gases released by spoiled food, causing the sensor to, for example, change its color, thus giving a clear visible signal if the food is fresh or not (Ruengruglikit et al., 2004). Active packaging, in turn, will, for example, release antimicrobials, flavors, colors, or nutritional supplements into the food if triggered by, for example, temperature or pH changes (Neethirajan and Jayas, 2011).

Active packaging is currently one of the most dynamic technologies used to preserve the quality, safety, and sensory properties of food, preventing oxidation, controlling and avoiding the formation of undesirable flavors and textures in the food. Most studies in this field have concerned active packaging developed with antimicrobial (Cerqueira et al., 2014; Lee, 2010; Medeiros et al., 2014; Zhou et al., 2010) and antioxidant (Barbosa-Pereira et al., 2013; Woranuch and Yoksan, 2013a) agents, which can be directly incorporated into packaging films. Encapsulating bioactive substances inside the package itself has also been a promising approach because it allows the release of active compounds in a controllable manner. This behavior has been shown not only by the structure of nanocoatings (Medeiros et al., 2012a, 2014) but also by the incorporation of food grade additives, generally recognized as safe (GRAS), nanomaterials as nanoparticles (NPs), nanoemulsions or micellar systems, and nanocomposites in food coatings (Brasil et al., 2012; Jiang et al., 2013; Kanmani and Rhim, 2014; Martiñon et al., 2014; Noronha et al., 2014), many of which are known as "nanoadditives."

9.3.2 Applications of Nanomaterials and Nanotechnologies

The main benefits offered by nanomaterials and by nanotechnologies in the development of food packaging are reflected in their actual applications, covering (Bradley et al., 2011): (i) innovation—It is the main driver for applications of nanomaterials and for the development of new products in the food packaging sector. These new products can provide greater choice and more convenience to consumers as well as supporting social change and lifestyles. It may also open new markets, promote economic growth and create employment; (ii) economy—The use of less material in the packaging synthesis, but with the same technical performance; (iii) increased protection and preservation of food—Better barrier properties can help maintaining food quality and increasing shelf life without the addition of chemical preservatives. This can potentially provide a supply of cheaper and more reliable foods, promoting better nutrition, and less food waste; and (iv) improved performance of biomaterials—Boosted by the incorporation of nanomaterials or by the use of nanotechnology tools to study the properties of materials and improve production methods, allowing the replacement of synthetic polymers by locally sourced biomaterials.

9.3.3 Naturally Occurring Nanomaterials

Edible packaging using nanotechnology needs materials at the nanometer scale compatible with human consumption. This is the reason why the three groups of naturally occurring substances presented below have preferably been used in this research area.

9.3.3.1 Polysaccharides

In general, polysaccharides have been used as thickening and gelling agents in processed foods and, in the last decade, frequently referred in research works aimed at producing food coatings. The building blocks of polysaccharides are sugars. Polysaccharides are generally high-molecular-weight chain-like structures with diameters typically of a few nanometers in size. They can thus be considered as naturally occurring nanomaterials. Due to their complex structure, individual polysaccharides often have distinct functional properties (Chaudhry et al., 2010).

9.3.3.2 Proteins

Proteins are essentially high-molecular-weight linear polymers composed of a chain of amino acids. The composition and the sequence of amino acids determine the local secondary structure within the protein as well as its overall tertiary structure. In aqueous solution, the folded structure of a protein tends to bury hydrophobic regions within the interior of the molecule and expose hydrophilic regions on the protein surface. Most of the food proteins obtained from plants, blood, or milk are globular structures, although the caseins are present in milk as complex colloidal structures, called casein micelles. These micelles can be dissociated to obtain individual casein proteins. Proteins such as myosin and collagen, which play specialized structural roles in meat tissue, are rod-like structures. Globular proteins are generally tenths of nanometers in size, and the diameters of rod-like proteins are a few nanometers in size. These molecules are also an example of natural nanomaterials (Chaudhry et al., 2010).

9.3.3.3 Lipids

Lipids are linear polymers less than 2 nm in thickness. Lipids are a group of naturally occurring molecules that include fats, waxes, fat-soluble vitamins (such as vitamins A, D, E, and K), phospholipids, and others. They have applications in cosmetic and food industries as well as in nanotechnology (Mashaghi et al., 2013). The structure of the phospholipid molecule generally consists of hydrophobic tails and a hydrophilic head; for example, the phospholipid lecithin has been used in formulations of NPs (Sonvico et al., 2006). Lipids can self-assemble into nanofilms and other nanostructures such as micelles, reverse micelles, and liposomes (Israelachvili, 2011).

9.3.3.4 General Properties of These Nanomaterials

One of the main challenges of using biopolymers as food packaging materials is to adapt their oxygen and water permeability to the food product requirements. It is well known that barrier properties are dependent on a large number of interrelated factors including chemical affinity between the permeant and the matrix, structural organization of polymer chains, hydrogen-bonding characteristics, degree of crosslinking, processing technology, and degree of crystallinity (Caner, 2011).

Being of a hydrophilic nature, the primary advantages of most polysaccharides when used in packaging materials are their structural stability and ability to slow down oxygen transmission. Proteins are, in general, very poor moisture barriers and unsuitable for controlling mass transfer of O_2, CO_2, and other gases, but their major advantage is their structural stability, which makes it possible, for example, to hold a required shape. Proteins and polysaccharides have poor water vapor barrier properties due to their hydrophilic nature but they exhibit very good oxygen permeability values under dry conditions (Fabra et al., 2013). However, under wet conditions, their oxygen barrier properties are poorer due to swelling and plasticization of the polymer in the presence of diffused water molecules (Fabra et al., 2010; Jagannath et al., 2003; Yam 2009). Lipids, mainly waxes and resins, are known to produce opaque and brittle films with very low water vapor permeability (WVP). Beeswax is one of the most effective lipid materials used to decrease WVP of edible films due to its high hydrophobicity and its ability to be in solid state at room temperature (Fabra et al., 2008). It is generally accepted that polymer blending may provide an effective route for biopolymer properties diversification.

9.3.4 NANOLAMINATES

A nanolaminated coating (or a nanolayered coating) consists of two or more layers of materials with nanometric dimensions, physically or chemically bound (Weiss

Nanotechnology in Edible Packaging

et al., 2006) and is obtained by alternated deposition of oppositely charged polyelectrolytes on the surface of a solid substrate, by dipping or spray. Its assembly is often achieved using the "electrostatic layer-by-layer (LbL) self-assembly" technique, which is mainly based in electrostatic interactions. Since the early 1990s of the twentieth century, with the first report by Decher in 1990, the LbL technique has been widely applied in the assembly of thin films and coatings. The process is basically based on the alternated adsorption of materials with opposite charges onto a substrate, dipping it into a solution, where the characteristics of different surfaces can be modified by coating of multiple nanolayers of different materials, allowing the precise control over the thickness and the interfacial properties of the coating (Decher and Schlenoff, 2003). As time goes on, the process adopted the name of conventional dipping LbL method in opposition to the spray LbL method.

A film is a thin skin formed, for instance, through *casting* of the biopolymer solution prepared separately from the food and subsequently applied to it, while a coating can be a suspension or an emulsion applied directly on the surface of the food, leading to the subsequent formation of a film. On the contrary, self-standing films have an extremely thin nature making them very fragile. Due to this reason, it is more likely that the nanolaminates are used as coatings on food surfaces instead of attached to them (Kotov, 2003). For example, in the framework of antimicrobial packaging, Rojas-Graü et al. (2007) working with essential oils and alginate apple puree have created edible food films by dipping that are able to kill *Escherichia coli*.

9.3.5 Electrostatic LbL Self-Assembly Technique

The electrostatic LbL (both dipping LbL and spray LbL) is a self-assembly technique, since the system adopts the most suitable configuration "on its own," according to thermodynamic factors (Whitesides and Grzybowski, 2002).

9.3.5.1 Dipping LbL Method

The conventional dipping LbL method, illustrated in Figure 9.1, requires three steps in the deposition process of each coating layer. The first one is the immersion of food for a short period of time and at a given temperature into a polyelectrolyte solution;

FIGURE 9.1 Schematic representation of application steps of a nanolayered coating with the dipping LbL method.

the second one consists of washing with water at the same pH of the polyelectrolyte solution for removal of the excess of polyelectrolyte; and the third one concerns with drying of the adsorbed layer. The procedure is subsequently repeated with an oppositely charged polyelectrolyte solution. The number of alternated layers of the coating is dependant on the desired purpose. This method is suitable for small pieces of food with various shapes and small surfaces.

9.3.5.2 Spray LbL Method

This method requires, in principle, the same steps of the above method. The main difference is that the deposition is made by spraying and not by dipping. However, there are several aspects that characterize it, some of which are further developed in the next section.

9.3.5.3 Advantages and Disadvantages of LbL Technique

9.3.5.3.1 Dipping LbL Method

The conventional dipping LbL method is frequently preferred by researchers for nanocoatings of fruits whose skin is generally a low-energy surface (values lower than 100 mN m^{-1}). The self-assembly process may also be governed by nonionic interactions such as, for instance, hydrogen bonds, hydrophobic interactions, van der Waals interactions, and biospecific interactions (e.g., specific enzyme/substrate, lectin/carbohydrate), influencing the stability, morphology, thickness, and film properties (Lvov et al., 1996; Villiers et al., 2011).

The necessary equipment for production of films by dipping LbL is simple when compared with other techniques or methods used for the synthesis of ultrathin films, regardless of the form and type of substrate used, which makes it cheaper and with great technological potential for large-scale application. With the equipment that is currently available in packing houses (facility where fruits and vegetables are received, selected, cleaned, and a light mono coating of [eventually natural] wax is applied to help the foods to retain moisture and enhance their appeal, prior to distribution to the market), this method could be used. The advantages of the dipping LbL method are (i) simplicity, both of the process and of the equipment needed, (ii) allowing coating of a wide variety of surfaces, (iii) availability of a wide variety of natural and synthetic polyelectrolytes, (iv) possibility of applying different bioactive substances, (v) flexibility of use of substrates with irregular shapes and sizes, and (vi) formation of stable coatings and control over the thickness of the layers.

However, the dipping LbL methodology presents practical limitations, mainly related to long periods of time for a complete layer adsorption and to the relatively small coating areas required for the thin film construction (Aoki et al., 2012).

9.3.5.3.2 Spray LbL Method

This method was developed by Schlenoff et al. (2000) to overcome the limitations of dipping LbL methodology. Currently, the spray LbL method is being applied to the fabrication of thin films of different kinds of materials, depending on the different applications (Aoki et al., 2014). Although it is suitable for big substrates with large surfaces but limited shapes, this method has been a practice commonly used for coating whole fruits and vegetables in the packing houses.

The spraying pressure value should be controlled to avoid negative consequences, for example, dispersion of suspensions of lipid vesicles can be difficult due to vesicles' fragility, which might collapse with high spray pressure. The average mass deposited increases with the spraying time, and the thickness of spray LbL films was found to be greater in comparison with dipping LbL film. In fact, the droplets do not cover completely the adjacent layer, and the electrostatic interactions between the oppositely charged groups are not the main driving forces involved in the growth of spray LbL films, contrarily to what happens with dipping LbL films (Aoki et al., 2012).

9.3.6 Control of Multilayer Characteristics

The composition, thickness, and properties of the multilayers formed around the food can be controlled by several manners, including (Decher and Schlenoff, 2003; Villiers et al., 2011; Weiss et al., 2006): (i) change of the type of adsorbed substance in the dipping solution, (ii) concentration of solution containing the substance to be adsorbed, (iii) variation in the total number of immersions, (iv) change in the immersion order of food into the various solutions, (v) change of environmental conditions used, such as pH, ionic strength, dielectric constant, and temperature, and (vi) drying process (temperature and relative humidity).

9.3.7 Main Factors Affecting the Assembly of Adsorbed Multilayer Films

In the electrostatic LbL self-assembly technique at the nanometer scale, there are three main factors that can affect the assembly, the thickness of the layers, and the stability of the complex: pH, ionic strength, and concentration of the polyelectrolyte solution (Ai et al., 2003).

9.3.7.1 pH

The pH of polyelectrolyte solutions should be selected, aiming at maintaining a high degree of polyion ionization to ensure an effective LbL self-assembly process. A requirement for multilayer formation is that the addition of an oppositely charged polyelectrolyte to a charged surface results in charge reversal of the surface. When meant for edible packaging applications, normally these LBL-based structures are constructed of polysaccharides or polysaccharides and proteins. The polysaccharides may be either polycations or polyanions, depending on their functional group. An anionic polysaccharide has a full negative charge at pH well above the pK_a value of its side groups (typically carboxyl [$pK_a \sim 4$] or sulfate groups [$pK_a \sim 1$]), but progressively loses its charge as the pH is decreased. For example, the polycation polysaccharide chitosan, composed of glucosamine residues bound via the glycoside bond, has a $pK_a \sim 6.5$, so the amino groups can be positively charged at values of pH around 3.0. Also in the case of protein molecules, the electrical charges change from positive to negative when the pH is altered from below to above their isoelectric point (pI). According to Ai et al. (2003), if the pH value of the coating solution is too close to the isoelectric point (pI) of the polyion used, the charge is not sufficient to support the assembly (at least 10% of pendant groups have to be ionized). In weak polyelectrolytes, highly charged polymer chains tend to be adsorbed as thin layers with flat chain conformation, whereas the less loaded polymer chains tend to be adsorbed as thicker structures, hook type (loop) (Yoo et al., 1998).

9.3.7.2 Ionic Strength

Since the first report in 1992 by Decher et al., it has been clearly demonstrated that through ionic strength adjustments, the thickness of the adsorbed layers can be fine-tuned at the molecular level. However, this approach is limited to a certain extent by the poor solubility of high-molecular-weight polyelectrolytes in solutions of high ionic strength (adding salt makes water a poorer solvent for polyelectrolytes) (Yoo et al., 1998). The phosphate-buffered saline solution at pH 7.4 is appropriate to maintain the charge of many polyanions and polycations, including polysaccharides, while also providing physiological ionic strength, which is important for protein or enzyme assembly. The thickness of each layer in the LbL film can be finely adjusted by changing the ionic strength of the solution, which in turn induces polymer coil formation. Normally higher ionic strengths lead to thicker films (Ai et al., 2003).

9.3.7.3 Concentration

The concentration of polyelectrolyte solutions can influence the construction of multilayers and the layers' thickness. The nature of the intervening element in the solution can affect the layer thickness, making it rough due to interpenetrations of the adjacent layers. This means that the thickness of a certain layer is dependent, in several degrees, on the type of biopolymer used. For systems containing weak polyelectrolytes (e.g., pectin and chitosan), the layer thickness is dependent on the concentration of the corresponding biopolymer, and this effect is particularly marked, for example, for pectin and weaker, for example, for chitosan (Marudova et al., 2005). Several works dealing with building nanomultilayer systems have typically used polysaccharide concentrations ranging between 0.1% and 0.5% (Carneiro-da-Cunha et al., 2010; Fu et al., 2005; Medeiros et al., 2014; Pinheiro et al., 2012a, 2012b; Radeva et al., 2006; Richert et al., 2004).

9.3.7.4 Other Factors

In the electrostatic LbL self-assembly process, the uniform spreading of polyelectrolyte solutions on the surface to be treated is a basic requirement. The hydrophobicity or hydrophilicity degree, both of substrate and of polyelectrolyte solution, will contribute not only to a more efficient spreading process but also to the desired saturation of the surface.

The surface characteristics of a film or coating depend mainly on the nature of the outer layer, but it may happen that underlying layers also have an influence in such characteristics.

The LbL process itself, with immersions and drying steps, may also affect the characteristics of a given coating. For instance, besides the two elements—glycosidic bonds and side groups—of utmost importance in the chemistry of polysaccharides, hydration, and chemical reactivity can also directly affect the polysaccharide's charge density. The polysaccharide-based polyelectrolyte multilayer films are among the most highly hydrated, and their swelling properties (i.e., their ability to change volume and thickness as the environmental conditions are changed) are dependent on assembly conditions. Hydration can impact film thickness, swellability, and diffusion of components. Consequently, when the assembly of films or coatings is done by alternated immersions into polyelectrolyte solution followed by intermediate drying steps, the changes in swelling between the dried and the wet states can be very high (Crouzier et al., 2010), and temperature, time, and the environment's relative humidity may interfere with the process.

9.3.8 Applications of Multilayers

Over the last years, a great number of scientific articles have been published reporting the efficiency of nanomultilayers as an alternative new technology for ensuring food safety, providing healthy food and also extend shelf life. Some examples are pointed out herein.

Chitosan and sodium alginate, two oppositely charged polysaccharides, were deposited on aminolyzed/charged polyethylene terephthalate (A/C PET) support and the nanolaminate obtained with five alternated layers with a total thickness of 121.28 nm, showed good barrier property to water vapor and highly functional mechanical properties with great potential for use as edible coatings or coatings for biomedical devices (Carneiro-da-Cunha et al., 2010).

Subsequently, Souza et al. (2015) showed that chitosan–alginate nanomultilayer edible coating extended the shelf life of fresh-cut mangoes up to 8 days. Lower values of mass loss, pH, malondialdehyde content, browning rate, soluble solid, microorganism proliferation, and higher titratable acidity were found in coated fresh-cut mangoes stored under refrigeration (8°C).

In another work, the interactions between two polysaccharides [κ-carrageenan (negatively charged) and chitosan (positively charged)] forming five nanolayers assembled on a PET support showed this nanolayer formation to be an exothermic process. It was shown to occur mainly due to electrostatic interactions existing between the two polyelectrolytes, although other types of interactions such as dipole–dipole, hydrogen bonds, and conformational changes of both polyelectrolytes may also have been involved, resulting in the formation of a stable multilayer structure. This nanocoating presented barrier and mechanical properties compatible for use in biomedical and food industry (Pinheiro et al., 2012a). In a subsequent work, the same research group evaluated the morphology and architecture of this nanolayered coating, incorporating methylene blue (MB) in different positions (layers) of the nanolayered coating, as model the compound as aiming at evaluating its loading and release behavior. The results revealed that the amount of MB loaded increased with the distance from the first layer, suggesting that the MB was able to diffuse into the κ-carrageenan/chitosan nanolayered coating and adhered to the surface of the layer immediately below it. However, depending on temperature and pH of the medium and on the position of MB incorporated on the nanolayered coatings, different diffusion mechanisms prevail (Pinheiro et al., 2012b).

Chitosan and fibroblast growth factor-2 (FGF-2) were assembled with starch by LbL on a silicon wafer and quartz glass support, and the process was monitored by a profilometer and scanning electron microscopy (SEM) (Cho et al., 2013). It was found that FGF-2 release rate could be controlled, changing the thermally induced crystallinity of these starch-based multilayer films.

Nanolaminates with five alternated layers of pectin (anionic polysaccharide) and chitosan (cationic polysaccharide) showed antimicrobial activity and low oxygen permeability. The combination of these properties was fundamental in the extension of "Tommy Atkins" mangoes' shelf life (Medeiros et al., 2012b). Also, an edible nanocoating with five alternated layers of κ-carrageenan, a polysaccharide with good barrier properties, and lysozyme, a protein with antimicrobial action, was applied directly on "Rocha" (*Pyrus communis* L.) whole and minimally processed peeled pears and contributed to extend their shelf life by 6 days (Medeiros et al., 2012a). A similar effect was reported for a nanocoating with five alternated layers of alginate and lysozyme, applied directly on "Coalho" cheese. The combination of the gas barrier properties and antibacterial properties of the coating had a positive effect on the physical and chemical parameters of the cheese, which led to an extension of its shelf life by 8 days (Medeiros et al., 2014). Table 9.1 summarizes these applications.

The electrostatic LbL self-assembly technique not only has proved efficient for multilayer nanocoatings application on food, but has also been successfully used to obtain nanoscale structures with incorporated active molecules, for subsequent incorporation into edible coatings.

TABLE 9.1
Nanolaminate Coatings Applications

Polyelectrolytes				
Polyanion	Polycation	Properties	Potential Use	References
Heparin	Chitosan	Antiadhesive and antibacterial	Surface modification of cardiovascular devices	Fu et al. (2005)
Alginate	Chitosan	Water vapor barrier	Coating for biomedical devices or edible coating	Carneiro-da-Cunha et al. (2010)
Alginate	Chitosan	Water vapor barrier and antimicrobial	Shelf-life extension of fresh-cut mango	Souza et al. (2015)
Alginate	Lysozyme	Antimicrobial and good barrier properties	Shelf-life extension of cheese	Medeiros et al. (2014)
κ-Carrageenan	Chitosan	Gas barrier	Biomedical and food industry	Pinheiro et al. (2012a)
Pectin	Chitosan	Antimicrobial activity and low O_2 permeability	Shelf-life extension of mango	Medeiros et al. (2012b)
κ-Carrageenan	Lysozyme	Good barrier properties	Shelf-life extension of pears	Medeiros et al. (2012a)
FGF-2	Chitosan	Controlled release	Therapeutic	Cho et al. (2013)

Note: FGF-2, fibroblast growth factor-2.

A delivery system for edible bioactive components must have some specific characteristics (McClements, 2010), such as (i) to be able to encapsulate a high enough amount of the bioactive compound with retention capacity during storage. That is, to have high loading capacity, encapsulation efficiency, and retention capacity; (ii) to contain some type of protector (e.g., chemical) to protect a labile bioactive compound from degradation so that it remains in its active state (e.g., to prevent oxidative degradation of encapsulated lipophilic compounds, such as ω-3 fatty acids, β-carotene, or lycopene); (iii) to be compatible with the food or drink where it will be incorporated, without causing any adverse effect on product appearance, texture, taste, flavor, and shelf life; (iv) to be resistant to environmental stresses to which the food or drink, where it is incorporated is exposed during production, storage, transport, and use. For instance, thermal processing, refrigeration, freezing, dehydration, exposure to light, or mechanical agitation; (v) to be produced entirely with ingredients GRAS, using processing methods that are legally acceptable in the countries of manufacture and marketing; (vi) the additional added value to the finished product containing the encapsulated bioactive components should be sufficient to compensate the additional costs associated with the encapsulation process (i.e., the ingredients and processes used to produce the delivery system must be cheap, reliable, and robust); and (vii) to be designed to control the release and/or absorption of the bioactive component at a specific location within the gastrointestinal tract, such as the mouth, stomach, small intestine, or large intestine.

9.4 ENCAPSULATION OF BIOACTIVE MOLECULES

An efficient way to incorporate functional molecules (e.g., vitamins, antioxidants, and flavor components) into films or coatings without losing their functions is through encapsulation techniques, increasing considerably their stability and thus granting to those films and coatings a high potential to allow expanding the market of products with high added value. Most encapsulation techniques currently used in the food industry are based on matrices formed by biopolymers such as polysaccharides, proteins, peptides, and lipids or lipid systems such as liposomes and micelles (Bouwmeester et al., 2009; García et al., 2010; McClements, 2010).

9.4.1 NANOPARTICLES

Nanoparticles are among the efficient nanosystems for targeted delivery of bioactive compounds. NPs are defined as nanostructures with three dimensions of the order of 100 nm or less. The concept of the NP comprises nanocapsules (NCs) and nanospheres (NSs), which differ from each other according to their composition and structural organization. NCs contain an oily core surrounded by a polymeric membrane and the active constituent can be adsorbed to the polymeric membrane and/or dissolved into the oily core. NSs are made only from a polymeric structure, where the active constituent is retained or adsorbed. It is the small size of NPs, which in combination with the chemical composition and surface structure, gives the unique features of NPs and great potential for many applications (Choi and Kwak, 2014).

In the packaging materials, NPs can also be applied as reactive particles. These so-called nanosensors are designed to respond to environmental changes (e.g., temperature or moisture in storage), degradation products of the foodstuffs, or contamination by microorganisms (Garcia et al., 2010).

Zhang et al. (2014) prepared with success thymol-loaded zein NPs with sodium caseinate and chitosan hydrochloride as electrosteric stabilizers using a simple and low-energy liquid–liquid dispersion method. The average particle size and surface electrical characteristics could be controlled through the mass ratios between zein to sodium caseinate and sodium caseinate to chitosan hydrochloride. Under the experimental conditions, thymol encapsulated in NPs had much stronger antimicrobial activity against *Staphylococcus aureus*, suggesting that colloidal NPs could potentially be used as all-natural delivery systems for antimicrobial agents in food, pharmaceutical, and agricultural industries.

The possibility of using eugenol-loaded chitosan NPs as an antioxidant for active bio-based packaging material was evaluated by Woranuch and Yoksan (2013b). Eugenol-loaded chitosan NPs were incorporated into thermoplastic flour (TPF)—a model bio-based plastic—through an extrusion process at temperatures above 150°C. The incorporation of 3% (w/w) of eugenol-loaded chitosan NPs reduced significantly the extensibility and the oxygen barrier property of TPF, provided antioxidant activity, and improved the water vapor barrier property. Moreover, TPF containing eugenol-loaded chitosan NPs presented superior radical scavenging activity and stronger reducing power compared with TPF containing free (not encapsulated) eugenol.

Nanocapsules of poly-ε-caprolactone containing α-tocopherol were successfully incorporated into biodegradable methylcellulose films by Noronha et al. (2014), since it is a promising biopolymer for active food packaging. The results showed that these films may be an advantageous alternative for food preservation and shelf life extension since they prevent the lipid oxidation of fatty foodstuffs.

Cerqueira et al. (2014) characterized polysaccharide-based films (κ-carrageenan and locust bean gum) without the incorporation of any compound (named as GA) and with the incorporation of free natamycin (named as GA-NA) and natamycin loaded in a smart delivery device consisting in poly(N-isopropylacrylamide) nanohydrogels (named as GA-PNIPA). The results showed that natamycin and natamycin-loaded PNIPA nanohydrogels can be successfully added to edible films without changing their main packaging properties. The authors concluded that since natamycin could be successfully released from polysaccharide-based films, the system could be used as active packaging ingredient when used free in the matrix or as smart packing when loaded with PNIPA nanohydrogels.

9.4.2 Nanoemulsions

Nanoemulsions, which are alternatives to micellar formulations, are nanometric preparations obtained by dispersion of two immiscible liquid phases with the help of an adequate emulsifier system. The nanoemulsions can be formed by emulsification methods (i.e., high energy, low energy, or a combination of both). Recent studies report β-carotene nanoemulsions prepared using a high-energy technique of emulsification–evaporation based on a 2^3 factorial design (Silva et al., 2011). The results showed that it is possible to obtain dispersions at nanoscale presenting a volume mean diameter ($D_{4,3}$) and a surface mean diameter ($D_{3,2}$) of 9 and 280 nm, respectively, immediately after the production of the particles showing a monomodal size distribution. These nanoemulsions showed good physical stability over 21 days of storage and may therefore be seen as a low cost alternative for producing β-carotene nanoemulsions by the food industry.

Liposomes have substantial potential to deliver bioactive compounds in foods. Panya et al. (2010) showed that by combining the inclusion of appropriate antioxidants such as rosmarinic acid and the deposition of a chitosan coating onto the surface of liposomes may significantly increase the oxidative and physical stability of liposomes.

The effect of chitosan on the stability of monodisperse-modified lecithin-stabilized soybean oil-in-water emulsion was investigated by Chuah et al. (2009). The results suggested that chitosan, when used under optimum conditions of preparation, may have the potential to be utilized as a functional ingredient to provide structurally and microbiologically stable emulsions.

Cruz-Romero et al. (2013) evaluated the antimicrobial activity of low- and medium-molecular-weight chitosan and organic acids (benzoic acid and sorbic acid and commercially available nanosized benzoic and sorbic acid solubilisate equivalents) and compared against commercial mixtures of organic acids used as meat coatings (Articoat DLP-02® and Sulac-01®). The results suggested that the molecular weight of the chitosan used affected its antimicrobial activity. Nanosized solubilisates of benzoic acid and sorbic acid had significantly higher antimicrobial effect than their non-nanoequivalents. These authors concluded that the solubilized nanosized structures can potentially be used as an antimicrobial in intelligent and active packaging applications, since less antimicrobial substances are required to impart the antimicrobial effect.

9.4.3 Bio-Nanocomposites

Since its starting in the nineteenth century and with the move toward globalization and the evolution of consumer preferences, food packaging (with the aid of nanotechnology) has also made great advances providing to foods a longer shelf life, with safety and quality, according to international standards (Silvestre et al., 2011). It is well known in the scientific world of food packaging that the use of edible and biodegradable polymers is often limited by problems associated with their performance (such as low resistance to breakage, weak barrier properties) and with their high costs (Azeredo and Mattoso 2008). In fact, polymer nanotechnology has provided new food packaging materials with adequate mechanical properties and improved barrier and antimicrobial properties, along with nanosensors for the detection and monitoring of the condition of foods during transportation and storage. Regarding the improvement of barrier performance, it is not restricted only to gases, such as oxygen and carbon dioxide, but also to, for example, ultraviolet (UV) rays, while also adding strength, rigidity, dimensional stability, and resistance to heat (Silvestre et al., 2011). Thus, a new class of materials represented by bio-nanocomposites with improved properties (barrier, mechanical, and thermal stability) has been considered as a promising means of enhancing the properties of biopolymer-based packaging materials (Azeredo et al., 2009; Kumar et al., 2010; Rhim et al., 2013).

A bio-nanocomposite can be considered a multiphase material derived from the combination of two or more components, including a matrix (continuous phase) and a discontinuous nanodimensional phase with at least one nanosized dimension (i.e., with less than 100 nm). The nanodimensional phase can be divided into three categories according to the number of nanosized dimensions. NSs or NPs have the three dimensions in the nanoscale. Both nanowhiskers (nanorods) and nanotubes have two nanometric dimensions, with the difference that nanotubes are hollow, while nanowhiskers are solid. Finally, nanosheets or nanoplatelets have only one nanosized dimension (Alexandre and Dubois, 2000). Most of the nanosized phases have a structural role, acting as reinforcements for improvement of the mechanical properties of the biopolymer matrix, provided that the matrix transfers the tension to the nanoreinforcement across the interface established between them. The incorporation of reinforcements of nanosize into biopolymer structures may open new possibilities for improving not only the properties but also the cost-price efficiency (Sorrentino et al., 2007).

Nanocomposite edible films have been developed by addition of cellulose nanofibers (CNF) as nanoreinforcement to mango puree-based edible films. The interactions of CNF with mango components (mainly pectin and/or starch), as well as the interactions among the nanofibers was effective in increasing tensile strength, suggesting the formation of a fibrillar network within the matrix that was also effective in improving water vapor barrier of the films (Azeredo et al., 2009).

Bacterial cellulose (BC) nanoribbons were used as reinforcement in chitosan films dissolved in acetic and lactic acid. The results showed a difference between the acids in their behavior and effect on the reinforcement. The acetic acid films presented a tensile strength about four times higher than the lactic acid films for the same reinforcement concentration. Neither the reinforced acetic acid films nor the matrix showed any significant inhibitory antimicrobial effects; however, the lactic acid films showed antimicrobial activity indicating that the BC nanoribbons had no important influence on this activity (Velásquez-Cock et al., 2014). Another study also suggests that $Mg(OH)_2$ nanoplatelets, in comparison with other NPs, have high antibacterial potency. As $Mg(OH)_2$ is a nontoxic material and has been widely used in food and medical industries, $Mg(OH)_2$ nanoplatelets have great potential for application as a new antibacterial material (Dong et al., 2011).

The nanoplatelets composed of clays or other silicate materials have been the most promising fillers at nanoscale for food packaging. The popularity of nanoclays in food contact applications derives from their low cost, effectiveness, high stability, and (alleged) benignity. The montmorillonite (MMT) clay [$(Na,Ca)_{0.33}(Al,Mg)_2(Si_4O_{10})(OH)_2 \cdot nH_2O$)], a soft 2:1 layered phyllosilicate clay comprised of highly anisotropic platelets separated by thin layers of water (Duncan, 2011), has been widely used in food packaging.

Bio-nanocomposite films based in soy protein isolate (SPI) and MMT were prepared using melt extrusion. Effects of the pH of the film-forming solution, MMT content, and extrusion processing parameters on the structure and properties of SPI–MMT bio-nanocomposite films were investigated. The arrangement of MMT in the soy protein matrix ranged from exfoliated at lower MMT content (5%) to intercalated at higher MMT content (15%). A significant improvement in mechanical properties, thermal stability, and WVP of the films with the addition of MMT was observed. The SPI–MMT film properties were significantly affected by the pH of film-forming solutions, MMT content and extrusion process parameters. The results showed the feasibility of using bio-nanocomposite technology to improve the properties of biopolymer films based in SPI (Kumar et al., 2010).

Formulations of polypropylene (PP)/clay (MMT) nanocomposites with different proportions of the PP, MMT, and maleated PP (MAPP) prepared by melt compounding, for food packaging, were also investigated. The nanocomposites revealed increased d-spacing of the MMT layers, indicating that the compatibility of neat PP and clay was improved by the addition of MAPP, and the intercalation and partial exfoliation of the layers. The use of clay increased the mobility distance of the gas molecules, leading to the oxygen permeability of neat PP being reduced by 26%–55% as well as to the improvement in the mechanical and thermal properties (Choi et al., 2011).

A composite coating of 1% chitosan film with 0.04% nano-silicon dioxide, applied by dipping on surface of whole jujube, led to a shelf life extension. After 32 days, the red index, decay incidence, weight loss, and respiration rate of the coated jujubes were lower in comparison with the control. The composite coating showed to be superior in preserving total flavonoid than chitosan coating alone. These findings suggested that coating jujubes with chitosan + nano-silicon dioxide can be a promising alternative, if nano-silicon dioxide is allowed for food use (Yu et al., 2012). A similar study was carried out with a novel chitosan/nano-silica hybrid film for the preservation quality of longan fruits (*Dimocarpus longan* Lour. cv. Shijia). The results showed that the chitosan/nano-silica coating had quite beneficial effects on physicochemical and physiological quality compared with other treatments by efficiently forming an excellent semipermeable film. The authors reported that this packaging material has the advantages of simple processing and industrial feasibility and could provide a promising alternative to improve the preservation qualities of fresh longan fruits during extended storage life (Shi et al., 2013).

A novel alginate/nano-Ag coating material was applied on shiitake mushroom (*Lentinus edodes*) and showed to have quite a beneficial effect on the physicochemical and sensory quality as well as more effective in reducing microbial counts in comparison with the control. According to the authors (Jiang et al., 2013), the coating could be applied for preservation of the shiitake mushroom to expand its shelf life and improve its preservation quality.

Active nanocomposite films were prepared by blending aqueous solutions of gelatin with different concentrations of silver NPs (AgNPs) using a solvent casting method. Gelatin/AgNPs nanocomposite films exhibited strong antibacterial activity against food-borne pathogens. Gelatin/AgNPs nanocomposite films are expected to have high potential as an active food packaging system to maintain food safety and to extend the shelf life of packaged foods (Kanmani and Rhim, 2014). Table 9.2 summarizes the examples of bio-nanocomposites previously referred.

9.5 TOOLS FOR CHARACTERIZATION OF NANOSYSTEMS

The most common techniques available to characterize nanosystems such as nanocoatings and NPs are briefly described below.

TABLE 9.2
Bio-nanocomposites for Food Packaging Applications

Matrix	Nanodimensional Phase	Improved Properties	References
Mango puree	Cellulose nanofibers	Mechanical property and WVP[a]	Azeredo et al. (2009)
Chitosan	Cellulose nanoribbons	Mechanical property	Velásquez-Cock et al. (2014)
Soy protein	Montmorillonite	Mechanical property, WVP,[a] and thermal stability	Kumar et al. (2010)
Polypropylene	Montmorillonite	Mechanical, oxygen permeability, and thermal stability	Choi et al. (2011)
Chitosan	Nano-silicon dioxide	Gas permeability	Yu et al. (2012) Shi et al. (2013)
Alginate	Nano-Ag	Antimicrobial property	Jiang et al. (2013)
Gelatin	Nano-Ag	Antibacterial property	Kanmani and Rhim (2014)

[a] Water vapor permeability.

9.5.1 Light Scattering

Light scattering techniques are particularly useful tools for exploring homo- or hetero-association of proteins and other biological macromolecules. With this technique, it is possible to change solution conditions (e.g., pH, ionic strength, and concentration); it is a noninvasive method, where the alteration of the solution association properties does not occur, and no dilution or fractionation of the sample that might alter the association is needed (Murphy, 1997).

The characterization of the size of particles, emulsions, or molecules, which have been dispersed or dissolved in a liquid can be obtained by dynamic light scattering (DLS). The Brownian motion of particles or molecules in suspension causes laser light to be scattered at different intensities. This DLS technique, which in older literature is also called photon correlation spectroscopy (PCS) and quasi-elastic light scattering (QELS), measures Brownian motion and hence the particle size using the Stokes–Einstein relationship. DLS thus has been widely used to determine the size distribution of NPs (Souza et al., 2014; Zhang et al., 2014), nanoemulsions (Silva et al., 2011), and polyelectrolyte solutions used in the construction of nanomultilayers (Carneiro-da-Cunha et al., 2010; Medeiros et al., 2012a; Zhang et al., 2014).

The size and polydispersity index are also essential analyses for characterization of NPs, since they influence important parameters such as loading, release, and stability of the compound encapsulated. It is known that the smaller the particle, the greater is the exposed surface area, which leads to a faster release of encapsulated drugs. Smaller particles also have an increased risk of aggregation during storage, being important the development of NPs with a low polydispersity index to achieve the maximum stability by a better control (and lesser dispersion) of particles' size. It is worth mentioning that the reproducibility of parameters such as stability and release is directly connected to a low polydispersity index (≤0.4), since a high polydispersity index means that there is no uniformity in the size distribution of the sample (Souza et al., 2014).

9.5.2 Zeta Potential

Zeta potential or electrokinetic potential is an important parameter of the electrical double layer and is a characteristic of electrical properties of solid/liquid and liquid/gaseous interfaces (Salopek et al., 1992). Generally, when all particles have a large positive or negative zeta potential (where the positivity and negativity are greater or lower than +30 and −30 mV, respectively), they will repel each other and the dispersion is stable. On the contrary, when the particles have low zeta potential values, there will be no sufficient force to prevent the particles from aggregating (Carneiro-da-Cunha et al., 2011; Souza et al., 2014). Sufficiently, high values of zeta potential are also essential to maximize electrostatic interactions between polyelectrolytes in a nanomultilayer structure (Pinheiro et al., 2012b). Zeta potential can be used to optimize the formulations of different systems, predicting long-term stability, through the value and sign of the charge of polysaccharides (Carneiro-da-Cunha et al., 2010; Medeiros et al., 2012a, 2012b, 2014; Soares et al., 2014), proteins (Medeiros et al., 2014; Yu et al., 2006), lipids (Audu et al., 2012), nanostructures (Soares et al., 2014; Souza et al., 2014), being an important, widely used tool, for example, to control the deposition of alternating cationic and anionic polyelectrolytes on a surface (Carneiro-da-Cunha et al., 2010, 2011).

9.5.3 Ultraviolet-Visible Spectroscopy

Ultraviolet-visible (UV-Vis) spectroscopy uses visible and UV light and refers to the transmission or absorption of light by the sample in the UV-Vis spectral region. Many molecules or compounds absorb UV or visible (Vis.) light. The absorbance of a solution increases as the attenuation of the beam increases. Different molecules or compounds absorb radiation of different wavelengths. UV-Vis spectroscopy has been used to verify whether the adsorption of biomolecules on nanolayered films is occurring, which is measured by changes in absorbance at specific wavelengths (corresponding to those of the adsorbed materials). Medeiros et al. (2014) followed the successive and alternate deposition of alginate and lysozyme layers at 266 nm, Carneiro-da-Cunha et al. (2010) confirmed the successive deposition of chitosan/alginate nanolayered film on A/C PET surface at 260 nm, and Aoki et al. (2014) monitored the growth of LbL films (dipping and spray) by measuring the increase in absorbance at the range from 190 to 1100 nm.

9.5.4 Contact Angle

The contact angle (θ) is the angle formed between the liquid/solid interface and the liquid/vapor interface. The contact angle quantifies the wettability of a solid surface by a liquid via the Young equation $\cos\theta = (\gamma_{SV} - \gamma_{SL}/\gamma_{LV})$, providing information regarding the bonding energy of the solid surface and surface tension of the droplet of liquid. It is conventionally measured using the liquid, resulting in values of the

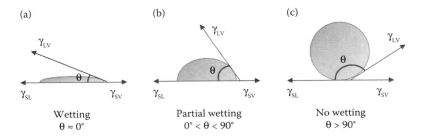

FIGURE 9.2 Schematic representation of the macroscopic characteristics of the wettability of different liquids on a surface.

surface tension for the solid–liquid (γ_{SL}), liquid–vapor (γ_{LV}), and solid–vapor (γ_{SV}) interfaces, which overall determine the energy required for adhesion, cohesion, and spreading. For a surface to be wet, the magnitude of the contact angle formed by the liquid and the solid surface must be nearly zero (Figure 9.2a), allowing the spreading of the liquid on the solid surface, or a partial wetting can occur when the magnitude of the angle is in the range of $0° < \theta < 90°$ (Figure 9.2b). Similarly, if a surface is not to be wet, the contact angle must be greater than 90° for the liquid to contract and pull away easily from the solid surface (Figure 9.2c) (Adamson and Gast, 1997).

This technique can be used to confirm the alternating deposition of polyelectrolytes in a specific surface (Carneiro-da-Cunha et al., 2010; Medeiros et al., 2012a, 2012b, 2014).

9.5.5 Nanoindentation

Nanoindentation also referred to as depth-sensing indentation (DSI) represents nowadays one of the principal techniques for the mechanical characterization of materials. The method monitors the penetration of an indenter into the material's surface, to induce local surface deformation during the application and release of a load (Rettler et al., 2013). The most advanced DSI instruments can produce indentations with depths of only a few tens of nanometers, most of them also offering the possibility of approaching an upper limit in the micron regime (Fischer-Cripps, 2011). This technique is emerging as a most valuable tool for the evaluation of the modulus, hardness, and creep enhancements upon incorporation of a filler into the biomaterials (Díez-Pascual et al., 2015), facilitating the measurement of the time-dependent and viscoelastic mechanical properties (Rettler et al., 2013). This methodology was used to assess the mechanical properties of a nanolayered alginate/chitosan coating by Carneiro-da-Cunha et al. (2010).

9.5.6 Electron Microscopy

Electron microscopy is the preferred method for visualization of structural details with nanometer-scale resolution (Liv et al., 2014). The great resolving power of electron microscopy is a result of the application of an electron beam with a wavelength well below the nanometer range. Optical microscopy, operating at wavelengths (λ)

in the range 400–800 nm and numerical aperture (NA) ~ 1, Abbe's diffraction limit $d = \lambda/(2 \cdot NA)$ gives a resolution of ~200 nm. Since the wavelength of an electron beam is much shorter than that of light, the resolution of an electron microscope is much higher. For an electron microscope working with an acceleration voltage of 100 kV, the value of the wavelength will be 0.0037 nm (Dudkiewicz et al., 2011). Two main imaging techniques commonly used for the analysis and characterization of nanostructures are the SEM and transmission electron microscopy (TEM).

9.5.6.1 Scanning Electron Microscopy

The principle of SEM is to use an electron beam of small diameter to explore the surface of a sample, point by point, in successive lines (Klang et al., 2013). The interaction between the sample and the electron probe produces various types of emissions, which are captured by different detectors placed in appropriate positions (Suga et al., 2014). The size, size distribution, and shape of nanomaterials can be directly acquired from SEM.

SEM is performed on dry samples; however, the process of drying and contrasting samples may cause shrinkage of the specimen and alter the characteristics of the nanomaterials. In addition, while scanned by an electron beam, many biomolecule samples that are nonconductive specimens tend to acquire charge and insufficiently deflect the electron beam, leading to imaging faults or artifacts. Coating an ultrathin layer of electrically conducting material onto the biomolecules is often required for this sample preparation procedure. Another disadvantage presented by the technique occurs during image interpretation, since biased statistics of size distribution of heterogeneous samples is unavoidable in SEM due to the small number of sample particles in the scanning region (Lin et al., 2014).

9.5.6.2 Transmission Electron Microscopy

In TEM, an incident electron beam is transmitted through a very thin foil specimen, during which the incident electrons interacting with the specimen are transformed to unscattered electrons, elastically scattered electrons, or inelastically scattered electrons. The scattered or unscattered electrons are focused by a series of electromagnetic lenses and then projected on a screen to generate an electron diffraction, amplitude-contrast image, a phase-contrast image, or a shadow image of varying darkness according to the density of unscattered electrons (Lin et al., 2014). TEM is used as one of the most powerful techniques for atomic scale characterization of structure and chemistry of solids, especially of nanostructured materials (Sharma et al., 2012), but can also be used for determining the particle size, dispersion, aggregation/agglomeration, and dynamic displacement of nanomaterials.

9.5.7 Atomic Force Microscopy

Atomic force microscopy (AFM) is a technique to obtain images and other information from a wide variety of samples, at extremely high (nanometer) resolution. AFM works by scanning a very sharp (end radius ca. 10 nm) probe along the sample surface, carefully maintaining the force between the probe and the surface at a set low level. Usually, the probe is formed by a silicon or silicon nitride cantilever with a sharp integrated tip, and the vertical bending (deflection) of the cantilever due to forces acting on the tip is detected by a laser focused on the back of the cantilever. The laser is reflected by the cantilever onto a distant photodetector. The movement of the laser spot on the photodetector gives a greatly exaggerated measurement of the movement of the probe. This setup is known as an optical lever. The probe is moved over the sample by a scanner, typically a piezoelectric element, which can make extremely precise movements. The combination of the sharp tip, the very sensitive optical lever, and the highly precise movements by the scanner with the careful control of probe-sample forces allows the extremely high resolution of AFM (Chaudhary and Sen, 2014). AFM is widely used to obtain images that show the topographic characteristics of a sample in three dimensions with high resolution and in real time (Aoki et al., 2012, 2014).

9.5.8 Confocal Laser Scanning Microscopy

In confocal laser scanning microscopy (CLSM), a laser beam passes through a light source aperture and then is focused by an objective lens into a small focal volume within or on the surface of a specimen. This technique has become an invaluable tool for a wide range of investigations in the biological and medical sciences for imaging thin optical sections in living and fixed specimens ranging in thickness up to 100 μm. Modern instruments are equipped with 3–5 laser systems controlled by high-speed acoustooptic tunable filters, which allow very precise regulation of wavelength and excitation intensity. Coupled with photomultipliers that have high quantum efficiency in the near-UV, visible, and near-infrared spectral regions, these microscopes are capable of examining fluorescence emission ranging from 400 to 750 nm. Instruments equipped with spectral imaging detection systems further refine the technique by enabling the examination and resolution of fluorophores with overlapping spectra as well as providing the ability to compensate for autofluorescence. Recent advances in fluorophore design have led to improved synthetic and naturally occurring molecular probes, including fluorescent proteins and quantum dots, which exhibit a high level of photostability and target specificity (Chaudhary and Sen, 2014). Silva et al. (2011) used this tool to confirm the presence of β-carotene in nanoemulsions placing a drop of the nanosystem directly on a slide without further treatment using the fluorescence excitation wavelength of 488 nm (green) with a 600× magnification (oil immersion).

CLSM is a scanning imaging technique in which the resolution obtained is best explained by comparing it with another scanning technique like that of the SEM. CLSM has the advantage of not requiring a probe to be suspended nanometers from the surface, as in an AFM, where the image is obtained by scanning with a fine tip over a surface.

9.6 NANOTECHNOLOGY: RISK, REGULATION, AND FUTURE TRENDS

Over the past few decades, the development of nanotechnology has generated innovations on production, processing, storage, transportation, traceability, safety, and security of food. However, many researchers have reported that nanosized particles frequently exhibit different properties from those found at the macroscale, because the very small sizes, in principle, would allow them to move through the body more freely than larger particles, once their high surface area increases their reactivity and allows a greater contact with cell membranes, as well as greater capacity for absorption and migration.

It is uncertain to what extent the use of these nanoproducts may or may not pose toxicity risks to human health. In general, it will remain so, until there is scientific information available and sufficient to characterize the harmful effects of some nanomaterials in humans and the environment. Therefore, nanomaterials may require a risk assessment, related to specific nanomaterials, and specific uses, which must be examined case by case, using relevant information (European Commission, 2012).

A few years ago, it was reported that edible films and coatings would be the packaging of the future. At the same time, much effort has been dedicated to preserving the environment following the awareness of the amount of nonessential plastic waste produced from packaging products and today, the tendency of businesses and consumers is to emphasize the importance of "being green" (Wei and Yazdanifard, 2013), and the growing demand by consumers for healthier and more ecological foods has driven researchers to develop new systems of packaging. To find the correct combination between the food product and the edible film/coating, which ensure the success of the technology has also been the general trend.

Nanomaterials and nanosystems are also being examined, offering the perspective to be integrated in new intelligent food packaging systems. It is expected that the next generation of intelligent packaging will make possible a better monitoring of the flow, safety, and quality of food products (Vanderroost et al., 2014).

ACKNOWLEDGMENT

M. G. Carneiro-da-Cunha expresses her gratitude to the Conselho Nacional de Desenvolvimento Científico e Tecnológico (CNPq) for research grants and fellowship.

REFERENCES

Adams, F. C., and Barbante, C. 2013. Nanoscience, nanotechnology and spectrometry. *Spectrochimica Acta B, 86*, 3–13.

Adamson, A. W., and Gast, A. P. 1997. *Physical Chemistry of Surfaces*. New York: John Wiley & Sons.

Ai, H., Jones, S. A., and Lvov, Y. M. 2003. Biomedical applications of electrostatic layer-by-layer nano-assembly of polymers, enzymes and nanoparticles. *Cell Biochemistry and Biophysics, 39*, 23–43.

Alexandre, M., and Dubois, P. 2000. Polymer-layered silicate nanocomposites: Preparation, properties and uses of a new class of materials. *Materials Science and Engineering, 28*, 1–63.

Andrade, R., Skurtys, O., and Osorio, F. 2013. Drop impact behavior on food using spray coating: Fundamentals and applications. *Food Research International, 54*, 397–405.

Aoki, P. H. B., Alessio, P., Volpati, D., Paulovich, F. V., Riul Jr., A., Oliveira Jr., O. N., and Constantino, C. J. L. 2014. On the distinct molecular architectures of dipping- and spray-LbL films containing lipid vesicles. *Materials Science and Engineering C, 41*, 363–371.

Aoki, P. H. B., Volpati, D., Cabrera, F. C., Trombini, V. L., Riul Jr., A., and Constantino, C. J. L. 2012. Spray layer-by-layer films based on phospholipid vesicles aiming sensing application via e-tongue system. *Materials Science and Engineering C, 32*, 862–871.

Arnaldi, S., Delgado G. C., Piccinni, M., and Poletti, P. 2011. *Nanomedicina. Entre políticas públicas y necesidades privadas.* Mexico: Ceiichunam.

Audu, M. M., Achile, P. A., and Amaechi, A. A. 2012. Phospholipon 90G based SLMs loaded with ibuprofen: An oral antiinflammatory and gastrointestinal sparing evaluation in rats. *Pakistan Journal of Zoology, 44*, 1657–1664.

Azeredo, L. H., and Mattoso, L. H. 2008. Nanotecnologia como ferramenta para melhorar o desempenho de embalagens comestíveis. http://www.bdpa.cnptia.embrapa.br/. (accessed on September 2, 2014).

Azeredo, H. M. C., Mattoso, L. H. C., Wood, D., Williams, T. G., Avena-Bustillos, R. J., and Mchugh, T. H. 2009. Nanocomposite edible films from mango puree reinforced with cellulose nanofibers. *Journal of Food Science, 74*, 31–35.

Barbosa-Pereira, L., Cruz, J. M., Sendón, R., Rodríguez, B. Q. A, Ares A, Castro-López M., Abad, M. J., Maroto, J., and Paseiro-Losada, P. 2013. Development of antioxidant active films containing tocopherols to extend the shelf life of fish. *Food Control, 31*, 236–243.

Bico, S. L. S., Raposo, M. F. J., Morais, R. M. S. C., and Morais, A. M. M. B. 2009. Combined effects of chemical dip and/or carrageenan coating and/or controlled atmosphere on quality of fresh-cut banana. *Food Control, 20*, 508–514.

Bouwmeester, H., Dekkers, S., Noordam, M. Y., Hagens, W. I., Bulder, A. S., and De-Heer, C. 2009. Review of health safety aspects of nanotechnologies in food production. *Regulatory Toxicology and Pharmacology, 53*, 52–62.

Bradley, E. L., Castle, L., and Chaudhry, Q. 2011. Applications of nanomaterials in food packaging with a consideration of opportunities for developing countries. *Trends in Food Science & Technology, 22*, 604–610.

Brasil, I. M., Gomes, C., Puerta-Gomes, A., Castell-Perez, M. E., and Moreira, R. G. 2012. Polysaccharide-based multilayered antimicrobial edible coating enhances quality of fresh-cut papaya. *LWT—Food Science & Technology, 47*, 39–45.

Brody, A. L. 2003. Nano, nano food packaging technology. *Food Technology, 57*, 52–54.

Caner, C. 2011. Sorption phenomena in packaged foods: Factors affecting sorption processes in package-product systems. *Packaging Technology & Science*, 24, 259–270.

Carneiro-da-Cunha, M. G., Cerqueira, M. A., Souza B. W. S., Carvalho, S., Quintas, M. A. C., Teixeira, J. A., and Vicente, A. A. 2010. Physical and thermal properties of a chitosan/alginate nanolayered PET film. *Carbohydrate Polymers, 82*, 153–159.

Carneiro-da-Cunha, M. G., Cerqueira, M. A., Souza, B. W. S., Teixeira J. A., and Vicente A. A. 2011. Influence of concentration, ionic strength and pH on zeta potential and mean hydrodynamic diameter of edible polysaccharide solutions envisaged for multinanolayered films production. *Carbohydrate Polymers, 85*, 522–528.

Cerqueira, M. A., Costa, M. J., Fuciños, C., Pastrana, L. M., and Vicente, A. A. 2014. Development of active and nanotechnology-based smart edible packaging systems: Physical–chemical characterization. *Food and Bioprocess Technology, 71*, 1472–1482.

Cerqueira, M. A., Lima, A. M., Teixeira, J. A., Moreira, R. A., and Vicente, A. A. 2009. Suitability of novel galactomannans as edible coatings for tropical fruits. *Journal of Food Engineering, 94*, 372–378.

Chaudhary, H. H., and Sen, D. J. 2014. Isothermal titration calorimetry, confocal laser scanning microscopy and atomic force microscopy in latest supramolecular ligand technology. *World Journal of Pharmaceutical Sciences, 3*, 341–363.

Chaudhry, Q., Castle, L., and Watkins, R. 2010. *Nanotecnologies in Food.* Cambridge: RSC Publishing.

Chaudhry, Q., Scotter, M., Blackburn, J., Ross, B., Boxall, A., Castle, L., Aitken, R., and Watkins, R. 2008. Applications and implications of nanotechnologies for the food sector. *Food Additives & Contaminants A*, 25, 241–258.

Cho, Y., Lee, J. B., and Hong, J. 2013. Tunable growth factor release from nano-assembled starch multilayers. *Chemical Engineering Journal*, 221, 32–36.

Choi, M. J., and Kwak, H. S. 2014. Advanced approaches of nano- and microencapsulation for foods ingredients. In H.-S. Kwak (Ed.), *Nano- and Microencapsulation for Foods* (pp. 95–116). Chennai, India: Wiley Blackwell.

Choi, R. N., Cheigh, C. I., Lee, S. Y., and Chung, M. S. 2011. Preparation and properties of polypropylene/clay nanocomposites for food packaging. *Journal of Food Science*, 76, 62–67.

Chuah, A. M., Kuroiwa, T., Kobayashi, I., and Nakajima, M. 2009. Effect of chitosan on the stability and properties of modified lecithin stabilized oil-in-water monodisperse emulsion prepared by microchannel emulsification. *Food Hydrocolloids*, 23, 600–610.

Coles, D., and Frewer, L. J. 2013. Nanotechnology applied to European food production: A review of ethical and regulatory issues. *Trends in Food Science & Technology*, 34, 32–43.

Crouzier, T., Boudou, T., and Picart, C. 2010. Polysaccharide-based polyelectrolite multilayers. *Current Opinion in Colloid & Interface Science*, 15, 417–426.

Cruz-Romero, M. C., Murphy, T., Morris, M., Cummins, E., and Kerry, J. P. 2013. Antimicrobial activity of chitosan, organic acids and nano-sized solubilisates for potential use in smart antimicrobially-active packaging for potential food applications. *Food Control*, 34, 393–397.

Decher, G., Hong, J. D., and Schmitt, J. 1992. Buildup of ultrathin multilayer films by a self-assembly process: III. Consecutively alternating adsorption of anionic and cationic polyelectrolytes on charged surfaces. *Thin Solid Films*, 210, 831–835.

Decher, G., and Schlenoff, J. B. 2003 Multilayer thin films: Sequential assembly of nanocomposite materials. In G. Decher, and J. B. Schlenoff (Eds.), *Multilayer Thin Films: Sequential Assembly of Nanocomposite Materials* (pp. 1–31). Germany: Wiley-VCH.

Díez-Pascual, A. M., Gómez-Fatou, M. A., Ania, F., and Flores, A. 2015. Nanoindentation in polymer nanocomposites. *Progress in Materials Science*, 67, 1–94.

Dong, C., Song, D., Cairney, J., Maddan, O. L., He, G., and Deng, Y. 2011. Antibacterial study of $Mg(OH)_2$ nanoplatelets. *Materials Research Bulletin*, 46, 576–582.

Dudkiewicz, A., Tiede, K., Loeschner, K., Loeschner, K., Jensen, L. H. S., Jensen, E., Wierzbicki, R., Boxall, A. B. A., and Molhave, K. 2011. Characterization of nanomaterials in food by electron microscopy. *Trends in Analytical Chemistry*, 30, 28–43.

Duncan, T. V. 2011. Applications of nanotechnology in food packaging and food safety: Barrier materials, antimicrobials and sensors. *Journal of Colloid and Interface Science*, 363, 1–24.

European Commission. 2011. Recomendação da Comissão de 18 de Outubro de 2011 sobre a definição de nanomaterial. (2011/696/UE). http://eur-lex.europa.eu/legal-content/PT/TXT/?uri=CELEX:32011H0696 (accessed on September 2, 2014).

European Commission. 2012. Communication from the Commission to the European parliament, the council and the European economic and social committee. Second Regulatory Review on Nanomaterials. Brussels, October 3, 2012. http://eur-lex.europa.eu/legal-content/EN/TXT/PDF/?uri=CELEX:52012DC0572&from=EN (accessed on January 27, 2015).

Fabra, M. J., Busolo, M. A., Lopez-Rubio, A., and Lagaron, J. M. 2013. Nanostructured biolayers in food packaging. *Trends in Food Science & Technology*, 31, 79–87.

Fabra, M. J., Talens, P., and Chiralt, A. 2008. Tensile properties and water vapour permeability of sodium caseinate films containing oleic acid-beeswax mixtures. *Journal of Food Engineering*, 85, 393–400.

Fabra, M. J., Talens, P., and Chiralt, A. 2010. Water sorption isotherms and phase transition of sodium caseinate-lipid films as affected by lipid interactions. *Food Hydrocolloid, 24,* 384–391.

Feynman, R. P. 2003. There's plenty of room at the bottom: An invitation to enter a new Field of Physics. In W. A. Goddard, D. Brenner, S. E. Lyshevski, and G. F. Iafrate (Eds.), *Handbook of Nanoscience, Engineering, and Technology* (pp. 3–12). Boca Raton, Florida: CRC Press LLC.

Fischer-Cripps, A. C. 2011. *Nanoindentation* (3rd ed.). New York: Springer.

Fu, J., Ji, J., Yuan, W., and Shen, J. 2005. Construction of anti-adhesive and antibacterial multilayer films via layer-by-layer assembly of heparin and chitosan. *Biomaterials, 26,* 6684–6692.

García, M., Forbe, T., and Gonzalez, E. 2010. Potential applications of nanotechnology in the agro-food sector. *Ciência e Tecnologia Alimentar, 30,* 573–581.

Huang, B., Cao, M. H., Nie, F., Huang, H., and Hu, C. W. 2013. Construction and properties of structure- and size-controlled micro/nano-energetic materials. *Defence Technology, 9,* 59–79.

Israelachvili, J. N. 2011. *Intermolecular and Surface Forces*. San Diego: Elsevier.

Jagannath, J. H., Nanjappa, C., Das Gupta, D. K., and Bawa, A. S. 2003. Mechanical and barrier properties of edible starch-protein-based films. *Journal of Applied Polymer Science, 88,* 64–71.

Jiang, T., Feng, L., and Wang, Y. 2013. Effect of alginate/nano-Ag coating on microbial and physicochemical characteristics of shiitake mushroom (*Lentinus edodes*) during cold storage. *Food Chemistry, 141,* 954–960.

Kaiser, H. 2014. Nano Food 2040, Nanotechnology in food, food processing, agriculture, packaging and consumption, state of science, technologies, markets, applications and developments to 2015 and 2040. http://www.hkc22.com/nanofood2040.html. (accessed on June 26, 2014).

Kanmani, P., and Rhim, J. W. 2014. Physicochemical properties of gelatin/silver nanoparticle antimicrobial composite films. *Food Chemistry, 148,* 162–169.

Klang, V., Valenta, C., and Matsko, N. B. 2013. Electron microscopy of pharmaceutical systems. *Micron, 44,* 45–74.

Kotov, N. A. 2003. Layer-by-layer assembly of nanoparticles and nanocolloids: Intermolecular interactions, structure and materials perspective. In G. Decher, and J. B. Schlenoff (Eds.), *Multilayer thin Films: Sequential Assembly of Nanocomposite Materials* (pp. 207–243). Germany: Wiley-VCH.

Kumar, P., Sandeep, K. P., Alavi, S., Truong, V. D., and Gorga, R. E. 2010. Preparation and characterization of bio-nanocomposite films based on soy protein isolate and montmorillonite using melt extrusion. *Journal of Food Engineering, 100,* 480–489.

Lee, K. T. 2010. Quality and safety aspects of meat products as affected by various physical manipulations of packaging materials. *Meat Science, 86,* 138–150.

Lin, P. C., Lin, S., Wang, P. C., and Sridhar, R. 2014. Techniques for physicochemical characterization of nanomaterials. *Biotechnology Advances, 32,* 711–726.

Liv, N., Lazić, I., Kruit, P., and Hoogenboom, J. P. 2014. Scanning electron microscopy of individual nanoparticle bio-markers in liquid. *Ultramicroscopy, 143,* 93–99.

Lvov, Y., Ariga, K., Ichinose, I., and Kunitake, T. 1996. Molecular film assembly via layer-by-layer adsorption of oppositely charged macromolecules (linear polymer, protein and clay) and concavalin A and glycogen. *Thin Solid Films, 285,* 797–801.

Martiñon, M. E., Moreira, R. G., Castell-Perez, M. E., and Gomes, C. 2014. Development of a multilayered antimicrobial edible coating for shelf life extension of fresh-cut cantaloupe (*Cucumis melo* L.) stored at 4°C. *LWT—Food Science and Technology, 55,* 341–350.

Marudova, M., Lang, S., Brownsey, G. J., and Ring, S. G. 2005. Pectin-chitosan multilayer formation. *Carbohydrate Research, 340,* 2144–2149.

Mashaghi, S., Jadidi, T., Koenderink, G., and Mashaghi, A. 2013. Lipid nanotechnology. *International Journal of Molecular Sciences*, 14, 4242–4282.

McClements, D. J. 2010. Design of nano-laminated coatings to control bioavailability of lipophilic food components. *Journal of Food Science*, 75, 30–42.

Medeiros, B. G. S., Pinheiro, A. C., Texeira, J. A., Vicente, A. A., and Carneiro-da-Cunha, M. G. 2012a. Polysaccharide/protein nanomultilayer coatings: Construction, characterization and evaluation of their effect on "Rocha" pear (*Pyrus communis* L.) shelf-life. *Food and Bioprocess Technology*, 5, 2435–2445.

Medeiros, B. G. S., Pinheiro, A. C., Carneiro-da-Cunha, M. G., and Vicente, A. A. 2012b. Development and characterization of a nanomultilayer coating of pectin and chitosan—Evaluation of its gas barrier properties and application on "Tommy Atkins" mangoes. *Journal of Food Engineering*, 110, 457–464.

Medeiros, B. G. S., Souza, M. P., Pinheiro, A. C., Bourbon, A. I., Cerqueira, M. A., Vicente, A. A., and Carneiro-da-Cunha, M. G. 2014. Physical characterisation of an alginate/lysozyme nano-laminate coating and its evaluation on "Coalho" cheese shelf life. *Food Bioprocess Technology*, 7, 1088–1098.

Murphy, R. M. 1997. Static and dynamic light scattering of biological macromolecules: What can we learn? *Current Opinion in Biotechnology*, 8, 25–30.

Neethirajan, S., and Jayas, D. S. 2011. Nanotechnology for the food and bioprocessing industries. *Food Bioprocess Technology*, 4, 39–47.

Noronha, C. M., Carvalho, S. M., Lino, R. C., and Barreto, P. L. M. 2014. Characterization of antioxidant methylcellulose film incorporated with α-tocopherol nanocapsules. *Food Chemistry*, 159, 529–535.

Panya, A., Laguerre, M., Lecomte, J., Pierre, V., Weiss, J., McClements, D. J., and Decker, E. A. 2010. Effects of chitosan and rosmarinate esters on the physical and oxidative stability of liposomes. *Journal of Agricultural and Food Chemistry*, 58, 5679–5684.

Pinheiro, A. C., Bourbon, A. I., Medeiros, B. G. S., Silva, L. H. M., Silva, M. C. H., Carneiro-da-Cunha, M. G., Coimbra, M. A., and Vicente, A. A. 2012a. Interactions between κ-carrageenan and chitosan in nanolayered coatings—Structural and transport properties. *Carbohydrate Polymers*, 87, 1081–1090.

Pinheiro, A. C., Bourbon, A. I., Mafalda, Q. M. A. C, Coimbra, M. A., and Vicente, A. A. 2012b. κ-Carrageenan/chitosan nanolayered coating for controlled release of a model bioactive compound innovative. *Innovative Food Science and Emerging Technologies*, 16, 227–232.

Radeva, T., Kamburova, K., and Petkanchin, I. 2006. Formation of polyelectrolyte multilayers from polysaccharides at low ionic strength. *Journal of Colloid and Interface Science*, 298, 59–65.

Rettler, E., Hoeppener, S., Sigusch, B. W., and Schubert, U. S. 2013. Mapping the mechanical properties of biomaterials on different length scales: Depth-sensing indentation and AFM based nanoindentation. *Journal of Materials Chemistry B*, 1, 2789–2806.

Rhim, J. W., Park, H. M., and Ha, C. S. 2013. Bio-nanocomposites for food packaging applications. *Progress in Polymer Science*, 38, 1629–1652.

Richert, L., Lavalle, P., Payan, E., Shu, X. Z., Prestwich, G. D., Stoltz, J. F., Schaaf, P., Voegel, J. C., and Picart, C. 2004. Layer by layer buildup of polysaccharide films: Physical chemistry and cellular adhesion aspects. *Langmuir*, 20, 448–458.

Roco, M. C. 2011. The long view of nanotechnology development: The national nanotechnology initiative at 10 years. *Journal of Nanoparticle Research*, 13, 427–445.

Rojas-Graü, M. A., Avena-Bustillos, R. J., Olsen, C., Friedman, M., Henika, P. R., Martín-Belloso, O., Pan, Z., and McHugh, T. H. 2007. Effects of plant essential oils and oil compounds on mechanical, barrier and antimicrobial properties of alginate–apple puree edible films. *Journal of Food Engineering*, 81, 634–641.

Ruengruglikit, C., Kim, H., Miller, R. D., and Huang, Q. 2004. Fabrication of nanoporous oligonucleotide microarrays for pathogen detection and identification. *Polymer Preprint, 45,* 526.

Salopek, B., Krasic, D., and Filipovic, S. 1992. Measurement and application of zetapotential. *Rudarsko-geoloiko-naftni zbornik, 4,* 147–151.

Schlenoff, J. B., Dubas, S. T., and Farhat, T. 2000. Sprayed polyelectrolyte multilayers. *Langmuir, 16,* 9968–9969.

Schummer, J. 2004. Interdisciplinary issues in nanoscale research. In D. Baird, A. Nordmann, and J. Schummer (Eds.), *Discovering the Nanoscale* (pp. 9–20). Amsterdam: IOS Press.

Sharma, P., Bhalla, V., Dravid, V., Shekhawat, G., Wu, J., Prasad, E. S., and Suri, C. R. 2012. Enhancing electrochemical detection on graphene oxide-CNT nanostructured electrodes using magneto-nanobioprobes. *Scientific Reports, 2,* 877.

Shi, S., Wanga, W., Liu, L., Wu, S., Wei, Y., and Li, W. 2013. Effect of chitosan/nano-silica coating on the physicochemical characteristics of longan fruit under ambient temperature. *Journal of Food Engineering, 118,* 125–131.

Silva, H. D., Cerqueira, M. A., Souza, B. W. S., Souza, B. W. S., Ribeiro, C., Avides, M. C., Quintas, M. A. C., Coimbra, J. S. R., Carneiro-da-Cunha, M. G., and Vicente, A. A. 2011. Nanoemulsions of β-carotene using a high-energy emulsification–evaporation technique. *Journal of Food Engineering, 102,* 130–135.

Silva, T. E. M., Premebida, A., and Calazans, D. 2012. Nanotecnologia aplicada aos alimentos e biocombustíveis: Interações sociotécnicas e impactos sociais. *Linc em Revista, 8,* 207–221.

Silvestre, C., Duraccio, D., and Cimmino, S. 2011. Food packaging based on polymer nanomaterials. *Progress in Polymer Science, 36,* 1766–1782.

Soares, P. A. G., Bourbon, A. I., Vicente, A. A., Andrade, C. A. S., Barros, W., Correia, M. T. S., Pessoa, A., and Carneiro-da-Cunha, M. G. 2014. Development and characterization of hydrogels based on natural polysaccharides: Policaju and Chitosan. *Materials Science & Engineering C, 42,* 219–226.

Sonvico, F., Cagnani, A., Rossi, A., Motta, S., Di Bari, M. T., Cavatorta, F., Alonso, M. J., Deriu, A., and Colombo, P. 2006. Formation of self-organized nanoparticles by lecithin/chitosan ionic interaction. *International Journal of Pharmaceutics, 324,* 67–73.

Sorrentino, A., Gorrasi, G., and Vittoria, V. 2007. Potential perspectives of bio-nanocomposites for food packaging applications. *Trends in Food Science & Technology, 18,* 84–95.

Souza, M. P., Vaz, A. F. M., Correia, M. T. S., Cerqueira, M. A., Vicente, A. A., and Carneiro-da-Cunha, M. G. 2014. Quercetin-loaded lecithin/chitosan nanoparticles for functional food applications. *Food Bioprocess Technology, 7,* 1149–1159.

Souza, M. P., Vaz, A. F. M., Cerqueira, M. A., Texeira, J. A., Vicente, A. A., and Carneiro-da-Cunha, M. G. 2015. Effect of an edible nanomultilayer coating by electrostatic self-assembly on the shelf life of fresh-cut mangoes. *Food and Bioprocess Technology, 8(3),* 647–654.

Suga, M., Asahina, S., Sakuda Y., Kazumori, H. Nishiyama, H., Nukuo, T., Alfredsson, V., Kjellman, T., Stevens, S. M., Cho, H. S., Cho, M., Han, L., Che, S., Anderson, M. W., Schüth, F., Deng, H., Yaghii, O. M., Liu, Z., Jeong, H. Y., Stein, A., Sakamoto, K., Ryoo, R., and Terasaki, O. 2014. Recent progress in scanning electron microscopy for the characterization of fine structural details of nano materials. *Progress in Solid State Chemistry, 42,* 1–21.

Takeuchi, M. T., Kojima, M., and Luetzow, M. 2014. State of the art on the initiatives and activities relevant to risk assessment and risk management of nanotechnologies in the food and agriculture sectors. *Food Research International, 64,* 976–981.

Vanderroost, M., Ragaert, P., Devlieghere, F., and Meulenaer, B. 2014. Intelligent food packaging: The next generation. *Trends in Food Science & Technology, 39,* 47–62.

Velásquez-Cock, J., Ramírez, E., Betancourt, S., Putaux,. J. L., Osorio, M., Castro, C., Gañán, P., and Zuluaga, R. 2014. Influence of the acid type in the production of chitosan films reinforced with bacterial nanocellulose. *International Journal of Biological Macromolecules, 69,* 208–213.

Verleysen, E., Temmerman, P. J., Doren, E. V., Francisco M. A. D., and Mast, J. 2014. Quantitative characterization of aggregated and agglomerated titanium dioxide nanomaterials by transmission electron microscopy. *Powder Technology, 258,* 180–188.

Villiers, M. M., Otto, D. P., Strydom, S. J., and Lvov, Y. M. 2011. Introduction to nanocoatings produced by layer-by-layer (LbL) self-assembly. *Advanced Drug Delivery Reviews, 63,* 701–715.

Wei, L. T., and Yazdanifard, R. 2013. Edible food packaging as an eco-friendly technology using green marketing strategy. *Global Journal of Commerce & Management Perspective, 2,* 8–11.

Weiss, J., and Gibis, M. 2013. Nanotechnology in the food industry. *Ernaehrungs Umschau International, 60,* 44–51.

Weiss, J., Takhistov, P. M. C., and Clements, D. J. 2006. Functional materials in food nanotechnology. *Journal of Food Science, 71,* 107–116.

Whitesides, G. M., and Grzybowski, B. 2002. Self-assembly at all scales. *Science, 295,* 2418–2421.

Woranuch, S., and Yoksan, R. 2013a. Preparation, characterization and antioxidant property of water-soluble ferulic acid grafted chitosan. *Carbohydrate Polymers, 96,* 495–502.

Woranuch, S., and Yoksan, R. 2013b. Eugenol-loaded chitosan nanoparticles: II. Application in bio-based plastics for active packaging. *Carbohydrate Polymers, 96,* 586–592.

Yam, K. L. 2009. Gas permeation of packaging materials. In K. L. Yam, (Ed.), *The Wiley Encyclopedia of Packaging Technology* (pp. 551–555). New York: John Wiley and Sons, Inc.

Yoo, D., Shiratori, S. S., and Rubner, M. F. 1998. Controlling bilayer composition and surface wettability of sequentially adsorbed multilayers of weak polyelectrolytes. *Macromolecules, 31,* 4309–4318.

Yu, S., Hu, J., Pan, X., Yao, P., and Jiang, M. 2006. Stable and pH-sensitive nanogels prepared by self-assembly of chitosan and ovalbumin. *Langmuir, 22,* 2754–2759.

Yu, Y., Zhang, S., Ren, Y., Li, H., Zhang, X., and Di, J. 2012. Jujube preservation using chitosan film with nano-silicon dioxide. *Journal of Food Engineering, 113,* 408–414.

Zhang, Y., Niu, Y., Luo, Y., Ge, M., Yang, T., Yu, L., and Wang, Q. 2014. Fabrication, characterization and antimicrobial activities of thymolloaded zein nanoparticles stabilized by sodium caseinate–chitosan hydrochloride double layers. *Food Chemistry, 142,* 269–275.

Zhou, G. H., Xu, X. L., and Liu, Y. 2010. Preservation technologies for fresh meat: A review. *Meat Science, 86,* 119–128.

10 Nanostructured Multilayers for Food Packaging by Electrohydrodynamic Processing

María José Fabra, Amparo López-Rubio, and José M. Lagaron

CONTENTS

Abstract 319
10.1 Introduction 320
10.2 Advantages in Multilayer Food Packaging Structures 321
10.3 Nanostructured Layers Produced by Electrospinning 322
    10.3.1 High Barrier Food Packaging Multilayer Structures 323
    10.3.2 Low Barrier Food Packaging Multilayer Structures 326
10.4 Conclusion 329
References 329

ABSTRACT

The increased interest in the development of biodegradable and renewable materials for food packaging applications has increased research on new strategies to improve their barrier performance. This chapter describes the potential of the electrohydrodynamic processing technique (EHDA) for the development of nanostructured layers of interest in food packaging. The most common process of the EHDA is the so-called electrospinning technique. Electrospinning is a simple, versatile, and efficient method to produce high-performance polymeric fibers with diameters ranging from the micro to the nanoscale. One of the most innovative applications of electrospinning is the fabrication of nanostructured multilayers with tailor-made physical properties such as barrier properties. For this application, this technique is very promising, since it allows to act as a functional layer and also as a tie layer, promoting adhesion between the constituting layers. This chapter compiles innovative strategies that make use of multilayer systems to achieve optimized barrier performance.

10.1 INTRODUCTION

Biopolymers obtained from natural resources are a promising alternative to non-biodegradable, petroleum-based plastics due to their environmentally friendly nature (Faruk et al., 2012; Jiménez et al., 2012; Tharanathan, 2003). Biodegradable materials (both renewable and nonrenewable) are usually classified into three main families. The first one includes polymers directly extracted from biomass, which include polysaccharides such as chitosan, starch, and cellulose, and proteins, such as gluten and zein. A second family makes use of either oil-based or biomass-derived monomers and applies classical chemical synthetic routes to obtain the final biodegradable polymer; this is the case, for instance, of polycaprolactones, polyvinyl alcohol, ethylene vinyl alcohol copolymers, and polylactic acid (PLA) (Arvanitoyannis et al., 1997; Busolo et al., 2010; Haugaard et al., 2001; Martínez-Sanz et al., 2012; Petersen et al., 1999; Thellen et al., 2013). The third family is currently being massively incorporated into the market and consists of polymers produced by natural or genetically modified microorganisms such as polyhydroxyalcanoates (PHA). One of the main disadvantages of polymers, in general, is that they are not impermeable materials and thus, they allow the passage of low-molecular-weight compounds which can compromise food quality and safety. And this is even more the case for biopolymers, as these materials present, in general, inferior barrier and thermal properties than their oil-based counterparts. Therefore, it is of great industrial interest to enhance their barrier properties while maintaining their inherently good properties like transparency and biodegradability (Bastiolo et al., 1992; Chen et al., 2003; Jacobsen and Fritz, 1996; Jacobsen et al, 1999; Koening and Huang, 1995; Park et al., 2002; Tsuji and Yamada, 2003). The most commercially viable materials at the moment are some biodegradable polyesters, which can be processed by conventional equipment. In fact, these materials are already used in a number of monolayer and also multilayer applications, particularly in the food packaging. Among the most widely researched thermoplastic, sustainable biopolymers for monolayer packaging applications are starch, PHA, and PLA (Phetwarotai et al., 2012; Shirai et al., 2013; Xiong et al., 2013). Specifically, starch and PLA biopolymers are undoubtedly the most interesting families of biodegradable materials, as they present an interesting balance of properties and have become commercially available, being produced in a large industrial scale. Of particular interest in food packaging is the case of PLA due to its excellent transparency and relatively good water resistance (Chen et al., 2003, 2012). The challenge for these specific biomaterials is to improve their properties so that they perform (in terms of barrier and thermal properties) like polyethylene terephthalate (PET) (Lagaron, 2011). There are also other materials extracted from biomass resources, such as proteins (e.g., zein), polysaccharides (e.g., chitosan), and lipids (e.g., waxes), with excellent potential as gas and aroma barriers (Fabra et al., 2008). The main drawbacks of these families of materials are their inherently high rigidity, difficulty of processing in conventional equipment, and for proteins and polysaccharides, the very strong water sensitivity arising from their hydrophilic character. This water sensitivity leads to a strong plasticization, deteriorating their excellent oxygen barrier characteristic (in dry state) as relative humidity (RH) and water sorption in the material increase (Fabra et al., 2010; Tian et al., 2010). This low water resistance of proteins and polysaccharides strongly handicap their use in food packaging applications. From an application point of view, it is of great relevance to diminish the water sensitivity of proteins and polysaccharides and to enhance the gas barrier properties and overall functionalities of thermoplastic biopolyesters to make them more adequate for food packaging applications.

Traditionally, the two most widely used approaches to improve the performance of biopolymers have been either the use of blends or the addition of dispersed nanoreinforcing agents to generate nanocomposites. Recently, these approaches have been extensively reviewed (Chivrac et al., 2009; Halley and Dorgan, 2011; Lagaron, 2011; Sanchez-Garcia et al., 2010; Sionkowska, 2011).

10.2 ADVANTAGES IN MULTILAYER FOOD PACKAGING STRUCTURES

As an alternative, the enhancement of biopolymers' properties can be tackled through the use of biolayering technologies such as monolayer biocoatings, multilayered biodegradable polymers, layer-by-layer bioassemblies, and through a novel methodology based on high-voltage spinning to generate interlayers; these interlayers are able to, on one hand, improve the barrier properties of the biopolymers, and on the other, tie together the different biopolymer layers serving as a natural adhesive (Carneiro-da-Cunha et al., 2010; Elsabee et al., 2008; Rhim et al., 2006). The biolayering approach is currently finding significant advantages when making use of nanofabrication or nanostructuration and is the main innovative subject of this chapter. The general principles of mass transport in polymeric materials have been summarized elsewhere (Paul and Bucknall, 1999). The corresponding modeling of the transport of low-molecular-weight components through multiphase polymer mixtures reveals that a laminate structure of the phase components in a series disposition regarding permeation direction should theoretically provide much better barrier properties than the blend/composite morphology disposition. Barrier properties of blends generally follow a relationship close to that proposed by Maxwell and extended by Roberson for particles of a low-barrier-phase homogeneously distributed in a high barrier continuous matrix (Paul and Bucknall, 1999). However, as mentioned earlier, the most efficient structure for a barrier material using two components is to dispose them in a layered form normal to the direction of the permeant flow (Lagaron et al., 2003; Paul and Bucknall, 1999), that is, in a multilayer form. However, one of the most important problems facing this type of structure is the adhesion between layers of diverse materials, when working with incompatible materials (such as PHA and hydrocolloids), a tie layer will most likely be required for adhesion purposes. To solve the adhesion problems of multilayer structures, an innovative route based on the incorporation of an intermediate electrospun layer has been recently developed in our group in which aliphatic polyesters and hydrophilic natural polymers such as proteins (both thermodynamically immiscible) are properly assembled (Fabra et al., 2013, 2014a, 2014b). Electrospinning is a physical process used for the formation of ultrathin fibers by subjecting the polymer solution to high electric fields. This technique relies on electrostatic forces to draw polymer solutions or melts into ultrathin fibers, which can be deposited as fibrous mats for diverse applications. Recently, various biopolymers have been electrospun, including soy protein (Vega-Lugo and Lim, 2008), chitosan (Vrieze et al., 2007), or zein (Torres-Giner et al., 2008). Further, the preliminary results obtained have demonstrated the potential of this multilayering technology for the development of fully renewable biopolyester-based systems of significant interest in food packaging applications.

10.3 NANOSTRUCTURED LAYERS PRODUCED BY ELECTROSPINNING

The electrospinning process was developed and patented at the beginning of the twentieth century, but in recent years, a renewed interest in the technique is taking place in line with the growing interest in nanotechnology. The so-called electrospinning processing technique is a simple, versatile, and efficient method to produce high-performance polymeric fibers with diameters ranging from the micro to the nanoscale (Sawicka and Gouma, 2006; Schiffman and Schauer, 2008). This technique basically requires a voltage power supply, a syringe containing the polymer solution with a needle, a pumping system, and a grounded collector screen. The electrospinning process is achieved by application of a high-voltage electric field that induces charges on the solution polymer surface. The hemispherical droplet at the tip of the needle stretches and a conical jet projection of the newly created fibers emerges. As the solvent evaporates before the jet reaches the collector, solid fibers are collected in the form of nonwoven mats. Several parameters affect the process and consequently, the final properties of the fibers. The electrospinning process is governed by solution properties (polymer concentration, viscosity, molecular weight, surface tension, and conductivity), process conditions (needle-to-collector distance, flow rate, and voltage), and ambient parameters (temperature and RH). All of them define the morphology and dimensions of the electrospun products, giving rise to fibers or beads (Fabra et al., 2014a; Torres-Giner et al., 2008)

Because of their very large area to volume ratio, ease of preparation and versatility in the design, electrospun ultrathin fibers are being considered in different areas such as filtration, nanotube reinforcement, sensors, catalysis, protective clothing, biomedicine, space applications, semiconductors, micro- or nanoelectronics, and in food packaging and encapsulation (Fabra et al., 2013; Lelkes et al., 2005; López-Rubio et al., 2009; López-Rubio and Lagaron, 2012; Ramakrishna et al., 2003). Specifically, in the food packaging area, a new route to develop high gas and vapor barrier multilayers combining biomass-derived nanocomposites and an electrospinning process has been very recently developed (Fabra et al., 2013). Such route generates high oxygen barrier structures, even at high RH, by combining layers of biopolyesters with layers based on proteins and polysaccharides. This proprietary technological approach and the adequate industrial instrumentation has become commercially available by the Fluidnatek® brand (BioInicia S.L., Valencia, Spain) and is based on creating intermediate or coating electrospun layers of biopolymers to enhance the barrier properties of other biopolymers while serving, at the same time, as natural tie or adhesive layers. This route has a number of advantages versus the multilayer biopolymer structures presented previously, such as:

- No need for tie layers: A challenge for the development of multilayer biobased structures is finding natural adhesives which do not compromise the biodegradability and/or renewability of the materials. Intermediate electrospun fiber layers of biopolymers have demonstrated to have the ability of acting as natural adhesives (Busolo et al., 2009).
- Tailor-made interlayers or coatings: Using the electrospinning technique, it is possible to tightly control the thickness of interlayers or coatings depending on the specific requirements in terms of barrier properties. Moreover, alternating layers of different biopolymers or even blends thereof can be electrospun as intermediate or coating barrier materials.
- Improved optical properties of the packaging materials: As the interlayers are nanofabricated, no changes in transparency or optical properties are virtually observed when developing the multilayer material. This is an important feature in the case of food packaging materials, as transparency is highly appreciated for most applications.
- Improved physical performance such as barrier properties since apart from the multilayer approach being the most efficient processing combination to enhance barrier properties, the nanostructuration of the biomaterials and the flexibility to combine different biopolymers and/or to introduce nanofillers during fabrication seem to allow more efficient blocking effects than expected by thicker individual layers.
- Virtually no impact on flexibility and higher crack resistance with no need for compatibilizers, since the layers are in nanostructured form and therefore, do not potentially affect the overall mechanical performance of the structure.

10.3.1 High Barrier Food Packaging Multilayer Structures

As described elsewhere, electrospun protein interlayers have the potential to generate multilayered biopolymeric structures with improved barrier properties in PLA (Busolo et al., 2009) and PHA films (Fabra et al., 2013, 2014a, 2014b). Concretely, oxygen permeability values were measured to drop by a maximum of 70% in both PHA and PLA multilayer systems by the addition of electrospun zein nanofibers, although the water vapor barrier properties were seen to depend on the polymer used as outer layers. As opposed to the PHA's material performance, PLA did not improve the water vapor permeability values of these films. Thus, Busolo et al. (2009) showed that water vapor permeability increased by 10% in PLA–zein multilayer systems as compared to control PLA films (without zein interlayer). On the contrary, the presence of a zein interlayer in polyhydroxybutyrate-co-valerate with a valerate content of 12% PHBV12 multilayer structures improved their water barrier properties by up to 39% or 93%, depending on the processing method used during the film formation, compression molding versus casting (Fabra et al., 2013). In these multilayer systems,

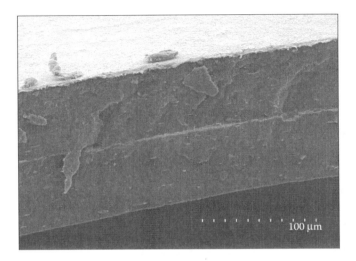

FIGURE 10.1 Scanning electron microscope image of a cryofractured section of the nanostructured multilayer systems based on biopolyesters (outer layer) containing electrospun zein interlayer (inner layer).

transparency was retained by the addition of zein interlayers. Figure 10.1 gives an example of the microstructure reached in this type of multilayer films. From the cross-section images, the zein interlayer could be easily identified as a very thin inner layer in comparison to the overall thickness of the multilayer, exhibiting strong adhesion to the outer matrices. Furthermore, as shown in Figures 10.2 and 10.3, zein fibers merged and aligned in parallel to the outer layers (favoring the formation of a

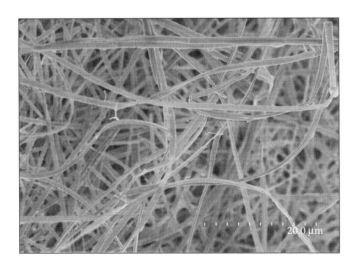

FIGURE 10.2 Scanning electron microscope image of the zein interlayers before the heat treatment used for multilayer assembly.

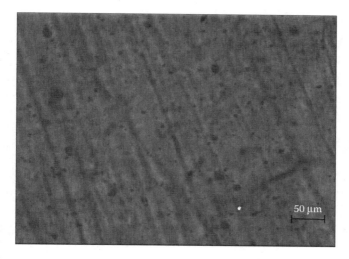

FIGURE 10.3 Optical microscopy image of the top view of zein nanofibers after the multilayer assembly, using polarized light.

continuous zein interlayer) after the heat-compression treatment used for multilayer assembly, which favored better adhesion between layers and reduced free volume for mass transport.

In a recent work, Fabra et al. (2014b) demonstrated that biopolyester–zein multilayers can be postulated as potential substitutes of PET for food packaging applications, since the addition of a zein interlayer improved water and oxygen permeability values of polyhydroxyalkanoate polymers, reaching permeability values similar to those obtained for PET. However, in an effort to find whether other proteins or polysaccharides provide a similar effect than the well-studied zein interlayers, Fabra et al. (2014a) presented for the first time a comparative study of the effect of incorporating electrospun whey protein isolate and pullulan nanofibers as interlayers on the physicochemical properties (optical, barrier, and mechanical properties) of polyhydroxybutyrate-*co*-valerate copolymer with 3% valerate content PHBV3-based multilayer films. It was observed that oxygen and water vapor barrier properties were greatly influenced by the morphology, thickness, and inherent barrier of the electrospun interlayer materials. Thus, zein and pullulan formed fibrillar structures which significantly contributed to improve barrier properties of the multilayer systems. On the contrary, electrospun whey protein isolate (WPI) formed bead microstructures (Figure 10.4) and did not improve oxygen and water barrier properties of these multilayer systems. The mechanical properties did not vary significantly by the addition of WPI or pullulan micro- or nanostructures. However, the addition of electropun zein nanofibers formed less stiff multilayer structures but with no higher elongation at break. Transparency of multilayer systems was slightly compromised by the addition of zein and/or pullulan interlayers, probably due to the light scattering generated by the arrangement of the biopolymeric components distributed throughout the inner layer.

FIGURE 10.4 Scanning electron microscope image of the WPI beads obtained by electrospinning.

Thus, the use of electrospun zein, whether pure or in blends with other hydrocolloids, and pullulan nanofibers as inner interlayers in biopolyester multilayer systems provided fully bio-based films of significant interest in food packaging, since they provide minimum changes in mechanical and optical properties but with enhanced barrier performance.

10.3.2 Low Barrier Food Packaging Multilayer Structures

The determination of the barrier properties of a biopolymer is crucial to estimate and predict the packaged product shelf life and the specific barrier requirements of a packaged system, which depends upon the food characteristics and the intended applications. For instance, the water vapor barrier properties for the packaged food product, whose physical and chemical deteriorations are related to its equilibrium moisture content, are of great importance for maintaining or extending its shelf life. In this sense, for fresh food, it is important to avoid dehydration, while for bakery or delicatessen, it is important to avoid water permeation.

As commented above, researchers are mainly focused on the development of biodegradable nanocomposites or multilayer structures with high barrier properties (Fabra et al., 2014a, 2014b; Rhim et al., 2013). However, there are some fresh products (e.g., mushroom and strawberry) with high respiration rates which require packaging films with high oxygen and carbon dioxide permeability values, but the permeability of the commonly used polymeric nonbiodegradable films is not sufficiently high and anaerobic conditions might occur. To avoid this problem, micro- and macroperforated films have been developed for some products (Farber et al., 2003).

In an effort to obtain highly permeable biopolymers (without the need of perforation), multilayer films based on plasticized wheat gluten or starch (inner layer) and biopolyester (outer layer) materials have been designed. These high permeable

FIGURE 10.5 Scanning electron microscope image of a cryofractured section of nanostructured plasticized WG multilayer systems containing biopolyesters fibers as outer layers.

biodegradable multilayer systems have been developed as an alternative to protect fresh products. It is well known that when proteins and polysaccharides are plasticized, water and oxygen permeability values increase, which could be a possibility to get the packaging requirements. Nevertheless, PHA's outer layer should be added to avoid swelling problems of the plasticized wheat gluten film. Wheat gluten was chosen due to its high potential as raw material for technical applications given their unique intrinsic properties, with numerous publications referring to their outstanding film-forming potential. Wheat gluten is a by-product of the wheat starch industry, that is, commercially available at low cost in large quantities. Wheat gluten–based materials are amorphous and can be prepared using either a solvent process (Gontard and Ring, 1996) or common thermoplastic processes such as extrusion (Hochstetter et al., 2006; Redl et al., 1999) or thermomolding (Angellier-Coussy et al., 2011; Gallstedt et al., 2004; Sun et al., 2008). Figure 10.5 gives an example of the microstructure reached in this type of multilayer films. Table 10.1 displays the mechanical properties of the nanostructured multilayer systems based on wheat gluten. The addition of biopolyester layers improved the mechanical properties of the plasticized wheat gluten films, especially Young's modulus (E), which increased up to ~40% for the multilayer structures containing greater amount electrospun fibers (Fabra et al., 2015b). However, wheat gluten multilayer nanostructures are less deformable and more flexible and elastic than high- or low-density polyethylene or polypropylene. Thus, further efforts should be focused on the development of new strategies to improve the tensile strength and the elongation at break of the outer layers and thus, of the overall multilayer structure. Table 10.1 also gathers the water vapor and oxygen permeability of the nanostructured multilayer systems based on wheat gluten. It is worth to note that no data could be obtained for the plasticized wheat gluten films which seem to be highly permeable. Probably the high amount of glycerol together

TABLE 10.1
Barrier (Water Vapor and Oxygen Permeability) and Mechanical Properties of WG Multilayer Structures, Measured at 0%–84% RH and 80% RH, Respectively

Inner Layer	Outer Layer		WVP 10^{14} (kg m Pa^{-1} s^{-1} m^{-2})	O$_2$ Permeability 10^{18} (m^3 m m^{-2} s^{-1} Pa^{-1})	TS (MPa)	EAB (%)
	Type of PHA's Fibers	mg PHA cm^{-2}	Measured at 0%–84% RH	Measured at 80% RH		
WG	—		16.02 (0.43)[a]	—	2.6 (0.1)[a]	136 (12)[a]
	PHBV3	0.5	4.80 (0.06)[b]	15.10 (2.42)[a]	6.1 (0.9)[b]	86 (19)[b]
		1.0	4.55 (0.25)[b]	4.96 (0.92)[b]	8.0 (1.8)[b]	32 (6)[c]
	PHB	1.5	4.65 (0.37)[b]	16.25 (2.77)[a]	4.9 (1.8)[b]	140 (4)[a]
		3.0	3.11 (0.64)[c]	8.95 (0.34)[c]	6.5 (0.8)[b]	47 (8)[d]
Low-density polyethylene[e]				21.5[e]	9–17[f]	500[f]
Polypropylene[e]				6.7[e]	42[f]	300[f]

Note: Mean value (standard deviation).
TS: tensile strength; EAB: elongation at break.
[a–d] Different superscripts within the same column indicate significant differences among multilayer structures and the plasticized wheat gluten film ($p < 0.05$).
[e] Films measured at 38°C and 0% RH (Lagaron, 2011).
[f] Films conditioned before and during test at 25°C, 50% RH (Bristol, 1986).

with the high RH used for the oxygen permeability measurements made the permeability analysis difficult with the equipment used. The preliminary results carried out in our research group (Fabra et al., 2015a) showed that nanostructured wheat gluten multilayer systems can be used in specific packaging applications, where low-barrier properties are required since the oxygen permeability values are in the same range as low-density polyethylene (Bristol, 1986).

10.4 CONCLUSION

This chapter demonstrates that nanostructured multilayers by electrohydrodynamic processing can be successfully used to develop tailor-made high and low barrier materials of interest in food packaging. These strategies have been used as highly efficient alternatives to nanocomposite films for improving physical and barrier properties of bio-based materials. The advantages of this nanotechnological process to constitute layers are mainly related to retaining transparency and enhancing barrier properties without significant changes in mechanical performance. The methodology used to develop the nanostructured multilayer systems can be potentially scaled-up, since the industrial instrumentation needed to develop the nanostructured layers are already commercially available.

REFERENCES

Angellier-Coussy, H., Gastaldi, E., Gontard, N., and Guillard, V. 2011. Influence of processing temperature on the water vapour transport properties of wheat gluten based agro-materials. *Industrial Crops and Products*, *33*, 457–461.

Arvanitoyannis, I., Psomiadou, E., Biliaderis, C. G., Ogawa, H., Kawasaki, H., and Nakayama, O. 1997. Biodegradable films made from low density polyethylene (LDPE), ethylene acrylic acid (EAA), polycaprolactone (PLC) and wheat starch for food packaging applications: Part 3. *Starch-Starke*, *49*, 306–322.

Bastiolo, C., Bellotti, V., Del Tredici, G. F., Lombi, R., Montino, A., and Ponti R. 1992. Biodegradable polymeric compositions based on starch and thermoplastic polymers. International Patent Application WO 92/19680. http://brevets-patents.ic.gc.ca/opic-cipo/cpd/eng/patent/2084994/summary.html, accessed July 15, 2013.

Bristol, J. H. 1986. Films, plastic. In M. Bakker (Ed.), *The Wiley Encyclopedia of Packaging Technology* (pp. 329–335). New York: John Wiley & Sons, Inc.

Busolo, M. A., Fernandez, P., Ocio, M. J., and Lagaron, J. M. 2010. Novel silver-based nanoclay as an antimicrobial in polylactic acid food packaging coatings. *Food Additives and Contaminants—Part A Chemistry, Analysis, Control, Exposure and Risk Assessment*, *27*(11), 1617–1626.

Busolo, M., Torres-Giner, S., and Lagaron, J. M. 2009. Enhancing the gas barrier properties of polylactic acid by means of electrospun ultrathin zein fibres. In *ANTEC, Proceedings of the 67th Annual Technical Conference*, Chicago, Illinois, Vol. 5, pp. 2763–2768.

Carneiro-da-Cunha, M. G., Cerqueira, M. A., Souza, B. W. S., Carvalho, S., Quintas, M. A. C., Teixeira, J. A., and Vicente, A. A. 2010. Physical and thermal properties of a chitosan/alginate nanolayered PET film. *Carbohydrate Polymers*, *82*, 153–159.

Chen, B.-K., Shih, C.-C., and Chen, A. F. 2012. Ductile PLA nanocomposites with improved thermal stability. *Composites Part A: Applied Science and Manufacturing*, *43*(12), 2289–2295.

Chen, C. C., Chueh, J. Y., Tseng, H., Huang, H. M., and Lee, S. Y. 2003. Preparation and characterization of biodegradable PLA polymeric blends. *Biomaterials*, 24, 1167–1173.

Chivrac, F., Pollet, E., and Avérous, L. 2009. Progress in nano-biocomposites based on polysaccharides and nanoclays. *Materials Science and Engineering: R: Reports*, 67, 1–17.

Elsabee, M., Abdou, E. S., Nagy, K. S. A., and Eweis, M. 2008. Surface modification of polypropylene films by chitosan and chitosan/pectin multilayer. *Carbohydrate Polymers*, 71(2), 187–195.

Fabra, M. J., López-Rubio, A., and Lagaron, J. M. 2013. High barrier polyhydroxyalcanoate food packaging film by means of nanostructured electrospun interlayers of zein prolamine. *Food Hydrocolloids*, 32, 106–114.

Fabra, M. J., López-Rubio, A., and Lagaron, J. M. 2014a. On the use of different hydrocolloids as electrospun adhesive interlayers to enhance the barrier properties of polyhydroxyalkanoates of interest in fully renewable food packaging concepts. *Food Hydrocolloids*, 39, 77–84.

Fabra, M. J., López-Rubio, A., and Lagaron, J. M. 2014b. Nanostructured interlayer of zein to improve the barrier properties of high barrier polyhydroxyalkanoates and other polyesters. *Journal of Food Engineering*, 127, 1–9.

Fabra, M. J., López-Rubio, A., and Lagaron, J. M. 2015a. Three-layer films based on wheat gluten and electrospun PHA. *Food and Bioprocess Technology*, 8(11), 2330–2340.

Fabra, M. J., López-Rubio, A., and Lagaron, J. M. 2015b. Effect of the film-processing conditions, relative humidity and ageing on wheat gluten films coated with electrospun polyhydryalkanoate. *Food Hydrocolloids*, 44, 292–299.

Fabra, M. J., Talens, P., and Chiralt, A. 2008. Tensile properties and water vapour permeability of sodium caseinate films containing oleic acid-beeswax mixtures. *Journal of Food Engineering*, 85, 393–400.

Fabra, M. J., Talens, P., and Chiralt, A. 2010. Water sorption isotherms and phase transitions of sodium caseinate–lipid films as affected by lipid interactions. *Food Hydrocolloids*, 24(4), 384–391.

Farber, J. N., Harrin, L. J., Parish, M. E., Beuchat, I. R., Susiow, T. V., Gorney, J. R., Garret, E. H., and Busta, F. F. 2003. Microbiological safety of controlled and modified atmosphere packaging of fresh and fresh cut produce. *Comprehensive Reviews in Food Science and Food Safety*, 2, 142–160.

Faruk, O., Bledzki, A. K., Fink, H.-P., and Sain, M. 2012. Biocomposites reinforced with natural fibres: 2000–2010. *Progress in Polymer Science*, 37(11), 1552–1596.

Gontard, N., and Ring S. 1996. Edible wheat gluten film: Influence of water content on glass transition temperature. *Journal of Agricultural and Food Chemistry*, 44, 3474–3478.

Gallstedt, M., Mattozzi, A., Johansson, E., and Hedenqvist, M. S. 2004. Transport and tensile properties of compression-molded wheat gluten films. *Biomacromolecules*, 5(5), 2020–2028.

Halley, P. J., and Dorgan, J. R. 2011. Next-generation biopolymers: Advanced functionality and improved sustainability. *MRS Bulletin*, 36, 687–691.

Haugaard, V. K., Udsen, A. M., Mortensen, G., Hoegh, L., Petersen, K., and Monahan, F. 2001. Food biopackaging. In C. J. Weber (Ed.) *Biobased Packaging Materials for the Food Industry—Status and Perspectives*. A report from the EU concerted action project: Production and application of biobased packaging materials for the food industry (Food Biopack), funded by DG12 under the contract PL98 4046. Available at http://www.biomatnet.org/publications/f4046fin.pdf.

Hochstetter, A., Talja, R. A., Helen, H. J., Hyvonen, L., and Jouppila, K. 2006. Properties of gluten-based sheet produced by twin-screw extruder. *LWT—Food Science and Technology*, 39(8), 893–901.

Jacobsen, S., Degée, P. H., Fritz, H. G., Dubois, P. H., and Jérome, R. 1999. Polylactide (PLA) a new way of production. *Polymer Engineering and Science*, 39, 1311–1319.

Jacobsen, S., and Fritz, H. G. 1996. Filling of poly(lactic acid) with native starch. *Polymer Engineering and Science*, 36, 2799–2804.

Jiménez, A., Fabra, M. J., Talens, P., and Chiralt, A. 2012. Edible and biodegradable starch films: A review. *Food and Bioprocess Technology*, 5(6), 2058–2076.

Koening, M. F., and Huang S. J. 1995. Biodegradable blends and composites of polycaprolactone and starch derivatives. *Polymer*, 36, 1877–1882.

Lagaron, J. M. 2011. *Multifunctional and Nanoreinforced Polymers for Food Packaging* (1st ed.). Cambridge, UK: Woodhead Publishing Limited.

Lagaron, J. M., Gimenez, E., Altava, B., Del-Valle, V., and Gavara, R. 2003. Characterization of extruded ethylene-vinyl alcohol copolymer based barrier blends with interest in food packaging applications. *Macromolecular Symposia*, 198, 473–482.

Lelkes, P. I., Li, M., Mondrinos, M. J., Gandhi, M. R., Ko, F. K., and Weiss, A. S. 2005. Electrospun protein fibres as matrices for tissue engineering. *Biomaterials*, 26, 5999–6008.

Lopez-Rubio, A., and Lagaron, J. M. 2012. Whey protein capsules obtained through electrospraying for the encapsulation of bioactives. *Innovative Food Science and Emerging Technologies*, 13, 200–206.

Lopez-Rubio, A., Sanchez, E., Sanz, Y., and Lagaron, J. M. 2009. Encapsulation of living bifidobacteria in ultrathin PVOH electrospun fibres. *Biomacromolecules*, 10, 2823–2829.

Martínez-Sanz, M., Lopez-Rubio, A., and Lagaron, J. M. 2012. Nanocomposites of ethylene vinyl alcohol copolymer with thermally resistant cellulose nanowhiskers by melt compounding (II): Water barrier and mechanical properties. *Journal of Applied Polymer Science*, 128(3), 2197–2207.

Park, E. S., Kim, M. N., and Yoon, J. S. 2002. Grafting of polycaprolactone onto poly(ethylene-co-vinyl alcohol) and application to polyethylene-based biodegradable blends. *Journal of Polymer Science, Part B: Polymer Physics*, 40, 2561–2569.

Paul, D. R., and Bucknall, C. B. 1999. *Polymer Blends. Volume 2: Performance*. New York: John Wiley & Sons, Inc.

Petersen, K., Nielsen, P. K., Bertelsen, G., Lawther, M., Olsen, M. B., Nilsson, N. H., and Mortensen, G. 1999. Potential of biobased materials for food packaging. *Trends in Food Science and Technology*, 10, 52–68.

Phetwarotai, W., Potiyaraj, P., and Aht-Ong, D. 2012. Characteristics of biodegradable polylactide/gelatinized starch films: Effects of starch, plasticizer, and compatibilizer. *Journal of Applied Polymer Science*, 126(1), E162–E172.

Ramakrishna, S., Huang, Z., Zhang, Y. Z., and Kotaki, M. 2003. A review on polymer nanofibres by electrospinning and their applications in nanocomposites. *Composites Science and Technology*, 63, 2223–2253.

Redl, A., Morel, M. H., Bonicel, J., Vergnes, B., and Guilbert, S. 1999. Extrusion of wheat gluten plasticized with glycerol: Influence of process conditions on flow behavior, rheological properties, and molecular size distribution. *Cereal Chemistry*, 76(3), 361–370.

Rhim, J. W., Mohanty, K. A., Singh, S. P., and Ng, P. K. W. 2006. Preparation and properties of biodegradable multilayer films based on soy protein isolate and poly(lactide). *Industrial and Engineering Chemistry Research*, 45, 3059–3066.

Rhim, J. W., Park, H. M., and Ha, C. S. 2013. Bio-nanocomposites for food packaging applications. *Progress in Polymer Science*, 38(10–11), 1629–1652.

Sanchez-Garcia, M. D., Lopez-Rubio, A., and Lagaron, J. M. 2010. Natural micro and nanobiocomposites with enhanced barrier properties and novel functionalities for food biopackaging applications. *Trends in Food Science and Technology*, 21, 528–536.

Sawicka, K., and Gouma, P. 2006. Electrospun composite nanofibres for functional applications. *Journal of Nanoparticle Research*, 8, 769–781.

Schiffman, J. D., and Schauer, C. L. 2008. A review: Electrospinning of biopolymer nanofibres and their applications. *Polymer Reviews*, 48(2), 317–352.

Shirai, M. A., Grossmann, M. V. E., Mali, S., Yamashita, F., Garcia, P. S., and Müller, C. M. O. 2013. Development of biodegradable flexible films of starch and poly(lactic acid) plasticized with adipate or citrate esters. *Carbohydrate Polymers*, *92*(1), 19–22.

Sionkowska, A. 2011. Current research on the blends of natural and synthetic polymers as new biomaterials: Review. *Progress in Polymer Science*, *36*, 1254–1276.

Sun, S. M., Song, Y. H., and Zheng, Q. 2008. Thermo-molded wheat gluten plastics plasticized with glycerol: Effect of molding temperature. *Food Hydrocolloids*, *22*(6), 1006–1013.

Tharanathan, R. N. 2003. Biodegradable films and composite coatings: Past, present and future. *Trends in Food Science and Technology*, *14*, 71–78.

Thellen, C., Cheney, S., and Ratto, J. A. 2013. Melt processing and characterization of polyvinyl alcohol and polyhydroxyalkanoate multilayer films. *Journal of Applied Polymer Science*, *127*(3), 2314–2324.

Tian, H., Wang, Y., Zhang, L., Quan, C., and Zhang, X. 2010. Improved flexibility and water resistance of soy protein thermoplastics containing waterborne polyurethane. *Industrial Crops and Products*, *32*(1), 13–20.

Torres-Giner, S., Gimenez, E., and Lagaron, J. M. 2008. Characterization of the morphology and thermal properties of zein prolamine nanostructures obtained by electrospinning. *Food Hydrocolloids*, *22*, 601–614

Tsuji, H., and Yamada, T. 2003. Blends of aliphatic polyesters. VIII. Effects of poly(L-lactide-co-ε-caprolactone) on enzymatic hydrolysis of poly(L-lactide), poly(ε-caprolactone), and their blend films. *Journal of Applied Polymer Science*, *87*, 412–419.

Vega-Lugo, A. C., and Lim, L. T. 2008. Electrospinning of soy protein isolate nanofibres. *Journal of Biobased Materials and Bioenergy*, *2*, 223–230.

Vrieze, S. D., Westbroek, P., Camp, T. V., and Langenhove, L. V. 2007. Electrospinning of chitosan nanofibrous structures: Feasibility study. *Journal of Material Science*, *42*, 8029–8034.

Xiong, Z., Yang, Y., Feng, J., Zhang, X., Zhang, C., Tang, Z., and Zhu, J. 2013. Preparation and characterization of poly(lactic acid)/starch composites toughened with epoxidized soybean oil. *Carbohydrate Polymers*, *92*(1), 810–816.

11 How to Evaluate the Barrier Properties for Edible Packaging of Respiring Products

Maria José Sousa-Gallagher

CONTENTS

Abstract 333
11.1 Introduction 334
11.2 Fresh Fruits and Vegetables 335
    11.2.1 Product Selection and Market Share 335
        11.2.1.1 Market Share of Selected Products 337
    11.2.2 Quality, Safety, and Storage Requirements 338
        11.2.2.1 Critical Quality Parameters, Level of Acceptance, and Safety Issues 338
        11.2.2.2 Optimal Storage Conditions 339
    11.2.3 Quantification of Physiological Properties 340
        11.2.3.1 Respiration Rate 340
        11.2.3.2 Transpiration Rate 342
11.3 Modified Atmosphere Packaging 343
    11.3.1 Current Packaging and Shelf Life Status 343
    11.3.2 Time–Temperature History at Various Points of Distribution Chain 345
    11.3.3 Opportunities for Niche Market Packaging Development 345
    11.3.4 Integrative MAP Modeling 346
11.4 Quantification of Barrier Properties: OTR, CTR, and WVTR 347
    11.4.1 Benchmark Packaging Properties 348
    11.4.2 Tailored Barrier Properties for Packaging 349
11.5 Conclusions 349
References 351

ABSTRACT

Modified atmosphere packaging (MAP) of fruits and vegetables is a technique that relies on the interplay between the respiratory metabolism of the product and the gas exchange kinetics through the package to achieve an optimum gas composition.

This can help extend shelf life and reduce product loss. The challenge would be to find the most appropriate packaging material (or the number of perforations required) to match the respiration rate for each specific fresh product. MAP as a technology has been studied for many years, but only recently efforts were made to integrate respiratory and film permeability data into predictive models for package gas compositions over time, as well as the effects of temperature. Integrative mathematical modeling for MAP has not only been used to determine the package gas composition but can also be used to determine the fresh produce packaging permeability needs for gases and water vapor. Integrative mathematical modeling of packaging system, which considers the quality by design (QbD) approach, was used to estimate the tailored oxygen transmission rate (OTR), carbon dioxide transmission rate (CTR), and water vapor transmission rate (WVTR) for the selected products that would result in the ideal packaging solution. These tailored barrier properties of packaging films were designed to match the optimal-modified atmosphere for the two selected products at the specified temperature. Both mushrooms and strawberries require packaging materials with high gas and water vapor permeability with the target range of OTR 47,501–71,251 (32,779–49,169) mL m^{-2} day^{-1} atm^{-1}, CTR 73,872–110,808 (22,275–33,413) mL m^{-2} day^{-1} atm^{-1}, and WVTR 425–638 (93–140) g m^{-2} day^{-1} atm^{-1} for packaging mushrooms (strawberries) at 5°C and 90% relative humidity. Engineering MAP design considering QbD approach can be not only used to define an effective MAP but also to determine material packaging permeability needs for fresh fruits and vegetables, and therefore maintain quality and safety and minimize product losses.

11.1 INTRODUCTION

The food packaging industry is constantly looking for new ways of preserving food products, maintaining quality and safety in an efficient manner to satisfy the needs of society and economy. The new packaging technologies play an important role in meeting the environment, society, and economic needs (sustainable packaging).

A packaging design can be evaluated based on performance (protection, convenience, communication, and containment), cost, and package environment (Yam and Lee, 2012) and should also sustain the conditions to which it is exposed during supply chain, so that produce quality attributes can be preserved. A package is therefore designed to protect and preserve the commodity from external contamination, and so, extend shelf life and maintain the quality and safety of packaged food, and at the same time to provide convenience at a minimal impact to the environment. To reduce the subjacent cost, it is crucial that product quality and safety are not significantly compromised; otherwise, we could potentially contribute to waste increase. To reduce postharvest losses, it is important to understand the causes related to fruits and vegetables deterioration and apply the appropriate and affordable technological procedures to maintain quality of the produce, delay senescence, and growth inhibition of many spoilage microorganisms.

Modified atmosphere packaging (MAP) technology offers the possibility to extend and preserve the shelf life of fresh fruits and vegetables, improve the product image, and at the same time reduce waste. However, if MAP is not properly designed, it can lead to water accumulation at the product surface, promoting product degradation and microbial growth, which compromises MAP objectives (Song et al., 2001, 2002). The low water vapor transmission rate (WVTR) of films commonly used for MAP of fresh fruits and vegetables, combined with the product metabolism (respiration and transpiration), rapidly brings about saturation (100% relative humidity, RH) of the package atmosphere. The saturated in-pack RH conditions along with temperature variation during supply chain can result in the condensation of water on the inside surface of the packaging film and on the contained produce, favoring microbial growth and discoloration.

Integrative mathematical modeling for MAP has been used to determine the gas composition packaging needs for fresh fruits and vegetables (Mahajan et al., 2006, 2007; Sousa-Gallagher and Mahajan, 2012, 2013) based on O_2 consumption and CO_2 production rate, that is, respiration rate of fresh produce. Xanthopoulos et al. (2012) developed a mathematical model to predict O_2, CO_2, N_2, and H_2O in perforated-mediated polymer packages during cold storage of strawberries accounting for respiration and transpiration. The model predictions were tested against published experimental data of O_2 and CO_2 concentrations in MAP showing a satisfactory agreement, but no validation was performed for H_2O. Mahajan et al. (2008) developed a mathematical model for transpiration rate (TR) of mushroom to represent the evolution of weight loss as a function of temperature and RH. A model development of strawberries transpiration and validation was integrated into the engineering packaging design (Sousa-Gallagher et al., 2013) and used to simulate the packaging requisites.

Therefore, integrative mathematical modeling for MAP can be used to determine the gas composition inside a package, but it is also very relevant to determine the packaging needs of fresh produce in terms of permeability to gases and water vapor (Sousa-Gallagher and Mahajan, 2012).

11.2 FRESH FRUITS AND VEGETABLES

All fresh fruits and vegetables are respiring bodies, whose freshness is maintained only when cells are alive and active (Vakkalanka et al., 2012). Due to high moisture content, richness in nutrients, tender nature, and active metabolism, they are susceptible to dehydration, mechanical damage, environmental stress, and microorganism contamination (Vakkalanka et al., 2012; Wu, 2010). Fresh produce are highly perishable and may be unacceptable for consumption if not handled properly after harvest (Wu, 2010).

11.2.1 PRODUCT SELECTION AND MARKET SHARE

Two fresh products were selected for product description specification; these products are whole mushrooms (*Agaricus bisporus*) and strawberries (cv. Elsanta). Ireland is one of the major producers of these two products (Figures 11.1 and 11.2).

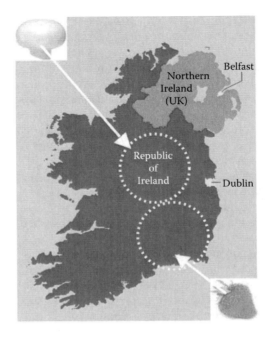

FIGURE 11.1 Fresh producers in Ireland.

FIGURE 11.2 Mushroom and strawberry production facilities in Ireland.

Most of the mushrooms and strawberries produced are sold as fresh product either locally or exported to the United Kingdom and other parts of Europe. However, short shelf life still poses difficulties in distribution and marketing of these products and thus, extending postharvest storage life and the development of improved packaging with biodegradable materials have been a constant quest. There is a need to improve packaging and shelf life to allow exporting these perishable products to the United Kingdom and European Union (EU), bringing economic advantage for the Irish and EU economy and potentially contributing to the waste minimization and added value to the agricultural products.

The white strain of *Agaricus* or button mushrooms is the most important variety in Ireland. These mushrooms vary in size from small to jumbo stuffer. They are very versatile and can be eaten fresh or cooked. They are low in calories, fat, cholesterol, and sodium free and high in riboflavin and fiber. They are also a good source of niacin, pantothenate, and copper. Mushrooms are highly perishable and their shelf life is 1–3 days at ambient temperature, due to high moisture content, high TR, and high respiration rate; therefore, mushrooms need a special packaging to keep their quality and freshness along the supply chain. At ambient temperature, cap is opened and colored, stem is elongated, and texture becomes soft and spongy, resulting in a depression of its commercial value. Currently, mushrooms are marketed in trays overwrapped with perforated polyvinyl chloride (PVC) stretchable packaging film with little or no atmosphere modification. Shelf life of 8 days in modified atmosphere (2%–5% O_2 and 3%–8% CO_2) at 3°C and a maximum of 14 days at 2°C in controlled atmosphere (5% O_2 and 10% CO_2) has been reported (Burton, 1991; De la Plaza et al., 1995). However, mushrooms are very sensitive to humidity levels, as high water levels favor microbial growth and discoloration; conversely, low water levels lead to loss of water (and thus economic value) and undesirable textural changes. Therefore, there is an urgent need to develop a packaging material with appropriate barrier properties for mushrooms.

Strawberries are highly perishable commodity products, due to high respiration and very high susceptibility to gray mold rot. Spoilage due to *Botrytis* infection is a major factor limiting quality of strawberries. It is one of the first visible attributes the consumer is confronted with in assigning quality to strawberries. Strawberries are typically packaged in the field and quickly transported to a cooling facility, where they are precooled by forcing cold air through the vents of the packaging. Strawberry producers must take into account a 3-day period of refrigerated, surface transport before the fruit is retailed. Both selected products are high value and perishable crops with high consumer demand and good export potential. However, there are many challenges, for example, shelf life, fungal decay, and moisture control and also both products are susceptible to moisture loss and microbial decay. The application of new biodegradable material in conjunction with MAP accompanied with low temperature storage could effectively retard the quality changes and extend shelf life of these two selected respiring products.

11.2.1.1 Market Share of Selected Products

Mushroom production in Ireland stands at about 69,000 ton, representing 8% of the total EU market which comprises 800,000 ton. Around 55% of the 170,000 ton, UK fresh mushroom retail market is supplied by Irish businesses. The mushroom industry task force in Ireland estimated that the industry is currently worth €124 million at the farm gate, generating €95 million in export income with direct employment of around 4000 people in the production and supply chain. Mushroom production is expanding in the Netherlands and Belgium, which currently produce a collective 330,000 ton and export to Britain, France, Germany, and Scandinavia. Entrepreneurial drive, high investment, and low labor costs have lifted Polish mushroom production to 150,000 ton, and its target markets are Germany and importantly Britain.

TABLE 11.1
Strawberry Production in Europe by Country

Strawberry Production in Europe (ton)

Poland	200,723
Italy	155,583
Germany	150,854
United Kingdom	87,200
France	43,541
The Netherlands	41,000
Belgium	40,000
Romania	21,233
Austria	19,363
Ireland	1400

Worldwide production of fresh strawberries in year 2008 was 3.9 million ton with a contribution of approximately 28% of world production from Europe. Among European countries, Spain is the leading strawberry producer with an annual production of 263,900 ton, while Ireland produces about 1400 ton of strawberries. The strawberry industry in Ireland is worth close to €30 million annually and is one of the areas of commercial horticulture that is rapidly expanding. This industry in Ireland could grow to be worth up to €50 million in 5 years time as consumer demand increases due to higher available incomes and a huge interest in eating healthy foods. Irish fresh market production is centered along the east coast with potential for development around large towns throughout the country. Multiple retail outlets account for 50% of fresh strawberry sales. Roadside stalls and farm gate sales account for 35% and greengrocers account for 15%. Irish strawberries are normally sold on the home market with producer groups successfully exporting onto the United Kingdom, when supply exceeds demand on domestic market. Other main strawberry-producing countries in Europe are presented in Table 11.1.

11.2.2 Quality, Safety, and Storage Requirements

11.2.2.1 Critical Quality Parameters, Level of Acceptance, and Safety Issues

During postharvest storage, the main processes related to mushrooms deterioration have been related to the development of the sporophore, such as breaking of the veil, elongation of the stipe, opening of the pileus, expansion of gill tissue, and spore formation. These phenomena, together with browning of the cap and gills, weight loss, and a general loss of appearance, have been considered negative characteristics and limit the shelf life of mushrooms. The details are given in Table 11.2.

For strawberries, spoilage due to *Botrytis* infection is a major factor limiting quality of strawberries. It is one of the first visible attributes the consumer is confronted with in assigning quality to strawberries. Other criteria include loss of texture attributes, browning, weight loss, and loss of aroma volatiles, and the details are shown in Table 11.2.

TABLE 11.2
Quality and Safety Parameters for Strawberry

Quality Parameter	Method of Measurement
Fungal decay	The presence of fungal decay is estimated visually and results are expressed as percent of number of fruits infected.
Weight loss	Using electronic balance, weight of mushrooms measured at regular intervals and percent weight loss is calculated.
Color	Standard colorimeter with Hunter Lab Color scale (L, a, b). A single index called chroma is useful to reflect color evolution.
Firmness	Standard texture profile analyzer is used.
Sensory evaluation	9-point hedonic scale is used. The sensory attribute used are overall appearance, aroma, color, and firmness.
Safety issues	Infection with *Botrytis cinerea* is a major limiting factor for keeping quality of strawberries.

11.2.2.2 Optimal Storage Conditions

There has been widely published literature on the optimum storage and packaging atmospheric conditions for the two selected products. However, it is to be noted that the products vary in their tolerance to low O_2 and elevated CO_2 levels; indeed there are differing reports on recommendations about the optimal atmospheric compositions for mushrooms. Considering the optimal conditions, packaging is designed in such a way that equilibrium modified atmosphere created inside the package should fall within the recommended gas atmospheric ranges for the given product. The range of recommended gas atmospheres, along with temperature and RH, for the selected products (strawberry and mushroom) is shown in Table 11.3.

TABLE 11.3
Optimal Storage and Packaging Conditions for the Selected Products

Optimum Conditions	Strawberry (cv. Elsanta)	Mushroom (*Agaricus bisporus*)
Optimum storage conditions	Temperature: 0–5°C Relative humidity: 90%–95%	Temperature: 0–5°C Relative humidity: 90%–95%
Optimal package atmosphere	O_2: 5%–10%[a] CO_2: 15%–20%	O_2: 3%–5%[b] CO_2: <12% (~10%)
Benefits of MAP	• Reduces the growth of *Botrytis cinerea* (gray mold rot) • Reduces respiration rate • Firmness retention	• Prevents cap opening • Preserves texture loss • Reduces browning of cap

[a] Kader (1992).
[b] Parentelli et al. (2007), Nichols and Hammond (1973).

11.2.3 QUANTIFICATION OF PHYSIOLOGICAL PROPERTIES

11.2.3.1 Respiration Rate

Fresh fruits and vegetables are living commodities which respire even after harvest, that is, they consume O_2 and produce CO_2. Due to the respiration process, an O_2 and CO_2 concentration gradient between the package head space and the storage environment is generated. Thus, a gas flow is activated through the packaging material to the surrounding atmosphere. If this gas flow is controlled using suitable barrier properties of packaging material, equilibrium will be achieved. At this equilibrium, the rate of gas exchange between the product and the packaging material equals, therefore the gas composition surrounding the product is maintained subsequently for the rest of the storage life. Extensive research has been done on the effect of the main influential parameters such as temperature, O_2, CO_2 on respiration rate of fresh fruits and vegetables. For the selected products, the following mathematical models were selected for quantifying respiration rate (R_{O_2} and R_{CO_2}, mL kg^{-1} h^{-1}) at various influential variables such as temperature, O_2, and CO_2.

Respiration rate model reported by Iqbal et al. (2009) for whole mushroom (*A. bisporus*) was considered in this work (Table 11.4). This model considers the effect of temperature, O_2, and CO_2 concentrations on both O_2 consumption and CO_2 production as shown in Equations 11.1 and 11.2.

$$R_{O_2} = \frac{\alpha_{ref}^{O_2} y_{O_2}}{\phi_{ref}^{O_2} + y_{O_2}\left(1 + \dfrac{y_{CO_2}}{\gamma_{ref}^{O_2}}\right)} \times e^{-(E_{O_2}/R) \times [(1/T)-(1/T_{ref})]} \quad (11.1)$$

$$R_{CO_2} = \frac{\alpha_{ref}^{CO_2} y_{O_2}}{\phi_{ref}^{CO_2} + y_{O_2}\left(1 + \dfrac{y_{CO_2}}{\gamma_{ref}^{CO_2}}\right)} \times e^{-(E_{CO_2}/R) \times [(1/T)-(1/T_{ref})]} \quad (11.2)$$

where
E_{CO_2} Activation energy constant for CO_2 (kJ mol^{-1})
E_{O_2} Activation energy constant for O_2 (kJ mol^{-1})
R Universal gas constant (kJ mol^{-1} K^{-1})

TABLE 11.4
Constant Coefficients for R_{O_2} and R_{CO_2} for Mushrooms

	$\alpha_{ref}^{O_2, CO_2}$ (mL kg^{-1} h^{-1})	E_{O_2, CO_2} (kJ mol^{-1})	$\phi_{ref}^{O_2, CO_2}$ (% v/v)	$\gamma_{ref}^{O_2, CO_2}$ (% v/v)	T_{ref} (K)
R_{O_2}	63.64	54.38	4.09	38.60	283.15
R_{CO_2}	54.68	56.04	3.18	57.90	283.15

Source: Iqbal, T. et al. 2009. *International Journal of Food Science and Technology,* 44(7), 1408–1414.

R_{CO_2} CO_2 production rate (mL kg^{-1} h^{-1})
$R_{CO_2,\text{ref}}$ Reference CO_2 consumption rate (mL kg^{-1} h^{-1})
R_{O_2} O_2 consumption rate (mL kg^{-1} h^{-1})
$R_{O_2,\text{ref}}$ Reference O_2 consumption rate (mL kg^{-1} h^{-1})
T Temperature (K)
T_{ref} Reference temperature (K)
y_{CO_2} Volumetric concentration of CO_2 at time t (% v/v)
y_{O_2} Volumetric concentration of O_2 at time t (% v/v)
α Michaelis–Menten equation constant corresponding to the maximal respiration rate (mL kg^{-1} h^{-1})
α_{ref} Michaelis–Menten equation constant corresponding to the maximal respiration rate at the reference temperature (mL kg^{-1} h^{-1})
ϕ Michaelis–Menten uncompetitive equation constant corresponding to the half of maximal respiration rate (% O_2 v/v)
ϕ_{ref} Michaelis–Menten uncompetitive equation constant corresponding to the half of maximal respiration rate at the reference temperature (% O_2 v/v)
γ Michaelis–Menten uncompetitive equation constant for CO_2 inhibition (% CO_2 v/v)
γ_{ref} Michaelis–Menten uncompetitive equation constant for CO_2 inhibition at the reference temperature (% CO_2 v/v)

Respiration rate model reported by Hertog et al. (1999) for fresh strawberries (cv. Elsanta) was considered in this work (Table 11.5). This model considers the effect of temperature, O_2, and CO_2 concentrations on both O_2 consumption and CO_2 production as shown in Equations 11.3 and 11.4.

The O_2 consumption rate (R_{O_2}) is

$$R_{O_2} = \frac{[O_2]}{K_{mO_2} \cdot \left(1 + \frac{[CO_2]}{K_{mcCO_2}}\right) + [O_2] \cdot \left(1 + \frac{[CO_2]}{K_{muCO_2}}\right)} \times Vm_{O_2,\text{ref}} \cdot e^{(E_{aVmO_2}/R_{\text{gas}})[(1/T_{\text{ref}})-(1/T)]} \quad (11.3)$$

TABLE 11.5
Parameters for Respiration Rate Model for Strawberry

Parameter	Value	Parameter	Value
$Vm_{O_2,\text{ref}}$	23.39	$Vm_{CO_2(f),\text{ref}}$	43.32
$EaVm_{O_2}$	74,826	$EaVm_{CO_2(f)}$	57,374
Km_{O_2}	2.63	$Kmc_{O_2(f)}$	0.056
Kmc_{CO_2}	+∞	$Kmc_{CO_2(f)}$	+∞
Kmu_{CO_2}	+∞	$Km_{CO_2(f)}$	1
RQ_{ox}	0.91		

Source: Hertog, M. L. A. T. M. et al. 1999. *Postharvest Biology and Technology, 15*, 1–12.

The CO_2 production (R_{CO_2}) is the simultaneous result of oxidative and fermentative processes:

$$R_{CO_2} = RQ_{ox} \cdot R_{O_2} + \frac{1}{\left(1 + \frac{[O_2]}{K_{mc_{O_2(f)}}} + \frac{[CO_2]}{K_{mcc_{O_2(f)}}}\right) \cdot K_{mc_{O_2(f)}} + 1} \times Vm_{CO_2(f),\text{ref}} \quad (11.4)$$

$$\cdot e^{(E_aVm_{CO_2(f)}/R_{\text{gas}})[(1/T_{\text{ref}})-(1/T)]}$$

where

$Vm_{O_2,\text{ref}}$	Maximum O_2 consumption rate (mL kg^{-1} h^{-1}) at reference temperature
T_{ref}	Reference temperature (10°C)
$EaVm_{O_2}$	Activation energy (J mol^{-1}) of rate constant Vm_{O_2}
Km_{O_2}	Michaelis constant for O_2 consumption (%)
Kmi_{CO_2}	Michaelis constant for inhibition of O_2 consumption by CO_2 (%), where $i = c$: competitive; $i = u$: uncompetitive
RQ_{ox}	Respiration quotient
$Vm_{CO_2(f),\text{ref}}$	Maximum CO_2 production rate (mL kg^{-1} h^{-1}) at reference temperature
$EaVm_{CO_2(f)}$	Activation energy (J mol^{-1}) of rate constant Vm_{CO_2}
$Km_{CO_2(f)}$	Michaelis constant for fermentative CO_2 production (%)
$Kmci_{(f)}$	Michaelis constant for competitive inhibition of fermentative CO_2 production by either O_2 or CO_2 (%)

11.2.3.2 Transpiration Rate

Fresh fruits and vegetables continue to lose water after harvest causing shrinkage and loss of weight. As mentioned before, mushrooms are one of the most perishable products and usually their shelf life is 1–3 days at ambient temperature. They are very sensitive to humidity levels, as high water levels favor microbial growth and discoloration and conversely, low water levels lead to loss of weight (and thus economic value), and undesirable textural changes. The vapor can buildup in the package, allowing spoilage bacteria to grow and causing mushrooms to become brown and spotted. Mushrooms are conventionally packed in plastic trays overwrapped with perforated PVC films and stored under refrigeration temperature. Mushroom dehydration is prevented due to higher humidity inside the packages but this humidity is too high to deter the microbial growth. High humidity, created due to the high TR of mushrooms, causes condensation inside the package as clearly seen underneath the film. Predicting TR, therefore, is helpful for estimating the shelf life of fresh mushrooms and designing its storage and packaging conditions.

Modified atmosphere packaging is a well-known technique used to extend the shelf life of fresh fruits and vegetables by slowing down respiration. Nonoptimal packaging design which does not account for product transpiration can lead to water accumulation at the product surface promoting microbial growth and sliminess, which impairs the objective of MAP. This is especially important for mushrooms and strawberries, as they have high TRs. The steady-state package RH depends on the relative rate of water loss by the packaged produce and water loss through the polymeric film. For modeling and predicting reliable methods to control RH in MAP, it is desirable to have accurate values of TR as a function of different storage conditions.

A TR mathematical model (Equation 11.5) for whole button mushrooms (*A. bisporus*) as developed by Mahajan et al. (2008) is described below.

$$\text{TR} = \frac{K_i}{24} \times (a_{w_i} - a_w) \times (1 - e^{-aT}) \quad (11.5)$$

where a_w is water activity of the surrounding atmosphere, RH/100, a_{w_i} the water activity of mushrooms (0.984), K_i the mass transfer coefficient (8.85×10^{-3}), a a constant coefficient (0.197), T the storage temperature (°C), and TR in mg cm^{-2} h^{-1}. To convert TR from mg cm^{-2} h^{-1} to g kg^{-1} h^{-1}, use $A_s = 0.029 \times (M)^{0.738}$, where A_s is the surface area of mushroom in cm^2 and M is mass of mushroom in mg.

Transpiration rate is an important physiological process that also affects the main quality characteristics of fresh strawberries. The TR model shown in Equation 11.5 with constant parameters estimated now for strawberry holds $K_i = 115.15$ and $a = 0.223$, with TR in g kg^{-1} h^{-1} as reported by Sousa-Gallagher et al. (2013).

11.3 MODIFIED ATMOSPHERE PACKAGING

11.3.1 Current Packaging and Shelf Life Status

A market survey was conducted on fresh strawberry and mushroom packages commercially available in four major supermarket retailers (A, B, C, D) in Ireland. The results showed that the mushrooms have been generally packed in rigid punnet-type tray (RPET) covered with PVC stretchable film. In some cases, mushrooms were packed in rigid tray (RPET) with heat sealed microperforated film. All packages studied were found to have weaker modification of package gas composition, which deviated from the optimal gas composition (3%–5% O_2 and <12% CO_2) for maintaining quality and extending shelf life of mushrooms. The gas composition was not modified because the packages had either too many perforations or perforations of various sizes, mainly made using cold needles. Therefore, the holes were randomly pierced, without a uniform shape or size, sometimes causing tearing of the film. MAP technology has a great potential for increasing shelf life and quality of mushrooms, but there was no evidence of application of MAP in the current market. However, if packaging design does not take into account product transpiration, it can lead to water accumulation or condensation. This is perhaps the reason for too many microperforations to avoid condensation and also anoxia. Remaining shelf life was 2–5 days in all mushroom packages. Pack density was 0.2–0.3 g mL^{-1}. Temperature/RH of chilling cabinet was 7–11°C/45%–77% RH (optimal 90%–95% RH).

Modified atmosphere packaging is a well-known technique for preserving fruits and vegetables for longer time. It relies on the modification of the atmosphere inside the package, achieved by the natural interplay between two processes, the respiration of the product, and the transfer of gases through the packaging film (Figure 11.3), which leads to an atmosphere richer in CO_2 and H_2O, and poorer in O_2 (Mahajan et al., 2006; Sousa-Gallagher and Mahajan, 2012, 2013). MAP is often also called as

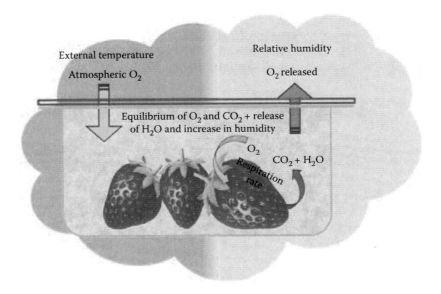

FIGURE 11.3 Principle of modified atmosphere packaging for fresh produce.

equilibrium MAP or passive MAP. In this system, atmosphere is generated naturally by product respiration rate.

The goal of MAP of fresh commodity is to create an equilibrium package atmosphere with %O_2 low enough and %CO_2 high enough to be beneficial to the produce and not injurious. This is accomplished through the proper balance of several variables that affect package atmosphere. For that it is necessary to define conditions that will create the best suited atmosphere for the extended storage of a given produce while minimizing the time required to achieve this atmosphere.

Market assessment for strawberry packages in Ireland showed that packaging styles have evolved over time. Many of the open-top pint baskets have been replaced by flow wrap packaging system for strawberry fruits. More recently, the rigid tray with heat sealable flexible film or clamshell with hinged lid is also being used. This latter design leads to less bruising and cutting of strawberries than pint baskets replacing the older systems. All packages studied were found to have gas composition close to air, that is, 21% O_2 and 0.03% CO_2 which deviated from the optimal gas composition (5%–10% O_2 and 15%–20% CO_2) for maintaining quality and extending shelf life of strawberries. The gas composition was not modified because all the packages had macroperforations of size 8 mm. The number of perforations varied from 3 to 22 for a given standard package size containing 300 g of strawberries. Although high O_2 and low CO_2 concentrations are not harmful, it does not help in gaining benefits of MAP technology for strawberries. Previous research clearly shows that MAP technology has a great potential for increasing shelf life and quality of strawberry; however, there was no evidence of application of MAP in the current market. Besides possible water accumulation or condensation in the package, as mentioned earlier, frequent temperature variations in distribution chain can also lead to condensation and strawberries being highly susceptible to condensation can potentially develop fungal infections. Remaining shelf life was 2–3 days in all the packages studied. Pack density was between 0.2 and 0.3 g of fruit mL^{-1} of pack volume which was low compared to other fresh-cut fruits (0.4–0.6 g mL^{-1}). Temperature of chilling cabinet was found to be between 10°C and 15°C which was much higher than the recommended range (0–5°C), whereas RH was found to be 40%, against recommended range of 90%–95% for strawberry fruit. This suboptimal temperature and RH might lead to heavy weight loss and faster quality degradation and thus low shelf life. Indeed, the retailers give a remaining shelf life of 2–3 days.

Despite many advantages of MAP, adoption of this technique has been rather slow because of lack of information on MAP principles, respiration rate data, and permeability data and how these parameters are affected by the real conditions during the supply chain, which lead to macroperforated packaging materials which are not allowing the full potential of MAP technology to be properly used.

11.3.2 TIME–TEMPERATURE HISTORY AT VARIOUS POINTS OF DISTRIBUTION CHAIN

Optimal storage temperature for mushroom is 0–5°C. However, the actual temperature could vary from 4°C to 20°C depending on the stage and duration of the distribution chain (Iqbal et al., 2009; Mohapatra et al., 2010). Optimal storage temperature for strawberry is 0–5°C. However, the actual temperature could vary from 4°C to 24°C (Table 11.6), depending on the stage and duration of the distribution chain as reported by Hertog et al. (1999).

However, it would be prudent to realistically evaluate the time and temperature conditions the product will likely encounter along the postharvest chain. It will then become possible to design packaging systems that can maintain optimum atmospheres and product quality throughout the postharvest handling chain.

11.3.3 OPPORTUNITIES FOR NICHE MARKET PACKAGING DEVELOPMENT

It is well established that MAP can help to increase the postharvest life of mushrooms and strawberries, but there was no evidence of application of MAP in the

TABLE 11.6
A Typical Chain for Strawberries

Location	Duration (h)	Temperature (°C)
Grower	24	12
Auction	16	4
Wholesale	4	24
Transport	20	10
Retail	72	16

Source: Hertog, M. L. A. T. M. et al. 1999. *Postharvest Biology and Technology, 15,* 1–12.

current market for these two products. However, MAP without engineering packaging design (e.g., considering product requirements, packaging permeability characteristics, and temperature fluctuation) can lead to an increase in product deterioration rate due to physiological damage caused by very low O_2 concentrations and condensation of water inside of lidding film.

Both mushrooms and strawberries have high respiration and TRs, requiring packaging films with high gas and water vapor permeability values. The existing packaging material used for mushrooms and strawberries was not permeable enough to compensate the high physiological requirements of these products.

For temperature fluctuations along the supply chain and if the packaging design does not take into account product transpiration, water accumulation, or condensation inside the package can occur leading to increased potential for microbial growth. Perhaps this is the reason why the market packages currently have macroperforations to avoid condensation and anoxia, but at the same time denying the MAP potential benefits.

Therefore, development of packaging materials with high gas and water vapor permeability that would allow reaching O_2 and CO_2 within the recommended range, and at the same time avoid moisture condensation inside the package is needed for fresh fruits and vegetables. Optimal packaging would ensure product quality and safety throughout the supply chain and extend products' shelf life. Biodegradability of packaging material is also a highly regarded characteristic by consumers because it adds value to the agricultural products and minimizes waste and impact of packaging on the environment.

11.3.4 INTEGRATIVE MAP MODELING

The major challenge would be to find the most appropriate packaging material (or the number of perforations required) to match the respiration rate for each specific fresh product. There is already enough information available on various aspects of MAP which could be compiled together to find the best packaging solutions for a particular fresh produce under a given set of processing and environmental conditions. By integrating various mathematical models available on respiration rate and TR with mass balance equations, the development phase of MAP can be shortened. Using packaging of fresh fruits and vegetables as an example, integrative mathematical modeling would simplify the design process that would help to identify the barrier properties required for a given product. The respiration rate models accounting for the effect of temperature can then be integrated into an overall mathematical model for mass transfer through the packaging to allow fresh produce processors to choose the packaging materials most suitable to the enclosed product (Sousa-Gallagher and Mahajan, 2012, 2013).

Modified atmosphere packaging is a technique used to extend the shelf life of fresh produce and has a preservative effect on the various product quality parameters by slowing down respiration, but it leads to water accumulation at the product surface, promoting microbial growth and sliminess, which impairs the objective of MAP. This is especially important for mushrooms and strawberries, as they have high TRs, coupled with frequent temperature variations in distribution chain. Hence, a careful analysis is required to measure and model the water loss rate of mushrooms in MAP conditions and includes this component in MAP engineering design. The steady-state package RH depends on the relative rate of water loss by the packaged produce and water loss through the polymeric film. For modeling and predicting reliable methods to control RH in modified atmosphere packages, it is desirable to have accurate values of water loss rate as a function of different storage conditions. Therefore, there is an urgent need for developing moisture regulating packaging film for fresh produce like mushrooms and strawberries.

11.4 QUANTIFICATION OF BARRIER PROPERTIES: OTR, CTR, AND WVTR

To estimate target oxygen transmission rate (OTR), carbon dioxide transmission rate (CTR), water vapor transmission rate (WVTR), knowledge of product respiration rate, TR, optimum storage, and packaging conditions for the given product, along with other parameters such as package size, product weight is required. Equations for barrier properties requirement can be computed by analyzing the mass balance of gases involved in going in and out of the MAP system. Assuming that there is no gas stratification inside the package and that the total pressure is constant, the differential mass balance equations that describe O_2 and CO_2 concentration and water vapor changes in a package containing a respiring product are given in Equations 11.6 through 11.8.

$$V_f \times \frac{d(y_{O_2})}{dt} = \frac{P_{O_2}}{e} \times A \times (y_{O_2}^{out} - y_{O_2}) - R_{O_2} \times M \tag{11.6}$$

$$V_f \times \frac{d(y_{CO_2})}{dt} = \frac{P_{CO_2}}{e} \times A \times (y_{CO_2}^{out} - y_{CO_2}) + R_{CO_2} \times M \tag{11.7}$$

$$V_f \times \frac{d(y_{H_2O})}{dt} = \frac{P_{H_2O}}{e} \times A \times (y_{H_2O}^{out} - y_{H_2O}) + TR \times M \tag{11.8}$$

where V_f is the headspace (free volume) in the package, y_{O_2}, y_{CO_2}, and y_{H_2O} are the volumetric fractions of O_2, CO_2, and H_2O, respectively, at time, P_{O_2}, P_{CO_2}, and P_{H_2O} are the permeability coefficients, e is the thickness of polymeric film, A is the cross-sectional area for the gas exchange, R_{O_2} and R_{CO_2} are the oxygen and carbon dioxide production rate, respectively, and M is the amount of the product.

At steady state, these equations are reduced to

$$y_{O_2}^{target} = y_{O_2}^{out} - \frac{R_{O_2} \times e \times M}{P_{O_2} \times A} \tag{11.9}$$

$$y_{CO_2}^{target} = y_{CO_2}^{out} + \frac{R_{CO_2} \times e \times M}{P_{CO_2} \times A} \tag{11.10}$$

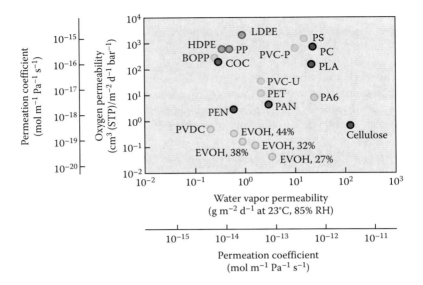

FIGURE 11.4 Permeability and permeation coefficient at 23°C of typical packaging plastics. (From Müller, K. 2011. *Kunststoffe International*, *101*(7), 62–67.)

$$y_{H_2O}^{target} = y_{H_2O}^{out} + \frac{TR \times e \times M}{P_{H_2O} \times A} \tag{11.11}$$

Equations 11.9 through 11.11 are coupled with the mathematical models for respiration rate and TR. These equations are then used to estimate the target OTR, CTR, and WVTR for the selected products that would result in the ideal packaging solution. The required permeability to O_2, CO_2, and WVTR that an ideal film must have if it is to be able to produce an optimal-modified atmosphere for the two selected products at specified temperature was calculated for market size packs as shown in Figure 11.4.

11.4.1 Benchmark Packaging Properties

Although the MAP industry has an increasing choice of packaging films, most packs are still constructed from four basic oil-based polymers: PVC, polyethylene terephthalate, polyproylene, and polyethylene for packaging of fresh produce (Figure 11.5). Polystyrene has also been used but polyvinylidene, polyester, and nylon have such low gas permeabilities that they would be suitable only for commodities with very low respiration rates. Biodegradable polymers are derived from replenishable agricultural feedstocks, animal sources, marine food processing industry wastes, or microbial sources. The most common biodegradable polymers used are made from cellulose and starches. Cellophane is the most common cellulose-based biopolymer. Starch-based polymers include amylose, hydroxylpropylated starch, and dextrins. Other starch-based polymers are polylactic acid, polyhydroxyalkanoate,

FIGURE 11.5 Typical commercial packs of strawberry and mushrooms.

polyhydroxybutyrate (PHB), and its copolymers with valeric acids (PHB/V). Made from lactic acid formed from microbial fermentation of starch derivatives, polylactide does not degrade when exposed to moisture.

Most of the data on film permeability are determined at a single temperature and RH. The film permeability data at realistic temperature between 0°C and 25°C and realistic RH between 85% and 95% must be determined. Therefore, when selecting polymer films for particular packaging applications, it is important that the film permeabilities are measured under the envisaged storage conditions.

11.4.2 Tailored Barrier Properties for Packaging

A combination of rigid tray covered by heat sealable film (option 2 of packaging specification) was selected for packaging of mushroom and strawberries (Figure 11.5).

Their respective tray size was $18.5 \times 12 \times 5$ cm^3 for mushroom and $17 \times 13 \times 6$ cm^3 for strawberry. These are the standard package sizes available in the retail market with product capacity of 275 and 300 g for whole mushrooms and fresh strawberries, respectively. The net breathable film area was 118 and 167 cm^2, considering the label area of 104 and 54 cm^2 as impermeable for mushrooms and strawberry package, respectively. The other details are shown in Table 11.7.

Equations 11.9 through 11.11 were used to estimate the tailored OTR, CTR, and WVTR for the selected products that would result in the ideal packaging solution. These tailored barrier properties of packaging films as determined from the mathematical model were designed to match the optimal-modified atmosphere for the two selected products at the specified temperature. The tailored properties are shown in Table 11.8 and the respective package atmosphere as predicted using integrative mathematical modeling of MAP.

11.5 CONCLUSIONS

A package is designed to protect and preserve the commodity and so extend shelf life, and at the same time to provide convenience at a minimal impact to the environment. It is also crucial that product quality and safety are not significantly compromised; otherwise we could potentially contribute to waste increase. Requirements

TABLE 11.7
Summary of Various Components Used in Integrative Modeling for MAP Engineering Packaging Design

Parameter	Strawberry	Mushroom
Storage temperature (°C)	5	5
Target RH (%)	90	90
Target O_2 (%)	10	5
Target CO_2 (%)	15	10
R_{O_2} (mL kg^{-1} h^{-1})	10.45	16.99
R_{CO_2} (mL kg^{-1} h^{-1})	9.67	16.46
Transmission rate (g kg^{-1} h^{-1})	0.27	0.95
Package format	Tray + Film	Tray + Film
Package dimensions (cm)	17 × 13 × 6	18.5 × 12 × 5
Film area (cm^2)	221	222
Label area (cm^2)	54	104
Net breathable area (cm^2)	167	118
Film thickness (μm)	25.4	25.4
Amount of product (kg)	0.300	0.275

TABLE 11.8
Tailored Range (from Modeling) Barrier Properties for Packaging the Selected Products at 5°C and 90% RH Considering 20% Variability in Produce Respiration Rate

Product	Target OTR	Target CTR	Target WVTR
	(mL m^{-2} day^{-1} atm^{-1})		(g m^{-2} day^{-1})
Strawberry	32,779–49,169	22,275–33,413	93–140
Mushroom	47,501–71,251	73,872–110,808	425–638

in terms of quality, safety, and physiological aspects for the selected products were reviewed and considered for the integrative packaging design.

Modified atmosphere packaging of fresh produce is a dynamic system and relies on the modification of the atmosphere inside the package, achieved by the natural interplay between two processes, the respiration of the product and the transfer of gases through the packaging film, which leads to an atmosphere richer in CO_2 and H_2O, and poorer in O_2.

Integrative mathematical modeling of packaging systems was used to estimate the tailored OTR, CTR, and WVTR for the selected products that would result in the ideal packaging solution. These tailored barrier properties of packaging films were designed to match the optimal-modified atmosphere for the two selected products at the specified temperature. Mushrooms and strawberries both require packaging materials with high gas and water vapor permeability with the target range of OTR 47,501–71,251 (32,779–49,169) mL m^{-2} day^{-1} atm^{-1}, CTR 73,872–110,808 (22,275–33,413) mL m^{-2} day^{-1} atm^{-1}, and WVTR 425–638 (93–140) g m^{-2} day^{-1} atm^{-1} for packaging mushrooms (strawberries) at 5°C and 90% RH.

REFERENCES

Almenar, E., Del-Valle, V., Hernandez-Munoz, P., Lagaron, J. M., Catala, R., and Gavara, R. 2007. Equilibrium modified atmosphere packaging of wild strawberries. *Journal of the Science of Food and Agriculture, 87*, 1931–1939.

Burton, K. S. 1991. Modified atmosphere packaging of mushrooms—Review and recent developments. In M. J. Maher (Ed.), *Science and Cultivation of Edible Fungi* (pp. 683–688). Rotterdam: Balkema.

De la Plaza, J. L., Alique, R., Zamorano, J. P., Calvo, M. L., and Navarro, M. J. 1995. Effect of the high permeability to O$_2$ on the quality changes and shelf life of fresh mushrooms stored under modified atmosphere packaging. In T. J. Elliott (Ed.), *Science and Cultivation of Edible Fungi* (pp. 709–716). Rotterdam: Balkema.

Hertog, M. L. A. T. M., Boerriter, H. A. M., van den Boogaard, G. J. P. M., Tijskens, L. M. M., and van Schaik, A. C. R. 1999. Predicting keeping quality of strawberries (cv. "Elsanta") packed under modified atmospheres: An integrated model approach. *Postharvest Biology and Technology, 15*, 1–12.

Iqbal, T., Rodrigues, F. A. S., Mahajan, P. V., and Kerry, J. P. 2009. Mathematical modelling of O$_2$ consumption and CO$_2$ production rates of whole mushrooms accounting for the effect of temperature and gas composition. *International Journal of Food Science and Technology, 44*(7), 1408–1414.

Kader, A. A. 1992. Modified atmospheres during transport and storage. In A. A. Kader (Ed.), *Postharvest Technology of Horticultural Crops, in Division of Agriculture and Natural Resources* (Publication 3311, pp. 85–92). Davis, University of California.

Mahajan, P. V., Oliveira, F. A. R., and Macedo, I. 2008. Effect of temperature and humidity on the transpiration rate of the whole mushrooms. *Journal of Food Engineering, 84*, 281–288.

Mahajan, P. V., Oliveira, F. A. R., Montanez, J. C., and Frias, F. 2007. Development of user-friendly software for design of modified atmosphere packaging for fresh and fresh-cut produce. *Innovative Food Science and Emerging Technology, 8*, 84–92.

Mahajan, P. V., Oliveira, F. A. R., Sousa, M. J., Fonseca, S. C., and Cunha, L. M. 2006. An interactive design of MA-packaging for fresh produce. In Y. H. Hui (Ed.), *Handbook of Food Science, Technology, and Engineering* (Vol. III, pp. 119-1–119-6). New York, NY: CRC Taylor & Francis.

Mohapatra, D., Bira, Z. M., Kerry, J. P., Frias, J. M., and Rodrigues, F. A. 2010. Postharvest hardness and colour evolution of white button mushrooms (*Agaricus bisporus*). *Journal of Food Science, 75*, E146–E152.

Müller, K. 2011. Faster results, permeation measurement of films. *Kunststoffe International, 101*(7), 62–67.

Nichols, R., and Hammond, J. B. W. 1973. Storage of mushrooms in pre-packs: The effect of changes in carbon dioxide and oxygen on quality. *Journal of the Science of Food and Agriculture, 24*, 1371–1381.

Parentelli, C., Ares, G., Corona, M., Lareo, C., Gámbaro, A., Soubes, M., and Lema, P. 2007. Sensory and microbiological quality of shiitake mushrooms in modified-atmosphere packages. *Journal of the Science of Food and Agriculture, 87*, 1645–1652.

Song, Y., Lee, D. S., and Yam, K. L. 2001. Predicting relative humidity in modified atmosphere packaging system containing blueberry and moisture absorbent. *Journal of Food Processing and Preservation, 25*, 49–70.

Song, Y., Vorsa, N., and Yam, K. L. 2002. Modelling respiration–transpiration in modified atmosphere packaging system containing blueberry. *Journal of Food Engineering, 53*, 103–109.

Sousa-Gallagher, M. J., and Mahajan, P. V. 2012. Quantification of mass transfer properties for engineering MAP design of fresh produce. In *Workshop: Environmental Microbiology and Biotechnology Conference (EMB12)*, April 11–13, 2012, Bologna, Italy.

Sousa-Gallagher, M. J., and Mahajan, P. V. 2013. Integrative mathematical modelling for MAP design of fresh-produce. Theoretical analysis and experimental validation. *Food Control, 29*(2), 444–450.

Sousa-Gallagher, M. J., Mahajan, P. V., and Mezdad T. 2013. Engineering packaging design accounting for transpiration rate: Model development and validation with strawberries. *Journal of Food Engineering, 119*, 370–376.

Vakkalanka, M., D'Sousa, T., Ray, S., and Yam, K. L. 2012. Emerging packaging technologies for fresh produce. In K. L. Yam, and D. S. Lee (Eds.), *Emerging Food Packaging Technologies: Principles and Practice* (1st ed., Vol. 1, pp. 109–133). Cambridge, UK: Woodhead Publishing Limited.

Wu, C. 2010. An overview of postharvest biology and technology of fruits and vegetables, 2–11. Available at http://ir.tari.gov.tw:8080/bitstream/345210000/2813/1/publication_no147_04.pdf, (accessed on February 9, 2015).

Xanthopoulos, G., Koronaki, E. D., and Boudouvis, A. G. 2012. Mass transport analysis in perforation mediated modified atmosphere packaging of strawberries. *Journal of Food Engineering, 111*, 326–329.

Yam, K. L., and Lee, D. S. 2012. Emerging food packaging technologies: An overview. In K. L. Yam, and D. S. Lee (Eds.), *Emerging Food Packaging Technologies: Principles and Practice* (1st ed., Vol. 1, pp. 1–9), Cambridge, UK: Woodhead Publishing Limited.

12 Edible Packaging for Fruits and Vegetables

Marta Montero-Calderón, Robert Soliva-Fortuny, and Olga Martín-Belloso

CONTENTS

Abstract 354
12.1 Characteristics and Preservation Requirements for Fresh and Processed Fruits and Vegetables 354
12.2 Fundamentals of Edible Packaging Technology Applied to Fruit and Vegetable Products 355
    12.2.1 Edible Packaging Formulation 355
        12.2.1.1 Polysaccharide-Based Materials 356
        12.2.1.2 Protein-Based Materials 361
        12.2.1.3 Lipids and Wax-Based Materials 361
        12.2.1.4 Coadjuvants and Functional Ingredients 361
    12.2.2 Physical Properties of Edible Films and Coatings for Application in Fruits and Vegetables 363
    12.2.3 Fruit and Vegetable Requirements to be Considered when Applying Edible Coatings 365
12.3 Edible Packaging Applications for the Development of Fruit and Vegetable Commodities with Improved Quality and Extended Shelf Life 366
    12.3.1 Edible Coating Applications for Controlling Mass Transfer Phenomena in Fruits and Vegetables 366
    12.3.2 Edible Coating Applications Designed for Internal Atmosphere Modification 368
    12.3.3 Edible Coatings Applied to Inhibit Microbial Spoilage and Suppress Physiological Disorders 369
    12.3.4 Edible Coatings Applied to Reduce Oxidative Reactions in Fresh and Processed Fruits and Vegetables 370
    12.3.5 Edible Coatings to Improve Texture-Related Properties 371
    12.3.6 Edible Coatings Applied to Improve Gloss and Shine on Fruits and Vegetables Surface 372
    12.3.7 Edible Coatings to Incorporate Active Ingredients 373
12.4 Industrial Implementation of Edible Packaging Alternatives 374
12.5 Regulatory Aspects 375
12.6 Future Prospects 376
References 376

ABSTRACT

Fresh fruits and vegetables have natural skins that provide protection against water loss, discoloration, and microbial spoilage, among other deleterious phenomena. However, this protection may be reinforced by applying coatings of edible materials on the fruit surface. Hence, the use of edible films and coatings is rapidly growing as consumers demand for fresh-like products increases and in response to this trend, technologists continue to look for new ways of reducing postharvest decay of fruit and horticultural produce. The economical importance of the edible coatings industry is increasing also because of the rise in the sales of fresh-cut products. In these products, peeling and cutting operations remove the natural protection, leading to deterioration and spoilage.

This chapter highlights the most significant applications regarding the use of edible coatings for preventing storage decay in fresh and fresh-cut products. Most recent applications will be reviewed. Furthermore, information on base materials, methods of industrial application, and regulatory aspects are considered with the view of providing a reference to both scientists and processors in the field.

12.1 CHARACTERISTICS AND PRESERVATION REQUIREMENTS FOR FRESH AND PROCESSED FRUITS AND VEGETABLES

Fruits and vegetables are very important for a healthy diet. They provide a wide range of nutrients in a large variety of shapes, colors, flavors, and aromas. Since they are made of living tissues, they continue their metabolic activities and eventually deteriorate during storage and transportation.

Fresh and minimally processed fruits and vegetables are very perishable products. Thus, cell wall degradation and increased transpiration phenomena rapidly occur upon manipulation or storage at abusive conditions. Loss of turgidity, softening and water migration, browning and discoloration, microbiological decay, loss of sugars, degradation of starch, and loss or modification of volatile compounds are some of the most remarkable changes leading to decreased commercial value (Kader, 2012). Hence, some simple but critical aspects, for example, temperatures during cooling and storage, must be controlled during postharvest and processing to delay deterioration and extend shelf life. In addition, preservation alternatives recommended for fresh-cut fruits and vegetables include disinfection, chemical or physical decontaminating treatments, modified atmospheres, and packaging selection to minimize damages along storage and transportation while they reach the consumer (Garrett, 2002; Garcia and Barrett, 2002).

Minimally processed fruits and vegetables are also composed of living tissues, so they have similar requirements than the whole fresh produce. However, minimally processed commodities have been washed, peeled, and cut, which deprive them from their natural protection, thus releasing fluids from cells and increasing contact area with the surrounding atmosphere. Because of this, minimally processed products need additional care to inhibit deleterious reactions and keep their fresh-like quality throughout the whole distribution and retail chains.

Edible films and coatings can help to preserve the quality of fresh and fresh-cut fruits and vegetables by providing a barrier that limits weight loss, gases exchange with the surrounding environment, and microbial growth while reducing the incidence of discoloration and browning phenomena, firmness decay, and aroma alteration. These edible layers may also be used as carriers of active ingredients which might help to preserve the product quality attributes and increase its shelf life.

Similarly, for fruits and vegetables subjected to further processing, edible coatings might also be a tool to minimize the impact of processing operations, as they may control mass transfer phenomena, which is crucial for keeping the sensory characteristics of the fresh-cut product, such as appearance, texture, and flavor.

12.2 FUNDAMENTALS OF EDIBLE PACKAGING TECHNOLOGY APPLIED TO FRUIT AND VEGETABLE PRODUCTS

The main purpose of packaging is to protect products against physical, chemical, and biological damage and to keep the product quality throughout storage. Edible films and coatings are not intended to replace non-edible synthetic polymers, which ensure prolonged storage. Instead, they can be used to improve food quality, extend shelf life, and improve the efficiency of the packaging material, while providing unique functionalities (Fabra et al., 2009a). Edible films and coatings are always in contact with the food surface. They may act as a barrier to moisture, gases, volatile compounds, or lipids, thus providing protection for a food commodity even once the primary package is opened (Fabra et al., 2009a). Edible films and coatings may be used to individually protect fresh and fresh-cut products to minimize undesirable changes during handling and storage. Novel edible packaging applications seek both quality maintenance and improvement of the health-related value of foods.

Some of the recognized contribution of edible films and coatings are

- Control of mass transfer phenomena
- Inhibition of microbial growth and suppression of physiological disorders
- Reduction in oxidative reactions
- Firmness preservation
- Improvement of the visual appearance of the product surface
- Incorporation of functional ingredients

All the above-mentioned advantages contribute to extend the shelf life of fruit and vegetable products (Rojas-Graü et al., 2011; Jianglian and Shaoying, 2013). Nevertheless, understanding the role of edible packaging requires considering both the coating composition and the characteristics of the product to be coated, as well as the interactions between the food and the packaging material.

12.2.1 EDIBLE PACKAGING FORMULATION

Edible packaging formulations can be cast as films used to wrap the product or as coatings formed directly on the product surface. Films may be formed through solidification of melted materials such as lipids, paraffin, and other waxes, casting by spreading protein dilute film solutions into a surface and let it dry under controlled conditions or by extrusion at high temperature, which reduces preparation times (Dangaran et al., 2009). Films are then applied to food pieces.

Edible coatings are formed by dipping, spraying, or panning depending on the characteristics of the foods to be coated and the physical properties of the coating (Andrade et al., 2012). Dipping includes submerging the product into the coating solution. After dipping, the excess of coating is allowed to drain, and the coated surfaces are usually dried at room temperature with or without forced air. The same authors pointed out the convenience of this method to achieve total coating and good uniformity around uneven surfaces, and rather thick coatings; however, the amount of coating solution per piece is difficult to control. In addition, produce contamination might take place if the coating operation is not carried out under proper hygienic conditions. For fresh-cut fruits and vegetables, this method has the advantage that cut pieces can be protected from mechanical damage, though care should be taken to avoid very thick coatings, since they will affect both the product fresh appearance, and metabolic activity of the tissues.

In contrast, spraying allows the application of a thin and uniform layer of coating. It can be applied in several steps to cover all parts of the product (top and bottom). Several layers of the same or different coating solutions may be applied. In addition, even rapid forming coatings like alginate combined with calcium chloride solutions can be sprayed (Andrade et al., 2012). These authors pointed out that coating spraying is usually easier to implement in industrial processes compared with dipping, because of the higher draining and drying times required when dipping.

On the contrary, panning is a process in which the product to be coated is placed into a rotating bowl, the coating solution is added while the product moves, and it is distributed over the surface of the food. In this method, the product has to be resistant to mechanical stresses caused by the movement, and better results are obtained for rounded or oval-shaped materials, which are easier to coat (Andrade et al., 2012). Although no reports were found for fruits and vegetables, this procedure could be interesting for frozen fruits, like strawberries and other berries, and a variety of frozen fruit pieces covered with chocolate, yogurt, or other coating to enhance their appearance, flavor, or nutritive value.

Film-forming solutions can be prepared using polysaccharides, proteins, or lipids as base components, alone or combined, and complemented with plasticizers, multivalent ions, surfactants, and other products which can improve their barrier characteristics (Tables 12.1 and 12.2).

12.2.1.1 Polysaccharide-Based Materials

Polysaccharide-based films and coatings are prepared with water-soluble, long-chain biopolymers. They exhibit unique colloidal properties, the ability to form strong gels which are insoluble upon the reaction with multivalent cations, for example, calcium, and provide excellent mechanical and structural properties, high selectivity to O_2 and CO_2, but poor moisture barrier, mainly due to their highly hydrophilic nature.

Polysaccharides used for edible coatings include those derived from plants (e.g., cellulose and derivatives, pectin, and starch), seaweed extracts (e.g., alginate and carrageenan), shell crustaceous extracts (e.g., chitosan), and polysaccharides secreted by bacteria (e.g., gellan gum) (Rojas-Graü et al., 2011). Although polysaccharides

TABLE 12.1
Main Compounds Acting as a Base in the Formulation of Edible Coatings for Fruits and Vegetables

Polysaccharide	Product	State	References
Starch	Strawberry	Whole	Ribeiro et al. (2007)
Cassava starch	Tomato	Whole	Barco Hernández et al. (2011)
	Papaya	Whole	Almeida Castro et al. (2013)
	Edible film		Pelissari et al. (2009)
			Auras et al. (2009)
Potato starch	Strawberry	Whole	Rehman et al. (2010)
Tapioca dextrin	Edible film		Trezza and Krochta (2001)
Pectin	Melon	Fresh-cut	Buitrago et al. 2012
	Strawberry	Whole	Rehman et al. (2010)
Carboxymethyl cellulose	Strawberry	Whole	Gol et al. (2013)
	Butternut squash	Fresh-cut	Ponce et al. (2008)
	Quince	Fresh-cut	Akbarian et al. (2014)
Hydroxypropyl methylcellulose	Mandarins	Whole	Contreras-Oliva et al. (2012)
	Strawberry	Whole	Gol et al. (2013)
	Apples	Fresh-cut	Perez-Gago et al. (2005)
	Characterization		Trezza and Krochta (2001)
Carboxymethyl cellulose	Pineapple	Processed	Talens et al. (2012)
Alginate	Pineapple	Fresh-cut	Azarakhsh et al. (2012), Montero-Calderón et al. (2008)
	Fuji apple	Fresh-cut	Rojas-Graü et al. (2007)
	Apples	Fresh-cut	Ghavidel et al. (2013)
	Melon	Fresh-cut	Buitrago et al. (2012)
	Peaches	Fresh-cut	Pizato et al. (2013)
Carrageenan	Strawberry	Whole	Ribeiro et al. (2007)
	Apples	Fresh-cut	Ghavidel et al. (2013)
Chitosan	Strawberry	Whole	Ribeiro et al. (2007)
	Strawberry	Whole	Gol et al. (2013)
	Star fruit	Whole	Mohamad Zaki et al. (2012)
	Mushroom	Fresh-cut	Eissa (2008)
	Butternut squash	Fresh-cut	Ponce et al. (2008)
	Tomato	Whole	Mustafa and Ali (2013)
	Strawberry	Whole	Rehman et al. (2010)
	Gala apple	Whole	Shao et al. (2012)
	Pineapple	Processed	Talens et al. (2012)
	Lotus root	Fresh-cut	Xing et al. (2010)
	Characterization		Pelissari et al. (2009)
Glucomannan (Konjac)	Guava	Fresh-cut	Sothornvit (2013)
Aloe vera	Kiwi	Fresh-cut	Benítez et al. (2013)
	Grapes	Whole	Valverde et al. (2005)
	Pineapple	Whole	Adetunji et al. (2012)

(*Continued*)

TABLE 12.1 (*Continued*)
Main Compounds Acting as a Base in the Formulation of Edible Coatings for Fruits and Vegetables

Polysaccharide	Product	State	References
Gellan	Pineapple	Fresh-cut	Azarakhsh et al. (2012)
	Fuji apple	Fresh-cut	Rojas-Graü et al. (2007)
Galactomannan	Acerola, cajá, mango, pitanga, seiguela, apple		Cerqueria et al. (2009), Lima et al. (2010)
Xanthan gum	Papaya	Fresh-cut	Cortez-Vega et al. (2013)
	Peaches	Fresh-cut	Pizato et al. (2013)
Tara gum	Peaches	Fresh-cut	Pizato et al. (2013)
Proteins			
Whey	Apples	Fresh-cut	Ghavidel et al. (2013), Perez-Gago et al. (2005)
	Pear	Whole	Javanmard (2013)
	Edible film		Trezza and Krochta (2001)
Casein (sodium caseinate)	Butternut squash	Fresh-cut	Ponce et al. (2008)
	Apples	Fresh-cut	Ghavidel et al. (2013)
	Characterization		Rezvani et al. 2013
	Pineapple	Processed	Talens et al. (2012)
Collagen	Mango, apple	Whole	Lima et al. (2010)
Gelatin	Pineapple	Fresh-cut	Bizura Hasida et al. (2013)
	Melons (cantaloupe)	Fresh-cut	Álvarez Arenas et al. (2012), Leonet et al. (2011)
Zein	Strawberry	Whole	Rehman et al. (2010)
	Characterization		Trezza and Krochta (2001)
Wheat gluten	Strawberry	Whole	Rehman et al. (2010)
Soy protein	Apples	Fresh-cut	Ghavidel et al. (2013)
Lipids and Waxes			
Carnauba wax	Apples	Whole	Alleyne and Hagenmaier (2000)
Beeswax	Mandarins	Whole	Contreras-Oliva et al. (2012)
	Apples	Fresh-cut	Perez-Gago et al. (2005)
	Edible film		Auras et al. (2009)
Paraffin	Cassava root	Whole	Medina et al. (2013)
	Characterization		• Auras et al. (2009)
Fatty acids and fatty acid esters	Apples	Fresh-cut	Perez-Gago et al. (2005)
	Characterization		Rezvani et al. (2013)
Shellac	Mandarins	Whole	Contreras-Oliva et al. (2012)
	Apples	Whole	Alleyne and Hagenmaier (2000)
	Characterization		Trezza and Krochta (2001)

TABLE 12.2
Main Additives Incorporated to Edible Coating Formulations for Application on Fruits and Vegetables

Plasticizers	Concentrations	Fruit or Vegetable	References
Sorbitol		Strawberry	Ribeiro et al. (2007)
Glycerol	0.75–5.0 (w/v), 1.0–15.0 (w/v)	Acerola, cajá, mango, pitanga, seriguela, and apple	Vanin et al. (2005), Auras et al. (2009), Cerqueira et al. (2009), Lima et al. (2010), Bosquez-Molina et al. (2010)
Propylene glycol			Vanin et al. (2005)
Ethylene glycol			Vanin et al. (2005)
Diethylene glycol			Vanin et al. (2005)
Glycerol monoestearate	0.75 (w/v)	Strawberry	Gol et al. (2013)
Surfactants			
Lecithin			Gol et al. (2013)
Fatty acids			
Fatty acid esters			
Antioxidant Agents			
Ascorbic acid	0.5–1.0 (w/v)	Apple and pear	Rojas-Graü et al. (2009), Leonet et al. (2011), Xing et al. (2010)
Citric acid	0.5–1.0 (w/v)	Apple	Rojas-Graü et al. (2009), Pizato et al. (2013), Xing et al. (2010)
Oxalic acid	0.5 (w/v)	Apple	Rojas-Graü et al. (2009)
Cysteine	0.1–0.5 (w/v)	Apple	Rojas-Graü et al. (2009)
4-Hexylresorcinol	0.005–0.02 (w/v)	Apple	Rojas-Graü et al. (2009)
N-acetylcysteine	1.0–2.0 (w/v)	Apple and pear	Rojas-Graü et al. (2009)
Glutathione	1.0–2.0 (w/v)	Apple and pear	Rojas-Graü et al. (2009)
Antimicrobial Agents			
Rosemary			Ponce et al. (2008), Ramos-García et al. (2012)
Sodium sorbate	0–400 ppm	Melon	Leonet et al. (2011)
Potassium sorbate	0.2 (w/v)	Strawberry	Rojas-Graü et al. (2009), Leonet et al. (2011)
Citric acid		Strawberry	Rojas-Graü et al. (2009), Pizato et al. (2013)
Lemongrass	0.7–1.5 (w/v)	Apple	Rojas-Graü et al. (2009)
Oregano	0.1–0.5 (w/v)	Apple	Rojas-Graü et al. (2009), Pelissari et al. (2009), Ramos-García et al. (2012)
Vanillin	0.3–0.6 (w/v)	Apple	Rojas-Graü et al. (2009)
Cinnamon	0.7 (w/v), 1:0.1 protein to lipid ratio	Apple, melon	Rojas-Graü et al. (2009), Atarés et al. (2010)

(Continued)

TABLE 12.2 (*Continued*)
Main Additives Incorporated to Edible Coating Formulations for Application on Fruits and Vegetables

Antimicrobial Agents	Concentrations	Fruit or Vegetable	References
Ginger essential oil	1:0.1 protein to lipid ratio		Atarés et al. (2010)
Clove	0.7 (w/v)	Apple, melon	Rojas-Graü et al. (2009), Ramos-García et al. (2012)
Cinnamaldehyde	0.5 (w/v)	Apple	Rojas-Graü et al. (2009)
Eugenol	0.5 (w/v)	Apple	Rojas-Graü et al. (2009)
Citral	0.5 (w/v)	Apple	Rojas-Graü et al. (2009)
Malic	2.5 (w/v)	Melon	Rojas-Graü et al. (2009)
Palmarosa	0.7 (w/v)	Melon	Rojas-Graü et al. (2009)
Chitosan	0.7 (w/v)	Melon	Rojas-Graü et al. (2009), Pelissari et al. (2009), Ramos-García et al. (2012)
Garlic oil	0.1 (v/v)		Pranoto et al. (2005)
Propolis (ethanolic extract of propolis)			Pastor et al. (2010)
Degradation of Organophosphorus Pesticides			
Zinc (complex of zinc II)		Chinese jujube	Wu et al. (2010)
Cerium (complex of cerium IV)		Chinese jujube	Wu et al. (2010)
Textural Enhancers			
Calcium chloride	1.0–10.0 (w/v)	Apple, pear, pineapple, melon	Rojas-Graü et al. (2009)
Calcium lactate	2.0 (w/v)	Apple, pear, pineapple, melon, grapefruit	Rojas-Graü et al. (2009), Moraga et al. (2009)
Calcium gluconate	1.0–5.0 (w/v)	Strawberry, raspberry	Rojas-Graü et al. (2009)
Nutraceuticals			
Calcium gluconate	1.0–5.0 (w/v)	Strawberry and raspberry	Rojas-Graü et al. (2009)
Vitamin E	0.2 (w/v)	Strawberry and raspberry	Rojas-Graü et al. (2009)
Ascorbic acid	1.0 (w/v)	Papaya and melon	Rojas-Graü et al. (2009), Leonet et al. (2011)
Bifidobacterium lactis	2.0 (w/v)	Apple and papaya	Rojas-Graü et al. (2009)

coatings moisture barrier are generally poor, they may partially control water loss. Adetunji et al. (2009) studied the influence of an *Aloe vera* gel applied on fresh pineapples on their weight loss and firmness throughout storage at $27 \pm 2°C$ and 50%–60% relative humidity (RH). Weight loss was reduced for pineapples coated with *A. vera* gel and firmness was better retained, as compared with uncoated fruits.

For deep fried foods, the use of polysaccharide edible coatings has been reported to reduce both water losses and fat intake of vegetable products during cooking, thus improving their nutritious value and reducing high blood cholesterol levels and coronary heart disease risks for the consumer (Porta et al., 2012).

12.2.1.2 Protein-Based Materials

Protein-based films and coatings have good nutritional value. Proteins can be used to stabilize emulsions and are good film-forming materials with excellent mechanical and structural properties. Protein films are relatively good barriers to O_2 and CO_2 but exhibit poor water vapor barrier properties (Fabra et al., 2009a). Proteins can be obtained from either animal or vegetable sources. The main proteins obtained from animal sources are whey proteins, caseins, gelatin, and collagen. Proteins obtained from plant sources include corn zein, wheat gluten, and soy protein, among others (Rojas-Graü et al., 2011).

12.2.1.3 Lipids and Wax-Based Materials

Lipid and wax-based materials are generally applied to improve resistance to water loss and dehydration of fruits during postharvest storage. Natural waxes include carnauba wax and beeswax, a mixture of wax esters, wax acids, and hydrocarbons with a viscoelastic behavior, whereas synthetic petroleum-based waxes include paraffin and polyethylene waxes (Fabra et al., 2009b; Maftoonazad and Badii, 2009). Other nonwaxy lipids that may be used in edible film, and coating formulations are oils from mineral and vegetable sources and oil derivatives such as fatty acids, fatty acid esters, and alcohols. However, lipid coatings usually have little structural integrity and therefore, they are often used in combination with proteins or polysaccharides, which reduces their water permeability.

12.2.1.4 Coadjuvants and Functional Ingredients

Some other compounds may be utilized to improve the properties of the film-forming solutions. A summary of the main compounds used in fruit- and vegetable-coating formulations is provided in Table 12.2. These compounds include:

- *Multivalent ions* are used to react with polysaccharide polymers and cross-link their chains forming a three-dimensional network, thus resulting in a strong insoluble gel (Rojas-Graü et al., 2011). Calcium ions, also act as texture enhancers. Therefore, calcium salts, for example, calcium chloride, calcium lactate, and calcium gluconate, are used to enhance the mechanical properties of the edible layers. Furthermore, the incorporation of calcium salts in the polymeric structure also presents interest from the point of view of texture preservation.
- *Antibrowning agents* may be used to protect against oxidative rancidity and enzymatic browning taking place in bruised fruit tissues (Rojas-Graü et al., 2009). Some examples of antibrowning agents used for fresh-cut fruits are sulfur-containing compounds such as *n*-acetylcysteine and glutathione, ascorbic acid, 4-hexilresorcinol, and tocopherols, and ascorbic, citric, and

oxalic acids. Most of these antibrowning agents are highly hydrophilic, thus increasing water vapor permeability of the cast layers and water loss of the product.

- *Antimicrobial agents* are used to inhibit aerobic microorganisms, molds, and yeasts and extend product shelf life. Potassium sorbate and chitosan are good examples of compounds with antimicrobial activity with application on fruits and vegetables. In addition, essential oils from cinnamon, clove, lemongrass, vanillin, and oregano, as well as organic acids, may be used to retard microbial proliferation on fruit and vegetable surfaces (Rojas-Graü et al., 2009).

- *Surfactants* are incorporated in coating formulations to reduce surface tension, improve wettability and spread coefficient of the coating on the product surface, and provide uniform thickness and contact with the food surface. They are also used as emulsifiers in composite-containing compounds of lipid nature. Some examples of surfactants with applications on coatings for fresh produce are lecithin, sorbitan monolaurate (Span 20), and polyoxyethylene sorbitan tioleate (Tween 85) (Trezza and Krochta 2000).

- *Plasticizers* are used to decrease the intermolecular attraction between adjacent polymeric chains, which in turn facilitates the penetration of gas molecules through the polymeric network. Plasticizers reduce tensile strength of the material and increase permeability to gas, solute, and water vapor (Rojas-Graü et al., 2007). Plasticizers are needed to improve flexibility, elongation, workability, and processability of polysaccharide- and protein-based films and coatings by increasing free volume, intermolecular spacing, as internal hydrogen bonding between polymer chains is reduced; the plasticizer molecules diffuse into the polymer and weaken the polymer-to-polymer van der Waals' forces, preventing the formation of a rigid network (Dangaran et al., 2009). Some examples of plasticizers are glycerol (GLY), sorbitol, ethylene glycol (ETG), diethylene glycol (DTG), polyethylene glycol, propylene glycol, manitol, and sucrose, as wells as some hydrophobic compounds derived from citric acid such as acetyltributyl citrate, triethyl citrate, and acetyltriethyl citrate (Andreuccetti et al., 2009; Vanin et al., 2005). Vanin et al. (2005) studied the influence of plasticizers on edible films characteristics (Figure 12.1). They found that puncture force (N) and deformation (%), water vapor permeability, and opacity of gelatin films varied with both concentration and type of plasticizer, when comparing GLY, polypropylene glycol (PPG), ETG, and DTG. Puncture force rapidly decreased with concentration for GLY and DTG, while puncture deformation was much more affected by concentration of GLY than for other plasticizer. On the contrary, their results showed that water permeability increased as DTG concentration increased and no effect was found in the opacity of the films for different plasticizer concentration.

- *Nutraceuticals* such as calcium gluconate, vitamin E, ascorbic acid, and plant extracts rich in other antioxidant compounds may be added to increase nutritional value of the coated product (Leonet et al., 2011; Rojas-Graü et al., 2009).

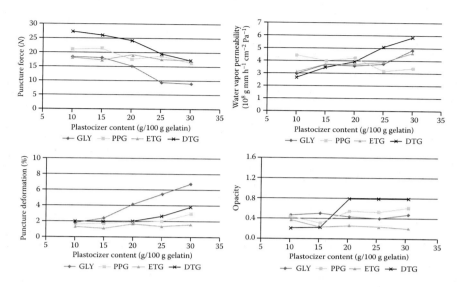

FIGURE 12.1 Effect of plasticizer type and concentration on the functional properties of gelatin-based films. (Adapted from *Food Hydrocolloids*, 19(5), Vanin et al., Effects of plasticizers and their concentrations on thermal and functional properties of gelatin-based films, 899–907, Copyright (2005), with permission from Elsevier.)

12.2.2 Physical Properties of Edible Films and Coatings for Application in Fruits and Vegetables

Mechanical and physical properties of edible packaging materials depend on their formulation and also on their interaction with the coated fruit, vegetable, or food system. Response to the film or coating may vary among products throughout handling and storage, for both fresh and minimally processed fruits and vegetables.

Surface and microstructure properties are relevant to develop film and coating applications for fruit and vegetable commodities (Ribeiro et al., 2007). Wettability indicates how well the coating-forming solution wets and uniformly spreads on the fruit surface. It takes into account the contact angle and liquid–vapor surface tension and is related with the thickness and uniformity of final layer. Adhesion or adherence of the coating to the fruit is also critical. Adherence should be as high as possible, which may be achieved in fruit and vegetable matrices with different base materials, for example, alginate and gellan for fresh-cut apples (Rojas-Graü et al., 2011). If adherence is not high enough, the coating will easily break during handling and storage. Cohesion, understood as a result of the strength of the forces between molecular components of the coating, will also strongly determine the characteristics of the edible layer. Furthermore, layer thickness has a direct relation with the optical properties of the film or coating and depends upon several properties such as composition of the coating formulation, particle size, and wettability (Rojas-Graü et al., 2009).

On the contrary, mass transfer between the product and the surrounding environment depends on temperature, as well as on the concentration gradients between the internal and external atmosphere of the product and the barrier properties of the film or coatings. Because respiration and transpiration phenomena are important in both fresh and fresh-cut commodities, water vapor resistance (WVR) and permeability to O_2 and CO_2 are often used as criteria for selecting the right film or coating for a certain product. A coating will impact the internal gas composition of a fruit or vegetable in a similar way as a modified atmosphere system act. Gas concentrations will change as a consequence of the respiratory activity of the product tissues and at the same time, due to the gas barrier characteristics of the coating. Therefore, barrier selectivity to CO_2 and O_2 may be desirable to reach appropriate internal concentrations for each gas (Rojas-Graü et al., 2007). WVR is another important property affecting transpiration and weight loss through storage. Water permeability of biopolymer layers is very sensible to changes in the coating formulation. For polysaccharide-based coatings, permeability is high and may be affected by the concentration of the polysaccharide, as well as by other components of the formulation (Rojas-Graü et al., 2011). When polysaccharide or protein coatings are developed for fruit and vegetable applications, the inclusion of a lipid is generally desirable to reduce the hydrophilic nature of the film, thus increasing its WVR.

Microstructure of edible films has an important effect on both mechanical properties and barrier properties. The stability of the emulsified solutions and the lipid particle size can have a direct effect on the resulting coating. Chiumarelli and Hubinger (2012) evaluated edible coatings formulated with cassava starch, GLY carnauba wax, and stearic acid and found that smaller lipid particles (<1.2 μm diameter) resulted in more stable solutions and emulsified coatings. High carnauba wax content resulted in larger and more heterogeneous droplets. Previously, Trezza and Krochta (2000) evaluated the gloss of a water-based whey protein isolate coating in contrast with shellac, zein, a water-based tapioca dextrin, and hydroxypropyl methylcellulose (HPMC) and found that the shape of dispersed phase particle size influenced coating gloss; solutions were stable for particle sizes smaller than 1.0 μm.

Optical properties are the result of the reflected, absorbed, and transmitted light through a specific layer. These are affected by both the internal and surface structure of the edible material, greatly depending on the film structure (droplet size and distribution of the film-forming dispersion), composition (lipid content and presence of surfactants), and external conditions, such as RH and storage time (Fabra et al., 2009a; Trezza and Krochta, 2000). For immiscible compounds, the particle size of the dispersed phase, the heterogeneity of its distribution, as well as the differences between refractive indexes of the different phases, greatly determine transparency of the edible layer. The gloss of edible coatings is generally reduced with surfactants and lipids (Fabra et al., 2009b). The incorporation of small particles into the coating-forming solution may also help to reduce its glossy appearance. In this regard, the larger the amount of surfactant, the lesser the gloss of the cast coating. Surface microstructure also affects particle size and gloss. Surface roughness refers to how smooth a surface is, and depends on particle size, angle of incidence of light, and changes in the refractive index (Barco Hernández et al. 2011; Villalobos et al., 2005). Villalobos et al. (2005) pointed out that roughness of banana peel is inversely related to its gloss. Some studies with several hydrocolloid-based films show that gloss increases linearly with polymer concentration.

All in all, the characteristics of the coating-forming solutions and as a consequence, those of the edible layer will be determined by their formulation, microstructure, and interaction with the fruit. Chiumarelli and Hubinger (2012) studied the stability, solubility, mechanical, and barrier properties of cassava starch–carnauba wax edible coatings, also containing GLY and stearic acid. They reported that density differences among components, particle size, and composition affected the stability of the solutions and the barrier properties of the coatings and optimized the formulations for a maximum stability using 3.0 g of cassava starch and 1.5 g of GLY (plasticizer) in 100 g of coating solutions, with a carnauba wax/stearic acid ratio of 0.2/0.8 g per 100 g of coating solution. They found that internal atmosphere and water resistance of the coating were most affected by the GLY content, while the ratio of carnauba wax–fatty acid ratio have a strong effect on the stability and lipid particle size of the coating solutions. Mechanical properties were also affected by the formulation, tensile strength varied from 0.252 to 2.138 MPa for solutions with lower cassava starch concentrations and higher lipid concentration, respectively.

12.2.3 Fruit and Vegetable Requirements to be Considered when Applying Edible Coatings

Fruits and vegetables are obtained from different parts of the plant, thus being really heterogeneous in terms of tissue structure, development stage, and metabolic rate. Some of them require limited O_2 availability to avoid internal browning reactions, such as the case of apple. Other produce can withstand high CO_2 concentrations, which contribute to their shelf life extension (e.g., strawberries), and yet others are very susceptible to water loss. These factors affect the selection of adequate edible materials. Taste and flavor imparted might be relevant, especially for certain products in which the coating is an integral part of the edible portion of the fruit or if the product is going to be consumed without skin removal.

Minimally processed fruits and vegetables also greatly differ in tissue structure, porosity, juiciness, turgidity, surface texture, firmness, and metabolic activity, therefore exhibiting uneven responses to edible films and coatings. In general, good adhesion and cohesion properties are required to achieve a fresh-like appearance of the cut pieces. Differences in the coating performance can be even larger considering the effect of maturity stage, product shape and size, processing procedures, and packaging and storage conditions (Zhao and McDaniel, 2005). The same is true for other processed fruit and vegetable products, for which edible packaging may be used to provide protection against mechanical bruises caused during processing, enhance the product quality attributes, reduce mass transfer phenomena, or incorporate functional ingredients. For composite foods, edible films and coatings may be applied even between different phases of the food preparation to increase their shelf life, which is generally dependent on the barrier efficiency of the edible material against the transfer of water and low-molecular-weight solutes (Karbowiak et al., 2007).

12.3 EDIBLE PACKAGING APPLICATIONS FOR THE DEVELOPMENT OF FRUIT AND VEGETABLE COMMODITIES WITH IMPROVED QUALITY AND EXTENDED SHELF LIFE

Edible coatings have great prospects regarding fruit and vegetable applications because they can provide a semipermeable barrier to water and other gases, in turn reducing weight losses and extending the fruit shelf life. Fruit and vegetable response to edible films and coatings depends on the attributes of the edible materials, the product characteristics and requirements, and their interaction. Because of the large differences among fresh, minimally processed, and other processed fruit and vegetable commodities, applications will be discussed separately throughout the following section.

12.3.1 Edible Coating Applications for Controlling Mass Transfer Phenomena in Fruits and Vegetables

Mass transfer processes, including water loss, gas exchange, as well as solutes migration, may be reduced and/or controlled through an appropriate coating formulation.

Water loss reduction is one of the major benefits recognized in edible packaging. Its impact on the quality of fruit and horticultural products is really high, as it causes shriveling and visual changes, which directly affects the consumer acceptance, and hence the product value and marketability. Packaging design for fresh and minimally processed fruits needs to consider these losses throughout the storage and distribution chain, thus standing as a major cause of economical losses.

Hydroxypropyl methylcellulose- and carboxymethyl cellulose (CMC)-based coatings incorporating 1% chitosan rendered good results regarding water loss reduction in strawberry stored during 8 days at 11°C and 70%–75% RH (Gol et al., 2013). Similar results were reported by Ribeiro et al. (2007) on strawberry for different coating formulations including starch- and carrageenan-based coatings and chitosan-based formulations. In another work, a triple layered chitosan beeswax edible coating significantly decreased water loss and senescence of strawberries (Velickova et al., 2013), due to changes in respiration rates and retardation of changes like color and texture. A chitosan-based coating also considerably reduced weight loss of Eksotika II papaya during 5 weeks of storage at 12°C and 85%–90% RH to less than a half of that of uncoated fruits, and additionally retarded firmness, color, soluble solid content, and titratable acidity changes, which were attributed to the effect of the modification of internal atmosphere (Ali et al., 2011).

Pectin, arabic, and xanthan gums mixed with candelilla wax as an hydrophobic phase and jojoba oil as plasticizers were evaluated in the coating formulations for green bell pepper by Ochoa-Reyes et al. (2013). The three coatings were effective to significantly reduce weight losses during storage (10 days, 25°C), as compared with uncoated peppers, with no differences among the three types of coatings. As well, no difference was found among other quality parameters.

For some fruits such as pineapple and watermelon, juices are difficult to retain. Montero-Calderón et al. (2008) found that juice leakage in fresh-cut pineapple (gold cultivar) coated with 1% alginate was significantly reduced in comparison with uncoated fruit purge, and Azarakhsa et al. (2012) reported reduced weight losses using both alginate- and gellan gum-based coatings in Josapine fresh-cut pineapple.

On the contrary, waxy coatings have been found to be very effective in retarding weight losses due to transpiration. Hu et al. (2012) found that a wax coating (FMC Sta-Fresh 2952) not only reduced weight loss in pineapples from 3.1% to 2.6% after 24 days of storage (21 days at 7°C, 3 days at 25°C), but also decreased blackheart symptoms, delayed changes in polygalacturonase and endo-1,4-glucanase activities associated to pectin degradation of cell walls and to the development of chilling injury symptoms, respectively, thus leading to a better preservation of the fresh-like firmness and flesh color.

Nevertheless, not all coating formulations may have a beneficial effect regarding water loss prevention. In fact, edible packaging could favor water loss. Arnon et al. (2014) compared a bilayer edible coating, made from CMC and chitosan in two steps and a commercial polyethylene-based wax. It was shown that, for Navel oranges, water loss was reduced after both types of coating; however, for other orange cultivars and grapefruits, water losses were larger as compared with those reported for uncoated control fruits.

Water loss also depends on temperature and RH conditions surrounding the product, since mass transfer phenomena are linked to vapor pressure differences between internal and external atmospheres. Hence, fruits or vegetables stored under low RH environments will generally lose weight more easily than those stored on higher RH conditions. However, most of the papers dealing with edible coating applications do not pay much attention to the effect of the different RH conditions. Indeed, information on the water vapor transmission rate of primary packages used to contain coated fresh or minimally processed fruits and vegetables are generally not mentioned in literature.

For both protein- and polysaccharide-based coatings, lipids may be added to the formulation to improve their water loss resistance. The lipid characteristics and particle size will affect not only the effectiveness of the coating to control water vapor loss but also the coating appearance. Zaki et al. (2012) and Hanani et al. (2012) pointed out that the moisture barrier efficiency of lipid-containing coatings depends on their physical state, as lipids with a high solid fat content are more efficient than melted lipids, though they are generally brittle and cannot form cohesive layers. Combination of these lipids with proteins or polysaccharides leads to coatings with better mechanical properties and increased permeability to respiration gases, though WVR decreases because of their hydrophilicity. These same authors formulated chitosan-based coatings incorporating stearin wax (SMP 54°C, IV 33) for star fruits by preparing stable emulsions with different chitosan/stearin ratios (C:S) and 0.1% Tween 80 as a surfactant. Water losses were minimal for C:S ratios of 0:1 but increased as the chitosan content became larger. Nonetheless, fruit firmness was better retained for coatings formulated with a 1:1 ratio, which was related to water loss control and to the beneficial effect of high CO_2 and low O_2 concentrations in the fruit internal atmosphere, which led to reduced enzymatic activities. Interestingly, Karbowiak et al. (2007) reported fat distribution changes during drying of edible films formulated with carrageenan and high-melting-point fat emulsions, leading to a reduction in the functional properties of the film. Because of this destabilization phenomenon, in which hydrophobic substances aggregate, authors stated that moisture barrier efficiency of edible films can only be improved if a homogeneous and continuous lipid layer is achieved throughout the hydrocolloid matrix.

Regarding fresh-cut commodities, Ferrari et al. (2013) studied the effect of osmotic dehydration (OD) and pectin-based edible coatings on the quality and shelf life of fresh-cut melon. In their study, the OD treatment was applied prior to the dipping with the coating formulation. They found that a calcium lactate dipping (0.5% solution for 15 min) prior to coating (1% pectin) of fresh-cut melon was effective in reducing weight loss for 14 days storage at 5°C. Osmotically coated melon pieces (40°Bx + 0.5% calcium lactate for 30 min at 30°C) exhibited similar water loss rates than uncoated fruits, although changes in OD fruits are expected to cause cell integrity because of turgor loss, tissue shrinkage, and cellular collapse.

Water vapor resistance of alginate- and gellan-based coatings on fresh-cut papaya and fresh-cut Fuji apples varied with formulation, as reported by Tapia et al. (2008) and Rojas-Graü et al. (2007). These authors observed an increase in the WVR as the oil content increased for both polysaccharide coatings, with a larger effect on gellan-based coatings. For fresh-cut "Piel de Sapo" melon, Oms-Oliu et al. (2008a) reported an increase in the WVR of polysaccharide-based coatings, changing from 10.48 s cm^{-1} (uncoated fruit) to 19.3, 20.6, and 23.35 s cm^{-1} for pectin, alginate, and gellan coatings, respectively. The changes were explained by the differences in polysaccharides structure and behavior, as well as by the interaction with the fruit surface.

12.3.2 EDIBLE COATING APPLICATIONS DESIGNED FOR INTERNAL ATMOSPHERE MODIFICATION

Edible coatings for fruits and vegetables may act as a selective barrier to gases required or produced by the metabolism of the living tissue. Thus, O_2 and CO_2 concentration inside the coated product will vary according to its respiration rate, the gradient of gas concentrations between internal and external environments, and the gas transmission rate of the coating material.

Different studies are presented in which internal gas concentrations are modified in a greater or lesser extent. For instance, Ali et al. (2011) found that CO_2 gas content in the central cavity of papaya fruits noticeably increased, while O_2 content decreased when a chitosan-based coating was used, in comparison with uncoated fruits. Moreover, changes in the internal atmosphere of the fruits can be even larger, when resins and waxy coatings are used. Therefore, care must be taken to avoid the generation of anaerobic conditions and excessive accumulation of volatile compounds inside the fruit, which may impinge on fresh fruit flavor. Changes in the internal atmosphere composition may eventually affect quality attributes of the product. Arnon et al. (2014) reported that the internal quality of a wax coating and a CMC/chitosan bilayer coating did not affect the total soluble solid contents and acidity values of mandarins, oranges and grapefruits, although consumers preferred the uncoated products due to further changes in the sensory properties. However, if properly formulated, a coating should preserve the fresh-like characteristics of fresh produce.

Coatings may be developed to improve the quality of dehydrated fruits and vegetables. Different edible formulations have been studied as a mean to reduce solute uptake during OD of fruits. Akbarian et al. (2014) proposed a coating made of CMC, pectin, and ascorbic acid to reduce shrinkage and color changes of quince fruit (*Cydonia oblonga*) slices. The use of the coating as a pretreatment for air drying reduced both shrinkage and discoloration of the dried product. Indeed, the positive effect was more noticeable when OD was carried out before air drying. Hence, the coating becomes a barrier which allows water removal but restricts the uptake of solutes. Similarly, Ferrari et al. (2013) reported a reduction in the respiration rate of fresh-cut melons treated with OD and calcium lactate dips. This decrease was attributed to the collapse of external cells, which can act as a barrier and limit O_2 diffusion into the tissue, as well as affect the sugar uptake, which in turn can cause structural changes in the cell wall and water availability for biological reactions.

Edible films can also be used as a barrier for moisture migration of semidried fruits in a composite food with crispy components, like mixed cereals, avoiding mass transfer between food components (Talens et al., 2012); or in frozen food matrices to reduce the effect of freezing or thawing operations. For instance, Han et al. (2004) found that chitosan-based coatings reduced 24% the drip losses of frozen-thawed strawberries.

12.3.3 EDIBLE COATINGS APPLIED TO INHIBIT MICROBIAL SPOILAGE AND SUPPRESS PHYSIOLOGICAL DISORDERS

Literature presents several examples of the beneficial effects of edible films regarding the postharvest quality of fruit and vegetable products. Waxing has been applied to reduce postharvest diseases and stress damage in many horticultural crops, being nectarine and strawberry two examples of products that may be preserved from decay with edible coating formulations (Hu et al., 2012). As an alternative, chitosan coatings have extensively been shown to inhibit microbial growth and reduce fruit decay rate of strawberries, raspberries, peaches, kiwifruits, Japanese pears, cucumbers, bell peppers, longan fruits, bananas, or mangoes (Gol et al. 2013; Janisiewicz and Conway, 2010). A bilayer CMC/chitosan coating has been shown to delay decay and mold occurrence up to 18 days at 2°C in Navel oranges (Arnon et al., 2014). On the contrary, an edible coating prepared with 10% gum arabic and 1.0% chitosan exhibited good antifungal effects for controlling anthracnose, a postharvest disease caused by fungus *Colletotrichum musae*, which causes up to 40% of marketable losses of banana during transportation and storage. A 100% suppression of mycelial growth and a 10% inhibition of conidial germination were achieved, thus controlling decay in around 80% of the fruits (Maqbool et al., 2010). Other benefits associated to the use of this edible coating were the reduction in weight loss and delay of ripening in terms of fruit firmness, soluble solid content, and titratable acidity.

Furthermore, Hernández-Muñoz et al. (2008) coated strawberries with a chitosan-based formulation and managed to reduce decay throughout storage at 10°C and 70% RH. They also reported that the use of calcium gluconate in the coating formulation increased strawberry firmness and calcium content. However, they observed a gradual loss of weight, and a reduction in firmness and external color for all coating formulations, as compared with uncoated fruits. This may be explained by the accelerated deterioration of the fruit at 10°C and by the high vapor pressure differences between the fruit and its surroundings (70% RH), which differ from those recommended for strawberries (0°C and 90%–95% RH). Perdones et al. (2012) achieved similar results but succeeded in reducing respiration rate and quality decay of strawberries without significantly affecting their physicochemical quality throughout storage at 5°C.

As well, some promising results were reported by Mantilla et al. (2013) for fresh-cut pineapple. They applied a multilayer coating in a five-step layer-by-layer deposition process, by sequentially dipping fruit pieces in calcium chloride, alginate with trans-cinnamaldehyde encapsulated with β-cyclodextrins, calcium chloride, pectin, and calcium chloride. Microbial growth was effectively reduced, especially for psychrothrophic microorganisms, yeasts, and molds, while at the same time, juice leakage and texture changes were reduced. Similarly, Sipahi et al. (2013) reported a significant reduction in the growth of yeasts and molds on fresh-cut watermelon stored at 4°C for 15 days, but no significant differences were observed between psychroytropic and coliform organisms.

Aloe vera–based coatings have also been shown to reduce microbial spoilage in minimally processed sliced Hayward kiwifruit (Benítez et al., 2013). Hussain et al. (2012) evaluated the effect of combining a CMC edible coating with gamma irradiation on inhibition of mold growth and quality retention of strawberry. The edible film was not effective in delaying strawberry decay and loss of visual quality, but the combination of 1.0% CMC coating with fruit irradiation effectively maintained the quality of the fruit, delaying the decay and appearance of the mold growth for up to 18 days.

Chitosan edible coatings (1%) have also been successfully applied for controlling microbial inhibition of Sweet Charley strawberries and at the same time, in reducing weight loss during storage at 7°C and 85% RH, compared to uncoated fruits (Rehman et al., 2010). Chitosan coatings led to less water losses than those obtained with other polysaccharides such as pectin and potato starch. However, the authors pointed out that water losses with any substance of polysaccharide nature were much larger than those observed with protein coatings such as zein and wheat gluten. Microbial loads were controlled over 21 days of storage at 7°C and 85% RH by using chitosan, zein, or wheat gluten coatings but not when pectin and potato starch formulations were used. In the latter case, microbial load was observed after 14 days of refrigerated storage.

12.3.4 Edible Coatings Applied to Reduce Oxidative Reactions in Fresh and Processed Fruits and Vegetables

Tapia et al. (2008) evaluated the application of alginate- and gellan-based coatings on fresh-cut papaya (*Carica papaya* L.) and found that coatings reduced ascorbic acid losses, with better results for alginate formulations due to the improved O_2 barrier properties. Oms-Oliu et al. (2008a) found that the use of polysaccharide-based edible coatings on pear wedges significantly reduced vitamin C loses. Indeed, when *N*-acetylcysteine and glutathione were added to the coating formulation, ascorbic acid levels were better preserved; in addition, total phenolic content of pear wedges increased and barely changed throughout storage. Oms-Oliu et al. (2008b) observed a reduction in vitamin C losses in "Piel de Sapo" melon coated with gellan gum-based formulations.

For fresh strawberries (*Fragaria ananassa*), Gol et al. (2013) reported a progressive decrease in total phenolic content and ascorbic acid, which were better retained when coated with 1% HPMC or 1% CMC coatings. The effect was enhanced when the formulation was enriched with 1% chitosan during the first 4 days of storage at 11°C. Hussain et al. (2012) found that ascorbic acid retention in strawberries (*Fragaria* sp. cv. Confitura) stored at 3°C and 80% RH was better achieved as CMC content increased from 0.5% to 1.0%.

On the contrary, anthocyanins content in strawberries increased during storage due to their synthesis; however, lower increments were observed for HPMC- and CMC-coated fruits stored at 11°C (Gol et al., 2013). Controversial results were reported in other works. Hence, the anthocyanins content of strawberries diminished during storage at 3°C, but reductions were smaller when CMC coatings were used (Hussain et al., 2012). These authors also evaluated gamma irradiation treatments applied to strawberries in combination with CMC coatings. They found that, when fruits were irradiated (2.0 kGy), anthocyanin content increased and at the same time, ascorbic acid and total phenolic content were better maintained along refrigerated storage at 3°C.

Serrano et al. (2006) studied the use of *A. vera* coatings to preserve functional properties of Crimson table grapes throughout refrigerated storage at 1°C up to 35 days. They reported that loss of functional compounds was reduced as compared with the uncoated fruits. Hence, the total antioxidant activity of both grape skin and pulp were better retained in the coated grapes. Similarly, total phenolics of skin and pulp, and ascorbic acid content of the pulp decreased throughout cold storage, but losses were much smaller for coated grapes. Furthermore, total anthocyanin content in the skin increased due to the ripening process; nevertheless, larger changes in uncoated fruits were attributed to a delay in fruit ripening as a consequence of the coating application.

Finally, Mahfoudhi et al. (2014) proposed the use of edible coatings prepared with almond gum tree exudate (10%) and gum arabic (10%) for shelf life extension of tomato fruits. They found that ascorbic acid content of coated and uncoated tomato samples significantly decreased during 20 days at 20°C and 80%–90% RH; however, changes occurred much faster in uncoated tomatoes reaching 12.28 mg/100 g at the end of the storage, compared with 18.12 and 18.41 mg/100 g, for gum arabic- and almond gum-coated tomatoes.

12.3.5 Edible Coatings to Improve Texture-Related Properties

Coating formulations may help to increase firmness of food products because of the mechanical protection provided by the biopolymer matrix, as well as the slowdown of texture deleterious processes during storage. In this regard, Zhou et al. (2011) evaluated a shellac resin-based coating regarding its ability to prevent texture losses in Huanghua pears through storage. The coating was able to preserve cell membrane integrity as a consequence of the little changes in the cell wall constituents. After 60 days of storage, shellac-coated pears better preserved brittleness and firmness and show lower losses of cellulose, hemicellulose, and pectin. They suggested that higher peroxidase activity and lower activity of cell-wall-degrading enzymes (e.g., pectinesterase, polygalacturonase, and cellulase) in the coated pears were related with the reduced changes in the integrity of the cell membranes and the depolymerization and degradation of cell wall constituents. When comparing a bilayer edible coating, made from CMC and chitosan in two steps with a commercial polyethylene-based wax, Arnon et al. (2014) found no differences between the firmness of coated and uncoated mandarins after 4 weeks of storage at 5°C. Coating application on other fruits provided a slight firmness increase in coated Star Ruby grapefruits stored at 10°C and a significant increase in coated Navel oranges at 5°C; however, a good correlation could not be drawn between firmness and water loss values.

Changes in texture have been reported to be inhibited in strawberries coated with 1% (w/v) CMC and 1% (w/v) HPMC, with or without chitosan (1% w/v), stored at 11°C and 70%–75% RH (Gol et al., 2013). Polygalacturonase, cellulose, pectin methyl esterase, and β-galactosidase (β-Gal) activities were better inhibited in coated strawberries than in the uncoated fruits, which correlates well with the decrease in fruit softening. In accordance with these results, texture scores were much higher (6.0–8.2) for coated than those received for uncoated fruit (1.9–2.5) in a sensory test.

Other studies have failed at preventing firmness through storage. Texture deterioration and reduction in total pectin depolymerization were observed in fresh-cut kiwi coated with *A. vera* (Benítez et al., 2013). As well, fresh-cut papaya firmness rapidly declined when a xanthan gum-based coating was used—as compared with uncoated fruit pieces (Cortez-Vega et al., 2013).

On the contrary, lipid-based coatings may also be applied with the objective of reducing abrasion damages during postharvest handling of fresh fruits and vegetables. In this regard, waxy coatings have been shown to decrease the incidence of decay caused by microbial infection at injury sites (Maftoonazad and Badii, 2009).

For fresh-cut fruits and vegetables, Ghavidel et al. (2013) found that alginate coatings (2% w/v) with GLY (1.5% v/v) improved firmness retention of apple wedges during 15 days of storage at 4°C, as compared with other coatings obtained from whey proteins, soy protein isolates, and carrageenan. However, the coated fruits received higher sensory scores for odor, taste, appearance, and acceptability than uncoated fruits regardless the applied formulation.

In addition, OD treatments and calcium lactate dips combined with a pectin coating contributed to the reduction in firmness losses of fresh-cut melon along storage (Ferrari, 2013). OD treatments favored the sensory acceptance of fresh-cut melon, which was attributed to the uptake of sugar, whereas calcium lactate dips were found to mask melon taste, thus reducing acceptability.

12.3.6 EDIBLE COATINGS APPLIED TO IMPROVE GLOSS AND SHINE ON FRUITS AND VEGETABLES SURFACE

Edible coatings are evaluated to replace commercial synthetic waxes. Some ingredients may be added to the coating formulation to improve the fruit gloss. This is especially interesting for fruits that will not be consumed with peel (e.g., citrus fruits). Shellac is a natural resin, used either as a complement or as a substitutive of waxes or of other lipids, with promising prospects regarding this issue. Contreras-Oliva et al. (2012) evaluated the effect of beeswax–shellac ratio on the quality of HPMC-coated mandarins. They found that fruit gloss was higher as the shellac content was increased; however, they reported that commercial waxes provided higher gloss, because of their lipid particle size, and highlighted that wax coatings need to be prepared as a microemulsion for obtaining smooth and glossy surfaces once the water phase evaporates. Arnon et al. (2014) applied a bilayer edible coating made from CMC and chitosan in two steps, as a substitute of commercial polyethylene-based wax, generally used to enhance appearance and gloss of citrus fruits. These authors found similar gloss for oranges and mandarins coated with both the CMC/chitosan bilayer coating and commercial wax; gloss, measured in a 0–10 scale, increased from 3–4 to 7–9 for mandarins and from 2.5 to 6–7 for oranges, with no significant differences between coatings. In contrast, the bilayer coating significantly improved the glossy appearance of Star Ruby grapefruits from 2.5 to 6 but not as much as with the commercial wax, which reached 8 in the gloss scale.

12.3.7 Edible Coatings to Incorporate Active Ingredients

Edible films or coatings may be used as a vehicle to incorporate active ingredients, including antimicrobial agents, flavor compounds, colorings, antioxidants, vitamins, and pharmaceutical substances (Karbowiak et al., 2007; Maqbool et al., 2011). This opens promising applications toward to use of edible coatings as carriers of active agents that help to stabilize the surface of fresh and fresh-cut fruits and vegetables, thus contributing to the extension of their shelf life.

Polysaccharides such as sodium alginate or chitosan provide an excellent matrix for the encapsulation of active compounds. For instance, Ramos-García et al. (2012) evaluated the use chitosan-based edible coating in combination with natural compounds to control *Rhizopus stolonifer* and *Escherichia coli* DH5α in fresh tomatoes. The first of these microorganisms causes large postharvest losses, as it rapidly spreads throughout the fruit, whereas the second can cause serious illness to the consumers. A coating formulation of 1% chitosan, 0.1% beeswax, and 0.1% lime essential oil inhibited *R. stolonifer* and *E. coli* growth on fresh tomatoes at both 12°C and 15°C. Some good results were also found for thyme essential oil (1%), which was effective in controlling *E. coli* DH5α.

Chitosan coatings have also been used in combination with biocontrol agents. Such is the case of the antagonistic *Candida saitoana*, which has proved to control green mold of oranges and lemons in a similar way that imazalil, a commercial chemical antifungal (Janisiewicz and Conway, 2010). The same agent has been shown to be effective in controlling the incidence of gray mold and blue mold on apples. Indeed, polysaccharide coatings incorporating other populations of antagonistic yeasts have also shown good results in reduction in fruit decay. Authors suggest that this may be due to the good distribution of the biocontrol agents within the coating matrix (Janisiewicz and Conway, 2010). Furthermore, coatings incorporating oregano essential oil have been suggested for controlling the growth of

R. stolonifer and *Aspergillus niger* in grapes (Dos Santos and Athayde Aguilar, 2012). Low concentrations of essential oil were found to enhance the antifungal effect of chitosan, causing changes in spores and mycelia, thus leading to the inhibition of fungal growth in liquid media.

12.4 INDUSTRIAL IMPLEMENTATION OF EDIBLE PACKAGING ALTERNATIVES

One of the main restrictions for the commercial application of films and coatings pertains to limitations about preparation and handling before packaging. Handling of whole fruits and vegetables is generally much easier than that of minimally processed produce, because drying times required are usually shorter in the first case. The structure and composition of the edible layer may also be critical for successfully implementing a certain application at an industrial level. Edible coatings may be formed as a single layer or as a superposition of multiple layers. Single layer coatings may combine hydrophilic and hydrophobic components in a heterogeneous structure to control mass transfer phenomena across the fruit. Homogeneous distribution of lipid particles emulsified in a polymeric aqueous phase is critical to control mass transfer phenomena. For multilayered coatings, each layer has a specific function, which complements with those of layers with distinct composition. On the contrary, the coating composition may determine the method of application. Lipids and waxes are usually allowed to solidify on the product after application and are therefore applied by spraying or dipping. Spray application is applicable to whole fruits and vegetables, provided that the product surface is regular enough to be coated uniformly. Otherwise, immersion is the simplest and inexpensive way of application for irregular surfaces, as well as for fresh-cut commodities. Mechanical properties of the coating formulation, especially those depending on surface tension, adhesion to the fruit surface and cohesiveness of the cast material are very important for obtaining uniform coatings. Therefore, the addition of surfactants and plasticizers is often required to avoid brittleness (Pavlath and Orts, 2009). On the contrary, the gelling and thickening properties of the hydrocolloids used as a coating base may greatly impact the sequence of operation. Namely, alginate- and gellan-based coatings are instantly formed as soon as the polysaccharides get in contact with multivalent metallic cations, for instance calcium ions (Ca^{2+}). The resulting surfaces are smooth and do not need to be dried, thus being ready to pack as soon the coating is cast. In contrast, many other coatings used for either fresh or minimally processed foods need to be dried by exposing them to a stream at a certain temperature for enough time, which brings up critical issues regarding the industrial application of the developed formulations.

More recently, electrospraying techniques have been proposed to produce thin and uniform coatings on fruit surfaces. In electrospraying, microdroplets are generated through the application of a potential difference over a droplet emerging from a nozzle. The droplets become charged and are adsorbed to the product surface as a result of electrostatic forces. The main benefit of this technique is the high transfer efficiency in comparison to conventional methods. Kahn et al. (2014) evaluated the use of electrospraying for the application of water-in-oil emulsion and chocolate

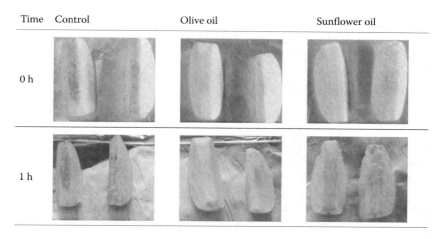

FIGURE 12.2 Influence of oil type on the browning of apple slices coated with water in oil emulsion by electrospraying method. (Adapted from *Innovative Food Science and Emerging Technologies*, 25, Khan et al., Anti-browning and barrier properties of edible coatings prepared by electrospraying, 9–13, Copyright (2014), with permission from Elsevier.)

coatings to prevent browning of apple slices. The feasibility of forming thin lipid-based coatings through this methodology was demonstrated. Furthermore, the effect of different oil types present in the emulsion formulation was evaluated regarding their effect on the browning delay of cut apples (Figure 12.2).

12.5 REGULATORY ASPECTS

Whenever they are an integral constituent of the edible portion of a fruit or vegetable, edible films and coatings must be regarded as food ingredients. Therefore, any compound and additive used in the formulation should be food grade and produced under strict hygienic conditions (Rojas-Graü et al., 2011). The U.S. Food and Drug Administration request the consideration of generally recognized as safe (GRAS) for any compound included in a coating formulation. Additives may be also used but within the specified limitations for each commodity. With some exceptions, polysaccharides such as cellulose, starch and their derivatives, seaweed extracts (e.g., agar), alginate and carrageenan, as well as hydrophobic substances (e.g., beeswax, carnauba wax, candelilla wax, stearic acid, and GLY) are considered to be GRAS substances. Regarding the European scope, edible coating components are generally regarded as food additives and listed within the list for general purposed. However, the use of specific compounds for coating purposes is regulated for each specific application (EC, 2008).

Another important issue regards the presence of allergens in the formulated edible layers for fruits and vegetables. This is relatively important in the case of coatings containing proteins from either animal or vegetable origin. Within this, milk, soybean, fish, peanut, nuts, and wheat proteins are those more closely related to the generation of intolerances and other allergenic reactions. Therefore, the presence of a known allergen or any ingredient which could contain traces of an allergen needs to be properly labeled. The use of cereal proteins may also pose an issue if the source contains gluten (e.g., wheat, barley, rye, and oat) because this could deprive celiac consumers of their dietary needs.

12.6 FUTURE PROSPECTS

Consumer demands are driving the development of new packaging materials with the attributes that contribute to better keep the quality of fresh and processed products in a sustainable way. The characterization of materials from sustainable agricultural sources, as well as from by-products and of the food processing industries broadens the panorama of the generation of new coating formulations with commercial application. Furthermore, the incorporation of micro- and nanostructured compounds opens the door for novel or improved functionalities, as well as for an enhanced release of active compounds to food matrices. Hence, much interest will be focused during the next years on the practical development of structures that can be used to improve the stability and availability of biologically active compounds, thus allowing the design of a new generation of high-quality, safe, and healthy foods.

REFERENCES

Adetunji, C. O., Fawole, O. B., Arowora, K. A., Nwaubani, S. I., Ajayi, E. S., Oloke, J. K., Aina, J. A., and Adetunji, J. B. 2012. Effects of edible coatings from *Aloe vera* gel on quality and postharvest physiology of *Ananas comosus* L. Fruit during ambient storage. *Global Journal of Science Frontier Research*, 12(5), 39–43.

Adetunji, C. O., Fawole, O. B., Arowora K. A., Nwaubani, S. I., Ajayi, E. S., Oloke, J. K., Majolagbe, O. M., Ogundele, B. A., Aina, J. A., and Adetuni, J. B. 2012. Effects of edible coatings from Aloe vera gel on quality and postharvest physiology of *Ananas comosus* (L.) fruit during ambient storage. *Global Journal of Science Frontier Research Bio-Tech and Genetics*, 12(5), 39–43.

Akbarian, M., Moayedi, F., Ghasemkhani, N., and Ghaseminezhad, A. 2014. Impact of antioxidant edible coatings and osmotic dehydration on shrinkage and colour of Quince dried by hot air. *International Journal of Biosciences*, 4(1), 27–33.

Ali, A., Muhammad, M. T. M., Sijam, K., and Siddiqui, Y. 2011. Effect of chitosan coatings on the physicochemical characteristics of eksotika II papaya (*Carica papaya* L.). Fruit during cold storage. *Food Chemistry*, 124, 620–626.

Alleyne, V., and Hagenmaier, R. D. 2000. Candelilla-shellac: An alternative formulation for coating apples. *HortScience*, 35(4), 691–693.

Almeida Castro, A., Reis Pimentel, J., Santos Souza, D., Vieira de Oliveira, T., and de Costa Oliveira, M. 2013. Estudio de la conservación de la papaya (*Carica papaya* L.) asociado a la aplicación de películas comestibles [Conservation of papaya fruit study (*Carica papaya* L.) using edible films]. *Revista Venezolana de Ciencia y tecnología de alimentos*, 25(2), 218–226.

Álvarez Arenas, C., Fermin, N., García, J., Peña, E., and Martínez, A. 2012. Evaluación del efecto de la aplicación de un recubrimiento comestible en melones (*Cucumis melo* L., var. *cantaloupe*) cortados y almacenados en refrigeración [Effect of an edible coating application on Fresh-cut cantaloupe quality (*Cucumis melo* L., var. *cantaloupe*) during cold storage]. *Saber Universidad de Oriente, Venezuela*, 25, 323–337.

Andrade, R. D., Sturkys, O., and Osorio, F. A. 2012. Atomizing spray systems for application of edible coatings. *Comprehensive Reviews in Food Science and Food Safety*, 11, 323–337.

Andreuccetti, C., Carvalho, R. A., and Grosso, C. R. F. 2009. Effect of hydrophobic plasticizers on funcional properties of gelatin-based films. *Food Research International*, *42*, 1113–1121.

Arnon, H., Zaitsev, Y., Porat, R., and Poverenov, E. 2014. Effects of carboxymethyl cellulose and chitosan bilayer edible coating on postharvest quality of citrus fruit. *Postharvest Biology and Technology*, *87*, 21–26.

Atarés, L., Bonilla, J., and Chiralt, A. 2010. Characterization of sodium caseinate-based edible films incorporated with cinnamon or ginger essential oils. *Journal of Food Engineering*, *100*, 678–687.

Auras, R., Arroyo, B., and Selke, S. 2009. Production and properties of spin-coated cassava-starch-glycerol-beeswax films. *Starch*, *61*, 463–471.

Azarakhsh, N., Osman, A., Ghazali, H. M., Tan, C. P., and Mohd Adzahan, N. 2012. Optimization of alginate and gellan-based edible coating formulations for fresh-cut pineapples. *International Food Research Journal*, *19*(1), 279–285.

Barco Hernández, P. L., Burbano Delgado, A. C., Mosquera Sánchez, S. A., Villada Castillo, H. S., and Navia Porras, D. 2011. Efecto del Recubrimiento a Base de Almidón de Yuca Modificado sobre la Maduración del Tomate [Effect of the use of a cassava modified starch based coating on tomato ripening]. *Revista Lasallista de Investigación*, *8*(2), 279–285.

Benítez, S., Achaerandio, I., Sepulcre, F., and Pujola, M. 2013. *Aloe vera* based edible coatings improve the quality of minimally processed "hayward" kiwifruit. *Postharvest Biology and Technology*, *81*, 29–36.

Bizura Hasida, M. R., Nur Aida, M. P., Zaipun, M. Z., and Hairiyah, M. 2013. Quality evaluation of fresh-cut "josapine" pineappple coated with hydrocolloid based edible coating using gelatin. In H. Abdullah, and M. N. Latifah (Eds.), *Proceedings 7th International Postharvest Symposium*, June 25, 2012, Kuala Lumpur (Malaysia). Acta Horticulturae (ISHS) *1012*, 1037–1041.

Bosquez-Molina, E., Tomás, S. A., and Rodríguez-Huezo, M. E. 2010. Influence of $CaCl_2$ on the water vapor permeability and surface morphology of mesquite gum based edible films. *LWT—Food Science and Technology*, *43*, 1419–1425.

Buitrago, C., Saavedra, A., and Pinzón, M. 2012. Conservación del Melón Cantaloupe (*Cucumis melo* L. var. *cantalulpensis*) Fresco Cortado recubierto y Almacenado en Atmósferas Modificadas [Quality evaluation of coated fresh-cut Cantaloupe melon (*Cucumis melo* L. var. *cantalulpensis*) stored under modified atmosphere conditions]. *Vitae*, *19*(1), 123–125.

Cerqueira, M. A., Lima, Á. M., Teixeira, J. A., Moreira, R. A., and Vicente, A. A. 2009. Suitability of novel galactomannans as edible coatings for tropical fruits. *Journal of Food Engineering*, *94*, 372–378.

Chiumarelli, M., and Hubinger, M. D. 2012. Stability, solubility, mechanical and barrier properties of cassava starch—carnauba wax edible coatings to preserve fresh-cut apples. *Food Hydrocolloids*, *28*, 59–67.

Contreras-Oliva, A., Rojas-Argudo, C., and Pérez-Gago, M. B. 2012. Effect of solid content and composition of hydroxypropyl methylcellolose–lipid edible coatings on physicochemical and nutritional quality of "oranules" mandarins. *Journal of Science Food Agriculture*, *92*, 794–802.

Cortez-Vega, W. R., Piotrovicz, I. B. B., Prentice, C., and Borges, C. D. 2013. Conservation of papaya minimally processed with the use of edible coatings based on xanthan gum. *Ciencias Agrarias*, *34*(4), 1753–1764.

Dangaran, K., Tomasula, P. M., and Qi, P. 2009. Structure and function of protein-based edible films and coatings. In M. E. Embuscado and K. C. Huber (Eds.), *Edible Films and Coatings for Food Applications* (pp. 25–56). New York: Springer.

Dos Santos, N. S., Athayde Aguiar, A. J., de Oliveira, C. E., Veríssimo de Sales, C., de Melo E Silva, S., Sousa da Silva, R., Stamfor, T. C., and Souza, E. L. 2012. Efficacy of the application of a coating composed of chitosan and *Origanum vulgare* L. essential oil to control *Rhizopus stolonifer* and *Aspergillus niger* in grapes (*Vitis labrusca* L.). *Food Microbiology*, *32*, 345–353.

EC Regulation No 1333/2008 of the European Parliament and of the Council. Retrieved from http://eur-lex.europa.eu/legal-content/EN/TXT/HTML/?uri=CELEX:02008R1333-20140625&qid=1397036760325&from=EN. (accessed on July 10, 2014).

Eissa, H. A. A. 2008. Effect of chitosan coating on shelf-life and quality of fresh-cut mushroom. *Polish Journal of Food and Nutrition Sciences*, *58*(1), 95–105.

Fabra, M. J., Talens, P., and Chiralt, A. 2009a. Microstructure and optical properties of sodium caseinate films containing oleic acid-beeswax mixtures. *Food Hydrocolloids*, *23*, 676–683.

Fabra, M. J., Jiménez, A., Atarés, L., Talens, P., and Chiralt, A. 2009b. Effect of fatty acids and beeswax addition on properties of sodium caseinate dispersions of films. *Biomacromolecules*, *10*, 1500–1507.

Ferrari, C. C., Sarantopoulos, C. I. G. L., Carmello-Guerreiro, S. M., and Hubinger, M. D. 2013. Effect of osmotic dehydration and pectin edible coatings on quality and shelf-life of fresh-cut melon. *Food and Bioprocess Technology*, *6*, 80–91.

Garcia, E., and Barrett, D. M. 2002. Preservative treatments for fresh-cut fruits and vegetables. In O. Lamikanra (Ed.), *Fresh-Cut Fruits and Vegetables. Science, Technology and Market* (pp. 267–304). Boca Raton: CRC Press.

Garrett, E. H. 2002. Fresh-cut produce: Tracks and trends. In O. Lamikanra (Ed.), *Fresh-Cut Fruits and Vegetables. Science, Technology and Market* (pp. 1–10). Boca Raton: CRC Press.

Ghavidel, R. A., Davoodi, M. G., Adib, A., Ahmad, F., Tenori, T., and Sheykholeslami, Z. 2013. Effect of selected edible coatings to extend shelf-life of fresh-cut apples. *International Journal of Agriculture and Crop Sciences*, *6*(16), 1172–1178.

Gol, N., Patel, P. R., and Rao, R. T. V. 2013. Improvement of quality and shelf-life of strawberries with edible coatings enriched with chitosan. *Postharvest Biology and Technology*, *85*, 185–195.

Han, C., Zhao, Y., Leonard, S.W., and Traber, M.G. 2004. Edible coatings to improve storability and enhance nutritional value of fresh and frozen strawberries (*Fragaria x ananassa*) and raspberries (*Rubus ideaus*). *Postharvest Biology and Technology*, *33*, 67–78.

Hanani, M. Z., Zahrah, M. S., and Zaibunnisa, A. H. 2012. Effect of chitosan-palm stearin edible coating on the postharvest life of star fruits (*Averrhoa carambola* L.) stored at room temperature. *International Food Research Journal*, *19*(4), 1433–1438.

Hernández-Muñoz, P., Almenar, E., Del Valle, V., Velez, D., and Gavara, R. 2008. Effect of chitosan coating combined with postharvest calcium treatment on strawberry (*Fragaria ananassa*) quality during refrigerated storage. *Food Chemistry*, *110*, 428–435.

Hu, H., Li, X., Chen, D., and Chen, W. 2012. Effects of wax treatment on the physiology and cellular structure of harvested pineapple during cold storage. *Journal of Agricultural and Food Chemistry*, *60*, 6613–6619.

Hussain, P. R., Dar, M. A., and Wani, A. M. 2012. Effect of edible coating and gamma irradiation on inhibition of mould growth and quality retention of strawberry during refrigerated storage. *International Journal of Food Science & Technology*, *47*, 2318–2324.

Janisiewicz, W. J., and Conway, W. S. 2010. Combining biological control with physical and chemical treatments to control fruit decay after harvest. *Stewart Postharvest Review*, *1*(3), 1–16.

Javanmard, M. 2013. Effect of edible coating based on whey protein and Zataria multiflora bioss extract on shelf life of "Shah Mive" pear (*Pyrus communis*). In H. Abdullah and M. N. Latifah (Eds.), *Proceedings 7th International Postharvest Symposium*, June 25, 2012, Kuala Lumpur (Malaysia). *Acta Horticulturae (ISHS) 1012*, 427–433.

Jianglian, D., and Shaoying, Z. 2013. Application of chitosan-based coating in fruit and vegetable preservation: A review. *Journal of Food Processing and Technology*, 4, 227–230.

Kader, A. A. 2002. Quality arameters of fresh-cut fruit and vegetable products. In O. Lamikanra (Ed.), *Fresh-cut Fruits and Vegetables. Science, Technology and Market* (pp. 11–20). Boca Raton: CRC Press.

Khan, M. K. I., Cakmak, H., Tavman, S., Shutyser, M., and Schroën, M. 2014. Anti-browning and barrier properties of edible coatings prepared by electrospraying. *Innovative Food Science and Emerging Technologies*, 25, 9–13.

Karbowiak, T., Debeaufort, F., and Voilley, A. 2007. Influence of thermal process on structure and functional properties of emulsion-based edible films. *Food Hydrocolloids*, 21, 879–888.

Leonet, M. J., Álvarez Arenas, C., and Bracho, N. 2011. Cálculo del Tiempo de Vida útil de Melones (*Cucumis melo* L.) Variedad Cantaloupe, Cortados, con Recubrimiento Comestible a Base de Gelatina, Mediante un Modelo de Superficie de Respuesta. *Saber. Universidad de Oriente*, 23(2), 127–133.

Lima, Á. M., Cerqueira, M. A., Souza, B. W. S., Santos E. C. M., Teixeira, J. A., Moreira, R. A., and Vicente, A. A. 2010. New edible coatings of galactomannans and collagen blends to improve the postharvest quality of fruits- influence on fruits gas transfer rate. *Journal of Food Engineering*, 97, 101–109.

Maftoonazad, N., and Badii, F. 2009. Use of edible films and coatings to extend the shelf-life of food products. *Recent Patents on Food, Nutrition & Agriculture*, 1, 162–170.

Mahfoudhi, N., Chouaibi, M., and Hamdi, S. 2014. Effectiveness of almond gum trees exudate as a novel edible coating for improving postharvest quality of tomato (*Solanum lycopersicum* L.) fruits. *Food Science and Technology International*, 20, 33–43.

Mantilla, N., Castell-Perez, M. E., Gomes, C., and Moreira, R. G. 2013. Multilayered antimicrobial edible coating and its effect on quality and shelf-life of fresh-cut pineapple (*Ananas comosus*). *LWT—Food Science and Technology*, 51, 37–43.

Maqbool, M., Ali, A., Alderson, P. G., Zahid, N., and Siddiqui, Y. 2011. Effect of a novel edible composite coating based on gum Arabic and Chitosan on biochemical and physiological responses of banana fruits during cold storage. *Journal of Agricultural and Food Chemistry*, 59, 5474–5482.

Maqbool, M., Ali, A., Ramachandran, S., Smith, D. R., and Alderson, P. G. 2010. Control of postharvest anthracnose of banana using a new edible composite coating. *Crop Protection* 29, 1136–1141.

Medina Jiménez, S., García Toro, L. M., Vélez Pasos, C., Alonso Alcalá, L., and Fernández Quintero, A. 2013. Comparative study of conservation of fresh cassava roots (*Manihot esculenta* Crantz) coated with natural wax and paraffin. *Informador Técnico (Colombia)*, 77(1), 17–21.

Montero-Calderón, M., Rojas-Graü, M.A., and Martín-Belloso, O. 2008. Effect of packaging conditions on quality and shelf-life of fresh-cut pineapple (*Ananas comosus*). *Postharvest biology and Technology*, 50, 182–189.

Moraga, M. J., Moraga, G., Fito, P. J., and Martínez-Navarrete, N. 2009. Effect of vacuum impregnation with calcium lactate on the osmotic dehydration kinetics and quality of osmodehydrated grapefruit. *Journal of Food Engineering*, 90, 372–379.

Mustafa, M. A., and Ali, A. 2013. Application of a chitosan based nanoparticle formulation as an edible coating for tomatoes (*Solanum lycopersicum* L.). In H. Abdullah and M. N. Latifah (Eds.), *Proceedings 7th International Postharvest Symposium*, June 25, 2012, Kuala Lumpur (Malaysia). *Acta Horticulturae (ISHS) 1012*, 445–452.

Ochoa-Reyes, E., Martínez-Vazquez, G., Saucedo-Pompa, S., Montañez, J., Rojas-Molina, R., de León-Zapata, M. A., Rodríguez-Herrera, R., and Aguilar, C. N. 2013. Improvements of shelf life quality of green bell peppers using edible coating formulations. *Journal of Microbiology, Biotechnology and Food Sciences*, 3(6), 2448–2451.

Oms-Oliu, G., Soliva-Fortuny, R., and Martín-Belloso, O. 2008a. Using polysaccharide-based edible coatings to enhance quality and antioxidant properties of fresh-cut melon. *LWT—Food Science and Technology*, *41*, 1862–1870.

Omms-Oliu, G., Soliva-Fortuny, R., and Martín-Belloso, O. 2008b. Edible coatings with antibrowning agents to maintain sensory quality and antioxidant properties of fresh-cut pears. *Postharvest Biology and Technology*, *50*, 87–94.

Pastor, C., Sánchez-González, L., Cháfer, M., Chiralt, A., and González-Martínez, C. 2010. Physical and antifungal properties of hydroxypropylmethylcellulose based films containing propolis as affected by moisture content. *Carbohydrate Polymers*, *82*, 1174–1183.

Pavlath, A. E., and Orts, W. 2009. Edible films and coatings: Why, what and how? In M. E. Embuscado and K. C. Huber (Eds.), *Edible Films and Coatings for Food Applications* (pp. 1–23). New York: Springer.

Pelissari, F. M., Grossmann, M. V. E., Yamashita, F., and Pineda, E. A. G. 2009. Antimicrobial, mechanical, and barrier properties of cassava starch-chitosan films incorporated with oregano essential oil. *Journal of Agricultural and Food Chemistry*, *57*, 7499–7504.

Perdones, A., Sánchez-González, L., Chiralt, A., and Vargas, M. 2012. Effect of chitosan-lemon essential oil coatings on storage-keeping quality of strawberry. *Postharvest Biology and Technology*, *70*, 32–41.

Perez-Gago, M. B., Serra, M., Alonso, M., Mateos, M., and del Río, M. A. 2005. Effect of whey protein- and hydroxypropyl methylcellulose-based edible composite coatings on color change of fresh-cut apples. *Postharvest Biology and Technology*, *36*, 77–85.

Pizato, S., Cortez-Vega, W. R., Andreghetto de Souza, J. T., Prentice-Hernández, D., and Borges, C. 2013. Effects of different edible coatings in physical, chemical and microbiological characteristics of minimally processed peaches (*Prunus persica* L. Batsch). *Journal of Food Safety*, *33*, 30–39.

Ponce, A. G., Roura, S. I., del Valle, C. E., and Moreira, M. R. 2008. Antimicrobial and antioxidant activities of edible coatings enriched with natural plant extracts: *In vitro* and *in vivo* studies. *Postharvest Biology and Technology*, *49*, 294–300.

Porta, R., Mariniello, L., Di Pierro, P., Sorrentino, A., Giosafatto, V. C., Rossi Marquez, G., and Esposito, M. 2012. Water barrier edible coating of fried foods. *Journal of Biotechnology and Biomaterials*, *2*(7), 1–3.

Pranoto, Y., Salokhe, V. M., and Rakshit, S. K. 2005. Physical and antibacterial properties of alginate-based edible film incorporated with garlic oil. *Food Research International*, *38*, 267–272.

Ramos-García, M. L., Bautista-Baños, S., Barrera-Necha, L. L., Bosquez-Molina, E., Alia-Tejacal, I., and Estrada-Carrillo, M. 2012. Compuestos antimicrobianos Adicionados en Recubrimientos Comestibles para Uso en Productos Hortofrutícolas [Evaluation of anti-microbial agents on edible coatings formulation for horticultural products]. *Revista Mexicana de Fitopatología*, *28*(1), 44–57.

Rehman, M. U., Saravanan, S., Mir, M. M., and Umar, L. 2010. Effects of carbohydrate and protein based edible coatings on quality of strawberry during storage. *SAARC Journal of Agriculture*, *8*(2), 1–10.

Rezvani, E., Schleining, G., Sümen, G., and Taherian, A. R. 2013. Assessment of physical and mechanical properties of sodium caseinate and stearic acid based film-forming emulsions and edible films. *Journal of Food Engineering*, *116*, 598–605.

Ribeiro, C., Vicente A. A., Teixeira, J. A., and Miranda, C. 2007. Optimization of edible coating composition to retard strawberry fruit senescence. *Postharvest Biology and Technology*, *44*, 63–70.

Rojas-Graü, M. A., Soliva-Fortuny, R., and Martín-Belloso, O. 2009. Edible coatings to incorporate active ingredients to fresh-cut fruits: A review. *Trends in Food Science and Technology*, *20*, 438–447.

Rojas-Graü, M. A., Soliva-Fortuny, R., and Martín-Belloso, O. 2011. Use of edible coatings for fresh-cut fruits and vegetables. In O. Martín-Belloso, and R. Soliva-Fortuny (Eds.), *Advances in Fresh-Cut Fruits and Vegetables Processing* (pp. 285–311). Boca Raton: CRC Press.

Rojas-Graü, M. A., Tapia, M. S., Rodríguez, F. J., Carmona, A. J., and Martín-Belloso, O. 2007. Alginate and gellan-based edible coatings as carriers of antibrowning agents applied on fresh-cut Fuji apples. *Food Hydrocolloids*, 21, 118–127.

Serrano, M., Valverde, J. M., Guillén, F., Castillo, S., Martínez-Romero, D., and Valero, D. 2006. Use of *Aloe vera* gel coating preserves the functional properties of table grapes. *Journal of Agricultural and Food Chemistry*, 54, 3882–3886.

Shao, X. F., Tu, K., Tu, S., and Tu, J. A. 2012. Combination of heat treatment and chitosan coating delays ripening and reduces decay in "Gala" apple fruit. *Journal of Food Quality*, 35, 83–92.

Sipahi, R. E., Castell-Perez, M. E., Moreira, R. G., Gomes, C., and Castillo, A. 2013. Improved multilayered antimicrobial alginate-based edible coating extends the shelf life of fresh-cut watermelon (*Citrullus lanatus*). *LWT—Food Science and Technology*, 51, 9–15.

Sothornvit, R. 2013. Effect of edible coating on the qualitiles of fresh guava. In H. Abdullah and M. N. Latifah (Eds.), *Proceedings 7th International Postharvest Symposium*, June 25, 2012, Kuala Lumpur (Malaysia). *Acta Horticulturae (ISHS) 1012*, 453–459.

Talens, P., Pérez-Masía, R., Fabra, M. J., Vargas, M., and Chiralt, A. 2012. Application of edible coatings to partially dehydrated pineapple for use in fruit-cereal products. *Journal of Food Engineering*, 112, 86–93.

Tapia, M. S., Rojas-Graü, M. A., Carmona, A., Rodríguez, F. J., Soliva-Fortuny, R., and Martín-Belloso, O. 2008. Use of alginate- and gellan-based coatings for improving barrier, texture and nutritional properties of fresh-cut papaya. *Food Hydrocolloids*, 22, 1493–1503.

Trezza, T. A., and Krochta, J. M. 2000. The gloss of edible coatings as affected by surfactants, lipids, relative humidity and time. *Food Engineering and Physical Properties*, 65(4), 658–662.

Trezza, T. A., and Krochta, J. M. 2001. Specular reflection, gloss, roughness and surface heterogeneity of biopolymer coatings. *Journal of Applied Polymer Science*, 79, 2221–2229.

Valverde, J. M., Valero, D., Martínez-Romero, D., Guillén, F., Castillo, S., and Serrano, M. 2005. Novel edible coating based on *Aloe vera* gel to maintain table grape quality and safety. *Journal of Agricultural Food Chemistry*, 53(20), 7807–7813.

Vanin, F. M., Sobral, P. J. A., Menegalli, F. C., Carvalho, R. A., and Habitante, A. M. Q. B. 2005. Effects of plasticizers and their concentrations on thermal and functional properties of gelatin-based films. *Food Hydrocolloids*, 19, 899–907.

Velickova, E., Winkelhausen, E., Kuzmanova, S., Alves, V., and Moldao-Martins, M. 2013. Impact of Chitosan-beeswax edible coatings on the quality of fresh strawberries (*Fragaria ananassa* cv. camarosa) under commercial storage conditions. *LWT—Food Science and Technology*, 52, 80–92.

Villalobos, R., Chanona, J., Hernández, P., Gutiérrez, G., and Chiralt, A. 2005. Gloss and transparency of hydroxypropyl methylcellulose films containing surfactants as affected by their microstructure. *Food Hydrocolloids*, 19, 53–61.

Wu, H., Wang, D., Shi, J., Xue, S., and Gao, M. 2010. Effect of the complex of zinc (II) and cerium (IV) with chitosan on the preservation quality and degradation of organophosphorus pesticides in Chinese jujube (*Zizyphus jujube* Mill. cv. Dongzao). *Journal of Agricultural and Food Chemistry*, 58, 5757–5762.

Xing, Y., Li, X., Xu, Q., Jiang, Y., Yun, J., and Li, W. 2010. Effects of chitosan-based coating and modified atmosphere packaging (MAP) on browning and shelf life of fresh-cut lotus root (*Nelumbo nucifera* Gaerth). *Innovative Food Science and Emerging Technologies*, 11, 684–689.

Zaki, N. H. M., Som, H. Z. M., and Haiyee, Z. A. 2012. Application of palm stearin-chitosan edible coating on star fruits (*Averrhoa carambola* L.). *The Malaysian Journal of Analytical Sciences*, *16*(3), 325–334.

Zhao, Y., and McDaniel, M. 2005. Sensory quality of foods associated with edible film and coating systems and shelf-life extension. In J. H. Han (Ed.), *Innovations in Food Packaging* (pp. 434–453). Amsterdam: Elsevier Academic Press.

Zhou, R., Li, Y., Yan, L., and Xie, J. 2011. Effect of edible coatings on enzymes, cell-membrane integrity, and cell-wall constituents in relation to brittleness and firmness of Huanghua pears (*Pyrus pyrifolia* Nakai, cv. Huanghua) during storage. *Food Chemistry*, 124, 569–575.

13 Edible Packaging for Dairy Products

*Óscar Leandro da Silva Ramos,
Ricardo Nuno Correia Pereira,
Joana T. Martins, and F. Xavier Malcata*

CONTENTS

Abstract 383
13.1 Introduction 384
13.2 Cheese 385
13.3 Packaging Requirements and Types of Cheese 385
    13.3.1 Hard Cheese 386
    13.3.2 Soft Cheese 387
    13.3.3 Fresh Cheeses 388
    13.3.4 Pasteurized Processed Cheese and Analogs 389
13.4 Novel Packaging Solutions: Edible Packaging 389
13.5 Edible Packaging: Recent Developments, Features, and Functional Properties 390
13.6 Edible Packaging Materials 392
    13.6.1 Polysaccharide-Based Packaging 392
    13.6.2 Protein-Based Packaging 392
    13.6.3 Lipid-Based Packaging 395
    13.6.4 Structuring Agent-Based Packaging 395
        13.6.4.1 Plasticizers 395
        13.6.4.2 Surfactants 395
13.7 Edible Packaging: Cheese Applications 396
13.8 Bioactive Agents for Enhanced Preservation 401
    13.8.1 Antimicrobials 401
    13.8.2 Antioxidants 404
13.9 Concluding Remarks 405
Acknowledgments 406
References 406

ABSTRACT

Dairy products, particularly several cheese varieties, are excellent sources of proteins, lipids, minerals, and vitamins, thus contributing to a balanced diet. Edible packaging (coatings and films) materials create a modified atmosphere surrounding the commodity similar to that achieved by controlled or modified atmosphere storage conditions. Edible films and coatings may potentially improve ripening control, quality, and safety during storage by constraining mass transfer and permeability to water vapor, oxygen, and carbon dioxide, and prevent organoleptic changes (e.g., texture, flavor, and aroma losses). Active compounds (i.e., antimicrobials, antioxidants, and nutrients) can also be incorporated into coatings' formulation to extend shelf life, preserve color, and improve the nutritional value of cheese. This chapter provides an overview of packaging requirements for preservation of different cheese varieties, as well of conventional synthetic packaging materials, and a review of the latest developments regarding application of edible packaging to cheese preservation.

13.1 INTRODUCTION

Dairy products, including milk, yoghurt, and cheese, have been uptaken worldwide for thousands of years and are still an important component of the diet—playing an essential role in meeting nutritional and functional requirements for maintenance of the human body. Cheese, in particular, is undoubtedly the most diversified and challenging group of dairy products—and an excellent source of proteins, lipids, essential minerals (such as calcium, magnesium, and phosphorus), and vitamins, as well as other functional or nutraceutical compounds—for example, bioactive peptides and conjugated linoleic acid. For this reason, production and marketing of cheese as a functional food have become increasingly important and are upholding a shift in cheese manufacture from optimization of quality to optimization of consumer benefit (Walther et al., 2008). The worldwide natural cheese production (except processed cheese) reached 20 million metric tons (MT) in 2011, experienced a growing trend in 2012 and 2013, in which European Union and United States accounted for 70% of the overall production in 2012. The cheese production is expected to show dynamic growth until 2020, with a forecasted production of ca. 25 million MT (PM Food & Dairy Consulting, 2014).

A great variety of cheeses can be found in the global market being usually produced from a narrow range of raw materials (i.e., ovine, caprine, bovine, and buffalo milks), using essentially similar processing aids (i.e., lactic acid bacteria, coagulant, and sodium chloride) and technological protocols. Cheese is naturally unstable and results from a "finely orchestrated series of consecutive and concomitant biochemical events that, if synchronized and balanced, lead to products with highly desirable aromas and flavors—but, when unbalanced, result in off-flavors and odors" (Fox and McSweeney, 2004, p. 5). Even when cheese is properly manufactured and stored, biochemical events occur due to gas exchange reactions promoted by microorganisms or enzyme activity. In addition to biological activity, the exposure to light and oxygen are key environmental factors that must be considered during storage and preservation of cheese (Robertson, 2012b). Scientists and technologists have explored methods to eliminate, or at least minimize, some of these critical issues through development of effective cheese packaging systems. Currently, packaging is seen as an imperative factor to protect and assure cheese quality (Cerqueira and Vicente, 2013), as well as in addressing consumer concerns. An ideal packaging system should increase cheese shelf life by following its dynamic changes without interfering with the ripening process. It is the way of finding an accurate balance between maturation, preservation, and cheese quality that makes cheese packaging design a challenging task. In this chapter, recent developments of edible films and coatings used in the packaging of cheese will be discussed by identifying main features and functional properties (i.e., bioactive agents for enhanced preservation). Other pertinent issues regarding cheese packaging requirements, preservation constrains, and conventional synthetic packaging materials will be addressed.

13.2 CHEESE

Production of the vast majority of cheese varieties may be subdivided into two well-defined stages: (i) manufacture of milk and (ii) ripening of the fresh curd milk. Definition of cheese quality is related with these stages, which involve different types of characteristics (Guinee and O'Callaghan, 2010): nutritional (i.e., contents of protein, fat, calcium, lactose, and sodium), physical (i.e., sliceability, crumbliness, hardness, springiness, and mouthfeel), chemical (i.e., intact casein and free fatty acid composition), heating (i.e., extent of flow, springiness, and browning), microbiological (i.e., absence of pathogens, toxic residues and foreign bodies, and conformity to approved levels of substances), and sensory (i.e., taste, aroma, texture, and appearance).

Quality patterns of modern cheese production are dependent on knowledge and control of the dynamic biochemical pathways (i.e., changes in cheese curd brought about by the action of industrial enzymes and complex fermentations) during ripening. Cheese ripening is considered a complex phenomenon (McSweeney and Fox, 1993) that entails biochemical, microbiological, structural, physical, and sensory changes during storage, which may have a strong effect on the quality of most cheese varieties (Guinee and O'Callaghan, 2010). It is not possible to review in this chapter the biochemistry involved in the ripening of all individual cheese varieties. However, different cheese varieties undergo common ripening reactions—such as glycolysis of residual sugars, proteolysis (hydrolysis of caseins to low-molecular-weight peptides and free amino acids), and lipolysis (hydrolysis of triacylglycerols). The quality of cheese ripening depends on crucial factors such as upholding of water activity to appropriate levels, and oxygen and carbon dioxide transfer across the packaging material at a rate compatible with respiration of the cheese microflora (Mathlouthi et al. 1994). Once cheese is not an inert product, these reactions continue to occur during storage and even after packaging (Poças and Pintado, 2010). Hence, cheese quality will be greatly influenced by packaging effectiveness toward factors such as temperature, humidity, light, and gas exchange.

13.3 PACKAGING REQUIREMENTS AND TYPES OF CHEESE

Classification schemes for the great multitude of cheeses are not a straightforward task, as many of these varieties are in fact variants from the same production process (Fox and McSweeney, 2004). Note that it is during the ripening phase that the characteristic flavor and texture of individual cheese varieties develop; hence classification schemes based on texture or rheological properties (that are very dependent of moisture content) are often chosen. It is therefore considered more useful to classify cheeses into the following classes: hard, soft, fresh, and processed (Cerqueira and Vicente, 2013; Poças and Pintado, 2010).

Packaging requirements are very much aligned with this kind of classification once moisture level of the cheese (on the surface and at the interior) will be dictated by the water vapor transmission rate (WVTR), and thus water vapor permeability (WVP) of the packaging material used. In addition, permeability to gases (i.e., oxygen, nitrogen, and carbon dioxide); level of opacity, which can avoid incidence of light thus reducing development of off-flavors due to oxidative processes; and potential migration of compounds from packaging to cheese or vice versa, are all to be taken into account. These are some of the factors that, in combination with surrounding humidity and temperature, should be considered in the selection of a cheese packaging material (Bekbölet, 1990; Cerqueira and Vicente, 2013; Robertson, 2012b). Despite these packaging requirements being interchangeable between the different classes of cheeses, some particularities should be highlighted.

13.3.1 Hard Cheese

This class of cheese includes very hard, hard, and semi-hard cheeses and is characterized by low moisture content (on a fat-free basis). Moisture content of very hard cheese is below 51%, while in hard and semi-hard ranges from 49% to 56%, and from 54% to 63%, respectively. Other features that characterize this class are as follows (van den Berg et al., 2004):

- Use of fresh pasteurized cow's milk that is partly skimmed (generally leading to at least 40% fat in dry matter of the final cheese)
- Milk clotting by means of a rennet
- Use of mixed-strain starters, consisting of mesophilic *Lactococci* and *Leuconostocs*, which generally produce carbon dioxide
- Acidification in the curd block, after separation of whey during pressing, holding, and early of brining
- Salting after pressing (usually in brine)
- Absence of an essential surface flora
- Maturation (for 4 weeks), thus permitting significant proteolysis

Cheeses such as Cheddar, Edam, Kasar, Port Salut, Gouda, and Mozarella, as well as those with eyes—such as Emmental and Gruyere—are included in this class. A common feature is that they are usually ripened in an anaerobic environment, under a plastic film that minimizes or even prevents occurrence of moisture evaporation. Therefore, in addition to variables such as temperature and time of ripening, the permeability of the packaging material is a critical issue for appropriate cheese maturation. For these, low permeability packages, for example, mineral wax-based coatings consisting of refined hard paraffin, and other constituents such as a blend of waxes, mineral oil (e.g., petroleum jelly), and microcrystalline waxes with various additives have been used (Poças and Pintado, 2010; Robertson, 2012b). These wax coatings play several roles: (i) barrier to oxygen, thus preventing aerobic ripening and mold growth; (ii) barrier to weight loss through avoiding moisture evaporation; and (iii) image differentiation—that is, yellow and white coatings are normally used for Gouda cheeses, while red wax is used in Edam cheeses. Since these monolayers may not provide all the required protection against gas and water vapor exchange, it is common to use coextruded materials instead—that is, polyamide (PA), polyethylene (PE), polyvinylidene chloride (PVDC), or polyethylene terephthalate (PET). These materials can be combined in laminates or multilayers, such as PA/PE, PE/PA/PE, and PE/PVDC/PE (Schneider et al., 2010). Water-based dispersions made from low-molecular-weight copolymers of ethylene, and polyvinyl acetate can be used as coating in combination with wax layers. These types of coatings are also associated with functional features, being used as carriers of several antifungal agents—that is, natamycin, calcium, and potassium sorbate (Poças and Pintado, 2010; Robertson, 2012b). The role of some of these bioactive agents for enhanced preservation is detailed in Section 13.8. Modified atmosphere packaging (MAP) and vacuum packaging, combined with specific permeability rates of the packaging material are also applied for portioned and sliced hard cheeses, once they are susceptible to high respiration rates—and consequently, deterioration. MAP and vacuum packaging are combined with ethylene vinyl alcohol (EVOH) and polyolefin-based materials; together, they block oxygen diffusion and act as moisture barrier (Poças and Pintado, 2010). Some of the drawbacks of packaging systems used for cheese maturation and sliced cheese derive from mold growth when the packaging is opened/damaged, and organoleptic defects—particularly in cheeses with a high fat content (i.e., appearance, taste and aroma), induced by vacuum or the gas mixture used in MAP packaging.

13.3.2 Soft Cheese

This class of cheese is characterized by a moisture content ranging from 61% to 69% on a fat-free basis, and the different variants can be distinguished between each other by the microorganisms involved in maturation: soft cheeses ripened by surface molds (e.g., Brie and Camembert) and blue vein soft cheeses, such as Gorgonzola, Stilton, Roquefort, Danablu, and Blue Wensleydale, ripened by internal (blue) mold. Surface mold–ripened soft cheeses are characterized by the presence of a felt-like coating of mycelia due to growth of molds on the surface (Spinnler and Gripon, 2004). In general, the microenvironment in soft cheese is heterogeneous with pronounced gradients of pH, salt, and water activity (Cantor et al., 2004). One of the most critical parameters during ripening of a packaged soft cheese is water activity. For instance, Camembert cheese water activity should range between 0.98 and 0.99 in the rind, which is particularly difficult to achieve, once cheese is normally stored without control of relative humidity (RH) (Mathlouthi et al., 1994). According to the nature of the ripening process, the softness level of the cheese and its respiration rate are critical issues for the choice of package material (Poças and Pintado, 2010). Packaging materials with low WVP minimize loss of moisture, while maintaining water activity in the range necessary for growth of microorganisms that contribute to cheese ripening. In the case of internal molded cheese, the packaging material should allow passage of oxygen to promote mold development in the curing channels of the cheese. However, Roquefort is usually wrapped with aluminum foil, polypropylene

(PP) films, or else thermoformed packages made from polyvinyl chloride, and polystyrene (PS). These impermeable materials allow the presence of carbon dioxide, which seems to minimize oxygen transfer and stimulate growth of *Penicillium roqueforti*—thus exchanging cheese ripening (Robertson, 2012b; Schneider et al., 2010). MAP and vacuum packaging can also be applied to blue vein cheeses, once they are slowly respiring cheeses. However, in the case of very soft cheeses, MAP technology is normally used—once vacuum packaging can affect cheese structure (Poças and Pintado, 2010).

Surface mold–ripened cheeses (e.g., Camembert and Brie) are fast respiring cheeses. Hence, packaging material, together with stage of development of surface molds, temperature, and atmosphere prevailing inside the package will determine its respiration rate. For these kinds of cheeses, the packaging material should balance permeability to oxygen and water vapor. Materials such as perforated films of orientated polypropylene (OPP) and combinations of OPP and paper are commonly used. Film perforations are crucial to control WVTR and their properties—that is, number, size, and density of perforations, which may vary according to level of the barrier required (Poças and Pintado, 2010; Robertson, 2012b). For example, Limburger-type cheeses are characterized by a high-water activity, and their packaging material is usually composed of coated paper as inner layer and cellophane, OPP or PET films, as outer layer, which allow retaining adequate levels of moisture inside the package. Camembert-type cheeses normally use perforated and laminated materials, with higher gas and WVP, thus ensuring appropriate mold growth, such as perforated lacquered cellophane laminated on paraffin-coated paper, perforated lacquered aluminum foils, aluminum foils laminated on low-density polyethylene (LDPE)-coated tissue paper, and PET films on cellulose paper, with PE and ethylene vinyl acetate lamination (Schneider et al., 2010).

13.3.3 Fresh Cheeses

Fresh cheeses are unripened cheeses characterized by very high moisture content (>80%), low salt concentration, high pH, and slow draining. These cheeses can be manufactured via coagulation of milk, cream, or whey using acid (i.e., lactic acid fermentation), a combination of acid and rennet, or a combination of acid and heat. Fresh cheeses can be divided into various categories, according to the method of coagulation (i.e., acid, acid rennet, or acid heat), consistency (i.e., paste, grainy, or gel like), and raw material used (i.e., milk or whey) (Schulz-Collins and Senge, 2004). This category includes cheeses such as Cottage, Mascarpone, Ricotta, Chèvre, Feta, Cream and whey cheese, Quark, and Petit Suisse. The shelf life of fresh cheese is very limited, because their inherent properties make them highly susceptible to microbial spoilage. Regarding packaging requirements, loss of moisture should be avoided once fresh cheese remains draining slowly. In particular, cream cheese, due to its relatively high fat content (~34%), is sensitive to photooxidation, so the packaging must provide protection against light transmission. Some fresh cheeses are also packaged under a low-oxygen-modified atmosphere to avoid quality deterioration (Poças and Pintado, 2010). Standard material packages used for these cheeses are plastic cups made of high-impact or high-density polyethylene (HDPE) and PP; these

Edible Packaging for Dairy Products

materials can also be coated with PVDC to improve barrier and pigmented with titanium dioxide to offer protection against light (Robertson, 2012b). High-barrier materials, such as PA/LDPE laminates, can also be used to provide a higher barrier to oxygen and modified atmosphere with the required levels of carbon dioxide in the headspace, between product and package (Poças and Pintado, 2010).

13.3.4 Pasteurized Processed Cheese and Analogs

The products in this group differ from the previous cheeses as they result from blending, heating, and shearing of mixtures of dairy and vegetable ingredients, such as skim milk, natural cheese, casein, caseinates, water, butter oil, and vegetable oils and proteins, among other ingredients (e.g., emulsifying salts and spices). Processed cheese is mainly composed by ingredients of dairy origin, in which natural cheese must account for more than 51% of the total final product. Analog cheese products do not necessarily include natural cheese; however, it may be added at small levels (~5%) to impart a cheesy flavor, if needed (Guinee et al., 2004). The high temperatures applied during thermal process (i.e., 70°C to 140°C, depending on the heating method), combined with the hot-filling process confer an extended shelf life to products. Moreover, sorbic and propionic acids may be added as preservatives to prevent mold growth, as well as nisin to prevent growth of anaerobic sporeformers, such as *Clostridia* spp. (Robertson, 2012b). However, an appropriate packing barrier to oxygen and light is still critical to prevent undesired oxidative reactions that lead to organoleptic flaws. Concerning packaging material, the most common systems used for processed cheese and analogs are mainly composed of aluminum foil, multilayer materials, and plastic laminates, as described by Poças and Pintado (2010):

- Squeezable nonbarrier tubes made of LDPE
- High-barrier tubes made of multilayer materials, containing EVOH as a barrier layer or metal tubes
- Cups made of PP, PET/LDPE or PS/EVOH/LDPE heat sealed with aluminum foil or plastic laminate
- Glass cups heat sealed with an aluminum foil plastic laminate or with an easy-open tinplate cap

One of the most traditional packages of processed cheese consists of triangular portions packaged by heat sealable, lacquered aluminum. This package contains also an opening device consisting of narrow strip of PET film (normally colored red to attract consumer's attention), sealed onto the inner side of the aluminum foil and extended several millimeters beyond the packaging material—to facilitate opening of a portion of cheese (Robertson, 2012b).

13.4 NOVEL PACKAGING SOLUTIONS: EDIBLE PACKAGING

Application of the aforementioned conventional synthetic packaging materials to cheese has been widely studied and applied by the industry depending on cheese variety and specific storage requirements (Cerqueira and Vicente, 2013). Major food retailers and consumers have been concerned with the waste generated, expenditure of scarce natural resources, and energy used in the manufacture of synthetic packaging materials (Tomasula, 2009). During recent decades, serious ecological problems, together with nonbiodegradability and nonedibility of petroleum-based films, have led food processors to turn their attention to renewable materials. Hence, an emergent area in food packaging has been on the rise encompassing edible films, coatings, or biodegradable packagings manufactured from fully renewable agricultural materials. Edible films and coatings can be obtained using simple or mixed materials, such as polysaccharides, proteins, lipids, and resins, in various forms—that is, coatings, single layer, bilayer, and multilayer. Edible films and coatings are not a replacement for nonedible, petrochemical-based packaging materials for extended food storage, because their biodegradability and the mechanical and physical properties do not last for a very long time. Depending on the material used, they can be highly sensitive to moisture and show poor water vapor barrier properties (i.e., cellulose and derivatives, starch and derivatives, gums, gelatin, zein, and gluten); despite of good water vapor barrier properties, they can be opaque, relatively inflexible and unstable (i.e., waxes, lipids or derivatives, and microbial polymers). However, they still have the potential to replace one or more polymeric film layers in multilayer packaging systems, and thus act as adjunct for improving overall food quality, extending shelf life, and exchanging economic efficiency of packaging materials (Gontard and Guilbert, 1994; Robertson, 2012a; Tomasula, 2009). The following sections will focus on detailed aspects related to application of edible films and coatings for cheese products, with or without incorporation of functional ingredients, and address the main advantages of edible films and coatings over traditional petrochemical-based polymeric packaging materials.

13.5 EDIBLE PACKAGING: RECENT DEVELOPMENTS, FEATURES, AND FUNCTIONAL PROPERTIES

Edible films and coatings offer alternative biodegradable packaging options without significant environmental impacts. Although edible packaging materials are not meant to fully replace synthetic packaging, they can be utilized on foods to meet challenges associated with stable quality, market safety, nutritional value, and economic cost. The main advantages and/or benefits of using edible films and coatings include: (i) limit moisture, aroma, and lipid migration between components and the outer environment; (ii) provide a gas barrier for controlling gas exchange between food and surrounding atmosphere; (iii) restrict exchange of volatile compounds; and (iv) protect foods from physical damage caused by mechanical impact, pressure, vibrations, and other mechanical factors, or susceptible ingredients from oxidation—thus extending shelf life throughout storage, while improving food quality (Ramos et al., 2012a; Rezvani et al., 2013). Edible films and coatings can also serve other purposes with associated high added value, viz. carrying of antimicrobial, antioxidant, and other nutraceutical ingredients; in such active packages, the intended primary barrier and mechanical properties of the films or coatings are not significantly compromised by said additives (Cerqueira et al., 2010b; Ramos et al., 2012a).

Edible packaging formulations are usually divided in two types: related with the form of application and with the type of film or coating. Film is a layer formed through solidification of melted materials, such as lipids, paraffin, and other waxes—or by casting, where a dilute film solution is spread on a surface and let to dry under controlled conditions before detachment. These stand-alone structures are prepared separately from the food and are later used to wrap food products (Dangaran et al., 2009; Ramos et al., 2012a). For an industrial scale production, techniques similar to those used to manufacture flexible plastic films are in general applicable to obtain edible and biodegradable films; these techniques are usually extrusion (or coextrusion for multilayer films), lamination, molding, and roll drying for solvent removal (Dangaran et al., 2009; Debeaufort et al., 1998). On the contrary, alternative methodologies have been developed to produce edible films at large scale: Lafargue et al. (2007) designed a dip-molding method for films based on a modified starch/carrageenan mixture, in which a gelled state is preferred for setting films on a surface (Lafargue et al., 2007), whereas Kozempel and Tomasula (2004) established a continuous process for calcium caseinate films, in which the forming solution is spread onto a moving belt, and then passed through a dryer (Kozempel and Tomasula, 2004).

Coating is defined as a thin layer of material obtained from the same film-forming solution, but applied directly on the surface of the food or between food components. Although removal of coatings may be possible, they are not typically designed to be disposed off separately from the coating material itself. Coatings are normally regarded as part of the final product, thus providing extra advantages regarding nutritional and quality aspects by incorporation of bioactive compounds therein (Ramos et al., 2012a).

The most common methodologies for application of coatings upon cheese surface encompass dipping, spraying, or brushing; the choice is highly dependent on the characteristics of the foods to be coated and the physical properties of the coating (Cerqueira and Vicente, 2013). Brushing method is frequently used in traditional cheeses wholesalers and usually requires several applications to cover all parts of the product (top and bottom)—and may accordingly lead to a less homogeneous coverage. In this method, the product should possess some resistance to mechanical stresses caused by brush movements—and better results are obtained for rounded or oval-shaped materials that are easier to coat (Andrade et al., 2012). Dipping is the most common lab-scale technique due to its simplicity and good coverage on uneven surfaces. However, this method leads to coating-solution dilution and a residual with a high quantity of coating materials, which often results in microorganism growth in the dipping tank (Andrade et al., 2012). Furthermore, the scale upgrade (viz. processing control and automation of continuous production) represents a major challenge. Spraying is another widely used technique for applying coatings. This methodology offers uniform coating, thickness control, and possibility for sequential applications that do not contaminate the solution. Moreover, it is usually easier to implement in industrial settings than dipping, because of the higher draining and drying times required (Andrade et al., 2012).

Zhonga et al. (2014) investigated the application of different edible coating materials (chitosan, sodium alginate, and soy protein isolate) and different coating application methods (dipping, enrobing, and spraying) on Mozzarella cheese. They showed that the spraying method produced thinner film with almost equal preservation abilities compared to dipping method, thus creating advantages for saving raw materials and better process control. Moreover, the authors concluded that the spraying method may lower production cost and help implementation of fully automatic production (Zhonga et al., 2014). This process is usually followed by drying in the case of aqueous products—which may need a drier to remove excess water or accelerate drying under room conditions—or by cooling for lipid-based coatings (Cerqueira and Vicente, 2013; Debeaufort et al., 1998).

13.6 EDIBLE PACKAGING MATERIALS

Edible packages can be manufactured from several materials via different processes, thus leading to final material exhibiting distinct properties. The edibility of the biomaterials used in the manufacture of this type of packaging may raise a problem, once the material to be used must be of food grade to ensure edibility (Brody, 2005; Pavlath and Orts, 2009). The main edible, renewable, and natural biopolymers intended for packaging are (obtained from) polysaccharides, proteins, lipids, and resins, which can be used solely, or in combination with each other, with or without addition of plasticizers and surfactants (Falguera et al., 2011). Biomaterials commonly used in the production of edible films and coatings for application on cheese are shown in Table 13.1.

The functionality and performance of edible packaging materials will greatly depend on their intrinsic barrier and mechanical properties, and their behavior will be affected by material composition, production process, and type of application. Polysaccharide and protein materials are characterized by high moisture permeability, and low oxygen and lipid permeability at low RH; however, their barrier and mechanical properties are compromised at high RH.

13.6.1 POLYSACCHARIDE-BASED PACKAGING

Polysaccharides were the first and still the most extensively studied materials used in development of edible packaging for cheese application. A variety of polysaccharides and derivatives thereof have been used for edible and biodegradable packaging—including those obtained from plants (e.g., pectin, galactomannans, starch and cellulose, and their derivatives), seaweed extracts (e.g., alginate and carrageenan), exoskeleton of crustaceans (e.g., chitosan), and microorganisms (e.g., gellan, pullulan, and dextran). They exhibit a wide variety of structures that allow development of packaging with distinct features, thus leading to a wide range of properties useful in food handling. They possess the ability to produce strong gels, which may be insoluble upon reaction with multivalent cations, and provide excellent mechanical and structural features. However, they have poor water vapor barrier due to a highly hydrophilic nature—which can be improved by addition of lipids (Ramos and Malcata, 2011).

13.6.2 PROTEIN-BASED PACKAGING

Proteins as basis for edible packaging have not been so extensively used as polysaccharides, but there is an increasing trend for their use in cheese packaging as can be inferred for recent publications in this field (Ramos et al., 2012b). Proteins can be useful to

TABLE 13.1
Examples of Antimicrobial Edible Films and Coatings Applied to Cheese

Bioactive Compound		Concentration	Film/Coating Material	Cheese Type	Effects/Results	References
Bacteriocin	Nisin	1000 IU cm^{-2}	Sodium caseinate	Semi-hard cheese (Babybel®)	*L. innocua* counts reduction (1.1 log CFU g^{-1}) in surface-inoculated cheese samples after 1 week (at 4°C) as compared to control samples	Cao-Hoang et al. (2010)
		50 IU g^{-1}	Galactomannan	Ricotta	*L. monocytogenes* growth was prevented for 7 days at 4°C	Martins et al. (2010)
Antibiotic	Natamycin		Chitosan	Semi-hard regional	Molds/yeasts population decrease 1.1 log (CFU g^{-1}) compared to uncoated cheese after 27 days of storage at 4°C	Fajardo et al. (2010)
		0%, 0.2%, 0.5%, 1%, 2%, and 4% per dry weight of cellulose	Cellulose	Gorgonzola	No significant growth of fungi at 2% and 4% natamycin, compared with uncoated cheese	de Oliveira et al. (2007)
Bacteriocin + antibiotic	Nisin and natamycin	9.25 mg natamycin dm^{-2} and 2.31 mg nisin dm^{-2} of film	Tapioca starch	Port Salut	Cheese coated showed *L. innocua* counts lower than 10 CFU mL^{-1} and *S. cerevisiae* counts of 1.5 cycles lower than uncoated cheese after 196 h of storage.	Ollé Resa et al. (2014)
Polysaccharide	Chitosan	2% (w/v)	Sodium caseinate	Cheedar	Mesophilic, psychrotrophic, and yeasts and molds populations significant reduced (2.0–4.5 log CFU g^{-1}) in coated cheese	Moreira et al. (2011)
		0.8% (w/v)	Chitosan/whey protein (CWP)	Ricotta	Mesophilic and psychrotrophic microorganisms viable numbers were significantly lower in CWP-coated than uncoated cheese	Di Pierro et al. (2011)

(*Continued*)

TABLE 13.1 (Continued)
Examples of Antimicrobial Edible Films and Coatings Applied to Cheese

Bioactive Compound		Concentration	Film/Coating Material	Cheese Type	Effects/Results	References
Plant extract	Olive leaf extract	1.5% (w/v)	Methylcellulose	Kasar	$S.\ aureus$ counts decreased 0.68 and 1.22 log cycle after 7 and 14 days of storage, respectively	Ayana and Turhan (2009)
	Linalool, carvacrol, or thymol	2.38% (w/w)	Starch	Cheddar	Linalool, carvacrol, or thymol reduced $A.\ niger$ population by 1.8, 2.0, and 2.2 log CFU g^{-1}, respectively, after 35 days of storage at 15°C	Kuorwel et al. (2014)
Enzyme	Lysozyme	11.7 mg g^{-1} film-forming solution	Zein	Kashar	Film prevented the increase of $L.\ monocytogenes$ counts in cheese for 8 weeks at 4°C	Ünalan et al. (2013)
		60% per dry weight of chitosan	Chitosan	Mozzarella	Chitosan–lysozyme films enhanced antimicrobial effect against $P.\ fluorescens$ and $L.\ monocytogenes$, and inhibited completely mold growth in cheese	Duan et al. (2007)
Organic acid	Lactic acid	6 g L^{-1} lactic acid and 0.25 g L^{-1} natamycin	Whey protein isolate	Semi-hard bovine cheese	Cheeses coated with antimicrobial coatings did not show growth (<100 CFU g^{-1}) of $Staphylococcus$ spp., $Pseudomonas$ spp., Enterobacteriaceae, yeasts, or molds, for at least 60 days at 10°C.	Ramos et al. (2012b)
		0.6% (w/v) lactic acid and 0.0125% (w/v) natamycin	Ovine whey protein concentrate	Semi-hard bovine cheese	Whey protein concentrate antimicrobial coatings prevented growth of $Staphylococcus$ spp., $Pseudomonas$ spp., Enterobacteriaceae, yeasts, and molds during 45 days at 11°C.	Henriques et al. (2013)

stabilize emulsions and are good film-forming materials essentially due to their gelation capacity. These materials produce indeed transparent packages with excellent mechanical and structural properties, but exhibit high permeability to moisture. Proteins can be obtained from animal (e.g., caseins, gelatin, whey proteins, and collagen) and plant (e.g., wheat gluten, corn zein, and soya) sources (Ramos and Malcata, 2011).

13.6.3 Lipid-Based Packaging

Among naturally occurring biopolymers, lipids have received the least attention—probably due to their poor film-forming capability and fairly weak mechanical strength. Lipids are generally used to improve barrier to water loss and dehydration, owing to their nonpolarity and hydrophobicity—and have been used mainly for fruits and meat products. The most common materials used include acetoglycerides, waxes (e.g., carnauba wax, bees wax, and paraffin wax), oils (e.g., mineral and vegetable), and surfactants (Maftoonazad and Badii, 2009).

13.6.4 Structuring Agent-Based Packaging

The formulation of edible and biodegradable films and coatings typically requires incorporation of structuring agents to some degree—for example, plasticizers to avoid brittleness and surfactants to stabilize the basic solution (Ramos and Malcata, 2011; Skurtys et al., 2010). These two functional compounds are discussed below to further extent.

13.6.4.1 Plasticizers

Plasticizers are added during formation of edible and biodegradable packaging—especially when polysaccharides or proteins are used as starting material, once the resulting structures are often brittle and stiff due to extensive interactions between the polymer molecules (Skurtys et al., 2010). Plasticizers are needed to improve packaging flexibility, extensibility, toughness, workability, and processability, by increasing chain mobility and intermolecular spacing—because internal interactions are reduced (Dangaran et al., 2009). On the contrary, plasticizer incorporation may lead to a marked increase in the diffusion coefficients for gases or water vapor, as well as a decrease in cohesion and tensile strength of the packaging material (Ramos and Malcata, 2011). The plasticizers most commonly used for biodegradable packaging are polyols (e.g., glycerol, sorbitol, PE glycol, mannitol, and sucrose), as well as some hydrophobic compounds derived from citric acid (e.g., acetyltributyl citrate, triethyl citrate, and acetyltriethyl citrate) (Andreuccetti et al., 2009; Ramos and Malcata, 2011).

13.6.4.2 Surfactants

Surfactants are useful to reduce surface tension and improve wettability and adhesion of the coating to the product surface. They can often act as emulsifiers, aimed at stabilizing an emulsion, being essential when lipid particles are present, as they are able to modify surface energy. The main surfactant compounds used for biodegradable packaging include lecithin, sorbitan monolaurate (known as Span 20), and polyoxyethylene sorbitan trioleate (known as Tween 85) (Skurtys et al., 2010).

13.7 EDIBLE PACKAGING: CHEESE APPLICATIONS

Application of synthetic packaging on food products has been thoroughly studied, and the great number of conventional synthetic solutions available in the market displaying good results in terms of food preservation has retarded implementation of food-grade alternative solutions. However, the use of edible packaging based on natural polymers and food-grade additives to extend shelf life of food products has attracted an increasingly greater attention, and numerous projects are under way in this field (Cerqueira and Vicente, 2013; Ramos and Malcata, 2011).

The most studied food products where edible packaging materials have been applied are fruits and vegetables. For instance, Ribeiro et al. (2007) tested the ability of polysaccharide coatings of starch, carrageenan, and chitosan to extend the shelf life of strawberries (*Fragaria ananassa*) (Ribeiro et al., 2007); Cerqueira et al. (2009b) used coatings composed of galactomannans from two different sources (*Caesalpinia pulcherrima* and *Adenanthera pavonina*) for application on five tropical fruits: acerola (*Malpighia emarginata*), cajá (*Spondias lutea*), mango (*Mangifera indica*), pitanga (*Eugenia uniflora*), and seriguela (*Spondias purpurea*) (Cerqueira et al., 2009b). Lima et al. (2010) likewise studied application of coatings based on mixtures of polysaccharide (galactomannan from *C. pulcherrima* and *A. pavonina*), collagen, and glycerol on mangoes and apples (Lima et al., 2010).

Their application on cheese has been hardly explored, and only a few studies exist to date. The main reason for this realization may be the fact that cheese has a complex composition that makes it biologically and biochemically dynamic—and consequently, inherently unstable (Ramos et al., 2012b). During cheese maturation and storage, biological reactions caused by microorganisms and enzymes take place that influence their texture, flavor, and odor properties. Additionally, cheese physical characteristics are strongly affected by several environmental factors during storage (e.g., light, RH, temperature, permeability to water vapor, oxygen, and carbon dioxide), which must be taken into account when selecting the material formulation for specific cheese coatings. All these factors affect not only cheese physical characteristics, but also evolution of flavor during storage (Fox and McSweeney, 2004; Robertson, 2006).

The environmental conditions prevailing throughout handling and storage often promote uncontrolled and extensive fungal and bacterial development on the cheese surface, which considerably reduces quality, causes economic losses, and may even lead to health problems on the consumer side (de Oliveira et al., 2007). Concerning unpackaged cheese, water loss depends on chemical properties of the cheese itself and on storage conditions. Although the shelf life of cheese can be extended via lowering the rates of water loss, application of a coating is considered a more effective approach toward this aim (Pantaleão et al., 2007). Therefore, recent research studies have attempted to apply edible coating materials toward cheese preservation, by controlling some of the aforementioned parameters. Overall, the use of edible packaging manufactured from different materials (e.g., chitosan, alginate, gellan, κ-carrageenan, galactomannan, and whey protein isolate) decreased weight and moisture loss of coated cheeses, improving textural and sensory properties as compared to uncoated cheeses (Altieri et al., 2005; Cerqueira et al., 2010a; Fajardo et al., 2010; Kampf and Nussinovitch, 2000). Both weight and moisture loss are related to the kinetics of water permeation through the coating used (Robertson, 2006). Examples of materials used in the production of edible packaging for cheese applications are presented in Table 13.1.

Kampf and Nussinovitch (2000) tested different edible coatings made from κ-carrageenan, alginate, and gellan as base material for semi-hard and dry white-brined cheeses. These authors showed that, for semi-hard cheese, all coatings reduced weight loss for 46 days of storage, with no significant differences between the various coatings, thus contributing to a better color and gloss, as well as to a softer and a less brittle texture, when compared to uncoated cheeses. In the case of dry white-brined cheese, all coating solutions reduced weight loss of cheese, but a significant advantage was observed for κ-carrageenan-based coating. These coatings contributed to a lower reduction in pH, thus leading to a higher quality of cheese. In addition, the application of coatings on cheese surfaces increased gloss, improved the textural and sensory properties, and did not affect taste of white-brined cheese in comparison with uncoated cheese (Kampf and Nussinovitch, 2000). Altieri et al. (2005) demonstrated that chitosan edible coatings significantly increased the shelf life of Mozzarella cheese by reducing coliform growth during storage compared to uncoated cheese. Moreover, sensory analysis performed in this cheese unfolded no significant differences between coated and uncoated cheese—thus corroborating the idea that chitosan at low concentrations (i.e., 0.075%) does not affect the sensory profile of cheese. This application in Mozzarella cheese contributed to maintenance of a better texture and consequently to a wider acceptability (Altieri et al., 2005). Cerqueira et al. (2010a) showed that semi-hard cheeses coated with galactomannan and chitosan edible coatings (as base material) decreased moisture loss by 2.5% and 1.9%, and weight loss by 3.8% and 3.1% at 4°C and 20°C, respectively, during 21 days of storage. In this study, reduction of oxygen consumption and carbon dioxide production rates were also observed for both coatings in comparison with uncoated cheese. Furthermore, the hardness and color changes of coated cheese decreased in relation to uncoated cheese, which brought clear advantages in terms of marketing and commercialization of cheese (Cerqueira et al., 2010a). On the contrary, Di Pierro et al. (2011) showed that coatings composed of a mixture of edible materials (i.e., chitosan and whey proteins) extended the shelf life of Ricotta cheese by reducing growth of microbial contaminants, delaying development of undesirable acidity, and maintaining texture without modifying the sensory characteristics for 30 days of storage at 4°C, when compared to uncoated cheese. Differences in visual appearance, texture, flavor, and odor were also not detected between coated and uncoated Ricotta cheese during sensory evaluation (Di Pierro et al., 2011). Ramos et al. (2012b) demonstrated that edible coatings prepared by a mixture of edible materials (i.e., whey protein isolate, glycerol, guar gum, sunflower oil, and Tween 20) as base matrix decreased weight and moisture loss, water activity, hardness, and color change of semi-hard cheese, whereas they did not affect pH, salt, and fat contents in comparison with uncoated cheese, throughout 60 days of storage (Figure 13.1).

These edible coatings together with several combinations of antimicrobial compounds (i.e., natamycin and lactic acid, natamycin and chitooligosaccharides [COSs], and natamycin, lactic acid, and COSs) prevented growth of pathogenic and other contaminant

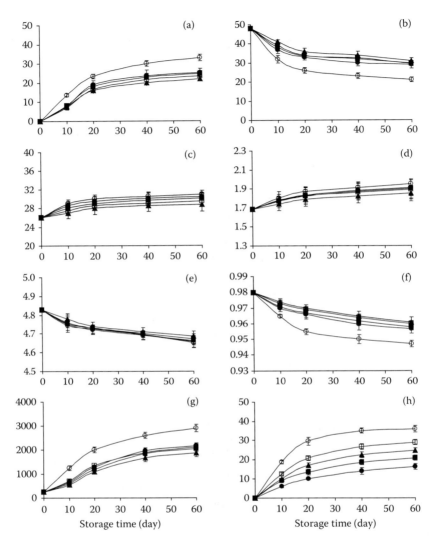

FIGURE 13.1 Values (average ± standard deviation) of (a) weight loss, (b) moisture, (c) fat, (d) salt contents, (e) pH, (f) water activity, (g) hardness, and (h) color of cheese coated with antimicrobial edible coatings—solution 1 (▲, 6 g L^{-1} lactic acid and 0.25 g L^{-1} natamycin), solution 2 (●, 20 g L^{-1} COS and 0.25 g L^{-1} natamycin), and solution 3 (■, 6 g L^{-1} lactic acid, 20 g L^{-1} COS, and 0.25 g L^{-1} natamycin), compared with cheese coated with commercial coating (□) and uncoated cheese (○), along 60 days of storage at 10°C and 85% RH.

microorganisms, while allowing regular growth of lactic acid bacteria throughout 60 days of storage. In turns, commercial nonedible coatings made from polyvinyl acetate used as control inhibited only yeasts and molds (Table 13.2) (Ramos et al., 2012b).

Efforts of R&D focused on development of novel edible packaging solutions for cheese preservation have boomed in the last decade, as concluded from the increasing

TABLE 13.2
Viable Cell Counts of Cheese Uncoated, and Coated with Antimicrobial Edible and Commercial Coatings

Microorganisms	Type of Coating Solution	Storage Time (days)				
		0	10	20	40	60
Lactococcus spp.	1	7.23 ± 0.22[a,a]	8.46 ± 0.44[b,a]	8.77 ± 0.32[b,a]	8.43 ± 0.38[b,a]	8.22 ± 0.43[b,a]
	2	7.36 ± 0.38[a,a]	8.54 ± 0.36[b,a]	8.89 ± 0.46[b,a]	8.63 ± 0.33[b,a]	8.35 ± 0.35[b,a]
	3	7.29 ± 0.43[a,a]	8.32 ± 0.44[b,a]	8.60 ± 0.35[b,a]	8.33 ± 0.23[b,a]	8.19 ± 0.35[b,a]
	Commercial	7.24 ± 0.20[a,a]	8.33 ± 0.29[b,a]	8.71 ± 0.35[b,a]	8.50 ± 0.35[b,a]	8.37 ± 0.41[b,a]
	None	7.33 ± 0.28[a,a]	7.51 ± 0.19[a,b]	7.79 ± 0.35[a,c]	7.58 ± 0.20[a,b]	7.41 ± 0.22[a,b]
Lactobacillus spp.	1	6.80 ± 0.33[a,a]	7.43 ± 0.34[b,a]	7.92 ± 0.37[b,a]	7.55 ± 0.35[b,a]	7.32 ± 0.43[b,a]
	2	6.63 ± 0.38[a,a]	7.68 ± 0.30[b,a]	8.16 ± 0.46[b,a]	7.75 ± 0.33[b,a]	7.52 ± 0.35[b,a]
	3	6.74 ± 0.43[a,a]	7.50 ± 0.45[b,a]	8.09 ± 0.23[b,a]	7.63 ± 0.41[b,a]	7.68 ± 0.31[b,a]
	Commercial	6.71 ± 0.20[a,a]	7.71 ± 0.20[b,a]	8.13 ± 0.41[b,a]	7.59 ± 0.38[b,a]	7.35 ± 0.50[b,a]
	None	6.49 ± 0.15[a,a]	6.81 ± 0.20[b,b]	7.29 ± 0.31[b,b]	6.89 ± 0.11[b,b]	6.53 ± 0.10[a,b]
Mesophiles	1	6.53 ± 0.22[a,a]	7.41 ± 0.33[b,a]	7.85 ± 0.41[b,a]	7.53 ± 0.28[b,a]	7.35 ± 0.27[b,a]
	2	6.46 ± 0.18[a,a]	7.33 ± 0.43[b,a]	7.97 ± 0.36[b,a]	7.48 ± 0.20[b,a]	7.26 ± 0.41[b,a]
	3	6.50 ± 0.23[a,a]	7.39 ± 0.41[b,a]	8.06 ± 0.32[b,a]	7.65 ± 0.34[b,a]	7.41 ± 0.38[b,a]
	Commercial	6.37 ± 0.19[a,a]	7.21 ± 0.30[b,a]	7.79 ± 0.39[b,a]	7.42 ± 0.19[b,a]	6.21 ± 0.26[b,a]
	None	6.40 ± 0.20[a,a]	8.31 ± 0.29[b,b]	8.90 ± 0.40[b,b]	8.51 ± 0.40[b,b]	8.20 ± 0.32[b,b]
Staphylococcus spp.	1	<2.00[a,a]	<2.00[a,a]	<2.00[a,a]	<2.00[a,a]	<2.00[a,a]
	2	<2.00[a,a]	<2.00[a,a]	<2.00[a,a]	<2.00[a,a]	<2.00[a,a]
	3	<2.00[a,a]	<2.00[a,a]	<2.00[a,a]	<2.00[a,a]	<2.00[a,a]
	Commercial	<2.00[a,a]	2.82 ± 0.31[b,b]	2.50 ± 0.29[a,b]	3.33 ± 0.31[b,b]	4.22 ± 0.43[c,b]
	None	<2.00[a,a]	2.82 ± 0.31[b,b]	3.97 ± 0.43[c,c]	4.82 ± 0.28[d,c]	6.02 ± 0.51[e,c]

(Continued)

TABLE 13.2 (Continued)
Viable Cell Counts of Cheese Uncoated, and Coated with Antimicrobial Edible and Commercial Coatings

Microorganisms	Type of Coating Solution	Storage Time (days)				
		0	10	20	40	60
Pseudomonas spp.	1	<2.00[a,a]	<2.00[a,a]	<2.00[a,a]	<2.00[a,a]	<2.00[a,a]
	2	<2.00[a,a]	<2.00[a,a]	<2.00[a,a]	<2.00[a,a]	<2.00[a,a]
	3	<2.00[a,a]	<2.00[a,a]	<2.00[a,a]	<2.00[a,a]	<2.00[a,a]
	Commercial	<2.00[a,a]	<2.00[a,a]	<2.00[a,a]	<2.00[a,a]	2.12 ± 0.32[b,b]
	None	<2.00[a,a]	<2.00[a,a]	2.32 ± 0.12[b,b]	2.95 ± 0.33[c,b]	3.72 ± 0.21[d,c]
Enterobacteriaceae	1	<2.00[a,a]	<2.00[a,a]	<2.00[a,a]	<2.00[a,a]	<2.00[a,a]
	2	<2.00[a,a]	<2.00[a,a]	<2.00[a,a]	<2.00[a,a]	<2.00[a,a]
	3	<2.00[a,a]	<2.00[a,a]	<2.00[a,a]	<2.00[a,a]	<2.00[a,a]
	Commercial	<2.00[a,a]	<2.00[a,a]	2.87 ± 0.35[b,b]	3.61 ± 0.40[c,b]	4.42 ± 0.21[d,b]
	None	<2.00[a,a]	2.25 ± 0.21[b,b]	3.75 ± 0.33[c,c]	4.82 ± 0.21[d,c]	6.12 ± 0.41[e,c]
Yeasts and molds	1	<2.00[a,a]	<2.00[a,a]	<2.00[a,a]	<2.00[a,a]	<2.00[a,a]
	2	<2.00[a,a]	<2.00[a,a]	<2.00[a,a]	<2.00[a,a]	<2.00[a,a]
	3	<2.00[a,a]	<2.00[a,a]	<2.00[a,a]	<2.00[a,a]	<2.00[a,a]
	Commercial	<2.00[a,a]	<2.00[a,a]	<2.00[a,a]	<2.00[a,a]	<2.00[a,a]
	None	<2.00[a,a]	3.82 ± 0.24[b,b]	4.65 ± 0.43[c,b]	5.72 ± 0.33[d,b]	6.62 ± 0.53[e,b]

Note: Viable cell counts (average ± standard deviation), in log (CFU g^{-1}), of cheese coated with antimicrobial edible coatings containing solution 1 (i.e., 6 g L^{-1} lactic acid and 0.25 g L^{-1} natamycin), solution 2 (i.e., 20 g L^{-1} COS and 0.25 g L^{-1} natamycin) and solution 3 (i.e., 6 g L^{-1} lactic acid, 20 g L^{-1} COS, and 0.25 g L^{-1} natamycin) compared with cheese coated with commercial coating and uncoated cheese, along 60 days of storage at 10°C and 85% RH.

[a, b, c, d, e] Means within the same line and column ([a,a], respectively), for each microorganism, labeled with the same letter, do not statistically differ from each other ($p > 0.05$).

number of publications resulting therefrom. Hence, it is a matter of time before such edible solutions become commercially available and only the higher costs associated with their scale-up have probably prevented so. The high potential of these edible packages associated with consumer demand for high quality and natural food products, coupled with the environmental pressure for eco-friendly packaging materials, will lead both the scientific community and the packaging industry to bet stronger in this promising alternative to synthetic packaging materials.

13.8 BIOACTIVE AGENTS FOR ENHANCED PRESERVATION

In recent years, active packaging has gained increased interest in food packaging. It is an innovative food concept appearing in response to growing consumer's demands and stricter market tendencies. Moreover, this type of packaging is an example of innovation going beyond the traditional functions of a package. This technology is based on the incorporation of agents into packaging systems that can be released into the food product or the surrounding environment to extend shelf life and sustain quality, safety, and sensory food characteristics (Pereira de Abreu et al., 2012).

An active agent can be incorporated in the packaging material or on its surface, in multilayer structures or particular elements associated with the packaging itself (Appendini and Hotchkiss, 2002; Rooney, 2005). This type of packaging system could bring several advantages compared with direct addition (e.g., spraying or immersion) of active compounds to food: (i) control migration, to maintain the functional agent's concentration at the surface; (ii) avoid loss of effectiveness due to dilution below active concentration by food matrix components; and (iii) lower amount of active compounds required to provide a given effect (Appendini and Hotchkiss, 2002; Cagri et al., 2004).

The list of active agents that can be added is wide (e.g., bacteriocins, fungicides, natural extracts, organic acids, and enzymes) as well as the nature of materials into which they can be included—such as edible (e.g., polysaccharides and proteins) and nonedible (e.g., papers and plastics) materials (Bastarrachea et al., 2011; Cagri et al., 2004; Ponce et al., 2008). Incorporation of active compounds into edible films and coatings could have a beneficial or adverse effect upon their functional properties (i.e., physical, mechanical, and biochemical features). Hence, it is critical to ascertain the impact of compound concentration, stability, chemical structure, degree of dispersion, and its interaction with the supporting polymer (Suppakul et al., 2003). An increasing number of published studies dealing with antimicrobial and antioxidant properties of some compounds—that is, bacteriocins, enzymes, salts, and plant extracts (Campos et al., 2010; Ponce et al., 2008; Sadaka et al., 2014)—have unlocked new opportunities for application of these compounds in edible coatings/films, and consequently to cheese products.

13.8.1 Antimicrobials

The food industry has struggled to reduce microbial growth rate or prevent food spoilage toward the development of safer food products with longer shelf life. Postprocess handling and moisture condensation on the surface of refrigerated (packaged) foods are two causes of microbial contamination of packaged foods (Ayala-Zavala et al., 2008; Zagory, 1999).

Packages with antimicrobial activity based on active agent released from the surface of the packaging material without migration to the food itself provides a promising form of active system applicable to food processing (Quintavalla and Vicini, 2002). The antimicrobial role can be played often (i) adding antimicrobial compounds directly into the packaging material, (ii) immobilizing antimicrobials onto the packaging polymer surface, and/or (iii) using polymers that are inherently antimicrobial (e.g., chitosan) (Kuorwel et al., 2011). Features of different types of antimicrobials, as well as the possibility to apply them in food products, have been extensively reviewed (Campos et al., 2010; Dhall, 2013).

Cheese products, especially fresh and semi-hard cheeses, provide excellent growth media for a variety of microorganisms, and they accelerate changes in aroma, color, and texture of cheese resulting in increased risk of food borne illness and shelf life reduction. Therefore, incorporation of antimicrobial compounds in the packaging materials has a significant potential to improve food safety and quality and extending the shelf life of cheese products (Campos et al., 2010). Several antimicrobial packaging films should be able to provide antimicrobial activity during or after cheese processing and can exert antimicrobial activity upon remaining microorganisms. Antimicrobial films may also act as an additional cheese postprocessing safety measure (Martins et al., 2010).

Several antimicrobial agents, for example, natural and/or chemical antimicrobials, antimicrobial polymers, biotechnological products, and gases were approved as food additives and may be incorporated in packaging systems for eventual application to food products. However, some antimicrobial agents can be classified as generally recognized as safe (GRAS) by Food Drug Administration (FDA), but not by European Food Safety Authority (EFSA), once these organizations have their own guidelines for food regulation (Han, 2005; Suppakul et al., 2003). Among antimicrobials permitted by FDA are (i) enzymes, such as lysozyme; (ii) bacteriocins, such as nisin; (iii) essential oils (EOs), such as thymol and carvacrol; (iv) weak organic acids, such as acetic, benzoic, and lactic; (v) antibiotics, such as natamycin; (vi) polysaccharides, such as chitosan; and (vii) peptides, such as lactoferrin (Quintieri et al., 2012; Sadaka et al., 2014).

In recent years, several studies have focused on applications of edible films and coatings containing natural antimicrobial compounds to cheese products as a treatment to reduce the deleterious effects raised by microorganisms (Table 13.1).

Enzymes (e.g., lysozyme) could be suitable for incorporation into edible films. Lysozyme is a food grade antimicrobial enzyme (E1105) with bacteriostatic, bacteriolytic, and bactericidal activity, particularly against Gram-positive bacteria (Duan et al., 2007). Medeiros et al. (2013) applied a nanolaminate coating of alginate/lysozyme in "Coalho" cheese, and microbiological analyses thereof were carried out for 20 days (Medeiros et al., 2013). The mesophilic and psychrotropic microbial counts and the visual evaluation of fungal contamination were lower on coated cheese than on its uncoated counterpart.

Bacterocins are one the most popular antimicrobial agents incorporated in coatings/films. Nisin, a natural polypeptide produced by *Lactococcus lactis*, possesses antimicrobial activity against a broad spectrum of Gram-positive bacteria, and it can be added to edible films. Scannell et al. (2000) applied cellulose coatings containing nisin and lacticin to Cheddar cheese. Nisin-adsorbed bioactive inserts reduced levels of *Listeria innocua* by 2 log cycles, and *Staphylococcus aureus* by 1.5 log cycles in cheese (Scannell et al., 2000). Martins et al. (2010) used galactomannan coatings containing nisin on Ricotta cheese. Nisin-added galactomannan coatings reduced *Listeria monocytogenes* population on cheese at least during 14 days of storage at 4°C (Martins et al., 2010).

As previously stated, film-forming agents, such as chitosan, possess antimicrobial activity, which is associated to its positively charged amino group. However, its antimicrobial activity depends on several features, including degree of acetylation, molecular weight, presence of other additives or food components, and target microorganism. Several authors (Aider, 2010; Cerqueira et al., 2009a) demonstrated that chitosan coatings inhibited growth of molds in semi-hard cheese.

Essential oils are composed of aromatic and volatile oil extracts obtained from aromatic and medicinal plant extracts and show strong antimicrobial activities (Sánchez-González et al., 2011). The antimicrobial activity of EOs has been consistently linked to presence of hydroxyl groups in their phenolic constituents (Kuorwel et al., 2014). EOs have been shown to possess antifungal activity against several microorganisms associated with cheese. Balaguer et al. (2013) evaluated the effectiveness, against food-contaminating fungi (*Penicillium expansum* and *Aspergillus niger*), of films made from wheat gliadins with cinnamaldehyde (GRAS flavoring substance) incorporated and provided evidence for their applicability in the design of active food packaging systems for cheese foodstuffs (Balaguer et al., 2013). These authors found fungal growth in control packaged cheese after 16 days of storage; however, no fungi were observed after 26 days of storage at 4°C, when the product was packaged with the active film.

Natamycin, a tetraene polyene macrolide, is a natural antifungal agent produced by *Streptomyces natalensis*. This compound is active against nearly all molds and yeasts but has no effect on bacteria. Natamycin is usually applied as a postproduction surface treatment for cheese to suppress fungal growth in the product during storage (Kallinteri et al., 2013). Yildirim et al. (2006) evaluate the effectiveness of using a casein-coating containing natamycin to prevent mold growth on Kashar cheese (Yildirim et al., 2006); their results confirmed that mold growth on cheese was suppressed for 1 month. Fajardo et al. (2010) assess the effectiveness of applying chitosan coating having natamycin on semi-hard cheese (Fajardo et al., 2010). Their results showed that mold growth decreased 1.1 log (CFU g^{-1}) in comparison with control (i.e., cheese without coating) after 27 days of storage.

Organic acids (e.g., acetic, benzoic, sorbic, citric, and lactic) and their salts (e.g., sorbates, benzoates, and propionates) are used to inhibit microbial growth and increase shelf life of a variety of food products. Addition of organic acids and their salts to edible films has been proposed as a way of minimizing surface microbial contamination (Kuorwel et al., 2011; Valencia-Chamorro et al., 2011). Ramos et al. (2012c) studied the effectiveness of a whey protein-based film with lactic acid and COS as antimicrobial agents against molds and bacteria on semi-hard cheeses made from cows' milk (Ramos et al., 2012c). They concluded that this system is very effective against bacteria (i.e., *Staphylococcus* spp. and Enterobacteriaceae) on the cheese surface.

It is expected that research will step forward in the direction of cost-effective and highly active antimicrobial combinations. Recently, studies have focused on the effectiveness of alginate coating combined with organic acids—potassium sorbate, sodium benzoate, calcium lactate, and calcium ascorbate, upon quality of Mozzarella cheese; potassium sorbate and sodium benzoate exchanged microbial inhibition on Mozzarella cheese (Lucera et al., 2014).

13.8.2 Antioxidants

A food antioxidant may be defined as any substance capable of delaying, retarding, or preventing development of rancidity or other flavor deterioration in food due to oxidation (Gordon, 2001). Moreover, such molecules can safely interact with free radicals and terminate the chain reaction before vital molecules are damaged (Pelli and Lyly, 2003).

Biological systems are subjected to a constant oxidative stress by reactive oxygen species (ROS) under normal physiological conditions, namely as a result of respiration (Morrissey and O'Brien, 1998; Weisburger, 1999). Several types of antioxidants exist: (i) inhibitors of free-radical oxidation reactions, (ii) inhibitors interrupting propagation of autoxidation chain reaction, (iii) singlet oxygen quenchers, (iv) synergists of proper antioxidants, (v) reducing agents, (vi) metal chelators, and (vii) inhibitors of pro-oxidative enzymes (Choe and Min, 2009; Pokorný, 2007). Some of the enzymatic antioxidants that provide a protection against ROS are superoxide dismutase, glutathione peroxidase, and catalase, further to numerous nonenzymatic small molecules distributed widely in biological systems and capable of scavenging free radicals. These molecules included lutathione, tocopherol (vitamin E), vitamin C, β-carotene, and selenium (Lee et al., 2003).

The deterioration of some foods has been associated with the oxidation of lipids and formation of undesirable secondary lipid peroxidation products. Lipid oxidation by ROS, such as superoxide anion, hydroxyl radicals, and H_2O_2, also decreases the nutritional value of lipid-containing foods, affecting their safety and appearance (Brewer, 2011).

In the food industry, many synthetic commercial antioxidants, such as butylated hydroxytoluene, butylated hydroxyanisole, tertiary butylhydroquinone, and propyl gallate, have been used to retard oxidation and peroxidation processes. However, recent studies suggested limitation of synthetic antioxidant use, owing to their toxicity (Jayathilakan et al., 2007). Additionally, consumer's preferences entail natural sources of antioxidant compounds, such as fruits and vegetables (Gramza and Korczak, 2005). As a result, there is a growing interest toward characterization of natural antioxidants, for instance plant extracts (e.g., tea, herbs) (Krishnaiah et al., 2011), peptides (Samaranayaka and Li-Chan, 2011), and food by-products (Balasundram et al., 2006). Some of these natural antioxidants can be added directly to cheese products. For example, Olmedo et al. (2013) demonstrated that oregano and rosemary EOs had a protective effect against lipid oxidation and fermentation in flavored cheese prepared with cream cheese base (Olmedo et al., 2013). In another work, addition of catechin at concentrations up to 500 ppm increased the total phenolic content and antioxidant activity of low-fat hard cheese, throughout a 90-day ripening period at 8°C (Rashidinejad et al., 2013). Huvaere et al. (2011) found that presence of green tea extract effectively reduces formation of lipid-derived aldehydes in light-exposed, reduced-fat soft cheese, possibly resulting in improved flavor stability (Huvaere et al., 2011).

The antioxidants content could decrease during storage, due to diffusion of the bioactive compounds and subsequent evaporation at the surface. To prevent such a decrease, edible coatings can be utilized as encapsulating matrices for many antioxidant bioactive compounds (Quirós-Sauceda et al., 2014). Antioxidants application on foods is, however, limited due to their impact on organoleptic food properties (e.g., EOs have a strong flavor) and their variable activity in foods due to interactions with food components (Burt, 2004).

Several efforts were made to test edible films and coatings with antioxidants for foodstuffs as fruits and vegetables, as protectors against the deleterious effects of oxygen, and via preventing lipid oxidation thus extending shelf life of food (Ponce et al., 2008; Sánchez-González et al., 2011). The use of antioxidants in films and coatings for preservation of cheese can therefore enhance stability of cheese against oxidation processes. However, information about their specific incorporation into edible films and coatings, and corresponding cheese application is rather limited. Ünalan et al. (2013) studied the incorporation of catechin and gallic acid into zein films on cold-stored (4°C) fresh Kashar cheese. Zein with phenolic compound-coated cheeses (50 mg g^{-1} coating solution) exhibited significantly lower lipid oxidation than uncoated samples over 35 days of storage (Ünalan et al., 2013). Hence, catechin and gallic acid mixtures are effective in preventing oxidative changes in cheese.

13.9 CONCLUDING REMARKS

Edible packaging has the potential to improve food quality and safety (i.e., increase shelf life expectancy, improve stock management, and reduce waste), thus addressing both retailers' and consumers' demands. These are the main justification why edible packaging systems (e.g., active packaging) are expected to play a key role in perishable food sectors, such as the dairy food industry. However, technical limitations and high costs associated with these technologies (for instance, expensive bioactive agents) have raised constraints to a more extensive implementation of edible packaging in the dairy industry.

In the area of active packaging, it is estimated that commercial applications of antimicrobial and antioxidant materials will grow so as to offer increased safety and extended shelf life. The use of active compounds (e.g., antimicrobials, antioxidants, and nutrients) derived from natural resources is also likely to continue growing, in addition to the incorporation of biodegradable packaging materials as carrier polymers. When active compounds are added to edible coatings and films, mechanical, sensory, and even functional properties can be dramatically affected. Studies on this topic are scarce, and more information is still required to develop novel edible packaging applications with improved functionality and performance.

Currently, nanotechnology can provide solutions and new functionalities in food packaging by increasing barrier properties (mechanical, chemical, and microbial), incorporate, and/or control release of active compounds, increase stability, and improve heat resistance (Neethirajan and Jayas, 2010). This technology offers higher opportunities in food packaging by extending shelf life, and leading to safer packaging and healthier food. It is expected that such nanostructures, incorporated in package materials, will perceive deterioration or contamination, and only after that will start releasing antimicrobials or antioxidants onto the food product (Cushen et al., 2012; Silvestre et al., 2011; Sung et al., 2013). Nanoencapsulation of active compounds (e.g., nisin) with edible materials may help control their release under specific conditions, thus protecting them from moisture, pH, or other physicochemical conditions—and enhancing their stability and activity (Imran et al., 2010). The use of nanolaminate packaging systems (two or more nanolayers of material)—in which the charged surfaces are coated with multiple nanolayers—also offers promising prospects (Weiss et al., 2006). These nanolaminate coatings could be applied to cheese products, as they are manufactured entirely from food-grade ingredients (e.g., proteins or polysaccharides), and could include such functional agents as antimicrobials, antioxidants, and enzymes (Medeiros et al., 2013; Weiss et al., 2006). Continued research and development efforts will contribute to a wider adoption of edible packaging in the dairy industry in the near future.

ACKNOWLEDGMENTS

Óscar Leandro da Silva Ramos, Ricardo Nuno Correia Pereira, and Joana T. Martins gratefully acknowledge their postdoctoral grants (SFRH/BPD/80766/2011, SFRH/BPD/81887/2011, and SFRH/BPD/89992/2012, respectively) from Fundação para a Ciência e Tecnologia (FCT, Portugal). All authors thank the FCT Strategic Project PEst-OE/EQB/LA0023/2013 and the Project "BioEnv—Biotechnology and Bioengineering for a sustainable world," REF. NORTE-07-0124-FEDER-000048, cofunded by Programa Operacional Regional do Norte (ON.2—O Novo Norte), QREN, FEDER.

REFERENCES

Aider, M. 2010. Chitosan application for active bio-based films production and potential in the food industry: Review. *LWT—Food Science and Technology, 43*(6), 837–842.

Altieri, C., Scrocco, C., Sinigaglia, M., and Del Nobile, M. A. 2005. Use of chitosan to prolong mozzarella cheese shelf life. *Journal of Dairy Science, 88*(8), 2683–2688.

Andrade, R. D., Skurtys, O., and Osorio, F. A. 2012. Atomizing spray systems for application of edible coatings. *Comprehensive Reviews in Food Science and Food Safety, 11*(3), 323–337.

Andreuccetti, C., Carvalho, R. A., and Grosso, C. R. F. 2009. Effect of hydrophobic plasticizers on functional properties of gelatin-based films. *Food Research International, 42*(8), 1113–1121.

Appendini, P., and Hotchkiss, J. H. 2002. Review of antimicrobial food packaging. *Innovative Food Science & Emerging Technologies, 3*, 113–126.

Ayala-Zavala, J. F., del-Toro-Sánchez, L., Alvarez-Parrilla, E., and González-Aguilar, G. A. 2008. High relative humidity in-Package of fresh-cut fruits and vegetables: Advantage or disadvantage considering microbiological problems and antimicrobial delivering systems? *Journal of Food Science, 73*(4), R41–R47.

Ayana, B., and Turhan, K. N. 2009. Use of antimicrobial methylcellulose films to control Staphylococcus aureus during storage of Kasar cheese. *Packaging Technology and Science, 22*(8), 461–469.

Balaguer, M. P., Lopez-Carballo, G., Catala, R., Gavara, R., and Hernandez-Munoz, P. 2013. Antifungal properties of gliadin films incorporating cinnamaldehyde and application in active food packaging of bread and cheese spread foodstuffs. *International Journal of Food Microbiology, 166*(3), 369–377.

Balasundram, N., Sundram, K., and Samman, S. 2006. Phenolic compounds in plants and agri-industrial by-products: Antioxidant activity, occurrence, and potential uses. *Food Chemistry, 99*(1), 191–203.

Bastarrachea, L., Dhawan, S., and Sablani, S. S. 2011. Engineering properties of polymeric-based antimicrobial films for food packaging: A review. *Food Engineering Reviews, 3*(2), 79–93.

Bekbölet, M. 1990. Light effects on food. *Journal of Food Protection, 53*(5), 430–440.

Brewer, M. S. 2011. Natural antioxidants: Sources, compounds, mechanisms of action, and potential applications. *Comprehensive Reviews in Food Science and Food Safety, 10*(4), 221–247.

Brody, A. L. 2005. Edible packaging. *Food Technology, 59*, 65–66.

Burt, S. 2004. Essential oils: Their antibacterial properties and potential applications in foods—A review. *International Journal of Food Microbiology, 94*(3), 223–253.

Cagri, A., Ustunol, Z., and Ryser, E. T. 2004. Antimicrobial edible films and coatings. *Journal of Food Protection, 67*(4), 833–848.

Campos, C. A., Gerschenson, L. N., and Flores, S. K. 2010. Development of edible films and coatings with antimicrobial activity. *Food and Bioprocess Technology, 4*(6), 849–875.

Cantor, M. D., van den Tempel, T., Hansen, T. K., and Ardö, Y. 2004. Blue cheese. In P. L. H. M. T. M. C. Patrick, F. Fox, and P. G. Timothy (Eds.), *Cheese: Chemistry, Physics and Microbiology* (Vol. 2, pp. 175–198). Amsterdam: Elsevier Academic Press.

Cao-Hoang, L., Chaine, A., Grégoire, L., and Waché, Y. 2010. Potential of nisin-incorporated sodium caseinate films to control Listeria in artificially contaminated cheese. *Food Microbiology, 27*(7), 940–944.

Cerqueira, M. A., Lima, A. M., Souza, B. W. S., Teixeira, J. A., Moreira, R. A., and Vicente, A. A. 2009a. Functional polysaccharides as edible coatings for cheese. *Journal of Agricultural and Food Chemistry, 57*, 1456–1462.

Cerqueira, M. A., Lima, Á. M., Teixeira, J. A., Moreira, R. A., and Vicente, A. A. 2009b. Suitability of novel galactomannans as edible coatings for tropical fruits. *Journal of Food Engineering, 94*(3–4), 372–378.

Cerqueira, M. A., Sousa-Gallagher, M. J., Macedo, I., Rodriguez-Aguilera, R., Souza, B. W. S., Teixeira, J. A., and Vicente, A. A. 2010a. Use of galactomannan edible coating application and storage temperature for prolonging shelf-life of "Regional" cheese. *Journal of Food Engineering, 97*(1), 87–94.

Cerqueira, M. A., Souza, B. W. S., Martins, J. T., and Vicente, A. A. 2010b. Improved hydrocolloid-based edible coatings/films systems for food applications. In A. Tiwari and F. Columbus (Eds.), *Polysaccharides: Development, Properties and Applications* (pp. 299–332). New York, NY: Nova Science Publishers.

Cerqueira, M. A., and Vicente, A. A. 2013. Shelf-life extension of cheese using edible packaging materials. In V. R. Preedy, R. R. Watson, and V. B. Patel (Eds.), *Handbook of Cheese in Health* (pp. 123–135). Wageningen, the Netherlands: Wageningen Academic Publishers.

Choe, E., and Min, D. B. 2009. Mechanisms of antioxidants in the oxidation of foods. *Comprehensive Reviews in Food Science and Food Safety, 8*, 345–358.

Cushen, M., Kerry, J., Morris, M., Cruz-Romero, M., and Cummins, E. 2012. Nanotechnologies in the food industry—Recent developments, risks and regulation. *Trends in Food Science & Technology, 24*(1), 30–46.

Dangaran, K., Tomasula, P. M., and Qi, P. 2009. Structure and function of protein-based edible films and coatings. In M. E. Embuscado and K. C. Huber (Eds.), *Edible Films and Coatings for Food Applications* (pp. 25–56). New York: Springer.

Debeaufort, F., Quezada-Gallo, J. A., and Voilley, A. 1998. Edible films and coatings: Tomorrow's packagings: A review. *Critical Reviews in Food Science and Nutrition, 38*(4), 299–313.

de Oliveira, T. M., de Fátima Ferreira Soares, N., Pereira, R. M., and de Freitas Fraga, K. 2007. Development and evaluation of antimicrobial natamycin-incorporated film in gorgonzola cheese conservation. *Packaging Technology and Science, 20*(2), 147–153.

Dhall, R. K. 2013. Advances in edible coatings for fresh fruits and vegetables: A Review. *Critical Reviews in Food Science and Nutrition, 53*(5), 435–450.

Di Pierro, P., Sorrentino, A., Mariniello, L., Giosafatto, C. V. L., and Porta, R. 2011. Chitosan/whey protein film as active coating to extend Ricotta cheese shelf-life. *LWT—Food Science and Technology, 44*(10), 2324–2327.

Duan, J., Park, S. I., Daeschel, M. A., and Zhao, Y. 2007. Antimicrobial chitosan-lysozyme (CL) films and coatings for enhancing microbial safety of Mozzarella cheese. *Journal of Food Science, 72*(9), M355–M362.

Fajardo, P., Martins, J. T., Fuciños, C., Pastrana, L., Teixeira, J. A., and Vicente, A. A. 2010. Evaluation of a chitosan-based edible film as carrier of natamycin to improve the storability of Saloio cheese. *Journal of Food Engineering, 101*(4), 349–356.

Falguera, V., Quintero, J. P., Jiménez, A., Muñoz, J. A., and Ibarz, A. 2011. Edible films and coatings: Structures, active functions and trends in their use. *Trends in Food Science & Technology, 22*, 292–303.

Fox, P. F., and McSweeney, P. L. H. 2004. Cheese: An overview. In P. L. H. M. T. M. C. Patrick F. Fox, and P. G. Timothy (Eds.), *Cheese: Chemistry, Physics and Microbiology* (Vol. 1, pp. 1–18). Amsterdam: Elsevier Academic Press.

Gontard, N., and Guilbert, S. 1994. Bio-packaging: Technology and properties of edible and/or biodegradable material of agricultural origin. In M. Mathlouthi (Ed.), *Food Packaging and Preservation* (pp. 159–181). New York, NY: Springer.

Gordon, M. H. 2001. The development of oxidative rancidity in foods. In N. Y. J. Pokorny, M. Gordon (Ed.), *Antioxidants in Food—Practical Applications* (pp. 7–21). Boca Raton. FL: CRC Press.

Gramza, A., and Korczak, J. 2005. Tea constituents (*Camellia sinensis* L.) as antioxidants in lipid systems. *Trends in Food Science and Technology, 16*(8), 351–358.

Guinee, T. P., Carić, M., and Kaláb, M. 2004. Pasteurized processed cheese and substitute/imitation cheese products. In P. L. H. M. T. M. C. Patrick F. Fox and P. G. Timothy (Eds.), *Cheese: Chemistry, Physics and Microbiology* (Vol. 2, pp. 349–394). Amsterdam: Elsevier Academic Press.

Guinee, T. P., and O'Callaghan, D. J. 2010. Control and prediction of quality characteristics in the manufacture and ripening of cheese. *Technology of Cheesemaking* (pp. 260–329). Oxford, UK: Wiley-Blackwell.

Han, J. H. 2005. Antimicrobial packaging systems. In J. H. Han (Ed.), *Innovations in Food Packaging* (pp. 80–107). Oxford, UK: Elsevier Academic Press.

Henriques, M., Santos, G., Rodrigues, A., Gomes, D., Pereira, C., and Gil, M. 2013. Replacement of conventional cheese coatings by natural whey protein edible coatings with antimicrobial activity. *Journal of Hygienic Engineering and Design, 3*, 34–47.

Huvaere, K., Nielsen, J. H., Bakman, M., Hammershøj, M., Skibsted, L. H., Sørensen, J., Vognsen, L., and Dalsgaard, T. K. 2011. Antioxidant properties of green tea extract protect reduced fat soft cheese against oxidation induced by light exposure. *Journal of Agricultural and Food Chemistry, 59*(16), 8718–8723.

Imran, M., Revol-Junelles, A.-M., Martyn, A., Tehrany, E. A., Jacquot, M., Linder, M., and Desobry, S. 2010. Active food packaging evolution: Transformation from micro- to nanotechnology. *Critical Reviews in Food Science and Nutrition, 50*(9), 799–821.

Jayathilakan, K., Sharma, G., Radhakrishna, K., and Bawa, A. 2007. Antioxidant potential of synthetic and natural antioxidants and its effect on warmed-over-flavour in different species of meat. *Food Chemistry, 105*(3), 908–916.

Kallinteri, L. D., Kostoula, O. K., and Savvaidis, I. N. 2013. Efficacy of nisin and/or natamycin to improve the shelf-life of Galotyri cheese. *Food Microbiology, 36*(2), 176–181.

Kampf, N., and Nussinovitch, A. 2000. Hydrocolloid coating of cheeses. *Food Hydrocolloids, 14*(6), 531–537.

Kozempel, M., and Tomasula, P. M. 2004. Development of a continuous process to make casein films. *Journal of Agricultural and Food Chemistry, 52*(5), 1190–1195.

Krishnaiah, D., Sarbatly, R., and Nithyanandam, R. 2011. A review of the antioxidant potential of medicinal plant species. *Food and Bioproducts Processing, 89*(3), 217–233.

Kuorwel, K. K., Cran, M. J., Sonneveld, K., Miltz, J., and Bigger, S. W. 2011. Antimicrobial activity of biodegradable polysaccharide and protein-based films containing active agents. *Journal of Food Science, 76*(3), R90–R102.

Kuorwel, K. K., Cran, M. J., Sonneveld, K., Miltz, J., and Bigger, S. W. 2014. Evaluation of antifungal activity of antimicrobial agents on Cheddar cheese. *Packaging Technology and Science, 27*(1), 49–58.

Lafargue, D., Lourdin, D., and Doublier, J.-L. 2007. Film-forming properties of a modified starch/κ-carrageenan mixture in relation to its rheological behaviour. *Carbohydrate Polymers, 70*, 101–111.

Lee, J., Koo, N., and Min, D. B. 2003. Reactive oxygen species, aging, and antioxidative nutraceuticals. *Comprehensive Reviews in Food Science and Food Safety, 3*, 21–33.

Lima, Á. M., Cerqueira, M. A., Souza, B. W. S., Santos, E. C. M., Teixeira, J. A., Moreira, R. A., and Vicente, A. A. 2010. New edible coatings composed of galactomannans and collagen blends to improve the postharvest quality of fruits—Influence on fruits gas transfer rate. *Journal of Food Engineering, 97*(1), 101–109.

Lucera, A., Mastromatteo, M., Conte, A., Zambrini, A. V., Faccia, M., and Del Nobile, M. A. 2014. Effect of active coating on microbiological and sensory properties of fresh mozzarella cheese. *Food Packaging and Shelf Life, 1*(1), 25–29.

Maftoonazad, N., and Badii, F. 2009. Use of edible films and coatings to extend the shelf-life of food products. *Recent Patents on Food, Nutrition & Agriculture, 1*, 162–170.

Martins, J. T., Cerqueira, M. A., Souza, B. W. S., Carmo Avides, M. D., and Vicente, A. A. 2010. Shelf life extension of Ricotta cheese using coatings of galactomannans from nonconventional sources incorporating Nisin against *Listeria monocytogenes*. *Journal of Agricultural and Food Chemistry, 58*(3), 1884–1891.

Mathlouthi, M., de Leiris, J. P., and Seuvre, A. M. 1994. Package coating with hydrosorbent products and the shelf-life of cheeses. In M. Matlouthi (Ed.), *Food Packaging and Preservation* (pp. 100–122). New York, NY: Springer.

McSweeney, P. L. H., and Fox, P. F. 1993. Cheese: Methods of chemical analysis. In P. F. Fox (Ed.), *Cheese: Chemistry, Physics and Microbiology* (pp. 341–388). Amsterdam: Elsevier Academic Press.

Medeiros, B. G. S., Souza, M. P., Pinheiro, A. C., Bourbon, A. I., Cerqueira, M. A., Vicente, A. A., and Carneiro-da-Cunha, M. G. 2013. Physical characterisation of an alginate/lysozyme nano-laminate coating and its evaluation on "Coalho" cheese shelf life. *Food and Bioprocess Technology, 7*(4), 1088–1098.

Moreira, M. R., Pereda, M., Marcovich, N. E., and Roura, S. I. 2011. Antimicrobial effectiveness of bioactive packaging materials from edible chitosan and casein polymers: Assessment on carrot, cheese, and salami. *Journal of Food Science, 76*(1), M54–M63.

Morrissey, P. A., and O'Brien, N. M. 1998. Dietary antioxidants in health and disease. *International Dairy Journal, 8*, 463–472.

Neethirajan, S., and Jayas, D. S. 2010. Nanotechnology for the food and bioprocessing industries. *Food and Bioprocess Technology, 4*(1), 39–47.

Olmedo, R. H., Nepote, V., and Grosso, N. R. 2013. Preservation of sensory and chemical properties in flavoured cheese prepared with cream cheese base using oregano and rosemary essential oils. *LWT—Food Science and Technology, 53*(2), 409–417.

Ollé Resa, C. P., Gerschenson, L. N., and Jagus, R. J. 2014. Natamycin and nisin supported on starch edible films for controlling mixed culture growth on model systems and Port Salut cheese. *Food Control, 44*, 146–151.

Pantaleão, I., Pintado, M. M. E., and Poças, M. F. F. 2007. Evaluation of two packaging systems for regional cheese. *Food Chemistry, 102*(2), 481–487.

Pavlath, A. E., and Orts, W. 2009. Edible films and coatings: Why, what, and how? In K. C. Huber and M. E. Embuscado (Eds.), *Edible Films and Coatings for Food Applications* (pp. 57–112). New York, NY: Springer.

Pelli, K., and Lyly, M. 2003. *Antioxidants in the Diet: A Flair-Flow Europe Synthetic Report on EU-Sponsored Research on Antioxidants* (pp. 1–28). Finland: VTT Biotechnology.

Pereira de Abreu, D. A., Cruz, J. M., and Paseiro Losada, P. 2012. Active and intelligent packaging for the food industry. *Food Reviews International, 28*(2), 146–187.

PM Food and Dairy Consulting 2014. In P. Mikkelsen (Ed.), *World Cheese Market 2000–2020*. Denmark.

Poças, M. F., and Pintado, M. 2010. Packaging and the shelf life of cheese. In G. L. Robertson (Ed.), *Food Packaging and Shelf Life: A Pratical Guide*. Boca Raton. FL: CRC Press.

Pokorný, J. 2007. Are natural antioxidants better—And safer—Than synthetic antioxidants? *European Journal of Lipid Science and Technology, 109*(6), 629–642.

Ponce, A. G., Roura, S. I., del Valle, C. E., and Moreira, M. R. 2008. Antimicrobial and antioxidant activities of edible coatings enriched with natural plant extracts: *In vitro* and *in vivo* studies. *Postharvest Biology and Technology, 49*(2), 294–300.

Quintavalla, S., and Vicini, L. 2002. Antimicrobial food packaging in meat industry. *Meat Science, 62*, 373–380.

Quintieri, L., Caputo, L., Monaci, L., Deserio, D., Morea, M., and Baruzzi, F. 2012. Antimicrobial efficacy of pepsin-digested bovine lactoferrin on spoilage bacteria contaminating traditional Mozzarella cheese. *Food Microbiology, 31*(1), 64–71.

Quirós-Sauceda, A. E., Ayala-Zavala, J. F., Olivas, G. I., and González-Aguilar, G. A. 2014. Edible coatings as encapsulating matrices for bioactive compounds: A review. *Journal of Food Science and Technology, 51*(9), 1674–1685.

Ramos, O. L., Fernandes, J. C., Silva, S. I., Pintado, M. E., and Malcata, F. X. 2012a. Edible films and coatings from whey proteins: A review on formulation, and on mechanical and bioactive properties. *Critical Reviews in Food Science and Nutrition, 52*(6), 533–552.

Ramos, O. L., Pereira, J. O., Silva, S. I., Fernandes, J. C., Franco, M. I., Lopes-da-Silva, J. A., Pintado, M. E., and Malcata, F. X. 2012b. Evaluation of antimicrobial edible coatings from a whey protein isolate base to improve the shelf life of cheese. *Journal of Dairy Science, 95*(11), 6282–6292.

Ramos, Ó. L., Santos, A. C., Leão, M. V., Pereira, J. O., Silva, S. I., Fernandes, J. C., Franco, M. I., Pintado, M. E., and Malcata, F. X. 2012c. Antimicrobial activity of edible coatings prepared from whey protein isolate and formulated with various antimicrobial agents. *International Dairy Journal, 25*(2), 132–141.

Ramos, O. S., and Malcata, F. X. 2011. Edible and biodegradable packaging for food storage. In G. E. Cohen and C. M. Levin (Eds.), *Food Storage* (pp. 135–169). Hauppauge, NY: Nova Science Publishers.

Rashidinejad, A., Birch, E. J., Sun-Waterhouse, D., and Everett, D. W. 2013. Effects of catechin on the phenolic content and antioxidant properties of low-fat cheese. *International Journal of Food Science & Technology, 48*(12), 2448–2455.

Rezvani, E., Schleining, G., Sümen, G., and Taherian, A. R. 2013. Assessment of physical and mechanical properties of sodium caseinate and stearic acid based film-forming emulsions and edible films. *Journal of Food Engineering, 116*(2), 598–605.

Ribeiro, C., Vicente, A. A., Teixeira, J. A., and Miranda, C. 2007. Optimization of edible coating composition to retard strawberry fruit senescence. *Postharvest Biology and Technology, 44*(1), 63–70.

Robertson, G. L. 2006. Packaging of dairy products. In G. L. Robertson (Ed.), *Food Packaging: Principles and Practice* (pp. 400–415). Boca Raton, FL: CRC Press.

Robertson, G. L. 2012a. Edible, biobased and biodegradable food packaging materials. In G. L. Robertson (Ed.), *Food Packaging: Principles and Practice,* (3rd ed., pp. 49–90). Boca Raton, FL: CRC Press.

Robertson, G. L. 2012b. Packaging of dairy products. In G. L. Robertson (Ed.), *Food Packaging: Principles and Practice* (3rd ed., pp. 509–544). Boca Raton, FL: CRC Press.

Rooney, M. L. 2005. Introduction to active food packaging technologies. In J. H. Han (Ed.), *Innovations in Food Packaging* (pp. 63–79). Oxford, UK: Elsevier Academic Press.

Sadaka, F., Nguimjeu, C., Brachais, C.-H., Vroman, I., Tighzert, L., and Couvercelle, J.-P. 2013. WITHDRAWN: Review on antimicrobial packaging containing essential oils and their active biomolecules. *Innovative Food Science & Emerging Technologies, 20*, 350.

Samaranayaka, A. G. P., and Li-Chan, E. C. Y. 2011. Food-derived peptidic antioxidants: A review of their production, assessment, and potential applications. *Journal of Functional Foods, 3*(4), 229–254.

Sánchez-González, L., Vargas, M., González-Martínez, C., Chiralt, A., and Cháfer, M. 2011. Use of essential oils in bioactive edible coatings: A review. *Food Engineering Reviews, 3*(1), 1–16.

Scannell, A. G. M., Hill, C., Ross, R. P., Marx, S., Hartmeier, W., and Arendt, E. K. 2000. Development of bioactive food packaging materials using immobilised bacteriocins Lacticin 3147 and Nisaplin. *International Journal of Food Microbiology, 60*, 241–249.

Schneider, Y., Kluge, C., Weiß, U., and Rohm, H. 2010. Packaging materials and equipment. In B. A. Law and A. Y. Tamime (Eds.), *Technology of Cheesemaking* (2nd ed, pp. 413–439). Oxford, UK: Wiley-Blackwell.

Schulz-Collins, D., and Senge, B. 2004. Acid- and acid/rennet-curd cheeses part A: Quark, cream cheese and related varieties. In P. L. H. M. T. M. C. Patrick F. Fox and P. G. Timothy (Eds.), *Cheese: Chemistry, Physics and Microbiology* (Vol. 2, pp. 301–328). Amsterdam: Elsevier Academic Press.

Silvestre, C., Duraccio, D., and Cimmino, S. 2011. Food packaging based on polymer nanomaterials. *Progress in Polymer Science, 36*(12), 1766–1782.

Skurtys, O., Acevedo, C., Pedreschi, F., Enronoe, J., Osorio, F., and Aguilera, J. M. 2010. *Food Hydrocolloid Edible Films and Coatings*. Hauppauge, NY: Nova Science Publishers.

Spinnler, H. E., and Gripon, J. C. 2004. Surface mould-ripened cheeses. In P. L. H. M. T. M. C. Patrick F. Fox and P. G. Timothy (Eds.), *Cheese: Chemistry, Physics and Microbiology* (Vol. 2, pp. 157–174). Amsterdam: Elsevier Academic Press.

Sung, S.-Y., Sin, L. T., Tee, T.-T., Bee, S.-T., Rahmat, A. R., Rahman, W. A. W. A., Tan, A.-C., and Vikhraman, M. 2013. Antimicrobial agents for food packaging applications. *Trends in Food Science & Technology, 33*(2), 110–123.

Suppakul, P., Miltz, J., Sonneveld, K., and Bigger, S. W. 2003. Active packaging technologies with an emphasis on antimicrobial packaging and its applications. *Journal of Food Science, 68*(2), 408–420.

Tomasula, P. M. 2009. 23—Using dairy ingredients to produce edible films and biodegradable packaging materials. In M. Corredig (Ed.), *Dairy-Derived Ingredients* (pp. 589–624). Cambridge, UK: Woodhead Publishing.

Ünalan, İ. U., Arcan, I., Korel, F., and Yemenicioğlu, A. 2013. Application of active zein-based films with controlled release properties to control *Listeria monocytogenes* growth and lipid oxidation in fresh Kashar cheese. *Innovative Food Science & Emerging Technologies, 20*, 208–214.

Valencia-Chamorro, S. A., Palou, L., del Río, M. A., and Pérez-Gago, M. B. 2011. Antimicrobial edible films and coatings for fresh and minimally processed fruits and vegetables: A review. *Critical Reviews in Food Science and Nutrition, 51*(9), 872–900.

van den Berg, G., Meijer, W. C., Düsterhöft, E. M., and Smit, G. 2004. Gouda and related cheeses. In P. L. H. M. T. M. C. Patrick F. Fox and P. G. Timothy (Eds.), *Cheese: Chemistry, Physics and Microbiology* (Vol. 2, pp. 103–140). Amsterdam: Elsevier Academic Press.

Walther, B., Schmid, A., Sieber, R., and Wehrmuller, K. 2008. Cheese in nutrition and health. *Dairy Science & Technology, 88*(4–5), 389–405.

Weisburger, J. H. 1999. Mechanisms of action of antioxidants as exemplifed in vegetables, tomatoes and tea. *Food and Chemical Toxicology, 37*, 943–948.

Weiss, J., Takhistov, P., and McClements, D. J. 2006. Functional materials in food nanotechnology. *Journal of Food Science, 71*(9), R107–R116.

Yildirim, M., Guleç, F., Bayram, M., and Yildirim, Z. 2006. Properties of Kashar cheese coated with casein as a carrier of natamycin. *Italian Journal of Food Science, 18*(2), 127–138.

Zagory, D. 1999. Effects of post-processing handling and packaging on microbial populations. *Postharvest Biology and Technology, 15*, 313–321.

Zhonga, Y., Cavenderb, G., and Zhao, Y. 2014. Investigation of different coating application methods on the performance of edible coatings on Mozzarella cheese. *LWT—Food Science and Technology, 56*(1), 1–8.

14 Edible Coatings and Films for Meat, Poultry, and Fish

Eveline M. Nunes, Ana I. Silva, Claúdia B. Vieira, Men de Sá M. Souza Filho, Elisabeth M. Silva, and Bartolomeu W. Souza

CONTENTS

Abstract 413
14.1 Introduction 413
14.2 Edible Coatings for Fresh and Frozen Fish 415
    14.2.1 Incorporation of Antimicrobial and Antioxidant Substances 416
    14.2.2 Coating Frozen and Refrigerated Fish 421
14.3 Edible Coatings on Meats and Poultry 423
    14.3.1 Quality of Meat and Poultry 423
14.4 Future of Food Packaging for Monitoring Quality in Fish, Meat and Poultry 425
References 426

ABSTRACT

Today, interest in the development of edible films and coatings for foods has increased due to consumer demands for high-quality foods and environmental concerns over the disposal of nonrenewable food packaging materials. The potential applications of edible films to control water transfer and to improve food quality and shelf life have received increasing attention from researchers and industry. Currently, technologies utilizing edible coatings directly in food raised a significant interest by the food industry. Several of those technologies are now being used on a commercial scale for an extensive range of food products; in this chapter, the applications of edible coatings to food products such as meat, poultry, and fish are discussed in detail.

14.1 INTRODUCTION

Consumers around the world demand for food of high quality, without chemical preservatives, and with an extended shelf life (Lin and Zhao, 2007). In the case of produce, maintaining high quality of food for long periods of time is difficult, since meat, poultry, and fish are composed of living tissues that undergo major changes due to their high water content. Water is an important component of meat, poultry, and fish, and its presence supports microbial growth.

Maintaining the quality of meat, poultry, and fish needs more attention upon processing due to the various factors involved. Thus, fish and/or processed cut meat and poultry shelf life have become a very important issue among food scientists and technologists in recent years. Edible films and coatings are a good complement for preservation of meat, poultry, and fish, since they can provide protection by serving as barriers to moisture migration, and limiting diffusion of gases important in food deterioration, such as O_2 or CO_2. Edible films and coatings have also been studied as potential carriers of additives used to preserve or even improve food quality. They can also enhance quality and appearance of a food product by preventing flavor and aroma migration and by providing structural integrity. Up to now, edible or biodegradable films have been used as carriers of many functional ingredients. Such ingredients may include antioxidants, antimicrobial agents, flavors, spices, and colorants which improve the functionality of the packaging materials by adding novel or extra functions, including immobilization of bacterial cells with antimicrobial activity (Bonilla et al., 2014; Emiroglu et al., 2010; Gialamas et al., 2010).

In the last years, food and packaging industries joined efforts to reduce the amount of synthetic food packaging materials, once environmental issues became relevant issue to consumers. The materials used to produce edible films or coatings applied in meat, poultry, or seafood can be classified as biodegradable and renewable materials. A schematic diagram of the types of materials and respective applications is shown in Figure 14.1.

The development of edible films or coatings based in natural biopolymers (i.e., proteins, polysaccharides, and their derivatives) provides a potential alternative to nonbiodegradable packaging materials. Several biopolymers have been used to develop eco-friendly food packaging materials. A significant proportion of research on these films has been made using biopolymers from renewable sources, that is, products or by-products derived from agriculture or from agroindustries. Usually, films based in biopolymers are highly sensitive to environmental conditions (e.g., relative humidity) and generally present low mechanical resistance (Kanatt et al., 2012).

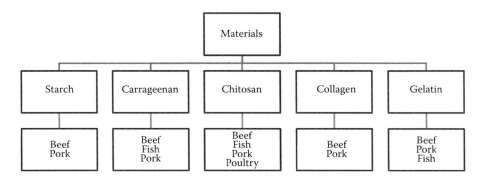

FIGURE 14.1 Materials used to produce edible films or coatings applied on meat, poultry, or seafood products.

14.2 EDIBLE COATINGS FOR FRESH AND FROZEN FISH

Seafood products are highly susceptible to quality deterioration caused by oxidation of their highly unsaturated fatty acids, catalyzed by the presence of high concentrations of hematin compounds and metal ions in the fish muscle (Decker and Hultin, 1992). Fish is an extremely perishable food compared with other fresh foodstuffs; quality is a complex concept involving a whole range of factors which for the consumer include, for example: freshness, safety, nutritional quality, availability, convenience and integrity, eating quality and the obvious physical attributes of the species, size, and product type. The state of freshness and quality of the end product are dependent on different biological and processing factors that influence the degree of various physical, chemical, biochemical and microbiological changes occurring *postmortem* in fish. Figure 14.2 depicts the relationship between quality and freshness, focusing on the different characteristics of freshness (Abbas et al., 2008).

Fish as a food is highly perishable, the quality and freshness of fish products are very affected by handling, processing, and storage methods. Moreover, seasonal condition, fishing grounds, and capture techniques affect the quality of this product. Storage techniques, including time/temperature histories, are a very important part of the production chain.

Cold storage and freezing are the normally employed methods for fish preservation. Development of frozen and processed fish products is hampered by the short shelf life of many seafood products. Freezing is a common conservation method used to control or decrease biochemical changes in fish that occur

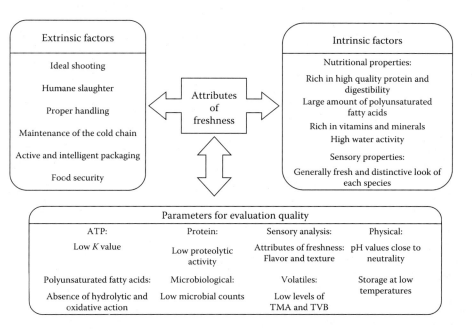

FIGURE 14.2 Relationship between quality and freshness, focusing on the different characteristics of freshness.

during storage. However, cold storage and freezing do not always completely suppress quality deterioration of seafood. Reactions leading to oxidative and enzymatic changes, and protein degradation may still proceed under chilled storage conditions (Sathivel, 2005).

Synthetic additives such as antioxidants, chelating agents, and antimicrobial compounds could be added to food products to improve their shelf life. Nevertheless, growing consumer interest in health and wellness is being reflected in increased demand for safer products. The preparation of any food product includes the addition of a number of ingredients that are not considered to be additives but clearly improve some properties of the food (e.g., maintaining quality, taste, flavor, or texture) and are originally intended as such. In this context, numerous studies are currently focused on using natural ingredients to enhance food quality and shelf life to avoid the use of synthetic preservatives. Biodegradable edible coatings applied on frozen foods can act as barriers to control moisture transfer and oxygen uptake (Debeaufort et al., 1998; Fan et al., 2009), aiming at improving overall food quality and prolonging storage life. Edible coatings have also been applied for preserving fresh fillets, and their preservative efficacy has reflected in reduced moisture loss, lipid oxidation, and growth of microorganisms in the tested fish (Souza et al., 2010).

14.2.1 INCORPORATION OF ANTIMICROBIAL AND ANTIOXIDANT SUBSTANCES

Over the past years, different changes have occurred in consumer demands for quality products, eating habits, and the retail food market. Consumers prefer foods that are minimally processed, contain fewer preservatives and additives, but maintain an unimpaired sensorial quality. In this sense, preservation of foods by application of technologies with minimized impact in the foods' quality can be an advantageous approach to solve many of the food spoilage-safety-related issues (Gialamas et al., 2010).

Active compounds and ingredients can be incorporated into packaging materials to provide several functions that do not exist in conventional packaging systems (Table 14.1). Active packaging may carry antioxidants, antimicrobial agents, and/or nutrients. Moreover, due to the health concerns of the consumers and environmental problems, current research in active packaging has focused on the use of natural preservatives in biodegradable packaging materials (Siripatrawan and Harte, 2010).

The compounds to be incorporated are of a different nature and may be limited to maintain quality and food safety or may even lead to a potential beneficial effect on consumer health. The compounds added more frequently are, among others, polyphenols from plants, chitosan, acids, antioxidant, and antimicrobial peptides (Lacey et al., 2012).

Chitosan has intrinsic antimicrobial activity and can still be used as a carrier for antimicrobials. The antimicrobial compounds incorporated into films and coatings can be gradually released from the polymeric matrix onto the surface of the food. Thus, the minimum inhibitory concentration needed for bacterial growth inhibition could be maintained for a longer time. Contrariwise, a direct incorporation of antimicrobials may be less effective due to rapid migration into the food bulk or to reaction with the food components (Iturriaga et al., 2012; Ye et al., 2008).

TABLE 14.1
Applications of Coatings, Films, and Active Compounds in Poultry, Meat, and Fish

Class	Food	Type	Material	Active Compound	References
Poultry	Cooked and uncooked	Film	Egg albumen	Fenugreek, rosemary, vitamin E	Armitage et al. (2002)
	Chicken breast	Coating	Chitosan	Oregano	Petro et al. (2012)
	Chicken breast	Coating	Whey protein isolate	Oregano or clove essential oils	Fernández-Pan et al. (2014)
	Chicken meat			Oregano essential oils	Chouliara et al. (2007)
Meat	Ground beef patties	—	—	Edible plants extracts	Kim et al. (2013)
	Ham	Film	Cellulose	Pediocin	Santiago-Silva et al. (2009)
	Pork	Coating	Pectin	Green tea extract	Kang et al. (2007)
	Roast beef	Coating	Chitosan	Acetic and lactic acids	Beverlya et al. (2008)
	Round beef steak	Film	Alginate	Nisin	Millette et al. (2007)
	Sausages	Film	Agar	Catechin	Ku et al. (2008)
Fish	Lingcod	Coating	Chitosan	Fish oil	Duan et al. (2010)
	Carp	Coating	Chitosan		Fan et al. (2009)
	Sardine fillets	Film	Ethylene vinyl alcohol Copolymer	Green tea extract, ascorbic acid, ferulic acid, quercetin	Dicastillo et al. (2012)
	Minimally processed fish products	Films	Gelatin, methyl cellulose, and their blend	Natural extracts	Iturriaga et al. (2012)
	Cod patties	Coating	Chitosan–gelatin blend		López-Caballero et al. (2005)
	Rainbow trout fillets	Coatings	Chitosan	Cinnamon oil	Ojagh et al. (2010)
	Fish fillet	Coating	Gluten, xanthan gum, wheat, and corn flours		Kilincceker et al. (2009)
	Salmon	Film	Chitosan	Nisin, sodium lactate, sodium diacetate potassium, sorbate, and sodium benzoate	Ye et al. (2008)
		Coating	Chitosan–tapioca starch	Potassium sorbate	Vásconez et al. (2009)
		Coating	Chitosan	—	Sathivel et al. (2007)
		Coatings	Chitosan	—	Souza et al. (2010)
		Film	Density polyethylene film	Barley husks	Pereira de Abreu et al. (2010)

Note: —, not found.

Fish flesh contains long-chain polyunsaturated fatty acids with a variety of health benefits. However, polyunsaturated fatty acids are highly susceptible to undergoing oxidation reactions and their composition varies largely between species. Antioxidants are added to food to delay the onset of oxidation or to slow down the rate of oxidation of the product. The most commonly used synthetic antioxidants in the food industry are butylated hydroxyanisole, butylated hydroxytoluene, and propylgalate (Pereira de Abreu et al., 2010; Siripatrawan and Harte, 2010). Although these synthetic antioxidants can effectively be used in food packaging because of high stability, low cost, and efficiency, there are significant concerns related to their toxicological aspects. Moreover, use of synthetic antioxidants is under strict regulation due to the potential health risk caused by such compounds (Siripatrawan and Harte, 2010), including the possibility of being carcinogenic.

Great interest has thus been shown regarding natural products because of the numerous benefits to health, such as the antioxidant effects of tea, rosemary, oregano, spices, herbs, clove, blueberries, mustard, red wine, and others (Pereira de Abreu et al., 2010). A promising trend in the last years involves incorporating natural extracts (essential oils [EOs], plant extracts, and their constituents) into packaging materials and into edible films and coatings for extending the shelf life of perishable and minimally processed foods as, for example, fish, meat, and fruits (Iturriaga et al., 2012). Plant extracts have received much attention as they contain high concentrations of phenolic compounds that possess strong antioxidant properties (Kanatt et al., 2012).

Naturally occurring antimicrobial compounds have good potential to be applied as food preservatives. EOs and other extracts from plants, herbs, spices, and some of their constituents have shown antimicrobial activity against different food pathogens and spoilage microorganisms. Extracts from rosemary, oregano, clove, thyme, and citrus fruit (e.g., lemon, orange, and grapefruit) are among the most studied natural antimicrobials for food applications (Burt, 2004; Corbo et al., 2008; Corrales et al., 2009; Cran et al., 2010; Del Nobile et al., 2009; Gutierrez et al., 2009). In fact, oregano and thyme have proven to be among the most active extracts, and citrus fruit extract has shown to effectively decrease the growth of specific fish spoilage bacteria (Iturriaga et al., 2012). It is well known that the antimicrobial activity of EOs is related to their chemical composition, mainly the phenolic components. However, the composition of EOs and the percentage of major components from a particular species of plant can differ depending on the harvesting season, geographical sources, and even on the function of a part from the same plant. For example, carvacrol can be found in quantities ranging from just traces to up to 80% in oregano EO. Therefore, results for antimicrobial activity are not always conclusive. EOs can be applied to the formulation of edible films, providing them with additional antioxidant and/or antimicrobial properties which can extend shelf life and reduce or inhibit food-borne pathogens (Gómez-Estaca et al., 2010). Incorporation of EOs into chitosan films or coatings, besides enhancing the film's antimicrobial and antioxidant properties, reduces water vapor permeability and slows down lipid oxidation of the product on which the film is applied (Ojagh et al., 2010). However, the use of EOs in foods could be limited because they would confer flavors and smells from those natural to the food in question, as in the case of fish (Gómez-Estaca et al., 2010).

Other extracts obtained from plants and fruits are composed of polyphenolic compounds such as flavonoids, which have also shown antimicrobial activity against a wide variety of microorganisms (Iturriaga et al., 2012). The beneficial effects of the phenolic compounds are the result from their ability to scavenge reactive oxygen and nitrogen species (Siripatrawan and Harte, 2010). The main advantage of EOs' application is their greater activity when compared with the effects of the individual active compounds, probably due to synergistic effects (Abdollahi et al., 2012).

Among the compounds added the most used are oils of clove (*Syzygium aromaticum* L.), fennel (*Foeniculum vulgare* Miller), cypress (*Cupressus sempervirens* L.), lavender (*Lavandula angustifolia*), thyme (*Thymus vulgaris* L.), herb of the cross (*Verbena officinalis* L.), pine (*Pinus sylvestris*), and rosemary (*Rosmarinus officinalis*) (Gómez-Estaca et al., 2010); oregano and rosemary extracts (Abdollahi et al., 2012; Gómez-Estaca et al., 2009); cinnamon oil (Ojagh et al., 2010); basil extracts (Suppakul et al., 2006); mint extract and pomegranate peel extract (Kanatt et al., 2012); extracts from barley husks (Pereira de Abreu et al., 2010, 2011a,b); lactic acid bacteria (LAB) (Chahad et al., 2012; Lacey et al., 2012; Gialamas et al., 2010); nanoclay (Abdollahi et al., 2012); fish oil (Duan et al., 2010); and generally recognized as safe (GRAS) antimicrobials (Ye et al., 2008).

Cinnamomum zeylanicum (L.), commonly known as cinnamon, is rich in cinnamaldehyde as well as β-caryophyllene, linalool, and other terpenes. Cinnamaldehyde is the major constituent of cinnamon leaf oil and provides the distinctive odor and flavor associated with cinnamon. It is used worldwide as a food additive and flavoring agent, and the Food and Drug Administration lists it as GRAS (Ojagh et al., 2010).

Basil (*Ocimum basilicum* L.) is a popular culinary herb and has been widely used as a food ingredient. Additionally, basil EOs and their principal constituents have been reported to be antimicrobial active against a wide spectrum of Gram-positive and Gram-negative bacteria, including important food-borne pathogens, multi-drug-resistant clinical bacteria, and yeasts and molds (Suppakul et al., 2006).

Green tea (*Camellia sinensis*), a nonfermented product, is a good source of polyphenolic compounds having strong antioxidant properties. The important polyphenolic compounds in tea leave include catechin, theaflavins, and thearubigins. Green tea catechins are proven to exhibit antimicrobial activity against some bacteria and have good antioxidant activity. Catechins, like quercetin, are bioflavonoids. They present a well-known antioxidant activity due to the catechol group, among other substituents. Green tea has been reported to delay the onset of lipid oxidation in various foods, including marine oil, soybean oil, and corn oil and dry-fermented sausage (López-De-Dicastillo et al., 2012; Siripatrawan and Harte, 2010). Recently, Cai et al. (2014) studied the effect of chitosan–ergothioneine coating (CHER) on the postmortem quality and shelf life of Japanese sea bass (*Lateolabrax japonicas*) stored at $4 \pm 1°C$ for 16 days. Four different treatments were used: control without coating, chitosan coating (CH), immersion in 0.3% ergothioneine (ER), and CHER with CH containing 0.3% ER coating. The results indicated that treatment with CHER coating inhibited increase in total volatile basic nitrogen, peroxide value, and thiobarbituric acid (TBA) value; maintained tissue hardness; and reduced microorganism counts, such as *Pseudomonas*, compared with control treatment. In addition, sea bass treated with CHER coating also exhibited a positive effect, causing low biogenic amine content, especially putrescine, cadaverine, and histamine. Sensory evaluation proved the efficacy of CHER coating by maintaining the overall quality of sea bass during the storage period (Cai et al., 2014).

Barley husk is a waste product obtained from the brewery industry. Chemical–biotechnological utilization of the different fractions provides added value to this residue. Lignin monomers and dimers have been observed to be active antioxidants. Extracts of phenolic compounds with high antioxidant activity can be obtained after prehydrolysis in acidic media followed by a delignification stage with NaOH. A suitable purification process leading to the removal of fractions with limited antioxidant activity enables use of lower amounts of natural extract to obtain the same effect as food antioxidant (Pereira de Abreu et al., 2011a).

Among alternative food preservation technologies, particular attention has been paid to biopreservation to extent the shelf life and to enhance the hygienic quality, minimizing the impact on the nutritional, and organoleptic properties of perishable food products such as seafood (Campos et al., 2012; Soomro et al., 2002). Biopreservation technology involves inoculating food with microorganisms, or their metabolites, selected for their antibacterial properties, which may be an efficient way of extending shelf life and food safety through the inhibition of spoilage and pathogenic bacteria without altering the nutritional quality of raw materials and food products. Inhibition of pathogenic microorganisms by LAB may be due to the effect of one or synergism between several mechanisms, such as competition for nutrients, lowering of pH, and production of lactic and acetic acids, hydrogen peroxide, and gas composition of atmosphere or production of antimicrobial substances such as bacteriocins (Gialamas et al., 2010). LAB have been historically linked with food fermentations, as acidification inhibits the growth of spoilage microorganisms. The antimicrobial activities of LAB have been known to play important roles in food fermentation and food preservation. In addition, lactic acid and other metabolic products produced by LAB contribute to the organoleptic and textural profile, as well as shelf life of the foods. The industrial importance of LAB is further evidenced by their GRAS status. Moreover, LAB have been observed which compete for nutrients or space with spoiling microorganisms through production of diacetyl and bacteriocins and should be of interest in marine fish and shellfish food biopreservation (Chahad et al., 2012). Among GRAS antimicrobials, a few can be cited due to their significance: nisin exerts rapid inhibitory effects against Gram-positive bacteria including *Listeria monocytogenes* in laboratory media or model food systems and in ready-to-eat (RTE) products including smoked salmon; sodium lactate (SL) is used as a flavor enhancer in meat and has also shown antilisterial effect in comminuted chicken, beef, cook-in bag roasts, and comminuted salmon model systems; sodium diacetate (SD) is approved as a direct ingredient for use in foods—at 0.1%–0.3%, SD can control growth of *L. monocytogenes* in meat; both potassium sorbate (PS) and sodium benzoate (SB) have been shown to inhibit growth of *L. monocytogenes* in artificial media, as well as in and on meat systems. The maximum legal limits for use in foods in the United States are 10,000 IU/g (0.025%) for nisin in RTE food, 4.8% for SL in meats, 0.25% for SD in meats, 0.1% for SB in foods, and 0.3% for PS in meat and fish products. The maximum limits allowed by the European Commission for these antimicrobials in foods are nisin (0.00125% or 500 IU/g for ripened cheese and processed cheese), PS (0.2%), SB (0.2%), while no upper limits are imposed for SL and SD (Ye et al., 2008).

14.2.2 Coating Frozen and Refrigerated Fish

Cold storage and freezing are the normally employed methods for fish preservation, but they do not completely inhibit biochemical reactions that lead to the quality deterioration of fish (Duan et al., 2010).

Fish generally spoil faster than do other muscle foods. Frozen storage is used to control or decrease biochemical changes in fish that occur during storage. Nevertheless, frozen storage does not completely inhibit microbial and chemical reactions that lead to quality deterioration of fish, as fish muscle is abundant in proteins and unsaturated fatty acids (Fan et al., 2009). Some authors attributed the process of lipid oxidation in frozen fish to the action of endogenous enzymes in each species. Endogenous fish enzymes may be active during frozen storage, even at −20°C. These enzymes may be influenced by a wide range of internal (enzyme content and composition) and external (intensity and feeding habits, temperature, and catching season) factors, which must be considered in the commercialization and storage of fish products (Pereira de Abreu et al., 2011a).

During frozen or cold storage, seafood products may develop surface drying and dehydration, which may lead to freezer burn, and may suffer from quality loss owing to oxidation or rancidity (Vanhaeckea et al., 2010). Although freezing is an effective method of preserving foods, some deterioration in frozen food quality occurs during storage. The final quality depends on the quality of the seafood at the time of freezing, as well as other factors during freezing, cold storage, and distribution, including the rate of freezing and thawing, storage temperature, temperature fluctuations, freeze–thaw abuse during storage, transportation, retail display, and consumption. Freezing and frozen storage can give a shelf life of more than 1 year if properly carried out. During frozen storage of shrimp and other shellfish products, the quality changes caused by oxidation, denaturation of proteins, sublimation, and recrystallization of ice crystals are predominant. These can result in off-flavors, rancidity, dehydration, weight loss, loss of juiciness, drip loss, and toughening, as well as microbial spoilage and autolysis (Gonçalves and Grindri Junior, 2009).

Oily fish, such as salmon, mackerel, and sardine, are rich in omega-3 fatty acids. As significant structural components of the phospholipids of cellular membranes, omega-3 fatty acids, particularly α-linolenic acid, eicosapentaenoic acid, and docosahexaenoic acid (DHA), are very important for human health. The inclusion of plentiful DHA in the diet not only prevents brain aging and Alzheimer disease but also improves the recovery from certain visual dysfunctions. American Heart Association recommends the intake of 1 g omega-3 fatty acids per day as a safe and effective way to obtain the heart health benefits (Duan et al., 2010).

Antioxidants are added to fish to delay the onset of oxidation or to slow down the rate of oxidation of the product (Pereira de Abreu et al., 2011b). Inhibition of the lipid oxidation process with antioxidant agents is of great importance in protecting foodstuffs that contain unsaturated fatty acids, from possible deterioration of quality.

In recent years, the food industry has used synthetic antioxidants. Nevertheless, the recent worldwide tendency to avoid or, at least, to decrease the use of synthetic additives has created a need to identify alternative, natural (possibly safer) sources of antioxidants for foods (Pereira de Abreu et al., 2011a).

Longer shelf life and better quality can be made possible by using different processing techniques such as freezing and appropriate combinations of these techniques. Several studies have shown that edible coatings made of protein, polysaccharide, and oil-containing materials help to prolong the shelf life and preserve the attributes of edible quality (Kilincceker et al., 2009).

CH has been well known for its good film-forming property, excellent compatibility with other substances, and broad antimicrobial activity against bacteria and fungi. The presence of a high density of amino groups and hydroxyl groups in its polymer structure makes CH a cationic polyelectrolyte, one of the few found in nature. CH possesses high positive charge on $-NH_3^+$ groups when dissolved in aqueous acidic solution, and therefore, it is able to adhere to or aggregate with negatively charged lipids and fats (Darmadji and Izumimoto, 1994; Gómez-Estaca et al., 2007; Kim and Thomas, 2007).

Lipid oxidation causes undesirable rancid off-flavors and potentially toxic products, which lead to the qualitative deterioration of fish. This is one of the major causes of quality deterioration in natural and processed foods. Oxidative deterioration is a large economic concern in the food industry because it affects many quality parameters such as flavor (rancidity), color, texture, and the nutritive value of foods. The free radical chain reactions in lipid oxidation are initiated by the attacks of oxygen to the double bond in fatty acids (Azhar and Nisa, 2006). Having good oxygen barrier properties, CH coatings applied on the surface of fish may act as a barrier between the fillet and its surroundings, thus slowing down the diffusion of oxygen from the surrounding to the surface of fillet and retarding lipid oxidation (Fan et al., 2009; Hershko et al., 1996). In addition, CH may reduce lipid oxidation by chelating ferrous ions present in fish proteins, thus eliminating their prooxidant activity or their conversion to ferric ion. The ability of CH to combine with lipid also plays a role in its antioxidative activity (Weist and Karel, 1992). Freezing temperatures do not prevent lipid oxidation in fish flesh. CH coatings might be sufficient to protect fish from lipid oxidation during the storage in refrigeration temperature (Duan et al., 2010).

Drip loss, during thawing, implies nutrient loss. Little drip loss occurs when the products are frozen quickly and stored properly, but if not, excessive drip loss can occur and render making the products unfit for consumption (Turan et al., 2003). Drip loss in frozen fish fillets is a complicated process, which may be caused by myosin aggregation during frozen storage, thus leading to muscle toughening and a loss in water-holding capacity. It has been reported that the usage of polyphosphate dips increases water-holding capacity of flesh and reduces drip and deterioration of the quality (Pigott and Tucker, 1990). In fact, CH coatings applied on the surface of fish might not prevent the loss in water-holding capacity of fish muscle during frozen storage, but they might reabsorb and hold the water expelled from the muscle during thawing and cause the reduction in drip loss (Duan et al., 2010). The most important advantages of this treatment are increase water-holding capacity, reduction in drip losses, nutrient retention, and improve texture and tenderness.

Successful inhibition of lipid oxidation and microbial growth in refrigerated rainbow trout fillet was possible with the application of an edible coating (CH and cinnamon oil), as together they kept the sensory characteristics within acceptable limits throughout storage. CH and cinnamon oil treatment could maintain trout fillet shelf life till the end of the storage period (day 16) without any significant loss of texture, odor, color, or overall acceptability and without significant microbial growth; showing that this type of active coating can be used as a safe preservative for fish under refrigerated storage (Ojagh et al., 2010).

Souza et al. (2010) evaluated the effect of CH coating on shelf life extension of fresh salmon (*Salmo salar*) fillets and showed that the success of edible coatings depends highly on their effective wetting capacity of the surfaces on which they are applied. In this context, solution with a spreading coefficient (Ws) of −4.73 mN m^{-1} was chosen to be subsequently analyzed and applied on fish fillets stored at 0°C for 18 days. These results demonstrated that fish samples coated with CH presented a significant reduction ($p < 0.05$) in the pH and K values after 6 days and in total volatile base nitrogen (TVB-N), trimethylamine (TMA), and TBA values after 9 days of storage, when compared with control samples. In terms of microbial growth, a slower increase was observed for coated fish, indicating that CH-based coatings were effective in extending for an additional 3 days the shelf life of salmon, not only helping in delaying the growth of microorganisms but also being an advantage in stabilizing chemical constituents, therefore reducing lipid oxidation (Souza et al., 2010).

14.3 EDIBLE COATINGS ON MEATS AND POULTRY

14.3.1 QUALITY OF MEAT AND POULTRY

Meat is the edible part of animals. Its chemical composition is, in descending order, water, protein, fat, other water-soluble organic materials, and water-soluble minerals. Meat proteins include sarcoplasmic, myofibrillar, and connective tissue proteins. Among the sarcoplasmic proteins are heme pigments and enzymes, which influence the color, smell, and structure of meat. Myofibrillar proteins and collagen are able to retain and hold water in the meat structure and to emulsify fat. Therefore, they influence the rheological properties of meat products. Mineral elements are present in enzymatic complexes and other structures that play an important biochemical role. They can affect the technological properties of meat, for example, water-holding capacity (Palka, 2007).

After slaughter, the various regulatory processes which prevent living meat from decomposing ceases to function. The most significant of these occurs as a result of circulatory failure, which causes the oxidation of muscle glycogen to cease and glycolysis to proceed. These quantities change significantly depending on type, age, sex, level of fattening, and part of animal carcass. The largest fluctuations are observed in the contents of water and lipids.

Lipid oxidation is a major cause of quality deterioration in meat and meat products because it leads to color alteration, off-flavor, and loss of nutrients, all of which are major determinants of meat quality (Kim et al., 2013). The color of poultry products is significantly affected by packaging because packaging can control endogenous and exogenous factors affecting the quality of products.

The increasing demand for meat and poultry products with an extended shelf life has intensified the search for new technologies using natural ingredients to enhance product quality and shelf life. In this context, edible films and coatings have received increased interest in recent years. Many authors have applied edible films and coatings on meat products, and several types of coatings and films have been tested in an attempt to maintain the quality of meat products, including fresh beef, pork, poultry, and also frozen products, such as beef and pork (Cagri et al., 2004; Cutter, 2006; Han, 2000; Oussalah et al., 2006; Stuchell and Krochta, 1995).

Emiroğlu et al. (2010) evaluated the antibacterial activity of thyme (*T. vulgaris*) and oregano (*Oreganum heracleoticum* L.) EOs incorporated in soy protein-based edible films against selected meat-associated bacteria and determined the effectiveness of these EO incorporated films in microbiological characteristics of fresh ground beef during refrigerated storage. Greater antimicrobial activities of soy edible films incorporated with EOs were demonstrated against *Staphylococcus aureus*, *Escherichia coli*, and *E. coli* O157:H7 (Emiroglu et al., 2010).

Zinoviadou et al. (2009) showed that 1.5% oregano EO incorporated in whey protein isolate (WPI) edible films resulted in 3.3 log reduction in the total viable count in fresh beef cuts when compared with the control at day 8 of refrigerated storage (Zinoviadou et al., 2009).

In a different type of application, Oussalah et al. (2004) reported that milk protein-based edible films containing 1.0% (w/v) oregano, 1.0% (w/v) chili, or a 1.0% oregano:chili (1:1) EOs mixture were applied on beef muscle slices to control the growth of pathogenic bacteria and increase the shelf life during storage at 4°C. Meat and film were periodically tested during 7 days for microbial and biochemical analysis; the results showed that oregano-based films stabilized lipid oxidation in beef muscle samples. Moreover, the films allowed a progressive release of phenolic compounds during storage (Oussalah et al., 2006).

In a recent study, Fernández-Pan et al. (2014) coated chicken breast with WPI edible coatings containing oregano or clove EOs as natural antimicrobials. During storage at 4°C for 13 days, the results showed that edible coatings containing 20 g kg^{-1} oregano EO increased the shelf life of chicken breast from 6 days (control) to 13 days, keeping total mesophilic aerobic, *Pseudomonas* spp. and LAB counts under the microbiological limits recommended for distribution and consumption. Those authors reported that the use of this kind of coatings would be considered as an emerging technology with the aim of extending the shelf life of refrigerated chicken breast (Fernández-Pan et al., 2014).

Armitage et al. (2002) evaluated the antioxidant activities of egg albumen coatings with natural antioxidants fenugreek, rosemary, and vitamin E in diced raw and diced cooked poultry breast meat. Coatings with added antioxidant exhibited the most significant effect against lipid oxidation in both raw and cooked samples (Armitage et al., 2002).

Petrou et al. (2012) examined the effect of natural materials, such as CH, oregano, and their combination, on the shelf life of modified atmosphere packaged chicken breast meat stored at 4°C. The results of this study indicated that the shelf life of chicken fillets can be extended using either oregano oil or CH, by approximately 6 days (samples treated with oregano oil 0.25% v/w, stored under modified atmosphere packaging [MAP]) and 15 days (samples treated with CH 1.5% w/v, stored under MAP) (Petrou et al., 2012).

Chouliara et al. (2007) studied the combined effect of oregano EO (0.1% and 1% w/w) and MAP (30% CO_2/70% N_2 and 70% CO_2/30% N_2) on shelf life extension of fresh chicken meat stored at 4°C. Based primarily on sensory data, results showed that shelf life of aerobically packaged fresh chicken meat was ca. 5 days. Addition of 0.1% oregano EO extended product shelf life by ca. 3–4 days, while MAP extended product shelf life by 2–3 days. The combination of both MAP and 0.1% oregano EO extended product shelf life by ca. 5–6 days. This extension is roughly equivalent to a 100% shelf-life extension (Chouliara et al., 2007).

Zinoviadou et al. (2009) investigated the potential of antimicrobial edible films based on WPI, incorporating different concentrations of oregano EO (0.5% and 1.5%) to extend the shelf life of refrigerated fresh beef. With coatings containing 1.5% of oregano, a significant reduction in total viable counts and *Pseudomonas* spp. during 12 days of storage was obtained. In addition, films containing 0.5% or 1% of oregano EO showed significant reductions against LAB but with 1.5% of oregano EO, the total inhibition was throughout the 12 days of storage (Zinoviadou et al., 2009).

In a different product, Siripatrawan and Noipha (2012) studied the possibility of using CH films incorporating green tea extract (CGT film) as active packaging for shelf life extension of pork sausages. It was found that samples wrapped with CGT film showed less changes in color, texture, TBA value, microbial growth, and sensory characteristics. Successful inhibition of lipid oxidation and microbial growth in the refrigerated pork sausages was also possible. The authors suggested that incorporation of green tea extract into CH films could enhance their antioxidant and antimicrobial properties and thus maintain the qualities and prolong the shelf life of the sausages (Siripatrawan and Noipha, 2012).

Bonila et al. (2014) studied edible films based in CH and prepared with and without basil or thyme EOs, with the aim of assessing their protective ability against lipid oxidation and their antimicrobial activity in pork meat products. The results showed that the incorporation of EOs increased O_2 permeability of the films when they have high moisture content. Further, CH films containing EOs effectively protected pork fat against lipid oxidation, which points to the specific chemical action of the oils' antioxidant compounds. The reduction in oxygen availability in coated minced meat led to expected changes in color associated with the conversion of myoglobin into metmyoglobin, which could decrease the products' acceptability. However, safety aspects were improved by CH films, increasing the product shelf life. On the contrary, the addition of EOs to CH films did not improve their antibacterial efficiency on meat products (Bonilla et al., 2014).

14.4 FUTURE OF FOOD PACKAGING FOR MONITORING QUALITY IN FISH, MEAT AND POULTRY

Recent improvements in the area of packaging for controlling quality in the fish, meat, and poultry industries include the introduction of active packaging, including antimicrobial films, the development of edible films, and coatings and the incorporation of nanotechnology or even biosensors.

In a review, Heising et al. (2014) discuss opportunities for monitoring directly or indirectly quality attributes of perishable packaged foods using intelligent packaging. Those authors explained that the information about the quality status of foods supplied by intelligent packaging can contribute substantially to the optimization of supply chain management. Expensive, highly perishable foods are the most important target group for intelligent packaging, because the intrinsic quality attributes of highly perishable foods change fast after processing and cause important economic losses. These packaging solutions will only be economically feasible, if the income from increased sales and/or reduced waste exceeds the increased costs of the package. When these issues have been tackled, intelligent packaging offers an enormous potential for commercial applications to improve supply product for consumers. In this perspective, the use of these packaging arise with a new promising tools for meat and fish preservation.

REFERENCES

Abbas, K. A., Mohamed, A., Jamilah, B., and Ebrahimian, M. 2008. A review on correlations between fish freshness and pH during cold storage. *American Journal of Biochemistry and Biotechnology*, 4(4), 416–421.

Abdollahi, M., Rezaei, M., and Farzi, G. A. 2012. Novel active bionanocomposite film incorporating rosemary essential oil and nanoclay into chitosan. *Journal of Food Engineering*, 111, 343–350.

Armitage, D. B., Hetiarachchy, N. S., and Monsoor, M. A. 2002. Natural antioxidants as a component of an egg albumin film in the reduction of lipid oxidation in cooked and uncooked poultry. *Journal of Food Science*, 67, 631–638.

Azhar, K. F., and Nisa, K. 2006. Lipids and their oxidation in seafood. *Journal of Chemical Society Pakistan*, 28(3), 298–305.

Bonilla, J., Vargas, M., Atarés, L., and Chiralt, A. 2014. Effect of chitosan essential oil film on the storage-keeping quality of pork meat products. *Food and Bioprocess Technology*, 7, 2443–2450.

Burt, S. 2004. Essential oils: Their antibacterial properties and potential applications in foods—A review. *International Journal of Food Microbiology*, 94, 223–253.

Cagri, A., Ustunol, Z., and Ryser, E. T. 2004. Antimicrobial edible films and coating. *Journal Food Protection*, 67(40), 833–848.

Cai, L., Li, X., Wu, X., Lv, Y., Liu, X., and Li, J. 2014. Effect of chitosan coating enriched with ergothioneine on quality changes of Japanese Sea Bass (*Lateolabrax japonicas*). *Food and Bioprocess Technology*, 7(8), 2281–2290.

Campos, A., Castro, P., Aubourg, S. P., and Velázquez, J. B. 2012. Use of natural preservatives in seafood. In A. McElhatton and P. J. do Amaral Sobral (Eds.), *Novel Technologies in Food Science, Integrating Food Science and Engineering Knowledge Into the Food Chain* (pp. 325–360). Santiago: Springer Science+Business Media.

Casariego, A., Souza, B. W. S., Cerqueira, M. A., Teixeira, J. A., Cruz, L., Díaz, R., and Vicente, A. A. 2009. Chitosan/clay films' properties as affected by biopolymer and clay micro/nanoparticles' concentrations. *Food Hydrocolloids*, 23, 1895–1902.

Chahad, O. B., Bour, M. E., Calo-Mata, P., Boudabous, A., and Barros-Velàzquez, J. 2012. Discovery of novel biopreservation agents with inhibitory effects on growth of foodborne pathogens and their application to seafood products. *Research in Microbiology*, 163, 44–54.

Chouliara, E., Badeka, A., Savvaidis, I. N., and Kontominas, M. G. 2007. Combined effect of oregano essential oil and modified atmosphere packaging on shelf-life extension of fresh chicken breast meat stored at 4°C. *Food Microbiology*, 24, 607–617.

Corbo, M. R., Speranza, B., Filippone, A., Granatiero, S., Conte, A., Sinigaglia, M., and Del Nobile, M. A. 2008. Study on the synergic effect of natural compounds on the microbial quality decay of packed fish hamburger. *International Journal of Food Microbiology*, 127, 261–267.

Corrales, M., Han, J., and Tauscher, B. 2009. Antimicrobial properties of grape seed extracts and their effectiveness after incorporation into pea starch films. *International Journal of Food Science and Technology*, 44, 425–433.

Cran, M. J., Rupika, L. A. S., Sonneveld, K., Miltz, J., and Bigger, S. W. 2010. Release of naturally derived antimicrobial agents from LDPE films. *Journal of Food Science*, 75, 126–133.

Cutter, C. 2006. Opportunities for bio-based packaging technologies to improve the quality and safety of fresh and further processed muscle foods. *Meat Science*, 74, 131–142.

Darmadji, P., and Izumimoto, M. 1994. Effect of chitosan in meat preservation. *Meat Science*, 38(2), 243–254.

Debeaufort, F., Quezada-Gallo, J. A., and Voilley, A. 1998. Edible films and coatings: Tomorrows packaginfs—A review. *Critical Reviews in Food Science and Nutrition*, 38, 299–313.

Decker, E. A., and Hultin, H. 1992. Lipid oxidation in muscle foods via redox iron. In Allen J. St. A. (Ed.), *Lipid Oxidation in Food* (pp. 33–54). Washington, DC: American Chemical Society.

Del Nobile, M. A., Conte, A., Bounocore, G. G., Incotonato, A. L., Massaro, A., and Panza, O. 2009. Active packaging by extrusion processing of recyclable and biodegradable polymers. *Journal of Food Engineering*, 93, 1–6.

Duan, J., Charian, G., and Zhao, Y. 2010. Quality enhancement in fresh and frozen lingcod (*Ophiodon elongates*) fillets by employment of fish oil incorporated chitosan coatings. *Food Chemistry*, 119, 524–532.

Emiroglu, Z. K., Yemis, G. P., Coskun, B. K., and Candogan, K. 2010. Antimicrobial activity of soy edible films incorporated with thyme and oregano essential oils on fresh ground beef patties. *Meat Science*, 86, 283–288.

Fan, W., Sun, J., Chen, Y., Qiu, J., Zhang, Y., and Chi, Y. 2009. Effects of chitosan coating on quality and shelf life of silver carp during frozen storage. *Food Chemistry*, 115, 66–70.

Fernández-Pan, I., Carrión-Granda, C., and Maté, J. I. 2014. Antimicrobial efficiency of edible coatings in the preservation of chicken breast fillets. *Food Control*, 36, 69–75.

Gialamas, H., Zinoviadou, K. G., Biliaderis, C. G., and Koutsoumanis, K. P. 2010. Development of a novel bioactive packaging based on the incorporation of *Lactobacillus sakei* into sodium-caseinate films for controlling *Listeria monocytogenes* in foods. *Food Research International*, 43, 2402–2408.

Gómez-Estaca, J., Lacey, A. L., López-Caballero, M. E., Gómez-Guillén, M. C., and Montero, P. 2010. Biodegradable gelatina-chitosan films incorporated with essential oils as antimicrobial agents for fish preservation. *Food Microbiology*, 27, 889–896.

Gómez-Estaca, J., Montero, P., Giménez, B., and Gómez-Guillén, M. C. 2007. Effect of functional edible films and high pressure processing on microbial and oxidative spoilage in cold-smoked sardine (*Sardina pilchardus*). *Food Chemistry*, 105(2), 511–520.

Gonçalves, A. A., and Grindri Junior, C. S. G. 2009. The effect of glaze uptake on storage quality of frozen shrimp. *Journal of Food Engineering*, 90, 285–290.

Gutierrez, J., Barry-Ryan, C., and Bourke, P. 2009. Antimicrobial activity of plant essential oils using food model media: Efficacy, synergistic potential and interactions with food components. *Food Microbiology*, 26, 142–150.

Han, J. 2000. Antimicrobial food packaging. *Food Technology*, 54, 56–65.

Hershko, V., Klein, E., and Nussinovitch, A. 1996. Relationships between edible coatings and garlic skin. *Journal of Food Science*, 61(4), 769–777.

Iturriaga, L., Olabarrieta, I., and Marañón, I. M. 2012. Antimicrobial assays of natural extracts and their inhibitory effect against *Listeria innocua* and fish spoilage bacteria, after incorporation into biopolymer edible films. *International Journal of Food Microbiology*, 158, 58–64.

Kanatt, S. R., Rao, M. S., Chawla, S. P., and Sharma, A. 2012. Active chitosan–polyvinyl alcohol films with natural extracts. *Food Hydrocolloids*, 29, 290–297.

Kilincceker, O., Dogan, I. S., and Kucuknev, E. 2009. Effect of edible coatings in the quality of frozen fish fillets. *LWT—Food Science and Technology*, 42, 868–873.

Kim, K. W., and Thomas, R. L. 2007. Antioxidative activity of chitosans with varying molecular weights. *Food Chemistry*, 101(1), 308–313.

Kim, S. J., Min, S. C., Shin, H. J., Lee, Y. J., Reum, C. A., Kim, S. Y., and Han, J. 2013. Evaluation of the antioxidant activities and nutritional properties of ten edible plant extracts and their application to fresh ground beef. *Meat Science*, 93, 715–722.

Lacey, A. M. L., López-Caballero, M. E., Gómez-Estaca, J., Gómez-Guillén, M. C., and Montero, P. 2012. Functionality of *Lactobacillus acidophilus* and *Bifidobacterium bifidum* incorporated to edible coatings and films. *Innovative Food Science and Emerging Technologies*, 16, 277–282.

Lin, D., and Zhao, Y. 2007. Innovations in the development and application of edible coatings for fresh and minimally processed fruits and vegetables. *Comprehensive Reviews in Food Science and Food Safety*, 6, 60–75.

López-De-Dicastillo, C., Gómez-Estaca, J., Catalá, R., Gavara, R., and Hernández-Muñoz, P. 2012. Active antioxidant packaging films: Development and effect on lipid stability of brined sardines. *Food Chemistry*, 131, 1376–1384.

Ojagh, S. M., Rezaei, M., Razavi, S. H., and Hosseini, S. M. H. 2010. Effect of chitosan coatings enriched with cinnamon oil on the quality of refrigerated rainbow trout. *Food Chemistry*, 120, 193–198.

Oussalah, M., Caillet, S., Salmiéri, S., Saucier, L., and Lacroix, M. 2006. Antimicrobial effects of alginate-based film containing essential oils for the preservation of whole beef muscle. *Journal Food Protection*, 69, 2364–2369.

Palka, K. 2007. Chemical composition and structure of foods. In Z. E. Sikorski (Ed.), *Chemical and Functional Properties of Food Components* (p. 14). London: CRC Press—Taylor & Francis Group.

Pereira de Abreu, D. A., Losada, P. P., Maroto, J., and Cruz, J. M. 2010. Evaluation of the effectiveness of a new active packaging film containing natural antioxidants (from barley husks) that retard lipid damage in frozen Atlantic salmon (*Salmo salar* L.). *Food Research International*, 43, 1277–1282.

Pereira de Abreu, D. A., Losada, P. P., Maroto, J., and Cruz, J. M. 2011a. Lipid damage during frozen storage of Atlantic halibut (*Hippoglossus hippoglossus*) in active packaging film containing antioxidants. *Food Chemistry*, 126, 315–320.

Pereira de Abreu, D. A., Losada, P. P., Maroto, J., and Cruz, J. M. 2011b. Natural antioxidant active packaging film and its effect on lipid damage in frozen blue shark (*Prionace glauca*). *Innovative Food Science and Emerging Technologies*, 12, 50–55.

Petrou, S., Tsiraki, M., Giatrakou, V., and Savvaidis, I. N. 2012. Chitosan dipping or oregano oil treatments, singly or combined on modifed atmosphere packaged chicken breast meat. *International Journal of Food Microbiology*, 156, 264–271.

Pigott, G. M., and Tucker, B. W. 1990. Adding and removing heat. In M. Dekker (Ed.), *Seafood Effects of Technology on Nutrition* (p. 359). New York: Marcel Dekker.

Sathivel, S. 2005. Chitosan and protein coatings affect yield, moisture loss, and lipid oxidation of pink salmon (*Oncorhynchus gorbuscha*) fillets during frozen storage. *Journal of Food Engineering*, 83, 366–373.

Siripatrawan, U., and Harte, B. R. 2010. Physical properties and antioxidant activity of an active film from chitosan incorporated with green tea extract. *Food Hydrocolloids*, 24, 770–775.

Siripatrawan, U., and Noipha, S. 2012. Active film from chitosan incorporating green tea extract for shelf life extension of pork sausages. *Food Hydrocolloids*, 27(1), 102–108.

Solomakos, N., Govaris, A., Koidis, P., and Botsoglou, N. 2008. The antimicrobial effect of thyme essential oil, nisin ant their combination against *Escherichia coli* 0157:H7 in minced beef during refrigerated storage. *Meat Science*, 80, 159–166.

Soomro, A. H., Masud, T., and Anwarr, K. 2002. Role of lactic acid bacteria (LAB) in food preservation and human health—A review. *Pakistan Journal of Nutrition*, 1, 20–24.

Souza, B. W. S., Cerqueira, M. A., Ruiz, H. A., Martins, J. T., Casariego, A., Teixeira, J. A., and Vicente, A. A. 2010. Effect of chitosan-based coatings on the shelf life of salmon (*Salmo salar*). *Journal of Agricultural and Food Chemistry*, 58, 11456–11462.

Stuchell, Y. M., and Krochta, J. M. 1995. Edible coatings on frozen ding salmon: Effect of whey protein isolate and acetylated monoglycerides on moisture loss and lipid oxidation. *Journal of Food Science*, 60, 28–31.

Suppakul, P., Miltz, J., Sonneveld, K., and Bigger, S. W. 2006. Characterization of antimicrobial films containing basil extracts. *Packaging Technology and Science*, 19, 259–268.

Turan, H., Kaya, Y., and Erkoyuncu, I. 2003. Effects of glazing, packaging and phosphate treatments on drip loss in rainbow trout (*Oncorhynchus mykiss* W., 1792) during frozen storage. *Turkish Journal of Fisheries and Aquatic Sciences*, 3, 105–109.

Vanhaeckea, L., Verbeke, W., and Brabander, H. F. 2010. Glazing of frozen fish: Analytical and economic challenges. *Analytica Chimica Acta*, 672, 40–44.

Weist, J. L., and Karel, M. 1992. Development of a fluorescence sensor to monitor lipid oxidation. 1. Fluorescence-spectra of chitosan powder and polyamide powder after exposure to volatile lipid oxidation products. *Journal of Agricultural and Food Chemistry*, 40(7), 1158–1162.

Ye, M., Neetoo, H., and Chen, H. 2008. Effectiveness of chitosan-coated plastic films incorporating antimicrobial in inhibition of *Listeria monocytogenes* on cold-smoked salmon. *International Journal of Food Microbiology*, 127, 235–240.

Zinoviadou, K. G., Koutsoumanis, K. P., and Biliaderis, C. G. 2009. Physico-chemical properties of whey protein isolate films containing oregano oil and their antimicrobial action against spoilage flora of fresh beef. *Meat Science*, 82, 338–345.

Index

A

AA, *see* L-(+)-ascorbic acid (AA)
Abietic acids, 129
Acacia senegal (*A. senegal*), 129
Acacia tree gum, *see* Gum Arabic
Acetic acids, 104
  films, 304
Acetylated monoglycerides, 134
A/C PET, *see* Aminolyzed/charged polyethylene terephthalate (A/C PET)
Active agent, 401
Active coatings, 99
  antimicrobial packaging, 104
  antioxidant effect of protein-based coatings, 103
  forces in edible protein-based coatings, 102
  on proteins and application on food products, 102
Active ingredients, 14
  edible coatings to incorporating, 373–374
Active materials, 99
Active packaging, 289, 292–293, 416
  materials, 104
Additives, 375
Adsorbed multilayer films assembly
  concentration, 298
  factors affecting, 297
  hydration, 299
  hydrophobicity, 298
  ionic strength, 298
  pH, 297–298
Adsorption of solutes, 193
AFM, *see* Atomic force microscopy (AFM)
Agar, 22, 49
  annual productions of dried agar seaweeds, 50
  applications for, 50–51
  double helices, 51
  idealized molecular structure of agarose, 51
  processing of, 49
Agaricus bisporus, *see* Whole mushrooms (*Agaricus bisporus*)
Agarose, 50, 51
Aggregation, 86–88
AgNPs, *see* Silver NPs (AgNPs)
Albumen, 85
Albumins, 85
Alcaligenes faecalis var. *myxogenes*, 60
Aldehydes, 91
Alginate, 42, 246
Alginic acid process, 42
Alkali cellulose, 28
All-polysaccharide composite films, 70
Allergens, 131
Aloe vera-based coatings, 370
α-Casein, 84
α-Helix, 84
α-Lactalbumin, 85, 249
α-Tocopherol, 302
2-amino-2-deoxy-β-D-glucopyranose (GlcN), 246
Amino acids, 83
Aminolyzed/charged polyethylene terephthalate (A/C PET), 299
Anionic gums, 40
Anogeissus latifolia tree, 130
Antibrowning agents, 361–362
Antimicrobial agents, 362
Antimicrobial and antioxidant substances incorporation, 416
  active packaging, 416
  barley husk, 420
  fish flesh, 418
  food preservation technologies, 420–421
  phenolic compounds, 419
Antimicrobial edible films and coatings, 393–394
  antimicrobial compounds in, 249–253
  with antimicrobial properties using chitosan, 253–254
  for cheese and other foods, 267–273
  for fruits and vegetables, 262–267
  lipids and waxes, 249
  for meat, poultry, and fish, 258–262
  physical properties of materials forming, 244–245
  polysaccharides, 245–247
  preparation methods for food active packaging purposes, 254–255
  proteins, 247–249
  viable cell counts of cheese uncoated, and coated with, 399–400
Antimicrobial edible packaging, 244; *see also* Edible packaging; Functional edible packaging
  antimicrobial compounds used in edible films and coatings, 249–253
  films and coatings with antimicrobial properties using chitosan, 253–258
  food applications of antimicrobial edible films and coatings, 258–272
  physical properties of materials forming antimicrobial edible films and coatings, 244–249

431

Antimicrobials, 220–221, 225–227, 267, 401
 film-forming agents, 403
 food industry, 401–402
 organic acids, 403–404
 packages with, 402
 packaging, 104
Antioxidants, 220–221, 225–227, 404–405, 421–422
AOAC, see Association of Official Analytical Chemists (AOAC)
Apple-based edible films, 252
A priori, 183
Arabinogalactan, 130
Area under the curve method (AUC method), 227
Aromatic amino acids, 159
Arrhenius-type relationship, 198
Ascorbic acids, 104
L-(+)-ascorbic acid (AA), 56
Association of Official Analytical Chemists (AOAC), 131
Astragalus gummifer (*A. gummifer*), 130
Atomic force microscopy (AFM), 90, 309–310
AUC method, see Area under the curve method (AUC method)

B

Bacterial cellulose (BC), 304
Bacterocins, 402–403
Balsams, 129
Barley husk, 420
Basil (*Ocimum basilicum* L.), 419
BC, see Bacterial cellulose (BC)
Beeswax, 124, 127, 294
Benchmark packaging properties, 348–349
β-Casein, 84
β-Lactoglobulin (β-Lg), 85, 249
β-Sheets, 84
β(1,4)-linked 2-acetoamido-2-deoxy-β-D-glucopyranose (Glc-NAc), 246
Bio-based materials, 3
Bio-based polymers, 10
Bio-nanocomposites, 289, 303
 composite coating, 305
 food packaging applications, 306
 multiphase material, 304
 polymer nanotechnology, 303
Bioactive agents for enhanced preservation, 401
 antimicrobials, 401–404
 antioxidants, 404–405
Bioactive coating, 216
Bioactive molecules encapsulation, 301
 bio-nanocomposites, 303–305
 nanoemulsions, 302–303
 nanoparticles, 301–302
Biodegradability, 105

Biodegradable edible coatings, 416
Biodegradable films, 105–106
Biodegradable materials, 154, 320, 348
Biodegradation, 105
Biological systems, 404
Bioplastics, 2
Biopolymers, 154, 244, 320, 414
 biopolymer-based packaging materials, 154
 films, 244
Blends of chitosan, 255, 256
Botrytis infection, 337, 338
Branched gums, 23
Branched regions, see Hairy regions
Brushing method, 170, 391

C

Calcium
 alginate process, 43
 calcium-induced gelation, 246
 ions, 361
Camellia sinensis, see Green tea (*Camellia sinensis*)
Camembert-type cheeses, 388
Candelilla wax, 124, 128
Carbon dioxide (CO_2), 85
Carbon dioxide transmission rate (CTR), 347
Carboxymethylcellulose (CMC), 17, 41–42, 366
 coating composite films, 161
Carnauba wax, 124, 128
Carob bean gum, see Locust bean gum
Carrageenan, 51, 246
 coating composite films, 161
 commercial sources, 51
 idealized structures, 53
 mechanical properties, 54
 principles, 52
Casein, 84, 91
 casein–lipid edible coatings, 158
 micelles, 294
CAT, see Catechin (CAT)
Catechin (CAT), 227
Cellophane, 348
Cellulose, 245
Cellulose gum, see Carboxymethylcellulose (CMC)
Cellulose nanofibers (CNF), 304
CH, see Chitosan (CH)
Cheese, 267, 384, 385
 antimicrobial edible films and coatings for, 267–273, 393–394
 contamination, 268
 edible packaging, 396–401
 fresh cheeses, 388–389
 hard cheese, 386–387
 packaging requirements and types, 385

Index

pasteurized processed cheese and analogs, 389
soft cheese, 387–388
Chemical(s), 218
   crosslinking, 90–92
   degradation, 163–164
   modification of proteins, 90
CHER coating, see Chitosan–ergothioneine coating (CHER coating)
Chickpea albumin extract (CPAE), 225
Chitin, 47, 223, 245–246
Chitooligosaccharides (COSs), 397–398
Chitosan (CH), 47, 223, 244–246, 253, 299, 303, 416, 419
   antimicrobial edible films and coatings preparation methods, 254–255
   coating, 159, 373–374, 419, 423
   DD in, 47
   edible coatings, 370
   factors affecting antimicrobial properties, 253–254
   films and coatings with antimicrobial properties using, 253
   films with, 260
   HA/PC–CH films, 49
   idealized structure, 48
   performance improvement, 255
   physical properties improvement, 255–257
   plastic wrap, 48
   use of chitosan films to deliver antimicrobial compounds, 257–258
Chitosan–ergothioneine coating (CHER coating), 419
Chitosan–gelatin films, 260
Chitosan–protein blends, 256
Cinnamaldehyde, 166
Cinnamomum zeylanicum (L.), 49
Citric acids, 104
Classic spraying systems, 169
CLSM, see Confocal laser scanning microscopy (CLSM)
CMC, see Carboxymethyl cellulose (CMC)
CNF, see Cellulose nanofibers (CNF)
CO_2P, see CO_2 permeability (CO_2P)
CO_2 permeability (CO_2P), 268
Coadjuvants, 361–363
Coated nonaromatic milled rice, 229
Coating(s), 289, 391
   frozen, 421–423
   operation time, 94
Cocoa butter, 126
Cocoa cake, 126
Coconut milk, 126
Coconut oil, 125–126
Coconut pulp (*Cocos nucifera*), 125
Collagens, 84
Colletotrichum musae (C. musae), 369
Commensal bacteria, 104

Complex conformations, gum polymers with, 61
   pectin, 61–64
   pullulan, 64–67
Composite films, 67, 122, 123
   all-polysaccharide composite films, 70
   blend films, 69
   insoluble composite films, 70–71
   rapid melt film composite, 70
   slow melt film composite, 70
   tensile and puncture strength and dissolution time, 68–69
Composite food, 183
   moisture transfer regulation in, 186
Compostability, 105
Confocal laser scanning microscopy (CLSM), 99, 310
Contact angle, 307–308
Contrasted organoleptic profiles, 186
COSs, see Chitooligosaccharides (COSs)
Covalent bonds, 86
CPAE, see Chickpea albumin extract (CPAE)
Critical moisture content, 204
Cross-links formations, 162
Crude oil, 126
Crystallinity, 195
Crystallization, 162
CTR, see Carbon dioxide transmission rate (CTR)
Curdlan, 59
   Alcaligenes faecalis var. *myxogenes*, 60
   Curdlan I, 60, 61
   Curdlan II, 60
   film studies, 60–61
   idealized structure of curdlan I, 61
Cutting, 126
Cyclodextrins, 196

D

DA, see N-deacetylation (DA)
Dairy products, 384
   cheese, 384, 385–389
DCMC, see Dialdehyde carboxymethyl cellulose (DCMC)
DD, see Degree of deacetylation (DD)
Degree of deacetylation (DD), 47
Degree of polymerization (DP), 17
Degree of substitution (DS), 27
Denaturation, 86–88
Depth-sensing indentation (DSI), see Nanoindentation
Derivative thermogravimetric curves (DTG curves), 98
Detection system, 191–192
DHA, see Docosahexaenoic acid (DHA)
Di-glycerides, 133
Dialdehyde carboxymethyl cellulose (DCMC), 92
Dialdehydes, 92

Diethanolamine, 89
Diethylene glycol (DTG), 362
Differential scanning calorimetry (DSC), 96
Diffusing/diffusion, 198
    coefficient, 189, 192
    molecule influence, 197–198
    substance, 193
Diffusivity, *see* Diffusion—coefficient
Diffusivity determination experimental setup and numerical treatment of data for, 192–193
Dipping, 94, 169, 269, 356, 391
Dipping LbL method, 295–296
    advantages and disadvantages, 296
Discoloration, 165
Dissolvable packaging, 12–13
DLS, *see* Dynamic light scattering (DLS)
Docosahexaenoic acid (DHA), 421
DP, *see* Degree of polymerization (DP)
Drip loss, 422
Dry processing, 93, 170
    of conventional polymers, 170
    extrusion, 171–172
    thermoforming, 171
    thermopressing, 171
Dry thermoplastic process, 166
DS, *see* Degree of substitution (DS)
DSC, *see* Differential scanning calorimetry (DSC)
DTG, *see* Diethylene glycol (DTG)
DTG curves, *see* Derivative thermogravimetric curves (DTG curves)
Durability of conventional plastics, 81
Dynamic light scattering (DLS), 306

E

E, *see* Young's modulus (E)
Edible barriers, 224
Edible coatings and films, 2, 13–14, 94, 103, 122, 156, 158–159, 162, 166, 169–170, 182–183, 215, 217, 218, 233, 244, 289, 355, 356, 390, 414; *see also* Protein-based films and coatings
    antioxidants and antimicrobials, functionality with, 226
    applications and functions of lipid-based, 139–141
    approaches, 201
    based on natural waxes, 138
    for controlling mass transfer, 366–368
    coupling diffusion and reaction, 205–206
    dimensioning based on mathematical modeling, 199
    flavors as bioactive compounds, functionality with, 228
    food preservation, 123
    for foods, 413
    formulation, 222–223
    for fresh and frozen fish, 415–423
    fruits and vegetables requirements to, 365
    hydrocolloids, 123
    to improving gloss and shine on, 372–373
    to improving texture-related properties, 371–372
    to incorporating active ingredients, 373–374
    to inhibit microbial spoilage, 369–370
    for internal atmosphere modification, 368–369
    lipids, 123, 124–126
    main functions of, 183
    materials, 414
    on meats and poultry, 423–425
    moisture barrier properties of hydrophilic films, 124
    optimization, 223
    other specific bioactive compounds, functionality with, 232
    permeation coupled with reaction, 204–205
    permeation in coating/film, 202–204
    physical properties of, 363–365
    predicting evolution of moisture content, 200, 201
    probiotics as bioactive compounds, functionality with, 228
    protein-based, 223–224
    to reducing oxidative reactions, 370–371
    resins, 123, 129–131
    to suppress physiological disorders, 369–370
    unsteady-state transfer from film into food, 205
    waxes, 123, 126–128
Edible gum film, 23
    casting, 24
    lab coater/dryer, 25
    method of manufacture and properties of, 23–25
    simple lab process for making gum films, 24
Edible packaging, 1, 23, 154, 216, 389; *see also* Functional edible packaging; Antimicrobial edible packaging
    advantages and limitations of functional compounds, 219–222
    antimicrobial edible films and coatings, 393–394
    bio-based materials, 3
    bioactive agents for enhanced preservation, 401–405
    cheese applications, 396–401
    companies with commercial edible packaging materials, 3
    edible bioactive coatings and films, 217
    edible films and coatings, 2, 390
    future perspectives, 5–6
    future trends, 233
    incorporating functional compounds, 218–219
    legislation and regulatory assessment, 4

Index

lipid-based packaging, 395
materials, 4, 392
polysaccharide-based packaging, 392
protein-based packaging, 392, 395
recent developments, features, and functional properties, 390–392
social, commercial, and scientific interest of, 2–3
structural matrix for functional compounds, 222–225
structuring agent-based packaging, 395
Edible packaging applications, 155
  fat barrier, 160
  hydrocolloids, 155
  intended applications, 157–158
  mechanical properties improvement of food, 160–161
  organic vapor barrier, 160
  oxygen barrier, 159
  potential applications, 156, 158
  properties, 156
  release systems of active compounds, 161
  sensory properties improvement, 161
  UV light barrier, 159–160
  water vapor barrier, 158–159
Edible packaging formulation, 355
  coadjuvants, 361–363
  edible coatings, 356
  functional ingredients, 361–363
  lipids-based materials, 361
  polysaccharide-based materials, 356, 361
  protein-based materials, 361
  wax-based materials, 361
Edible packaging stability, 162
  chemical degradation, 163–164
  enzymatic degradation, 164–165
  film matrix reorganizations, 162–163
  microbial growth, 165–166
EFSA, *see* European Food Safety Authority (EFSA)
"Egg-box" model, 246
Egg white, *see* Albumen
EHEC, *see* Enterohemorrhagic *E. coli* (EHEC)
Elaeis guineensis, *see* Palm fruit (*Elaeis guineensis*)
Electrokinetic potential, *see* Zeta potential
Electron microscopy, 308–309
Electrospinning, 321–322
  advantages, 323
  high barrier food packaging multilayer structures, 323–326
  low barrier food packaging multilayer structures, 326–329
  nanostructured layers produced by, 322
  processing technique, 322
Electrospraying, 374–375
"Electrostatic layer-by-layer self-assembly" technique, 295

advantages and disadvantages of LbL technique, 296–297
dipping LbL method, 295–296
spray LbL method, 296
Electrostatic spraying, 272
Endogenous fish enzymes, 421
Enrobing, 272
Enterohemorrhagic *E. coli* (EHEC), 258
Enzymatic browning, 165
Enzymatic degradation, 164–165
Enzymatic reactions, 199
Enzymes, 402
EOs, *see* Essential oils (EOs)
Epidermis, 263
Equilibrium MAP, 344
ER, *see* Ergothioneine (ER)
Ergothioneine (ER), 419
Essential oils (EOs), 104, 251, 402, 403, 418
  hydrophobicity, 104
ETG, *see* Ethylene glycol (ETG)
Ethylcellulose, 30–31
Ethyl chloride, 30
Ethylene glycol (ETG), 362
Ethylene vinyl alcohol (EVOH), 387
EU, *see* European Union (EU)
European Bioplastics, 10
European Food Safety Authority (EFSA), 123, 402
European Union (EU), 34, 336
EVOH, *see* Ethylene vinyl alcohol (EVOH)
Extended, fivefold helix structure, gum polymer with Xanthan gum, 56–59
Extended, sixfold helix structures, gum polymers with, 59
  curdlan, 59–61
Extended, threefold helix structures, gums polymers with, 49
  agar, 49–51
  Carrageenan, 51–54
  Gellan, 54–56
External parameters influence, 198–199
Extruder, 171
Extrusion, 97, 171–172

F

FAO, *see* Food and Agriculture Organization (FAO)
Fat barrier, 160
FDA, *see* Food and Drug Administration (FDA)
Fenugreek (*Trigonella foenum-graecum*), 31–32
Fenugreek gum, 32
Fibroblast growth factor-2 (FGF-2), 299
Fibrous proteins, 84, 247
Fick's laws, 189, 192
  first law, 189, 204
  second law, 189, 199

Film-forming
  agents, 403
  comparison of relative viscosities, 18–21
  factors, 15–17
  gum polymers interaction with water, 15
  HPC, 15
  hydrocolloids, 14
  materials, 267
  properties, 14, 22–23
  structural conformation, 15–16
  structural features, 17, 22
Films, 2, 295
  with chitosan, 260
  matrix reorganizations, 162–163
  organic vapor transfer, 160
  with pectin, 260, 261
  with proteins, 261–262
Films composition and structure, 194
  crystallinity, 195
  nanostructure voluntarily creation, 196
  plasticizers, 195
  posttreatments, 197
  raw material chemistry, 194–195
  solvents, 196
  structural defects, 196–197
Fish, antimicrobial edible films and coatings for, 258–262
Fish, fresh and frozen
  antimicrobial and antioxidant substances incorporation, 416, 418–421
  applications of coatings, films, and active compounds in, 417
  coating frozen, 421–423
  edible coatings for, 415
  freezing, 415–416
  future of food packaging for monitoring quality in, 425–426
  refrigerated fish, 421–423
  relationship between quality and freshness, 415
Flavors, 221, 227–229
Flocculation process, 27
Fluidized-bed processing, 272
Food
  bioplastics, 11
  coatings, 13–14
  degradation mechanism, 183–184
  industry, 401–402, 404
  preservation, 288
  preservation technologies, 420–421
  spoilage mechanism, 183–184
Food and Agriculture Organization (FAO), 129
Food and Drug Administration (FDA), 4, 13, 123, 402
Food packaging, 154, 155, 217
  active packaging, 292–293
  antimicrobial compounds used in edible films and coatings, 249–253
  applications of multilayers, 299–301
  applications of nanomaterials and, 293
  control of multilayer characteristics, 297
  "electrostatic layer-by-layer self-assembly" technique, 295–297
  factors affecting adsorbed multilayer films assembly, 297–299
  industry, 334
  innovations in, 215
  nanolaminates, 294–295
  nanotechnology in, 292
  naturally occurring nanomaterials, 293–294
  physical properties of materials forming antimicrobial edible films and coatings, 244–249
Food packaging multilayer structures
  high barrier, 323–326
  low barrier, 326–329
Formaldehyde, 91
Fourier transform InfraRed–attenuated total reflectance analysis (FTIR-ATR analysis), 192, 193
Fourier transform infrared spectroscopy (FTIR spectroscopy), 86
Fragaria ananassa, *see* Strawberries (*Fragaria ananassa*)
Freezing, 50, 415–416, 421
Fresh cheeses, 388–389
Frozen storage, 421
Fructose, 164
Fruits and vegetables, 335; *see also* Modified atmosphere packaging technology (MAP technology)
  additives to edible coating formulations, 359–360
  characteristics and preservation requirements for fresh and processed, 354–355
  compounds in edible coatings formulation, 357–358
  controlling mass transfer, edible coating for, 366–368
  critical quality parameters, 338
  edible packaging applications, 366
  edible packaging formulation, 355–363
  fresh producers in Ireland, 336
  fundamentals of edible packaging technology, 355
  improving gloss and shine, edible coatings to, 372–373
  improving texture-related properties, edible coatings to, 371–372
  incorporating active ingredients, edible coatings to, 373–374
  industrial implementation of edible packaging alternatives, 374–375
  inhibit microbial spoilage, edible coatings to, 369–370

Index

internal atmosphere modification, edible coating for, 368–369
level of acceptance, and safety issues, 338
optimal storage conditions, 339
physical properties of edible films and coatings, 363–365
product selection and market share, 335–338
quality and safety parameters for strawberry, 339
quantification of physiological properties, 340–343
reducing oxidative reactions, edible coatings to, 370–371
regulatory aspects, 375–376
requirements to edible coatings, 365
suppress physiological disorders, edible coatings to, 369–370
Fruits, antimicrobial edible films and coatings for, 262–267
Fruit waxing, 182
FTIR-ATR analysis, *see* Fourier transform InfraRed–attenuated total reflectance analysis (FTIR-ATR analysis)
FTIR spectroscopy, *see* Fourier transform infrared spectroscopy (FTIR spectroscopy)
Functional compounds, 215–217
advantages and limitations, 219–220
antioxidants and antimicrobials, 220–221
direct surface application, 217
disadvantages, 218
flavors, 221
incorporating, 218–219
probiotics, 221–222
structural matrix of edible packaging, 222–225
Functional edible packaging, 225; *see also* Antimicrobial edible packaging; Edible packaging
antioxidants and antimicrobials, 225–227
flavors, 227–229
other functional compounds, 230–231
probiotics, 229–230
Functional foods, 216
development, 217–218
Functional ingredients, 361–363
Fungi, 250

G

G-block polymer, 43, 44
GA-natamycin (GA-NA), 302
GA-poly(*N*-isopropylacrylamide) (GA-PNIPA), 302
GA, *see* Gallic acid (GA)
Galactomannans, 36, 247
Gallic acid (GA), 90, 227, 302

γ-Casein, 84
Gas exchange, excessive restriction of, 159
Gas transport properties, 268
Gelatin, 82, 84, 247–248
Gelation, 22
Gelidium amansii (*G. amansii*), 49, 50
Gellan, 22, 54
edible gellan films, 56
idealized structure, 55
manufacturing process for, 54
Gelling, 22–23
gums, 11
Generally recognized as safe (GRAS), 103, 123, 252, 293, 375, 402, 419
Ghatti gum, 130–131
Glc-NAc, *see* β(1,4)-linked 2-acetoamido-2-deoxy-β-D-glucopyranose (Glc-NAc)
GlcN, *see* 2-amino-2-deoxy-β-D-glucopyranose (GlcN)
Gliadins, 85
films, 250, 252
Globular proteins, 247, 294
Globulins, 85
Gloss and shine, edible coatings to improving, 372–373
Glucose, 164
Glutaraldehyde, 91, 92
crosslinked protein faster, 91
GLY, *see* Glycerol (GLY)
Glycerol (GLY), 362
glycerol-plasticized gelatin films, 92
Glyoxal-incorporated probes, 92
Gracilaria, 49
Gram-negative bacteria, 253
Gram positive bacteria, 253
Gram positive fungi, 253
GRAS, *see* Generally recognized as safe (GRAS)
Gravimetry, 191
Green tea (*Camellia sinensis*), 419
Group 2 bioplastics, 11
Guar gum, 32–34
Guar splits, 33
Guluronic acid, 43
Gum Arabic, 129–130
Gum polymers, 11, 13
with complex conformations, 61–67
with extended, fivefold helix structure, 56–59
with extended, sixfold helix structures, 59–61
with extended, threefold helix structures, 49–56
Gum polysaccharides
film-forming properties of, 14–17
gum polymers with complex conformations, 61–67
gum polymers with extended, sixfold helix structures, 59–61

Gum polysaccharides (*Continued*)
  gum polymer with extended, fivefold helix structure, 56–59
  gums polymers with extended, threefold helix structures, 49–56
  ionic gum polymers with extended, twofold helix structures, 40–49
  nonionic gum polymers with extended twofold helix structures, 25–40
  source, chemical structure, and film-forming properties of, 25
Gum tragacanth, 130

H

Hairy regions, 63
HA/PC–CH films, 49
Hard cheese, 386–387
Hard resins, 129
HDPE, *see* High-density polyethylene (HDPE)
Heat treatment, 96
Henry's law, 189
Hepatitis A, 250
High-density polyethylene (HDPE), 388
High-molecular weight glutenins, 85
High barrier food packaging multilayer structures, 323
  biopolyester–zein multilayers, 325
  electrospun protein interlayers, 323
  electrospun zein, 326
  multilayer systems, 323–324
  optical microscopy image of zein nanofibers, 325
  scanning electron microscope image of zein interlayers, 324
High set curdlan gel, 60
Homogalacturonans block, 62
HPC, *see* Hydroxypropyl cellulose (HPC)
HPMC, *see* Hydroxypropyl methylcellulose (HPMC)
Hydration, 299
Hydrocolloids, 14, 123, 155, 255
Hydrogen bonding, 87
Hydrophilic compounds, 89
Hydrophilic plasticizers, 134
Hydrophobic amino acid residues, 85
Hydroxypropyl cellulose (HPC), 15, 28–30
  miscibility, 30
Hydroxypropyl methylcellulose (HPMC), 28–30, 364
  coating, 169
  HPMC-based coatings, 366
  idealized twofold helix structure, 29
  mechanical strength and moisture barrier properties of, 29
Hylon VII solutions, 255

I

Initial moisture content, 204
Insoluble composite films, 70–71
Integrative MAP modeling, 346–347
Intelligent packaging, 292
Intermolecular hydrogen bonds, 16
Internal atmosphere modification, edible coating for, 368–369
International Union of Pure and Applied Chemistry (IUPAC), 129
Intramolecular hydrogen bonds, 16
Inulin, 24
Ionic charge on molecule, 17
Ionic gum polymers with extended, twofold helix structures, 40; *see also* Nonionic gum polymers with extended twofold helix structures
  chitosan, 47–49
  CMC, 41–42
  PGA, 45–47
  sodium alginate, 42–45
Ionic strength, 298
ι-Carrageenan, 52, 53
Irreversible intermolecular interactions, 86
Isoelectric point, 86, 94, 297
Isomeric carboxylic acids, 129
IUPAC, *see* International Union of Pure and Applied Chemistry (IUPAC)

J

Jojoba liquid wax, *see* Jojoba oil
Jojoba oil, 128
Jojoba shrub (*Simmondsia chinensis*), 128

K

κ-Carrageenan, 52, 53
κ-Casein, 84
Karaya gum, 130
KGM, *see* Konjac glucomannan (KGM)
Kibbling machines, 35–36
Knife coating, *see* Tape casting
Konjac glucomannan (KGM), 38, 40
Konjac gum, 38
  characteristics, 39
  idealized structure, 39
  potential disadvantage, 40
  process, 38
Kunststoffe International, 348

L

LAB, *see* Lactic acid bacteria (LAB)
Lab coater/dryer, 25
Lactic acid bacteria (LAB), 419

Index

Lactic acids, 104
Lactobacillus acidophilus (*L. acidophilus*), 44, 272
Lactoferrin, 252
Lactose, 251
Lateolabrax japonicas, *see* Sea bass (*Lateolabrax japonicas*)
Layer-by-layer (LbL), 295
LBG, *see* Locust bean gum (LBG)
LbL, *see* Layer-by-layer (LbL)
LC, *see* Lipid coating (LC)
LCA, *see* Life cycle assessment (LCA)
LDPE, *see* Low-density polyethylene (LDPE)
Legislation, 4
Life cycle assessment (LCA), 106
Light scattering, 306–307
Limburger-type cheeses, 388
Linear anionic gums, 17
Linear function of time, 192
Linear gums, 16
Lipid-based coatings, 372
Lipid-based materials, 185
Lipid-based packaging, 395
Lipid coating (LC), 29
Lipid oxidation, 422, 423
    products, 163
Lipids-based materials, 361
Lipids, 123, 124, 224, 249, 294, 367
    applications, 134, 138
    cocoa butter, 126
    coconut oil, 125–126
    in edible films, 124
    using in edible coatings and films, 131, 132
    effect of lipids in water vapor permeability, 135–137
    palm oil, 126
    plasticizer properties, 134
    properties, 132
    sunflower oil, 125
    water vapor barrier, 133–134
Liposomes, 303
Locust bean gum (LBG), 22, 35
    idealized structure, 37
    molecular weight, 36
    processing, 35–36
    WVP, 37, 38
Long-chain triglycerides, 133
Low-density polyethylene (LDPE), 90, 388
Low barrier food packaging multilayer structures, 326
    barrier and mechanical properties of WG multilayer structures, 328
    preliminary results, 329
    wheat gluten, 327
Low pH foods, 222
Low set curdlan gel, 60

M

M-block polymer, 44
Maillard reaction, 92, 163–164
    between proteins and sugars, 93
Maillard reaction products (MRPs), 92
Maleated PP (MAPP), 305
MAP, *see* Modified atmosphere packaging (MAP)
MAPP, *see* Maleated PP (MAPP)
Mass transfer
    actor to control mass transfer phenomena, 184–185
    assessment of edible films and coating performance and compatibility, 186–188
    food spoilage and degradation mechanisms, 183–184
    functions of edible films and coatings as regulating agent, 183
    group of researchers, 188
    materials, 185, 186
    moisture transfer regulation in composite food product, 186
    occurring in coating/food system, 184
Mass transfer
    basics on, 189–190
    detection system, 191–192
    diffusing molecule influence, 197–198
    diffusivity, 191
    edible coating applications for controlling, 366–368
    experimental setup and numerical treatment for effective diffusivity determination, 192–193
    external parameters influence, 198–199
    factors affecting mass transfer parameters, 194–199
    films composition and structure, 194–197
    mass transfer parameters measurement, 190–194
    parameters measurement, 190
    permeation, 190–191
    sorption, 193–194
Mastic gum, *see* Mastic resin
Mastic resin, 131
MB, *see* Methylene blue (MB)
MC, *see* Methyl cellulose (MC)
MCC, *see* Microcrystalline cellulose (MCC)
Meat
    antimicrobial edible films and coatings for, 258–262
    applications of coatings, films, and active compounds in, 417
    edible coatings on, 423
    future of food packaging for monitoring quality in, 425–426
    quality, 423–425

MEFs, *see* Moderate electric fields (MEFs)
Melanoidins, 92
Melt viscosity, 172
Mesophilic phase, 105
Mesquite gum, 130
Mesquite tree (*Prosopis* spp.), 130
Methyl cellulose (MC), 26, 245
    DS, 27
    flocculation process, 27
    molecular structure, 26
    tensile strength of films, 28
Methylene blue (MB), 299
Methylol compounds formation, 91
Metric tons (MT), 384
Michaelis–Menten equation, 199
Microbial growth, 165–166, 370
Microbial spoilage, edible coatings to inhibits, 369–370
Microcrystalline cellulose (MCC), 29
Microdroplets, 374–375
Microencapsulation of antimicrobial compounds, 253
Microstructure of edible films, 364
Milk proteins, 248–249
Mineral elements, 423
Minimal processing operations, 263
MMT clay, *see* Montmorillonite clay (MMT clay)
Modeling approaches, 187, 188
Moderate electric fields (MEFs), 87
Modified atmosphere packaging (MAP), 334, 387, 425
    benchmark packaging properties, 348–349
    current packaging and shelf life status, 343–345
    integrative MAP modeling, 346–347
    mathematical modeling for, 335
    opportunities for niche market packaging development, 345–346
    quantification of barrier properties, 347
    tailored barrier properties for packaging, 349
    time–temperature history, 345
Moisture
    loss, 134, 138
    transfer regulation in composite food product, 186
Molecular quantification, 191
Molecular weights (Mw), 16, 17, 52, 62, 245
Mono-glycerides, 133
Monolayer packaging applications, 154
Montmorillonite clay (MMT clay), 304
MRPs, *see* Maillard reaction products (MRPs)
MT, *see* Metric tons (MT)
Multilayer
    advantages in multilayer food packaging structures, 321–322
    applications, 299–301
    material development, 171
Multivalent ions, 361
Muscle processing method, 165
Muscle soluble proteins, 165
Mushrooms, 336
Mw, *see* Molecular weights (Mw)
Myofibrillar proteins, 423
Myoglobin, 159

N

N-deacetylation (DA), 245
NA, *see* Numerical aperture (NA)
Nanocapsules (NCs), 301
Nanocomposite edible films, 304
Nanoindentation, 308
Nanolaminates, 294–295, 300
    coatings applications, 300
Nanolayers, 253
Nanomaterials, 291–292
    applications, 293
    naturally occurring, 293–294
Nanometer, 289
Nanoparticles (NPs), 293
Nanosensors, 301–302
Nanospheres (NSs), 301
Nanostructure voluntarily creation, 196
Nanosystems, tools for characterization, 305
    AFM, 309–310
    CLSM, 310
    contact angle, 307–308
    electron microscopy, 308–309
    light scattering, 306–307
    nanoindentation, 308
    UV-Vis spectroscopy, 307
    zeta potential, 307
Nanotechnology
    active packaging, 292–293
    applications of multilayers, 299–301
    applications of nanomaterials and, 293
    control of multilayer characteristics, 297
    "electrostatic layer-by-layer self-assembly" technique, 295–297
    expectations of forthcoming developments, 291
    factors affecting adsorbed multilayer films assembly, 297–299
    in food packaging, 292
    knowledge level with scientific and economic impacts, 290–291
    nanolaminates, 294–295
    nanomaterials, 291–292
    naturally occurring nanomaterials, 293–294
    research area with multidisciplinary approaches and materials, 289–290
    risk, regulation, and future trends, 310–311
Natamycin, 251, 252, 403
Natural additives, 250
Natural compound, 250
    aromatic compounds, 221

Index

Natural resins, 129
Natural waxes, edible coatings and films based on, 138
NCs, *see* Nanocapsules (NCs)
Niche market packaging development, opportunities for, 345–346
Nisin, 252
Nonbiodegradable materials, 82
Noncovalent interactions, 86
Nonenzymatic browning development, 164
Nonionic gum polymers with extended twofold helix structures, 25; *see also* Ionic gum polymers with extended, twofold helix structures
    ethylcellulose, 30–31
    Fenugreek, 31–32
    Guar gum, 32–34
    HPC, 28–30
    HPMC, 28–30
    Konjac gum, 38–40
    Locust bean gum, 35–38
    methylcellulose, 26–28
    Tara gum, 34–35
Nonpathogenic microbes, 104
Nonpolar lipids, 123
Norovirus, 250
NPs, *see* Nanoparticles (NPs)
NSs, *see* Nanospheres (NSs)
Nuclear magnetic resonance spectroscopy investigation, 192
Numerical aperture (NA), 309
Nutraceuticals, 231, 362
Nutrients, 231

O

O_2P, *see* O_2 permeability (O_2P)
O_2 permeability (O_2P), 268
Ocimum basilicum L., *see* Basil (*Ocimum basilicum* L.)
OD, *see* Osmotic dehydration (OD)
Opacity, 268
OPP, *see* Orientated polypropylene (OPP)
Optical lever, 310
Organic acids, 403
Organic vapor barrier, 160
Orientated polypropylene (OPP), 90, 388
Osmotic dehydration (OD), 368
OTR, *see* Oxygen transmission rate (OTR)
Ovalbumin, 85
Oxidative deterioration, 422
Oxidative reactions, edible coatings to reducing, 370–371
Oxygen (O_2), 85
    barrier, 159
Oxygen transmission rate (OTR), 347

P

PA, *see* Polyamide (PA)
Package, 1, 334
Packaging, 1
Packaging materials production, methods for, 166, 167
    dry processing, 170–172
    edible coatings, 169–170
    solvent casting method, 166
    tape casting, 168–169
    wet processing, 168
Palm fruit (*Elaeis guineensis*), 126
Palm oil, 126
Panning, 272, 356
Paraffin, 249
Passive MAP, 344
Pasteurized processed cheese and analogs, 389
PCS, *see* Photon correlation spectroscopy (PCS)
PE, *see* Polyethylene (PE)
Pectin, 61, 246–247
    edible films, 64
    films with, 260, 261
    homogalacturonan moieties, 63
    idealized structure, 63
    process details, 62
    properties of, 62
PEFs, *see* Pulsed electric fields (PEFs)
PEG400, *see* Polyethylene glycol 400 (PEG400)
Peptide bonds, 83
Permeability, 189, 190
Permeation, 190–191
PET, *see* Polyethylene terephthalate (PET)
Petroleum-based packaging materials, 2
PGA, *see* Propylene glycol alginate (PGA)
pH, 297–298
PHA, *see* Polyhydroxyalkanoates (PHA)
PHB, *see* Polyhydroxybutyrate (PHB)
Phenolic acid, 90
Phenolic compounds, 419
Photobleaching technique, 191
Photon correlation spectroscopy (PCS), 306
Physiological disorders, edible coatings to suppressing, 369–370
Physiological properties quantification, 340
    respiration rate, 340–342
    transpiration rate, 342–343
Phytochemicals, 218
Pimaric acids, 129
PLA, *see* Polylactic acid (PLA)
Plant extracts, 418
Plasticizers, 13, 163, 195, 362, 395
    effect, 88
    properties, 134
Plasticizing, 88–90
Polyamide (PA), 387
Polyethylene (PE), 10, 387

Polyethylene glycol 400 (PEG400), 28
Polyethylene terephthalate (PET), 320, 387
Polyhydroxyalkanoates (PHA), 10, 154, 320
Polyhydroxybutyrate (PHB), 349
Polylactic acid (PLA), 10, 154, 171, 320
Polymer
    nanotechnology, 303
    structure modifications, 162
Polyphenols, 220
    compounds, 419
    oxidase, 165
Polypropylene (PP), 10, 305, 388
Polypropylene glycol (PPG), 362
Polysaccharide-based coatings, 134, 367
Polysaccharide-based films, 302
Polysaccharide-based materials, 356, 361
Polysaccharide-based packaging, 392
Polysaccharides, 223, 245, 253, 293, 373
    alginate, 246
    carrageenans, 246
    cellulose and derivatives, 245
    chitin/chitosan, 245–246
    edible films/coating for meat, poultry, and fish products, 261
    galactomannans, 247
    pectin, 246–247
    starch, 245
Polystyrene (PS), 348, 388
Polyvinyl alcohol (PVA), 30
Polyvinyl chloride (PVC), 337
Polyvinylidene chloride (PVDC), 387
Posttreatments, 197
Potassium sorbate (PS), 420
Poultry
    antimicrobial edible films and coatings for, 258–262
    applications of coatings, films, and active compounds in, 417
    edible coatings on, 423
    future of food packaging for monitoring quality in, 425–426
    quality, 423–425
PP, *see* Polypropylene (PP)
PPG, *see* Polypropylene glycol (PPG)
Principal preparation technologies, 196
Probiotics, 218–219, 221–222, 229–230
Product selection and market share, 335
    Botrytis infection, 337
    market share of selected products, 337–338
    mushrooms and strawberries, 336
Propylene glycol alginate (PGA), 45–47
Protective coating/film, 216
Protein-based films and coatings, 82; *see also* Edible coatings and films
    applications of protein-based materials, 99–106
    composition and structure of proteins, 83–85
    edible coatings/films, 223–224
    fibrous proteins, 84
    gelatin, 82
    modification of protein-based materials, 85–92
    processing of protein-based materials, 92–99
    Soy protein, 85
    Whey protein, 83
Protein-based materials, 361
    active coatings, 99–104
    aggregation, 86–88
    applications of protein-based materials, 99–106
    biodegradable films, 105–106
    chemical crosslinking, 90–92
    denaturation, 86–88
    FTIR spectra of soy protein-based film-forming solutions, 95
    ideal life cycle for protein-based films, 105
    modification of, 85–92
    physicochemical properties, 94
    plasticizing, 88–90
    preparation and characterization of, 100–101
    processing in solution, 94–97
    processing methods and parameters, 93
    processing of, 92–99
    temperature, 96
    thermomechanical processing, 97–99
Proteins, 223, 247, 293–294
    edible films/coating for meat, poultry, and fish products, 262
    films, 159, 261–262
    gelatin, 247–248
    milk proteins, 248–249
    networks, 162
    protein-based coatings, 367
    protein-based packaging, 392, 395
    protein-coated materials, 168
    protein–food forces, 102
    protein–protein forces, 102
    secondary structures, 86
    solubility reduction, 164
    sources, 168
    soy protein, 248
    wheat gluten, 248
    zein protein, 248
Proteolysis, 165
Provitamin A carotenoids, 159
PS, *see* Polystyrene (PS); Potassium sorbate (PS)
Pseudomonas elodea (*P. elodea*), 54
Pullulan, 64
    addition, 67
    basic structure, 65
    films, 66
    idealized structure, 66
    process for, 65
Pullularia pullulans (*P. pullulans*), 64
Pulp, 30
Pulsed electric fields (PEFs), 87

Index

PVA, *see* Polyvinyl alcohol (PVA)
PVC, *see* Polyvinyl chloride (PVC)
PVDC, *see* Polyvinylidene chloride (PVDC)

Q

Quasi-elastic light scattering (QELS), 306

R

RAMAN spectroscopy, 193
Rancidity, 159
Rapid melt film composite, 70
Raw material chemistry, 194–195
Reactive oxygen species (ROS), 404
Ready-to-eat products (RTE products), 420
Refined carrageenans, 54
Refrigerated fish, 421
    antioxidants, 421–422
    CH coating, 423
    freezing, 421
    lipid oxidation, 422
Regulatory assessment, 4
Relative humidity (RH), 96, 155, 190, 198–199, 320, 360, 387
Release systems of active compounds, 161
Residual proteolytic enzymes, 164
Residues, 84
Resins, 123, 129
    Ghatti gum, 130–131
    Gum Arabic, 129–130
    Gum tragacanth, 130
    Karaya gum, 130
    mastic resin, 131
    Mesquite gum, 130
    vanilla oleoresin, 131
Respiration rate, 340–342
Retrogradation, 162
RG, *see* Rhamnogalacturonan (RG)
RH, *see* Relative humidity (RH)
Rhamnogalacturonan (RG), 63
Rigid punnet-type tray (RPET), 343
Room temperature (RT), 15
ROS, *see* Reactive oxygen species (ROS)
RPET, *see* Rigid punnet-type tray (RPET)
RT, *see* Room temperature (RT)
RTE products, *see* Ready-to-eat products (RTE products)

S

Salicylic acid, 98, 172
Saran wrap, 13
Sausage casing, 44, 45
SB, *see* Sodium benzoate (SB)
Scanning electron microscopy (SEM), 59, 90, 299, 309
Schiff base, 92
SD, *see* Sodium diacetate (SD)
Sea bass (*Lateolabrax japonicas*), 419
Seafood products, 415
Secondary structures, 84
Self-assembly process, 296
SEM, *see* Scanning electron microscopy (SEM)
Sensory characteristics, 161
Silicate materials, 304
Silver NPs (AgNPs), 305
Simmondsia chinensis, *see* Jojoba shrub (*Simmondsia chinensis*)
SL, *see* Sodium lactate (SL)
Slow melt film composite, 70
Smart packaging, 292
SME, *see* Specific mechanical energy (SME)
Smooth region, *see* Homogalacturonans block
Sodium alginate, 42, 299
    annual production, 43
    coating composite films, 161
    idealized structure, 43
    sausage casing, 44, 45
    structural blocks, 44
    WVP, 45
Sodium benzoate (SB), 420
Sodium diacetate (SD), 420
Sodium lactate (SL), 420
Soft cheese, 387–388
Soft resins, 129
Solubility, 189, 254
Solution casting, 92, 97
Solvents, 196
    casting, 96, 166
Sorbic acids, 104
Sorption, 193–194
Soy protein, 85, 248
    films, 90
    soy protein-based edible film, 248
Soy protein isolate (SPI), 46, 248, 304
Specific mechanical energy (SME), 98
SPI, *see* Soy protein isolate (SPI)
Spray coating, 169
Spraying, 94, 169, 272, 391
Spray LbL method, 296
    advantages and disadvantages, 296–297
Spread casting, *see* Tape casting
Starch-based polymers, 348–349
Starch, 172, 245
    edible films, 162
Sterculia genus, 130
Steric groups, 17
Strawberries (*Fragaria ananassa*), 336, 371
    chain for, 345
    optimal storage and packaging conditions for, 339
    quality and safety parameters for, 339
Structural conformation, 15–16

Structural defects, 196–197
Structuring agent-based packaging, 395
Sunflower oil, 125
Sunflower seeds, 125
Surface mold-ripened cheeses, 388
Surface roughness, 364–365
Surfactants, 362, 395
Sushi wrap, 23
Synthetic additives, 416
Synthetic plastic materials, processability and versatility of, 154
Synthetic plastics, 10
Synthetic resins, 129

T

Tape casting, 168–169
Tara gum, 34–35
TBA, see Thiobarbituric acid (TBA)
TBARS, see Thiobarbituric acid reactive substance (TBARS)
TEAC, see Trolox equivalent antioxidant capacity (TEAC)
TEM, see Transmission electron microscopy (TEM)
Temperature, 96, 198
Tetrasaccharide, 56
Texture-related properties, edible coatings to improving, 371–372
TGA, see Thermogravimetric analysis (TGA)
Thawing, 50
Thermal stability of proteins, 96
Thermoforming, 171
Thermogravimetric analysis (TGA), 97
Thermomechanical processing, 97
    CLSM analysis, 99
    DTG curves, 98
    extrusion, 97
    SME, 98
Thermoplastic ether–ester elastomer (TPC-ET), 10
Thermoplastic flour (TPF), 302
Thermoplastic processing, 97
Thermoplastic sustainable biopolymers, 154
Thermopressing, 171
Thickening, 22–23
Thin film, 2
Thiobarbituric acid (TBA), 419
Thiobarbituric acid reactive substance (TBARS), 54, 103
Three-dimensional "egg box" molecular conformation, 44
Time–temperature history, 345
TMA, see Trimethylamine (TMA)
Total volatile base nitrogen (TVB-N), 423
TPC-ET, see Thermoplastic ether–ester elastomer (TPC-ET)
TPF, see Thermoplastic flour (TPF)
TR, see Transpiration rate (TR)

Tragacantic acid, 130
Transmission electron microscopy (TEM), 96, 309
Transparent films, 87
Transpiration rate (TR), 335, 342–343
"Trial-and-error" approach, 183, 187, 199
Triethanolamine, 89
Triglycerides, 133
Trigonella foenum-graecum, see Fenugreek (*Trigonella foenum-graecum*)
Trimethylamine (TMA), 423
Trolox equivalent antioxidant capacity (TEAC), 227
TVB-N, see Total volatile base nitrogen (TVB-N)
Type A gelatin, 94
Type B gelatin, 94

U

Ultrasonic waves, 96
Ultraviolet-visible spectroscopy (UV-Vis spectroscopy), 307
Ultraviolet light (UV light), 90, 91
    barrier, 159–160
    rays, 303
Unordered random coil structures, 84
UV-Vis spectroscopy, see Ultraviolet-visible spectroscopy (UV-Vis spectroscopy)
UV light, see Ultraviolet light (UV light)

V

Vanilla oleoresin, 131
Vanilla orchid (*Vanilla planifolia*), 131
Vanilla planifolia, see Vanilla orchid (*Vanilla planifolia*)
Vegetables, antimicrobial edible films and coatings for, 262–267
Viscosity measurements, 62

W

Water, 88, 99
    loss, 367
    vapor barrier, 133–134, 158–159
    vapor transmission, 158
Water vapor permeability (WVP), 29, 45, 90, 96, 133, 156, 225, 247, 249, 294, 386
    of gellan films, 56
Water vapor resistance (WVR), 364
Water vapor transmission rate (WVTR), 335, 347, 386
Wax-based materials, 361
Waxes, 123, 126, 249
    Beeswax, 127
    Candelilla wax, 128
    Carnauba wax, 128
    cutting, 126
    Jojoba oil, 128

Index

Waxy coatings, 367
Wet processing, 168
Wettability, 268, 363
Wheat gluten, 85, 171, 248, 327
Whey protein, 83, 84–85
    coating, 161
Whey protein concentrates (WPCs), 85
Whey protein isolate (WPI), 85, 87, 325, 424
    coating, 168
Whole mushrooms (*Agaricus bisporus*), 335–336
WikiCells, 11–12
WikiPearls, 11–12
Window of optimal mass transfer properties, 183
WPCs, *see* Whey protein concentrates (WPCs)
WPI, *see* Whey protein isolate (WPI)
Wrappers, 13–14
WVP, *see* Water vapor permeability (WVP)
WVR, *see* Water vapor resistance (WVR)
WVTR, *see* Water vapor transmission rate (WVTR)

X

Xanthan gum, 56
    idealized structure, 58
    process, 57
    synergistic interactions, 59
    three dimensional structure, 58
Xanthomonas campestris (*X. campestris*), 57

Y

Yeast, 250
Young's modulus (E), 327

Z

Zein coatings, 103–104
Zein protein, 248
Zeta potential, 307